Introducing Geographic Information Systems with ArcGIS®

Featuring GIS Software from Environmental Systems Research Institute

7 chapters
in 28 days

Michael Kennedy

University of Kentucky

WILEY

JOHN WILEY & SONS, INC.

Library of Congress Cataloging-in-Publication Data:

Kennedy, Michael, 1939–
    Introducing Geographic Information Systems with ArcGIS: featuring GIS software from
    environmental systems/Michael Kennedy.
        p. cm.
    Includes bibliographical references (p.   ) and index.
    ISBN-10 0-471-79229-2 (paper/cd) – ISBN-13 978-0-471-79229-1 (paper/cd)
    1. Geographic information systems.   2. ArcGIS.   I. Title.
    G70.212.K47 2006
    910.285—dc22

                                                                        2006000886

Printed in the United States of America

10 9 8 7 6 5 4 3 2 1

Dedicated to the memory of Evan Kennedy, who had every gift but that of years.

Contents

Contents

Contents

Contents

Contents

Contents

Contents

Contents

Contents

Contents

Contents

Contents

Contents

Contents

Contents

Foreword

by Jack Dangermond

Introducing Geographic Information Systems with ArcGIS offers a unique approach to GIS instruction. In it, Michael Kennedy recreates his time-tested methods of teaching GIS in the classroom in a step-by-step guidebook to GIS. Students on a journey to learn GIS with Professor Kennedy may feel like he is taking the journey with them, offering them his sage advice each step of the way. Professor Kennedy cares deeply for his students, and the detail of this care and years of teaching GIS come through in this book. In it, he walks students through the multitude of questions that come up daily in the classroom. His goal is to help students understand GIS concepts and learn GIS skills. It takes a master teacher to map GIS knowledge, making it clear to students and enabling them to gain confidence in their growing skills.

Once GIS students have learned the basics, the next step is to learn how to analyze spatial data and identify problems and create solutions. Learning to analyze spatial data moves students beyond exploration, beyond locating places on maps, and helps them create maps that guide better decisions.

All of us learn GIS skills in different ways. Some people are visual learners, some are auditory learners, and some need a hands-on approach. As the learning styles of students in general vary, so do the learning needs of students of GIS. Some students will need classroom study with conversations and time to process information about GIS concepts, spatial data, geodatabases, map projections, attribute tables, feature classes, datasets, and building maps, while others need only a guidebook with clear graphic illustrations. So, a variety of approaches to teaching GIS will help ensure that the increasing number of students worldwide have opportunities to gain GIS skills in ways that best suit their needs.

GIS is becoming part and parcel of the daily work lives of most people in many fields, from architects to zoologists, from academia to the business world, from city planning to national and international spatial data portals. Teachers are now taking on the essential task of opening the door for students to learn GIS. In *Introducing Geographic Information Systems with ArcGIS*, Professor Kennedy opens such a doorway for students to learn the skills basic to understanding GIS and to prepare students to make our communities better places.

Preface[1]

The purpose of Introducing Geographic Information Systems with ArcGIS is threefold.

1. To acquaint the reader with the central concepts of GIS and with those topics that are required to understand spatial information analysis.

2. To provide the person who works the exercises either (a) a considerable ability to operate important tools in the ArcGIS software or (b) a demonstration of other capabilities of the software.

3. To lay a basis for the reader to go on to the advanced study of GIS or to the study of the newly emerging field of GIScience, which might be described as the scientific examination of the technology of GIS and the fundamental questions raised by GIS.

Introducing Geographic Information Systems with ArcGIS is meant to serve as a text book for a standard one-semester course. It is suitable for a university, college, technical school, or advanced high school course, meeting for three hours per week. Between two and five additional hours per week are required for laboratory work, depending on the capabilities and computer experience of the students. The text may also be used for self-study.

The book, and any course taught from it, depend on having ESRI's ArcGIS Desktop and Workstation software, version 9.0, 9.1, or higher, available. The assumption is that the students will have access to full the ArcInfo package offered to colleges and universities under the generous site license agreement that ESRI offers to educational institutions. For more information about this program, point your browser at: http://www.esri.com/industries/university/education/faqs.html. However, if ArcInfo is not available, many of the exercises can be done with the ArcView level of ArcGIS, available to students with a free, one-year license.

While the author is impressed with the ArcGIS software (and with the aims of ESRI of being a force for conservation, preservation, and sustainable development worldwide), this book is not meant as a promotional text for ESRI. Like all large software packages, ArcGIS has its shortcomings, limitations, and bugs. When these arise in the process of working through the exercises, they are candidly pointed out to the reader. All bugs have been reported to ESRI, and most have been repaired or are scheduled for repair. By the way, the ESRI support staff is excellent: responsive and friendly.

---

[1]If this text is used in a classroom/laboratory setting, this preface is for the instructor and may be skipped by students. If you are using the book to learn GIS on your own you should probably read it.

Preface

The function of GIS software is to make a computer think it's a map—a map with characteristics that let the user analyze it, display its elements in a variety of ways, and use it for decision making. This text is oriented more toward preparing the student for doing analysis with GIS, rather than display, mapping, or standard data processing.

Contents of Introducing Geographic Information Systems with ArcGIS

❏ Part I: Basic Concepts of GIS

   ❏ Chapter 1: Some Concepts that Underpin GIS (and introduction to ArcCatalog)

   ❏ Chapter 2: Characteristics and Examples of Spatial Data (and introduction to ArcMap)

   ❏ Chapter 3: Products of a GIS: Maps and Other Information

   ❏ Chapter 4: Structures for Storing Geographic Data (and introduction to ArcToolbox and Workstation)

   ❏ Chapter 5: Geographic and Attribute Data: Selection, Input, and Editing (and introduction to ArcScene and ArcGlobe)

❏ Part II: Spatial Analysis and Synthesis with GIS

   ❏ Chapter 6: Analysis of GIS Data by Simple Examination

   ❏ Chapter 7: Creating Spatial Data Sets Based on Proximity, Overlay, and Attributes

   ❏ Chapter 8: Spatial Analysis Based on Raster Data Processing (and introduction to Spatial Analyst)

   ❏ Chapter 9: Other Dimensions, Other Tools, Other Solutions (and introductions to 3-D Analyst, Historical Data, Address Geocoding, Network Analyst, and Linear Referencing)

In my view, the pedagogical theme of a first course should be breadth, with depth in vital areas. The text covers virtually all the general GIS capability that ArcGIS has to offer. Vector and raster storage, analysis, and synthesis are, of course, discussed extensively, with many examples and exercises for the student. Other areas receive less attention, such as 3-D GIS, time and GIS, network analysis (path finding and allocation), surface creation, spatial analysis, statistical and numerical analysis, model builder, GIS & GPS, and so on. In some later instances, the exercises are primarily demonstrations of the capabilities of the ESRI software, but, in my opinion, a student in a first course needs to get at least a glimpse of almost all of what GIS can do. Omitted from the text is most of customization, programming, and the more esoteric capabilities of geodatabases, which I believe belong in a second course. Also not included is GIS on the Internet and the issues related to large, enterprise implementations of GIS. To mention it again: the thrust of the text is to lay a foundation from which the reader can move toward doing analysis and synthesis with GIS.

The emphasis in terms of data structure is on geodatabases. However, extensive use is made of shapefiles and coverages, since most existing GIS data sets are in these formats. The student will become comfortable with switching and converting among the various formats. Another reason for using all three formats is that, at this stage of ArcGIS development, there are operations that can be done with coverages that cannot be performed with geodatabases.

ArcMap, ArcCatalog, ArcToolbox, ArcScene, and ArcGlobe are all explored in considerable detail. ArcInfo Workstation is introduced. Enough of command-line ArcInfo Workstation is used to make the student aware of its existence and its capability to perform operations that are cumbersome or impossible with the point and click software. This is a book that creates knowledge for the student that is realistic and at least touches on virtually all the ArcGIS capabilities and products.

In the four years the text has been under development, most of the exercises in the book have been performed by scores of students. All of the exercises have been tested and they work, both from a technical and pedagogical standpoint.

In terms of time required to do the exercises, most students will require:

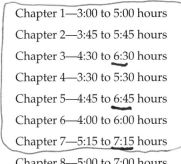

Chapter 1—3:00 to 5:00 hours
Chapter 2—3:45 to 5:45 hours
Chapter 3—4:30 to 6:30 hours
Chapter 4—3:30 to 5:30 hours
Chapter 5—4:45 to 6:45 hours
Chapter 6—4:00 to 6:00 hours
Chapter 7—5:15 to 7:15 hours
Chapter 8—5:00 to 7:00 hours
Chapter 9—4:45 to 6:45 hours

Handwritten annotations:
wk 2 +3
wk 4 +5
wk 6 +7
wk 8 + 9
wk 10+ 11
wk 12 + 13
wk 14

We have 35 days so we have 6 days per chapter

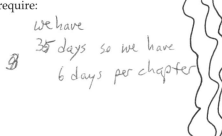

Theory and Practice

Of the myriad of GIS textbooks available, some are long on theory but don't train the student, while the rest are pretty much manuals on how to use software, but don't promote an understanding of what lies behind the mechanics. So frequently GIS is taught either with texts that teach only theory and leave it to the instructor to select software and data to illustrate points or taught with manuals and demonstrations.

The book is unusual, if not unique, in that it serves both as a general introduction to GIS (serving an education function) and a manual on ArcGIS software (serving a training function). This is accomplished by dividing each chapter into an

❑ Overview section, and a

❑ Step-by-Step section

The Overview section is descriptive. It is a top-down discussion of theory and ideas relating to GIS.

The Step-by-Step section is prescriptive. It operates in a sequential fashion—do this, then this, then this. Here the student learns about and practices ArcGIS. There are more than 60 exercises in the book, not counting the 9 review exercises. Almost 60 percent of the book consists of step-by-step instructions on how to use ArcGIS software.

All the data sets for the exercises are on the CD-ROM that accompanies the book.

Teaching with This Book

The contents of the following folders on the CD must be available for downloading by students:

❏   IGIS-Arc—the primary source of data sets for the exercises

❏   IGIS-Arc_AUX—a source for datasets occasionally needed for exercises

❏   IGIS_with_ArcGIS_FastFactsFile_Checklists—a combination chapter summary and set of Fast Facts File prompts

❏   IGIS_with_ArcGIS_Selected_Figures—full-color versions of some figures in the text that suffer from black-and-white reproduction

If you are an instructor, you should consider copying the four folders above from the CD-ROM to a location on a network where the students can access their contents.

Exercises are roughly put into categories of length or difficulty, with such notes as "Warm-up" (least effort), "Project" (greater effort), and "Major Projects" (most effort).

Some warnings:

Do not use any of the sample databases on the CD-ROM for anything other than tutorial purposes. Most of the data is old. Much of it has been modified for instructional purposes.

For students who aren't paying attention, the exercises will get harder and harder because it is expected that they will learn (or be able to quickly find) how to perform operations that they have performed before. The "hand holding" diminishes as the chapter numbers increase.

Exercise 5-8 is a cooperative exercise for eight to twenty-four students. Preparation and management on the part of the instructor is a really good idea.

If you, as an instructor, are quite sure that your students will not need more than the most basic knowledge about coverages and shapefiles, you can have them skip considerable portions of Chapter 4. You should read the sections on coverages yourself and, perhaps in lecture sessions, supplement the student's knowledge of the coverage concepts that apply to geodatabases.

If you serve ArcGIS, or even just its license manager, over a network, you should thoroughly test the process. Also, in Chapter 8, the unsupported CellTool is used. Students may not be able to install it, so someone from network services will have to be involved.

More Resources for the Instructor

If you are an instructor who is using this text, you are encouraged to register on the website www.wiley .com/college/kennedy. There you will find advice on how to use the book to its fullest potential. Included there are answers to the questions posed in the text, sample assignments with blanks for the students to complete, test data for some assignments, and suggestions of how to use the text—avoiding some pitfalls that lurk, especially when the datasets are served across a network. The Instructor's Guide there can be a valuable resource for those teaching with this text. Also look at the folder IGIS_with_ArcGIS_Instructor's_Guide on the CD-ROM.

Concepts, Devices, and Techniques that Underlie the Philosophy of the Book

How can one textbook touch on almost all of GIS when it takes thousands of pages of manuals to do this? Two ways:

❑ There are few figures, and, compared to the standard computer manual, there are few screen shots. When a student follows the instructions, he or she sees the proper screens. When a figure can be better understood by the use of color, the figure is available on the CD-ROM in the folder IGIS_with_ArcGIS_Selected_Figures. Such figures are designated in the text reference with three asterisks. For example, "See Figure 8-4***."

❑ As the student progresses through the later chapters, the exercises do not contain detailed instructions. The students are expected to be able to do steps that were explained in detail earlier. For example, in early chapters, detailed instructions are given for finding or changing a property of a data set or data frame. In later chapters, the students will simply be told to take that action. When students can't either remember or find out how to perform an action that has been previously detailed, teachers should take it as a clue that the students are simply going through the motions of executing the software tools and that learning is not really taking place.

I believe that students learn best by doing—while observing and recording what it is they are doing. Students are asked to develop a Fast Facts File in which they record what it is they have learned about the software. This is a computer file that they keep open during their work sessions, both for adding new material and ascertaining how to do a particular procedure that they have used previously but cannot remember. They periodically revise and augment this file. Then, at the end of the course, they have their own reference manual for the software. I have used this technique for some years now, and it pays dividends. Some students who have graduated and now work in the GIS field tell me they take their Fast Facts File with them and maintain it in their new positions. One failure of other workbooks and web-based courses is that, while students can go through the exercises and even pass a test at the end, they simply cannot operate the software when handed a new exercise. Now with ten-plus years of teaching GIS with the Overview-Step method behind me, insisting that students make a Fast Facts File to provide themselves a guide through the very complex GIS software, I'm convinced that the not-always-popular-with-the-students Fast Facts File is more than worth the trouble.

The exercise material is project oriented; students learn the software as needed for the particular project at hand. So rather than learning everything about labeling features at one time or everything about selecting, the students learn as they complete projects and record what has been learned in their Fast Facts Files, which are later reorganized. At the risk of losing adoptions and sales, please let me candidly point out that this textbook does not serve very well as reference material. The idea behind the book is to make things click in the students' brains, to promote comprehension of concepts, not to serve as a reference guide to the software. However, the diligent students—indeed even those who follow the instructions—will emerge from the course with their own reference guides, done in a style suitable for each student: the Fast Facts File. Some students resist creating the file, so I make it count for 5 percent of their grade. Further, the Fast Facts File will be a reference document that the learners can maintain and upgrade in future months and years. A student of mine of a decade ago came to my class to give a guest lecture. He was in charge of the GIS program of a state unit. He brought his Fast Facts File with him to show to the class. Over the years, he had updated it many times.

The text is workbook-like in that there are blanks in the text which the students are asked to complete, showing that they have performed and comprehended a particular section. This also serves as a mechanism for letting the instructors know how students are progressing. The Web site www.wiley.com/college/kennedy contains forms with these blanks, in context, so an instructor can copy and paste the material into assignment sheets. Student progress can be monitored using these assignments.

The text is set up so students can work at their own pace, respecting different learning styles and speeds of the students. For example, some students create entries in their Fast Facts Files with each step. Others make two passes through the material.

In a few places in the text, students are asked to record the names of menu choices or tabs in windows. Of course this information could have been printed for them, but having them write it in reinforces the words and the concepts behind them in the students' minds. It is also a modest protection against software changes (e.g., addition of menu or tab items).

The last exercise in each chapter is a checklist that can serve the students in development of the Fast Facts File. The students are given prompts that they fill in. The prompts appear in the text and also on the CD-ROM in the folder IGIS_with_ArcGIS_FastFactsFile_Checklists so they are available in machine readable form to the students. This allows students to copy the prompts into their Fast Facts Files and complete them.

The book simulates a teacher: sometimes a lecturer, standing in front of students, imparting information or giving directions. More often the instructor is a colleague, sitting beside the student, making suggestions, prompting, and, occasionally, making mistakes that he or she and the student rectify. One might describe the tone of the book as conversational. I believe the most important thing, after correctness, is engaging the student. I believe economy in writing is important. But sometimes additional words can set a tone. I actually use several tones in the text to provide variety. I change pace. I change style. I change attitude. I change the level of formality.

The writing style, for the most part, is informal—to convey the idea that the author is closely involved with the student, guiding her or him. Humor is used, but sparingly. Irony is used, but sparingly.

The ArcGIS software is so complex that there is no way to explore it "depth first." We must look at an overview. The book attempts to teach, or at least demonstrate, the major capabilities of ArcGIS. As you can tell from the weight of ESRI manuals, compared with the size of this text (which also serves as a general GIS theory text), I could hardly cover even a large portion in detail. However, the student will come away with considerable facility with the software and will know how to find and use additional capabilities.

Finally, I believe it is important to emphasize that computer is not a black box. An educated GIS professional should have some understanding of what makes a computer tick. So there is some general material on computers and representation of information, especially as they impact answers from a GIS.

Acknowledgments

This book was something of a family affair. My daughter Heather Kennedy provided help with the 3-D material[2]. My son Alex Kennedy carefully worked all the exercises in all chapters, making corrections and providing insightful observations. He also helped collect some of the GPS data.

Jack Dangermond, for his encouragement in general and writing the Foreword in particular, and ESRI for allowing use of numerous datasets and figures. In fact, this text was originally conceived of as a new edition of *Understanding GIS—The Arc/Info Method (UGIS-tiam)* reworked for the ArcGIS point-and-click version of the software, beginning with ArcGIS 8.x. It has clearly grown way beyond that first idea, with the addition of textual matter dealing with GIS itself, discussion of other capabilities of the software besides vector-based GIS site selection (such as a discussion of Spatial Analyst, the addition of material on several of the other important extensions to the software, and the major emphasis on geodatabases). Teachers who in the past used *UGIS-tiam* (last published almost a decade ago) will, however, recognize the site selection problem of that text (a laboratory to do research in "Aquaculture") as the Wildcat Boat Testing Facility of this text, albeit highly modified.

I am indebted to Dr. Michael Goodchild for writing the Afterword. As mentioned before, the aim of this textbook is to provide a general introduction to GIS and prepare students to use ArcGIS primarily for analysis. Some of those students will want to go on to study GIScience, and I commend Dr. Goodchild's Afterword to them.

Many colleagues and friends contributed to bringing this book to fruition. In particular, I want to thank:

Christian Harder, founding publisher, *ESRI Press*, for suggesting and encouraging the development of the text.

Gary Amdahl, an early, helpful editor with *ESRI Press*.

Randy Worch of ESRI, for support and encouragement over the years.

Damian Spangrud, ArcGIS product manager for ESRI, for being the person I could always count on when I had difficult questions about the software.

Ken Bates, extension specialist with Kentucky State University, for working through many of the exercises with the ArcView level of ArcGIS.

Dan Carey, Ph.D., of the Kentucky Geological Survey and the University of Kentucky, for his thorough reading.

I greatly appreciate the help of Richard Greissman, who facilitated my finding the time to write.

Demetrio Zourarakis of the Division of Geographic Information, Commonwealth Office of Technology for Kentucky-wide data.

The staff of ESRI technical support—friendly and helpful people too numerous to mention.

---

[2]She is author of *Data in Three Dimensions: A Guide to ArcGIS 3D Analyst* (Onward Press, 2004, ISBN 1-4018-4886-9) and editor of *The ESRI Press Dictionary of GIS Terminology* (2001, ISBN 1-879102-78-1)

Preface

Jim Harper and Amy Zarkos of John Wiley and Sons, who facilitated the editing process in spite of the author's frequent lack of organization and tardiness.

Richard Peal of Publishers' Design and Production Services, Inc. in Sagamore Beach, MA, who looked after the figures, again against complications created by the author.

Teaching assistants Chris Blackden and Amber Ruyter worked through the exercises and helped both students and me over rough spots.

Scores of students at UK had early drafts of the text inflicted on them. In particular Rebecca McClung, who reviewed the text exhaustively and helped prepare the index, and Travis Searcy, who combed through the later chapters.

I also wish to thank:

Taylor and Francis (and CRC Press) for allowing me to use portions of my textbook *The Global Positioning System and GIS* (ISBN 0-415-28608-5)

Chris Kimball of Digital Data Services, 10920 West Alameda Avenue, #206, Lakewook, CO (www.usgsquads.com, 303-986-6740) for the data behind Figure 2-3: Frankfort, MI and Crystal Lake DOQQs.

The Lexington Herald Leader for the photograph of the water facility on the Kentucky River.

Sue McCowan, account manager, GIS Markets of Tele Atlas, for San Francisco Street data for the Network Analyst exercise.

I'm appreciative of the University of Kentucky and its College of Arts and Sciences and Department of Geography, for providing the opportunity to develop the text.

Introduction

A geographic information system (GIS) software package is basically a computer program designed to make a computer think that it's a map. This new sort of map is a dynamic entity, designed to assist people in making decisions. Such decisions might be as simple and short range as determining an efficient way to get from place A to place B. Or as complex as designing a light rail transportation system for a city or delineating flood planes. The difference between a paper map and a GIS map is that the latter exhibits "intelligence." You can ask it a question and get an answer.

Geographic information systems are transforming all the activities and disciplines that formerly used maps as the basis for decision making. It's about time. Most fields of human endeavor have long since been heavily impacted by the digital computer; in fact it's hard to think of one that hasn't. Fifty years have gone by since computers began changing accounting, census taking, physical sciences, and communication, to name a few. Even the field of music has been altered. Most of these "nonspatial" fields already couched their problems in terms of discrete symbols (such as A, r, 5, and $) that are easily converted to the binary language (using only the symbols 0 and 1) that the computer understands. The spatial fields such as geography, planning, and land use management, had to stick with maps because, while maps also use symbols, they are not so neat and tidy as to fit on the key of a keyboard. A symbol for a road might be three feet long! Determining how to efficiently represent the real-world environment in the memory of a computer turned out to be quite a challenge. So until a decade or so ago, those who relied on maps usually could not use computers effectively as the primary source of data from which to work.

Why has computer-based GIS come to influence how decisions are made about land use planning, navigation, resource allocation, and so on? First, the shortcomings of maps for decision making are many. Second, computers have become greatly faster, bigger (in terms of memory size), and cheaper. And, third, we have developed sophisticated data structures and learned how to efficiently program computers to represent the huge, almost infinitely detailed environment that we live in. So those of you who are just now beginning to learn about GIS are not pioneers, but if you enter the field now, I bet that in a decade you will feel like a pioneer because the field is growing so rapidly. You are off on a great adventure!

PART I

Basic Concepts of GIS

Some Concepts That Underpin GIS

OVERVIEW

IN WHICH you begin to understand the rather large and complex body of ideas and techniques that allow people to use computers to comprehend and design the physical environment. And in which you use ESRI's ArcCatalog to explore geographic data.[1]

You Ask: "What Is GIS About?"

The poem, "The Blind Men and the Elephant," tells the story of six sightless men who approach an elephant, one by one, to satisfy their curiosity.

> It was six men of Indostan
> To learning much inclined
> Who went to see the Elephant
> (Though all of them were blind)
>
> That each by observation
> Might satisfy his mind.
>
> The First approached the Elephant,
> And happening to fall
> Against his broad and sturdy side,
> At once began to bawl,
> "God bless em! but the elephant
> Is very like a WALL!"
>
> The Second, feeling of the tusk
> Cried: "Ho! what have we here
> So very round and smooth and sharp?
> This wonder of an Elephant
> Is very like a SPEAR!"
>
> The Third approached the animal,
> And, happening to take
> The squirming trunk within his hands,
> Thus boldly up and spake:
> "I see," quoth he, "the Elephant,
> Is very like a SNAKE!"

---

[1]ESRI is the Environmental Systems Research Institute. ESRI makes the software, ArcGIS, which you will use in this text to understand the concepts of Geographic Information Systems.

The Fourth reached out an eager hand,
And felt about the knee
"What most this wondrous beast is like
Is mighty plain," quoth he:
" 'Tis clear enough the Elephant
Is very like a TREE!"

The Fifth, who chanced to touch the ear,
Said: "E'en the blindest man
Can tell what this resembles most;
Deny the fact who can,
This marvel of an Elephant
Is very like a FAN!"

The Sixth no sooner had begun
About the beast to grope,
Than seizing on the swinging tail
That fell within his scope,
"I see," quoth he, "the Elephant
Is very like a ROPE!"

And so these men of Indostan
Disputed loud and long,
Each in his own opinion
Exceeding stiff and strong
Though each was partly in the right
And all were in the wrong!

[. . .]²

Excerpted from "The Blind Men and the Elephant"
(based on a famous Indian legend)
John Godfrey Saxe
American Poet (1816–1887)

You Ask Again: "What Is GIS About?"

Poet Saxe's lines could apply to geographic information systems (GIS) in that relating to the subject may well depend on your point of view. Asking what GIS³ is about is sort of like asking, "What is a computer about?" The capabilities of GIS are so broad and its uses so pervasive in society, geography, and the technical world in general that a short, meaningful description is impossible. But for starters, here is a generic definition of GIS that you might find in a dictionary:

---

²[. . .] indicates an omission.
³In this text "GIS" stands for "a geographic information system" or for the plural "geographic information systems" depending on the context.

> *A geographic information system is an organized collection of computer hardware and software, people, money, and organizational infrastructure that makes possible the acquisition and storage of geographic and related attribute data, for purposes of retrieval, analysis, synthesis, and display to promote understanding and assist decision making.*

To better understand one facet of GIS, consider how you might use the technology for a particular application. Solve the following site selection problem:

Exercise 1-1 (Project)

Finding a Site by Manual Means

Wildcat Boat Company is planning to construct a small testing facility and office building to evaluate new designs. They've narrowed the proposed site to a farming area near a large lake and several small towns. The company now needs to select a specific site that meets the following requirements:

❑ The site should not have trees (to reduce costs of clearing land). A regional agricultural preservation plan prohibits conversion of farmland. The other land uses (urban, barren, and wetlands) are also out. So the land cover must be "brush land."

❑ The building must reside on soils suitable for construction.

❑ A local ordinance designed to prevent rampant development allows new construction only within 300 meters of existing sewer lines.

❑ Water quality legislation requires that no construction occur within 20 meters of streams.

❑ The site must be at least 4000 square meters in size to provide space for building and grounds.

Figure 1-1 is a key to the following maps. It shows the symbols for land cover, soil suitability, streams, and sewers.[4]

Figure 1-2 is a map showing land cover in the area where the site must be chosen. Different crosshatch symbols indicate different types of land cover; the white area in the northern area of the map is water. The land cover codes and categories are as follows:

LC CODE	LC TYPE
100	Urban
200	Agriculture
300	Brush land
400	Forest
500	Water
600	Wetlands
700	Barren

---

[4]These maps are also available on the CD-ROM that accompanies the text. They are the image files: Key_to_maps.jpg, Landcover.jpg, Soil_suitability.jpg, and Streams_&_Sewers.jpg located in the folder IGIS_Arc_AUX.

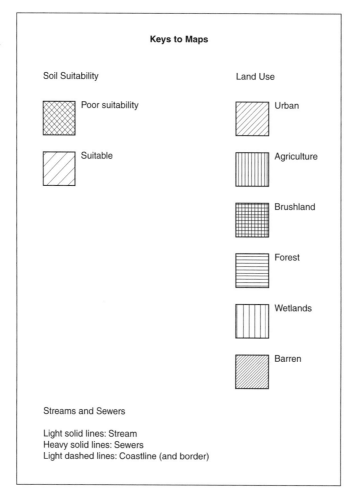

**FIGURE 1-1   Key to maps of the Wildcat Boat facility area
area maps**

Figure 1-3 is a soil suitability map. Lines separate soils of different types. The different soils are categorized as suitable or unsuitable for building. Therefore, you will see the same symbol on both sides of a
dividing line, indicating that while such soil types may be different, their suitability is the same.

Figure 1-4 shows natural streams in the area and human-made sewer lines.

You may use scissors, xerography, a computer-based drawing program, a light table, and any other tools
to solve the problem.

You are asked to present a map that shows *all the areas* where the company could build while meeting
the requirements stated previously. Make your map the same scale and size as those maps provided on
the CD-Rom. Outline in red the areas that meet the requirements. You don't need to produce a high-quality

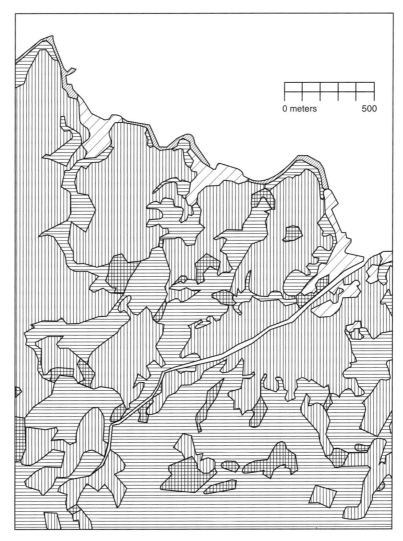

FIGURE 1-2   Land Cover

cartographic product. The main object here—indeed the object of this textbook—is to analyze geographic data. While making maps is important it is not the primary focus of this book.

Write a brief description (100 to 200 words, preferably typed or computer printed) of the procedure you used to make the map.

The problem is much easier than it might otherwise be because the maps provided cover exactly the same area, have the same underlying assumptions regarding the shape and size of Earth, are at the same scale, and use the same projection of Earth's sphere onto the flat plane of the map. These benefits are often not available in the real world, where you frequently need a considerable amount data preparation to solve such a problem.

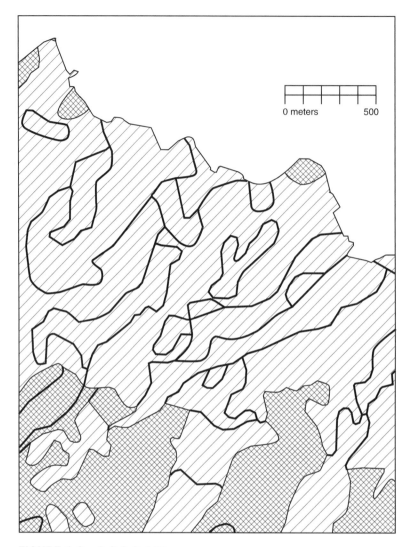

FIGURE 1-3 Soil Suitablilty

More of What GIS Is About

Completing Exercise 1-1 showed how GIS can help you solve one kind of problem. There are many others. Computer-based GIS not only serves the purpose of traditional maps but also helps you perform activities that involve spatial analysis, even without maps. Understanding conditions that occur in the vicinity of the Earth's surface are important in building structures, growing crops, preserving wildlife habitat, protecting ourselves from natural disasters, navigating from one point to the next, and a myriad of other activities.

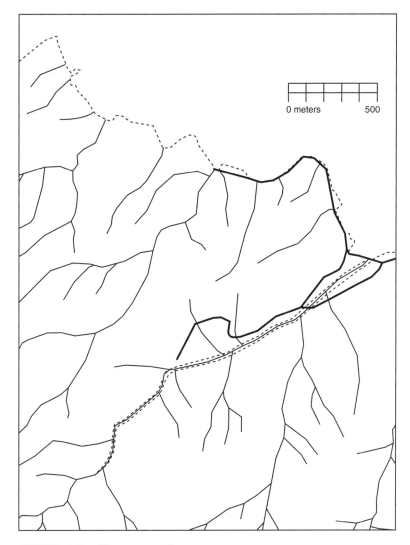

FIGURE 1-4   Streams and Sewers

GIS has many uses:

1   *Land use*—Helps determine land cover, zoning, environmental impact analysis, locational analysis, and site analysis.

2   *Natural environment*—Identifies, delineates, and manages areas of environmental concern, analyzes land-carrying capacity, and assists in environmental impact statements.

3   *Energy*—Examines costs of moving energy, determines remaining available energy reserves, investigates the efficiency of different allocation schemes, reduces waste, reduces heat pollution, identifies

12 USes of GIS   9

areas of danger to humans and animals, assesses environmental impacts, sites new distribution lines and facilities, and develops resource allocation schemes.

Human resources—Plans for mass transit, recreation areas, police unit allocation, and pupil assignment and analyzes migration patterns, population growth, crime patterns, and welfare needs. It also manages public and government services.

Areas of environmental concern—Facilitates identification of unique resources, manages designated areas, and determines the relative importance of various resources.

Water—Determines floodplains, availability of clean water, irrigation schemes, and potential and existing pollution.

Natural resources—Facilitates timber management, preservation of agricultural land, conservation of energy resources, wildlife management, market analysis, resource allocation, resource extraction, resource policy, recycling, and resource use.

Agriculture—Aids in crop management, protection of agricultural lands, conservation practices, and prime agricultural land policy and management.

Crime prevention, law enforcement, criminal justice—Facilitates selection of sites or premises for target-hardening attention, establishment of risk-rating procedures for particular locations, tactical patrol allocation, location selection for crime prevention analysis, crime pattern recognition, and selection of areas or schools for delinquency prevention attention.

Homeland security and civil defense—Assesses alternative disaster relief plans, needs for stockpiling of foods and medical supplies, evacuation plans, and the proper designation of disaster relief areas.

Communications—Facilitates siting of transmission lines, location of cellular equipment, and education.

Transportation—Facilitates alternative transportation plans, locational analysis, mass transit, and energy conservation.

Next Steps: Seemingly Independent Things You Need to Know

Before we launch into the theory and application of GIS, let's look ahead at the remaining text in the Overview of this chapter. You may know some or all of this material already, depending on your background. To use GIS effectively, you should know something about several topics that may seem unrelated at first glance. The next few sections briefly review the relevant aspects of the following:

❑ Determining where something is: coordinate systems

❑ Determining where something is: latitude and longitude

❑ Geodesy, coordinate systems, geographic projections, and scale

❑ Projected coordinate systems

❑ Geographic vs. projected coordinates: which to use

❑ Two projected coordinate systems: UTM and state plane

- ❏ Coordinate transformations
- ❏ Physical dimensionality
- ❏ Global positioning systems
- ❏ Remote sensing
- ❏ Relational databases
- ❏ Another definition of GIS
- ❏ Computer software: in general
- ❏ Computer software: ArcGIS in particular

Determining Where Something Is: Coordinate Systems

Cartesian Coordinate Systems

A coordinate system is a way of determining where points lie in space. We are interested in two-dimensional (2-D) space and three-dimensional (3-D) space. In general, it takes two numbers to assign a position in 2-D space and three numbers in 3-D space.

Coordinates may be thought of as providing an *index* to the locations of points in space, and hence to the features that these points define.

To make a 2-D Cartesian[5] coordinate system, draw two axes (lines) that cross at right angles on a piece of paper. The point at which they cross is called the origin. Arrange the page on a horizontal table in front of you so that one line points left-right and the other toward and away from you. The part of the line from the origin to your right is called the positive x-axis. The line from the origin away from you is called the positive y-axis. Mark each axis in equal linear units (centimeters, say) starting at the origin, as shown in Figure 1-5. Now, a pair of numbers serves as a reference to any point on the plane. The position $x = 5$ and $y = 3$ (shorthand: (5,3)) is shown. The sheet of paper is the x-y plane.

Create a 3-D Cartesian coordinate system from the 2-D version. Imagine a vertical line passing through the origin; call it the z-axis; the positive direction is up. Now you can reference any point in three-dimensional space. The point $x = 5$, $y = 3$, and $z = 4$ (5,3,4) is shown in Figure 1-6.

This is called a right-hand coordinate system. The thumb, forefinger, and middle finger represent the positive axes x, y, and z, respectively. With your right hand outstretched, arrange those three digits so that they are roughly mutually orthogonal—that is, with 90° angles between each pair. Point your thumb to the right and your forefinger away from you. Now, unless you are in considerable pain, your middle finger will be pointing up.[6]

---

[5]Rene Descartes, who lived from 1596 to 1650, made major contributions to both mathematics and philosophy. Descartes is credited with integrating algebra and geometry, by inventing the coordinate system that (almost) bears his name.

[6]It's best not to practice this exercise where other people can see you.

Cartesian Coord. Systems

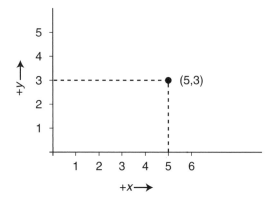

FIGURE 1-5   2-D Cartesian coordinate system

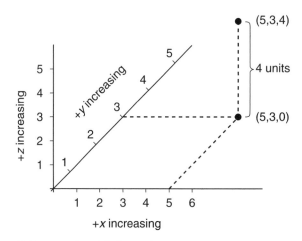

FIGURE 1-6   3-D Cartesian coordinate system

Spherical Coordinate Systems

A spherical coordinate system is another way to reference a point in 3-D space. It also requires three numbers. Two are angles and the third is a distance. Consider a ray (a line) emanating from the origin. The angles determine the direction of the ray. See Figure 1-7.

The latitude-longitude graticule (a gridded reference network of lines encompassing the globe) is based on a spherical coordinate system. As usual, referencing navigation and Earth location issues requires a

Spherical Coord. Systems

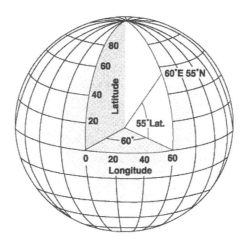

**FIGURE 1-7   Spherical coordinate system,
showing latitude 55° North and longitude
60° East**

different system from the one mathematicians use in more abstract systems.[7] The origin is considered to be the center of the Earth. The equator serves as intersection of the x-y plane and the hypothetical sphere of the Earth. To determine the coordinates of a point, one angle (latitude) is measured from the x-y plane. The other angle (longitude) is contained in the x-y plane and is measured from the meridian that passes through Greenwich, England. The third number in a mathematical spherical coordinate system is the distance along the ray from the origin to the point. When added to the latitude-longitude system, altitude is usually defined instead to be the distance to the point along the ray from mean (average) sea level (MSL) or from a gravity-defined pseudo-ellipsoid used with the NAVSTAR Global Positioning System, to be discussed shortly.

By using three numbers, you can determine and communicate the position of any point on Earth. Of course, a set of parameters must qualify these numbers. Any given point on the surface has probably been addressed by dozens of different sets of numbers, based on the parameters (e.g., units) of the coordinate system chosen. The parameters must match when you combine GIS data.

Determining Where Something Is: Latitude and Longitude

A fundamental principle underlies all geography and GIS: Most things on Earth don't move (or move very slowly) with respect to each other. Therefore, we can talk about the *position* of something embedded in or attached to the ground and know that its position won't change (much). It seems like a straightforward idea, but position confuses a lot of people when it is described as a set of numbers.[8]

---

[7]For example, in the two dimensional Cartesian plane, a mathematician will measure angles starting with the positive x-axis (east) as zero and increasing counterclockwise, to 360 degrees (which is, again, zero). The navigator (think: compass) or cartographer will also use 360 degrees to represent a full circle, but measures clockwise from the positive y-axis (north).

[8]Descriptions of points aren't always just numbers. A possibly apocryphal "meets and bounds" description of a point in Kentucky a couple of centuries ago was "Two tomahawk throws from the double oak in a northerly direction."

Latitude/Longitude

Let's suppose in 1955 somewhere in the United States you (or your parents, or their parents) drove a substantial metal stake or pin vertically into solid ground. Now consider that the object, unless disturbed by humans or natural forces such as erosion or earthquake, would not have moved with respect to the planet since then.[9] In other words, it is where it was, and it will stay there. Three numbers—latitude, longitude, and altitude—could identify the location of the object in 1955. But over the last half century, teams of mathematicians and scientists (skilled in geodesy) developed other sets of numbers to describe exactly the same spot where your object now resides. *The actual position of the object didn't change, but additional descriptions of where it is have been created.*

Ignoring the matter of altitude for the moment, suppose the object had been driven into the ground at latitude 38.0000000° (North) and longitude 84.5000000° (West) according to the calculations done before 1955 that indicated the location of the center of the Earth, its shape, and the location of its poles. Most people and organizations in the United States in 1955 used the North American Datum of 1927 (NAD27) to estimate the latitude-longitude graticules, based on parameters of the ellipsoid determined by Clarke in 1866.[10]

The datum described as the World Geodetic System of 1984 (WGS84) offers the most recent, widely accepted view of the center of the Earth, its shape, and the location of its poles. The ellipsoid of WGS84 is virtually identical to the GRS80 ellipsoid.[11] In the coterminous states of the United States, this datum is virtually identical (within millimeters) to the North American Datum of 1983 (NAD83), though they result from different calculations.

According to the latitude-longitude graticule (WGS84), the object previously described would be at latitude 38.00007792° and longitude 84.49993831°. The difference might seem insignificant, but it amounts to about 10 meters on the Earth's surface. Or consider it this way: According to NAD83, a second object placed in the ground at 38°N latitude and 84.5°W would be 10 meters away from the first one. Does that sound like a lot? People have exchanged gunfire in disputes over smaller distances. Given a latitude and longitude, a GIS must know the datum that is the basis for the numbers. Hundreds of datums exist, and many countries have their own.

Geodesy, Coordinate Systems, Geographic Projections, and Scale

First, a disclaimer: This text does not pretend to cover in detail such issues as geodetic datums, projections, coordinate systems, and other terms from the fields of geodesy and surveying. Nor will the text rigorously define most of these terms. Simply knowing the definitions will mean little without a lot of study. Many textbooks and Web pages are available for your perusal. These fields, concepts, and principles may or may not be important in your use of GIS, depending on your project. But the datum, projection, coordination system designations, and measurement units must be identical when you combine GIS or map information. If not, your GIS project may well produce inaccurate results.

---

[9]Well, it hasn't moved much. If it was on the island of Hawaii, it has moved northwest at about 4 inches per year. If you are unfortunate enough to be in a place where there was an earthquake, it might well have moved and not returned to its original position.

[10]Based on a monument on Meades Ranch, in Kansas; the Clarke 1866 ellipsoid was meant to minimize the error between the itself and the geoid, in the United States.

[11]GRS80 is a global geocentric system based on the ellipsoid adopted by the International Union of Geodesy and Geophysics (IUGG) in 1979. GRS80 is the acronym for the Geodetic Reference System of 1980.

How we apply the mathematically perfect latitude-longitude graticule to points on the ground depends partly on human's understanding of the shape of the Earth. This understanding changes the more we learn. Geodesy is the study of the shape of the Earth and the validity of the measurements humans make on it. It deals with such issues as spheroid and datum. You don't have to know much about geodesy to use a GIS effectively, provided your data are all based on the same spheroid and datum (and projection and units, as we will see later). It is the application of geodesic knowledge that caused the differences in the coordinates of the object that was put into the ground 50 years ago. The object hasn't moved. We simply have a better idea of the location of the object relative to the latitude-longitude graticule.

Projected Coordinate Systems

For many reasons, it's often not convenient to use latitude and longitude to describe a set of points (perhaps connected by straight lines to make up a coastline or a country's boundaries) on the Earth's surface. One is that doing calculations using latitude and longitude, say, determining the distance between two points, can involve complex operations such as products of sines and cosines. For a similar distance calculation, if the points are on the Cartesian x-y plane, the worst arithmetic hurdle is a square root.

Latitude and longitude measures for many geographic applications do not work well for several aspects of mapmaking. Suppose you plot many points on the Earth's surface—say, along the coastline of a small island some distance from the equator—on a piece of ordinary graph paper, using the longitude numbers for x-coordinates and latitude numbers for y-coordinates. The shape of the island would look strange on the map compared to how it would appear from an airplane. You would not get useful numbers if you measured distances or angles or areas on the plot. This is due to a characteristic of the spherical coordinate system: The length of an arc of a degree of longitude does not equal the length of an arc of a degree of latitude. They are almost equal near the equator, but the difference grows as you go further north or south from the equator. At the equator a degree of longitude translates to about 69.17 <u>miles</u>. Very near the North Pole a degree of longitude might be 69.17 <u>inches</u>. (A degree of latitude, in contrast, varies only between about 68.71 miles near the equator and 69.40 miles near the poles.)

For relatively small areas, mathematically projecting the spherically defined locations onto a plane provides a good solution to problems associated with calculation and plotting. Geographic projection might be thought of as imagining a process that places a light source inside a transparent globe that has features of the Earth inscribed on it. The light then falls on a flat piece of paper (or one that is curved in only one direction and may be unrolled to become flat).[12] The shadows of the features (say, lines or areas) will appear on the paper. Applying a Cartesian coordinate system to the paper offers the advantages of easy calculation and more realistic plotting. However, distortions are inherent in any projection process; most of the points on the map will not correspond exactly to their counterparts on the ground. The degree of distortion is greater on maps that display more area. Accuracy will suffer when you flatten a curved surface and convert a three-dimensional coordinate system to a two-dimensional one.

After constructing geographic data sets according to latitude and longitude, based on some agreed-upon spheroid (such as GRS80) and datum (such as WGS84), you decide how to represent them graphically for viewing. At one time, cartographers went directly from the latitude-longitude description of an area

---

[12]This description is a sort of a cartoon to describe a map projection. A map projection is actually a mathematical transformation that "maps" points on the globe to points on a plane; the process may be quite complex depending on the projection; a single light source at the center of the globe does not suffice to explain it.

Projected Coord. Systems

or feature to a graphical portrayal on paper. This usually involved "projecting" the data from a three-dimensional (3-D) spherical coordinate system to a two-dimensional (2-D) Cartesian one, and setting a scale: A certain number of units on a map represented that number of units on the ground. Using GIS changes this. As I commented earlier, latitude-longitude is the most fundamental and accurate way to represent spatial data. So large areas—those that are most subject to distortion by being projected—may best be left in latitude-longitude coordinates, and subareas projected to other coordinate systems when needed. Computers are really good at doing the complex computations necessary to project data.

The larger the area projected, the greater the tendency for things not to be where the coordinates say they are.

Four considerations for viewing and analysis are size, shape, distance, and direction. Myriad projections have been invented. Many distort size, shape, distance, or direction, and some preserve one or two of them.

Regarding linear units, GIS differs from cartography in that matters of scale may be left until the very end. The position defining numbers in the database should be real-world coordinates—not scaled coordinates. That is, GIS "maps" are stored in ground units rather than map units. Besides being a more fundamental way to store data, this makes it possible to easily make maps of any desired scale on the computer monitor or on paper. Scale is only a consideration when you measure on a physical map. Computers take the worry out of determining scale accuracy. Because of the vast computational power of a computer, there is no difficulty in scale conversion. The days of: "Let's see, this distance is 5.3 inches on the map, and one inch is 12 miles, so the distance is about 64 miles" are over. Scale is a minor concept in GIS—one used only on final output.

Geographic vs. Projected Coordinates: A Comparison

❑ Advantages of the spherical coordinate system: You can represent any point on the Earth's surface as accurately as your measurement techniques allow. The system itself does not introduce errors.

❑ Disadvantages of a spherical coordinate system: You will encounter complex and time-consuming arithmetic calculations in determining the distance between two points or the area surrounded by a set of points. Latitude-longitude numbers plotted directly on paper in a Cartesian coordinate system result in distorted—sometimes greatly distorted—figures.

❑ Advantages of a projected coordinate system on the Cartesian plane: Calculations of distances between points are trivial. Calculations of areas are relatively easy. Graphic representations are realistic, provided the area covered is not too large.

❑ Disadvantages of a projected coordinate system on the Cartesian plane: Almost every point is in the wrong place, though maybe not by much. All projections introduce errors. Depending on the projection, these errors are in distances, sizes, shapes, or directions.

Matching parameters is paramount in combining GIS data sets![13]

---

[13]The parameters of data sets may not always match but may be close enough so that any error introduced is trivial. In many instances and places, for example, NAD 83 and WGS 84 match within centimeters. But you must research carefully to know when you can use data sets that don't match exactly. The ESRI manual *Understanding Map Projections* can help.

Geog vs Projected Coordinates

Two Projected Coordinate Systems: UTM and State Plane

A coordinate system called Universal Transverse Mercator[14] was developed based on a series of 60 projections onto semi-cylinders that contact the Earth along meridians. (To consider, for example, one of these projections, imagine a sheet of paper bent so that it becomes a half cylinder. Then with the axis of the cylinder oriented in an east-west direction—hence, the term *transverse*—the paper is brought into contact with a globe along the meridian designating 3° longitude. Then the surface of the Earth between 0° and 6° is projected onto the paper). UTM projections are further subdivided into areas, called *zones*, covering 6° of longitude and, for most zones, 8° of latitude. A coordinate system is imposed on the resulting projection such that the numbers

❏ Are always positive

❏ Always increase from left to right

❏ Always increase from bottom to top

The representation of our previously discussed object (at 38°N and 84.5°) in the UTM coordinate system, when that system is based on WGS84, is a "northing" of 4,208,764.4636 meters and an "easting" of 719,510.3358 meters. The northing is the distance to the point, in meters[15], from the equator, measured along the surface of an "Earth" that has no bumps. The easting is somewhat more complicated to explain because it depends on the zone and a coordinate system that excludes negative numbers. Consult a textbook on geodesy or cartography, or review the thousands of Web pages that come up when you type the words

```
UTM "coordinate system" "Transverse Mercator"
```

into an Internet search engine (e.g., www.google.com).

A version of the UTM coordinate system is based on NAD27. In this case, our object would have different coordinates: Northing 4,208,550.0688 and easting 719,510.6393. This produces a difference of about 214 meters. If you combined WGS84 UTM data with NAD27 UTM data, the locations they depict would be close enough to cause trouble.[16]

Each state of the United States has State Plane Coordinate Systems (SPCSs) developed, originally in the 1930s, by the U.S Coast and Geodetic Survey. These systems are based on different projections (usually, Transverse Mercator or Lambert Conformal Conic), depending on whether the state is mostly north-south (like California) or mostly east-west (Tennessee). The units represent international feet, survey feet, or meters, depending on decisions made by the state itself.[17] Zone boundaries frequently follow county

---

[14]The idea of the Mercator projection was developed in 1568 by Gerhardus Mercator a Flemish geographer, mathematician, and cartographer.

[15]In fact, the meter was originally defined as one ten-millionth of the distance from the equator to the North Pole, along a meridian that passed through Paris.

[16]Comparing these coordinates with the WGS 84 UTM coordinates, you see that virtually all of the difference is in the north-south direction. While true for this particular position, it is not true in general.

[17]The meter is the standard unit of length in most places in the world. Two different lengths of "foot" are defined in terms of the meter. A "U.S. Survey Foot" is one in which a meter is considered to be 39.37 inches, exactly; the other sort of foot is the international foot, where an inch is 0.0254 meters, exactly. The survey foot and the international foot are almost, but not exactly, the same length.

UTM and State Plane System

boundaries. The coordinate system(s) used in one state are not applicable in neighboring states. Nor can you apply the SPCS of one zone to areas in the state only a short distance away in another zone. Furthermore, the difference between NAD 27 and NAD 83 (WGS 84) can be startlingly large. In Kentucky, for example, 38.0000000° (North) and longitude 84.5000000° (West) would translate into a northing of 1,568,376.1900 feet and an easting of 182,178.3166 feet when based on WGS84. However, when the basis is NAD27, the coordinates are 1,927,939.8692 and 182,145.9821, which makes a difference of some 68 miles!

Why the large differences in projected coordinate systems based on NAD27 versus those based on WGS84? Because those responsible for the accuracy of other coordinate systems took advantage of the development of WGS84—a worldwide, Earth-centered, latitude-longitude system—to correct or improve those projected coordinate systems.

State plane coordinate systems are generally designed to have a scale error maximum of about 1 unit in 10,000. Suppose you calculated the Cartesian distance (using the Pythagorean theorem) between two points represented in a state plane coordinate system to be exactly 10,000 meters. Then, with a perfect tape measure, pulled tightly across an idealized planet, you would be assured that the measured result would differ by no more than 1 meter from the calculated one. The possible error with the UTM coordinate system may be larger: 1 in 2500.

Coordinate Transformations

Coordinate transformation of a geographic data set is simply taking each coordinate pair of numbers in that data set and changing it to another pair of numbers that indicates exactly the same spot on the Earth's surface, but using a different system of assigning coordinates.

Let's make up a coordinate system for a garden delineated by lines between stakes. Suppose the origin (0,0) is at the southwest corner. Now drive several stakes in the ground so that if you passed a string around them they would outline a polygon. Suppose you use a surveyor's tape calibrated in survey feet to measure the Cartesian coordinate for each stake. Stake 0 is at the origin (0,0). Stake 1 is 33 feet east and 0 feet north. (i.e., 33,0). Stake 2 is at (33,10). And so on. See Figure 1-8

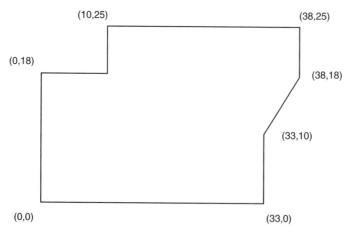

FIGURE 1-8   **Locations of garden stakes (coordinates in feet)**

Coord. Transform

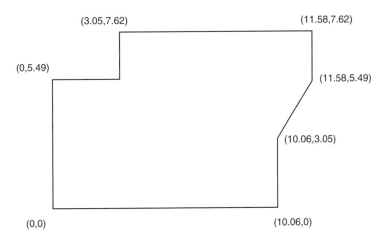

FIGURE 1-9   Locations of garden stakes (coordinates in meters)

The garden grows and becomes overgrown. Next year someone locates the origin and wants to find the locations of the original stakes. He also has a surveyor's tape, but it is calibrated in meters instead of feet. He asks you to provide the data on where the stakes are in meters. In this case, the transformation of coordinate systems is easy: You simply multiply each number by a conversion factor (the number of meters per survey foot, which is 0.3048). So Stake 1's coordinates are (10.06, 0). Those of Stake 2 are (10.06, 3.05). And so on. See Figure 1-9.

The stakes are in the same place that they were before. Their positions are simply referred to by a different set of numbers. Geographic coordinate or datum transformation is nothing more or less than this, except that the mathematical operations applied to each number are usually more involved.

Physical Dimensionality

All matter exists in four dimensions—roughly characterized as left-right, toward-away, up-down, and time. The first three are conceptually clumped together and called spatial dimensions. Time is treated as a dimension here because we want to talk about "position" in space and time.[18]

Any physical object occupies space and persists in time. This is true of the largest object in the universe and the smallest atom. However, when the measure of one or more dimensions of an object is small, or insignificant, with respect to the measure of others or is tiny with respect to its environment, it is useful to describe and represent objects by pretending that they occupy fewer than three spatial dimensions. For example, a sheet of paper can be considered a pseudo-two-dimensional object; it has thickness (up-down), but that dimension is miniscule compared with left-right and toward-away. A parking meter in

---

[18]In physics, time is treated as the fourth dimension because it is inexorably bound up with the other three according to Einstein's theories (special in 1905 and general in 1916) of relativity. This connection need not concern us except for some esoteric technical matters related to time on global positioning system satellites (for one, time on a satellite runs faster because it is farther from Earth, where there is less gravity).

Physical dimensionality

19

a city could be considered a pseudo-zero-dimensional object, because the measures of all its dimensions are insignificant with respect to its environment. (The parking meter would certainly not be considered a spatially zero-dimensional object by its designer, manufacturer, or the driver who uses it. Therefore, the pseudo-dimensionality of an object depends on its context.)

It is important to consider the pseudo-dimensionality of both an object and the field (the space) in which it resides. A point on a line is a zero-dimensional object in a one-dimensional space. A line segment on a plane is a one-dimensional object in a two-dimensional space. A polygon is a two-dimensional object in a two-dimensional space. A point in a volume is a zero-dimensional object in a three-dimensional space. The dimensionality of the object must be less than, or equal to, the dimensionality of the space in which it resides.

This issue of dimensionality comes into play here because great economies of computer storage are achievable if an object is considered to have fewer dimensions than it actually does. For example, describing a fire hydrant as a complex three-dimensional spatial object would be an arduous task and involve many numbers and much text. If it is considered a zero-dimensional object, its location can be described (very precisely, but still inexactly) with three numbers, perhaps representing the latitude, longitude, and altitude of some point on the hydrant or, perhaps, its center of mass.

The spatial dimensions of a spatial pseudo-zero-dimensional object are insignificant with respect to the context or environment in which it resides. A theoretical geometric point is the prototypical example. Examples of such zero-dimensional objects on maps or in a GIS could be streetlights, parking meters, oil wells, census tracts, or cities—depending on the spatial extent of the other objects or features in the database.

A spatial pseudo-one-dimensional object or feature has no more than one significant dimension or is made up of essentially one-dimensional objects. A straight-line segment is the prototypical example. In terms of making a one-dimensional object up from component parts, consider the example of a telephone wire strung from one pole to the next to the next and so on. It is considered a one-dimensional object, even though the poles may not be in a straight line, so that the phone line zigzags over a two-dimensional field. (The line itself sags and exists therefore in a three-dimensional context; but these facts notwithstanding, it can be simplified or generalized into a one-dimensional feature—saving us lots of computer storage, processing time, and conceptual complication.) Roads, school district boundaries, pipes for fluid transportation, and contour lines on a topographic map are all examples of one-dimensional entities.

You can disregard the up-down dimension in a spatial-pseudo-two-dimensional object. Plain areas such as voting districts and soybean fields are examples. Such pseudo-areas lack vertical components—variability in altitude—that can be important. The true surface area of hilly terrain may be considerably underrepresented by the plane area within the borders of the plane figure that represents it.

You must consider all dimensions of a spatial three-dimensional object in order to represent it. Three-dimensional features are volumes—say, a coal seam or a building.

Relative to each other, positions of features that we record with a GIS do not move, or do not move quickly, with respect to each other. We want to know how conditions change or objects move. The representation of the objects may be zero-, one-, two-, or three-dimensional. At its most complex, GIS would involve all four dimensions. Most GIS operations and data sets are two-dimensional; the data are assumed to pertain to moments in time, or stable phenomena or conditions over a period of time. Three-dimensional GIS could involve either three spatial dimensions (such as representing the volume of a limestone quarry) or two spatial dimensions and varying time (showing historical changes in a land-scape). Rarely will you use four-dimensional GIS.

Results from a GIS are always only an approximation of reality. One of the reasons, among others to be discussed later, is that we simplify objects by reducing their dimensionality.

Global Positioning Systems

A global positioning system (GPS) is a satellite-based system that provides users with accurate and precise location and time information. Using NAVSTAR GPS, you can determine locations on Earth within a few meters, and, with more difficulty and expense, within a few centimeters or better. Timing within 40 billionths of a second (40 nanoseconds) is easily obtained. Timing within 10 nanoseconds is possible.

The U.S. Department of Defense operates NAVSTAR GPS in cooperation with the U.S. Department of Transportation. The acronym stands for "Navigation System Timing and Ranging Global Positioning System." Informally, it is Navigation Star.

The Russian GLONASS (Global Navigation Satellite System) operates similarly. Concerns about U.S. control over NAVSTAR led Europe to begin development of its own independent Galileo system in 2002.

A GPS receiver, which "remembers where it has been," is becoming a primary method of providing data for GIS. For example, if you drive a van with a GPS antenna on its roof along a highway, recording data every, say, 50 feet, you will develop an accurate and precise map of the location of the highway. The NAVSTAR GPS is discussed in Chapter 5.[19]

Remote Sensing

Remote sensing probably started with photographs taken from balloons in the 1840s. The first automated system (not requiring humans to be with the sensors) may have been in the 1890s in Europe when cameras programmed to take pictures at timed intervals were strapped to pigeons!

Evelyn Pruitt probably introduced the modern use of the term "remote sensing" in the mid-1950s when she worked as a geographer/oceanographer for the U.S. Office of Naval Research (ONR). Remote sensing uses instruments or sensors to capture the spectral characteristics and spatial relations of objects and materials observable at a distance, typically from above them. Using that definition, everything we observe is remotely sensed. More practically, something is sensed remotely when it is not possible or convenient to get closer.

We can categorize remote sensing for GIS many ways. Data can be taken from aircraft or satellite "platforms." The energy that the sensor "sees" can come from the objects or areas being examined as a result of radiation emanating from them (caused by the sun or other heat or light sources) or from radiation bounced off them by an energy source associated with the sensor (e.g., radar). The images produced may be developed on film or produced by digital sensors. Satellites in geosynchronous orbits (hanging in a single spot over the equator) or near polar orbits can see different areas of the Earth as it turns. Chapter 2 offers examples of remotely sensed data.

---

[19]See also the author's textbook *The Global Positioning System and GIS*, 2002, Taylor & Francis, 375 pages.

GPS & Remote Sensing

Relational Databases

GIS relies heavily on databases of text and numbers. The relationship between such information and geography is discussed later in the chapter. For now, you simply need to know how text and number data sets are stored.

For this discussion, a *database* is a collection of discrete symbols (numbers, letters, and special characters) located on some physical medium with at least one principal underlying organization or structure. An old-fashioned library card catalog is an example of a database with a single underlying structure: an alphabetical list of authors; the medium is 3×5 index cards; and the data are the symbols on the cards describing books and their locations in the library. Many libraries have substituted computer-based catalogs with the advantages of searching and finding not only authors but titles and subjects as well, so a number of organizing themes may underpin a database.

Another example is a reel of magnetic computer tape that has recorded the most common type of soil found in specific acreage in a county. The tape is the physical medium, the codes assigned to a soil type constitute the data, and the location of each acre—as understood from the position of each datum on the tape—could be the underlying structure.

Existing general purpose databases usually:

❑ Result from some sort of project; some individual or team constructs it—frequently going to considerable effort.

❑ Need to be *updated* (modified and corrected as time progresses) if they are to continue to be of value.

❑ Contain errors *regardless* of size, care of construction, simplicity of data, or quality of physical medium used.

❑ Serve a function when allied with some process. The function may be as simple as supplying a telephone number (structure—alphabetical by name; medium—cheap bound paper; data—phone numbers, written quite small; process—looking up a name, finding the adjacent number). The function served by the database might also be quite sophisticated—supporting far-reaching land use decisions, for example.

Numerous databases are used to solve problems at all levels of government. Access to these databases is achieved through the use of referencing schemes. Following are some examples:

Referencing Scheme	*Examples of Data Contained*
Names of people	Salary, social security number, medical history, criminal record
Auto license plate number	Color, owner, serial number
Street address	House value, lot size
Job title	Person employed, duties, salary
Transaction number	Money received, paid, transferred, invested
Events	Schedules, orders, crimes, accidents

Many schemes exist for presenting or storing data. Suppose parents want to find a name for a newborn child. A list of potential names in random order might be provided. It would probably be more useful if the list were divided into girl's names and boy's names. Another improvement could be to alphabetize the list, or present it in terms of current popularity of names.

Relational database

As a second example, consider how you might store telephone numbers associated with names and addresses. You could order this list alphabetically so you could easily find a phone number, knowing the name. Or you could order the list numerically by telephone number so you could find a name, given the number.

A different sort of data structure is "hierarchical." You could use this approach to store the names and positions of people in a corporation or the military according to who reports to whom. Likewise you could store voting districts within cities within counties within states. In the computing world this "tree-structured directory" approach is used to organize the folders and files contained on a hard disk drive on a computer. Any folder on the disk may store other folders and files.

Many schemes exist to store information in the memory of a computer or on its secondary devices, such as disk drives or tapes. The primary method used to store large amounts of information is called the "relational database," or RDB, developed by E. F. Codd.[20] The software is described as a RDBMS (Relational Database Management System). The idea is simple: use a two-dimensional table format in which the rows relate to entities (objects, people, things in general), while the columns relate to attributes (characteristics, properties) of entities. The intersection of a given row and a given column is a cell, containing a value, which defines the particular attribute of the particular entity. See Figure 1-10.

Relational Database Nomenclature

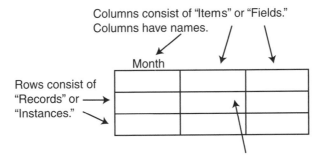

FIGURE 1-10   Components of a relational database table

You can use a database to store names, occupations, and pay schedules of employees, as in Figure 1-11.

Another example shows part of a database of automobiles registered in a state. Each row would represent one car; each column would represent one property of cars. See Figure 1-12.

Here is some terminology: The structure that contains the entity, the row, is also called a *tuple* or *record*. The structure that contains the attributes, the column, is also a *field* or an *item*.

---

[20]Edgar F. Codd was the originator of the relational approach to database management, beginning in 1970. For this work he was the recipient of the Association for Computing Machinery's (ACM) Turing award in 1981.

Relational Database Nomenclature

Garden-Variety Database

Name	Occup	Pay_Sched
Smith	Welder	Hourly
Jones	Accountant	Bi-Monthly
Adams	Attorney	Monthly

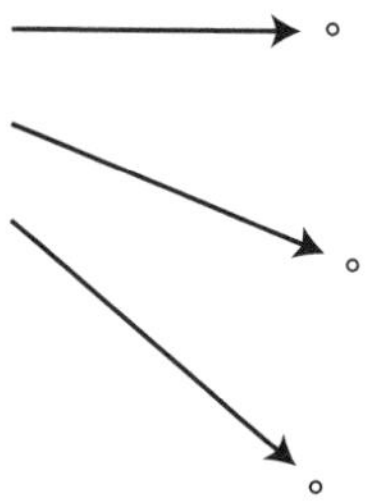

FIGURE 1-11   Trivial example of a relational database table

Mfgr	Num_Doors	VIN	Color	Weight
Porsche	2	123XXX	Silver	2300
Porsche	2	887ABC	White	2100
Toyota	4	9880123	Grey	2350
Honda	4	456789	Blue	2999

FIGURE 1-12   A RDB table of automobiles

All the cells of a given column must contain the same sort of value. Some common ones are as follows:

❑ *Character*—Any valid ASCII characters in a string of almost any length.

❑ *Short integer*—Can range from slightly less than –32,000 to slightly more than +32,000.

❑ *Long integer*—Can range from slightly less than negative 2 billion to slightly more that positive 2 billion.

❑ *Floating-point number*—Can have an exponent as small as 10^{-38} and as large as 10^{38} and you can count on six significant digits.

❑ *Double-precision floating-point number*—Can have an exponent as small as 10^{-308} and as large as 10^{308} and you can count on 16 significant digits

❑ *Boolean*[21]—A value that is either true or false.

---

[21]Named after George Boole, mathematician.

Getting Information from a Relational Database: Queries

Relational databases are designed to give you information. You can obtain the information by selecting a subset of records from the total set by writing an expression that is a mixture of attribute names, arithmetic and logical operators, and values.[22] For a trivial example, suppose research has found that gray cars that weigh less than 2,500 pounds put their occupants at greater risk than average. You want to select those records from the statewide automobile database. You might first get all the records of the cars that are gray.

    SELECT: COLOR = 'Gray'

Given that subset, you might then write

    SELECT: WEIGHT < 2,500

Given this sub-subset of records, you could perhaps write letters to the owners of those cars, making them aware of the danger they face.

Languages to get subsets of records can provide flexibility and efficiency. For example, to do both of the preceding operations with one expression, you might write

    SELECT: COLOR = 'Gray' AND WEIGHT < 2500

Suppose also that the research study showed increased danger to those occupying cars that were built in 1985 or before, regardless of color or weight. You might add to the selection above by saying:

    ADDSELECT: YEAR <= 1985 (ADDSELECT means add to the current set of selected
    records.

Or you may use a single query to select all the records you want at once:

    SELECT: (COLOR = 'Gray' AND WEIGHT < 2500) OR YEAR <= 1985

Note the use of parentheses to indicate the order in which operations are done.

Economies in Relational Databases

In theory, the information in a relational database system could reside in a single table. This isn't the best policy and in reality may not even be possible. A relational database usually consists of a set of tables that relate to one another—thus the word "relational."

Each relational database table must contain a column that has a unique identifier for each record in the table. This is known as the key field. If the relational database references people, the key might be social security number, employee number, or student number. In the case of automobiles, the key code might be the vehicle identification number (VIN).

---

[22]There are many relational database software products and several different languages used to query them. What is shown in the following is not a specific query language but rather a generic representation of such a language. ArcGIS may be used with several relational database products, each with a somewhat different query language.

Queries   Key field

To illustrate why multiple tables might be used, suppose in our relational database it is possible for one person to own more than one automobile. If the entire database is all in one table, then the names and addresses of each multicar owner must be repeated, which increases the amount of storage required. If a multicar owner gets a new street address, several records in the database table would have to be changed. Further, some of the data about owners might be located in another database and might be private. The elegant answer to these problems is to have several tables that contain the information, divided to minimize the repetitions. In the automobile example, it might be that only the owner's identification number is stored in the record of the car itself. This number could be the key field in the database table containing information about the owner. If some of the owner information is private, it could be stored in a separate table as well, with the same key field. The database could be set up so that this table could not be accessed by the automobile table.

Relational database tables that meet certain requirements of efficiency are said to be in First Normal Form (1NF), Second Normal Form (2NF), and so on. The higher the number, the more efficient the database.

In summary, it is useful to partition the information in a relational database into a number of tables. Such partitioning can:

❏ Reduce the amount of redundant data stored.

❏ Aid in updating the database.

❏ Reduce the chances of inconsistency and instability in the database.

❏ Aid in protecting private or sensitive data.

Languages have been developed for retrieving information from a relational database. One is the Structured Query Language (SQL), developed by the American National Standards Institute (ANSI). You use it to execute queries of a database. Vendors of relational database management systems also have developed proprietary languages.[23] ESRI products interface with a number of RDBMS from various sources.

Relational Databases and Spreadsheets

A relational database table may look a lot like an electronic spreadsheet, such as Microsoft Excel. There are some important differences:

❏ A relational database is structured, with rows and columns strictly defined. In a spreadsheet you can put anything anywhere.

❏ In a relational database, the headings of columns are not stored in cells of the relational database. The column headings (attribute names) are known to the database software and are displayed, but are not part of the data. In a spreadsheet, column headings occupy cells.

❏ A relational database is a logical object about which conjectures can be made and theorems proved. An entire branch of computer science is devoted to work with relational databases.

---

[23]Some names of past and present relational database management system software products are Access, dBASE, Informix, Ingres, Oracle, SQL Server, and Sybase.

Another Definition of GIS

Each record in a relational database references something: a person, object, idea, equation, feature, subject—some unique entity. In a GIS, the entity is usually a spatial three-dimensional feature such as a parking meter, lake, railroad, and so on.[24] These 3-D features are almost always reduced to abstract objects of fewer dimensions. The parking meter is a point, the railroad is a set of lines, and the lake is an area bounded by a sequence of lines.

Usually, a record in a relational database references a person or object without respect to its current location. A subject of a record in a relational database frequently moves around—like cars or people. But when the subjects of a relational database are fixed in space, the position, or positions, of the feature may be included in the description of the feature along with the attributes. In one sense, the location of the object becomes one of its attributes.

Points

Here, for example, is a map and part of a relational database that together describe parking meters in a town. Each meter has a unique number and is thus the key field. Parking meters (each of which has a latitude and longitude specification) correspond one-to-one with the records in the database. See Figure 1-13.

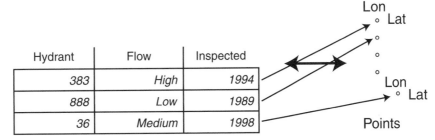

Hydrant	Flow	Inspected
383	High	1994
888	Low	1989
36	Medium	1998

FIGURE 1-13   Marriage of a point geographic database and a relational database table

Let's consider another definition of a GIS: From a computer software point of view, a GIS is the marriage of a (geo)graphical database and an attribute database (frequently a relational database).

Lines

As I mentioned earlier, a GIS can store and analyze pseudo-one-dimensional entities such as roads. Here each RDB table record would refer to segments of a roadway between intersections. The attributes might be number of lanes, highway number, street name, pavement type, length, and so on. See Figure 1-14.

---

[24]In raster-based GIS, the entity is a set of square areas that share common values. This approach will be discussed extensively in Chapter 8.

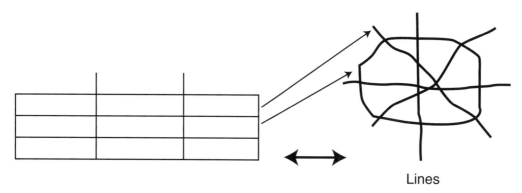

Lines

FIGURE 1-14   Marriage of a line geographic database and a relational database table

Again: A GIS is the marriage of a (geo)graphical database and an attribute database.

Areas

A GIS needs to be able to store information about areas, as well as points and lines. For example, such areas might be ownership parcels. Each parcel would be delineated by the lines determined by a surveyor. Those lines define an area. A record would exist in the relational database that would correspond to the area. The attributes in the record might be owner's name, tax identification number, area of the parcel, and perimeter of the parcel. See Figure 1-15.

Did I mention? *A GIS is the marriage of a (geo)graphical database and an attribute database.*

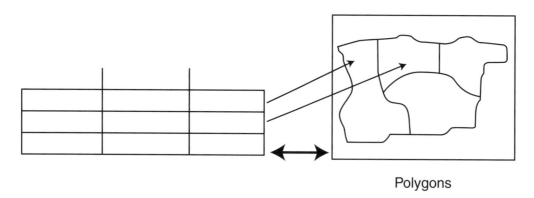

Polygons

FIGURE 1-15   Marriage of a polygon geographic database and a relational database table

Geo + Attrib. databases = GIS

Computer Software: In General

Computers, considered at the most fundamental level, do only three things:[25] get input, manipulate data, and produce output; that is, read bits, stir bits, and write bits. What tells the computer what to do? In the early days of computers, more than a half century ago, each individual instruction to a computer came from outside, one at a time. This process was soon automated so that external media, like perforated paper tape or magnetic tape, contained the instructions. Then several scientists[26] got the idea of placing the instructions in the store of the computer itself. This major breakthrough in computer development allowed the computer to execute one group of instructions and then, on the basis of testing bits in its memory (e.g., is number "a" larger than number "b"?), execute another group of instructions in a different place in the memory. This let the computer simulate "reasoning" and "decision making." This concept of a *stored program* revolutionized computer use.

The modern computer holds an immense amount of data and stores a large number of programs. If a given program is being used at a given time, it exists in the "fast" electronic memory of the machine. A program or data set not being used usually resides in "slow" disk (usually electromechanical) memory. Dealing with large numbers of diverse elements requires a management scheme. With computers this takes the form of an *operating system* that allows the user to execute specific programs when desired, connects the computer to other computers (e.g., by way of the Internet or other network), makes copies of files, and performs many housekeeping activities.

Writing sets of instructions to be stored in the computer's memory is known as *programming*. Developing and marketing *programs* (software packages) is an immense business. The operating system that your computer uses is a software package. The word processor used to write this text is a software package. And the GIS you are about to learn, known as ArcGIS from ESRI, is a software package, or rather a suite of software packages.

Computer Software: ArcGIS in Particular

ArcGIS is an integrated GIS that consists of three principal parts:

> *ArcGIS Desktop* software, which is an integrated suite of advanced GIS applications
>
> ArcSDE *Gateway*, which is an interface for managing geodatabases in a database management system (DBMS)
>
> ArcIMS software, an Internet-based GIS for distributing spatial data and services.

ArcGIS has software extensions such as ArcPad, which provides GIS capability for palmtop or handheld computers running under the Windows CE operating system.

ArcGIS is called a *scalable product* because it allows for deployment of GIS of many different "sizes." A user can choose to have only a system running on a single computer—a *personal GIS*—that allows viewing and simple selection of spatial data without capabilities of editing or extensively analyzing that data. At the other end of the scale, a *multiuser GIS* can serve an entire company or governmental agency. This *enterprise GIS* allows all the capabilities of the software. Options also include systems of intermediate functionality known as *workgroup GIS* and *departmental GIS*.

---

[25]Fundamentals that we'll discuss further in Chapter 6.
[26]John von Neumann, J. Presper Eckert, John Mauchly, Arthur Burks, Maurice Wilkes, and others.

For many products, the price charged depends on the features and utility that the product provides. When one manufacturer produces many different products, users must learn how to combine and package them to provide the correct degree of utility required (you may not need a school bus if a sedan will do). This is also true of ESRI products.

Software Focus of This Textbook

In this introductory text, we confine ourselves largely to ArcGIS Desktop. The other two products pertain primarily to managing GIS data for large applications or organizations and to distribute GIS capability and data to users over networks. Our emphasis is rather on the analysis, synthesis, and display of spatial and attribute information.

ArcGIS Desktop is itself divided in a number of different ways. It has three main software packages:

❑ ArcCatalog

❑ ArcMap

❑ ArcToolbox

ArcCatalog is mainly an operating system for geographic data. ArcCatalog allows you to set up shortcuts to reach particular files of data quickly and easily and examine those files visually and textually. You can use ArcCatalog to rename and copy spatial data sets.

ArcMap lets you put together graphic and geographic elements to make sophisticated maps and interact with them. It allows you to obtain information from those maps and the attributes associated with their components, by using a variety of processing methods.

You use ArcCatalog or ArcMap to access Arc Toolbox, which provides immense geo-processing and analysis capability. It has tools to do 3-D analysis, spatial analysis, and spatial statistical analysis, to convert from one spatial data paradigm to another, and to convert spatial data into a myriad of geographic projections.

These three packages are fundamental to ArcGIS. But you have probably noticed many other Arc*Xxxxxx* terms floating around. As a scalable product, ArcGIS desktop allows you to purchase what you need. The different levels of capability are bundled into four different packages. In terms of increasing utility (and cost) they are as follows:

ArcReader (free)
ArcView
ArcEditor
ArcInfo

Knowing the capabilities of each is not important for our immediate purposes. You should know that some of these terms are recycled from earlier ESRI product names. In particular, ArcView 9 is a different animal than ArcView 3.x. ArcView 9 is a level of capability of ArcGIS Desktop; ArcView 3.x is an older, but still supported and useful, independent software package.

ArcCatalog, ArcMap ArcToolbox

Some Concepts That Underpin GIS

Developing a Fast Facts File for the Information You Learn

To make this textbook work well for you, I strongly recommend that you create and maintain a Fast Facts File—a computer text file in which you can record what you learn so that it is at your fingertips. It will serve you whenever you need to know a particular bit of information or perform an operation. Much of what you do you will put into your own fast memory (contained in your cranium), but there will be a number of facts that your fast memory may not contain when you need them. Here's where the Fast Facts File comes in. It is a computer-based equivalent of a loose-leaf notebook that you continually revise and update. There's where you should put procedures and concepts you might forget after a couple of weeks of doing work other than GIS.

For example, you might use the file to note techniques for changing symbology (colors and symbols that represent features on a map), which are addressed at various different points in the text and in the software. You can note the techniques down as you work with them and reorganize them later. The computer-searchable file helps you find what you need when you need it, even if you fall behind in organizing.

Periodically reorganize your notes. Occasionally print out the file and put it in a notebook. Periodically back up the file onto a flash drive, CD-ROM, or other computer. As you progress through this text, you will develop your own little book that will aid you in this course and thereafter. And, because you write it, it will be organized in a way that meets your needs—both as a tutorial and a reference volume.

‾✓‾ **1.** Use a text editor (e.g., WordPad) or word processor (e.g., Microsoft Word) to create the Fast Facts File. Put your name and other contact information in it. Initially include your computer account name (not your password—put that somewhere else), how to start various software programs, and so on. The file will evolve as you go on. For the moment, start the file and keep it open. It should at least contain the following:

Name: _____*Fast Facts File*_____

User or Logon Identifier: _____

Password hint or secure location: _____

Do not save the file yet.

---

[1]The length of time most students will require to complete the exercises in the Step-by-Step sections can be found in the Preface.

Understanding the File Structure for the Exercises

You will be working primarily with two major folders or directories. The first of these folders will be IGIS-Arc. It will contain the data for all the exercises. It will start off as, and will remain, a direct copy of the IGIS-Arc folder that is on the CD-ROM in the back of the book. If you are in a class, operating off a network, your instructor will probably have put this folder somewhere in the network file structure. If you are using this book on your own, you should load the IGIS-Arc folder that is on the CD-ROM directly onto your local hard drive. In either case, the IGIS-Arc directory should be protected so that changes cannot be made to it.

2. Locate or create the IGIS-Arc folder. It may be in the root folder of a hard drive on your computer, or it may be several levels down in a hierarchy. Whenever it is referred to in this textbook, it will be represented as:

[__] IGIS-Arc

The symbols [__] might represent something as simple as C:\

Or it might be a long path such as:

U:\ABCNet\GIS_Students\GIS401

Write the path associated with [__], in your computer system, below:

[__] means _____

The second folder you will use to store the work that you do. It will be specifically yours; your initials will be made part of the folder name. For example, if your name were John W. Stephenson, the folder would be called IGIS-Arc_JWS. In the next step, you will create this folder.

3. Decide where on the computer or network you want to store your work. Use the operating system of the computer to make a new folder by navigating to and selecting the appropriate drive or path, clicking the File menu, and clicking Folder when it shows up under New. Once the folder is created, change its name to

IGIS-Arc_*YourInitialsHere*

(for example, IGIS-Arc_JWS).

In this textbook the simple blank, ___, will be the disk drive or path to the folder where you keep your work.

Write the path associated with ___ below:

___ means _____

4. Save your Fast Facts File in the folder you have just created, giving it the name `FastFactsFile`. In other words, the full path and filename for your new Fast Facts File will be as follows:

___IGIS-Arc_*YourInitialsHere*\FastFactsFile.someextension

To recap: when I use the designation

[___]IGIS-Arc

(note the square brackets), I mean the place where the data sets on the CD-ROM have been placed. You should not change any of these folders or files. When I use the designation

___IGIS-Arc_*YourInitialsHere*

(note the lack of square brackets and the presence of *YourInitialsHere*), I am referring to the location that contains your work.

Exercise 1-3 (Minor Project)

Getting Set Up with ArcGIS

A click or two of the mouse should give you access to ArcGIS. This should provide easy access to three major components: ArcCatalog, ArcMap, and Arc.

You may reach a component by clicking an icon on the desktop, or by navigating to them by clicking Start, then Programs>, then ArcGIS, then the component—or via some other way prescribed by your instructor.

___ **1.**[2] Find the name or icon for ArcCatalog, *right*-click it and choose Properties. In the ArcCatalog *Target* Properties window, the Target should be something like:

C:\Program Files\ArcGIS\Bin\ArcCatalog.exe Is it? ___*Yes*___ . If not, write of the target here:

_____

Find and write the targets for ArcMap and Arc (located under ArcInfo Workstation).[3]

C\Program Files\ArcGIS\Bin\ArcMap .exe
C\Program Files\ArcGIS\Bin\ArcCatalog. exe

___ **2.** Start ArcCatalog.

ArcCatalog serves as sort of an operating system for GIS. You manage data with it. The Step-by-Step part of this chapter is largely an introduction to Arc Catalog.

___ **3.** Dismiss Arc Catalog by selecting Exit from its File menu.

___ **4.** Start ArcMap. Agree to start using ArcMap with A new empty map.

---

[2]The line to the left of each step number provides a space where you can put a check mark when you have completed the numbered step.

[3]As you work through the text you will find blanks in which to enter requested information. Being able to do this lets you know you are on the right track. Rather than writing in the book you might want to use notebook paper or type into a document file.

ArcMap is the software package that performs primary and major GIS operations, such as making maps and analyzing spatial data. The Step-by-Step part of Chapter 2 serves as the introduction to ArcMap. ArcMap and ArcCatalog both allow you to get to ArcToolbox, which is used primarily for data conversion and many more advanced GIS operations.

5. Dismiss ArcMap by clicking the "X" in the red box in the upper right corner of the window.[4]

6. Start Arc, which is one of the modules within ArcInfo Workstation.

The programs Arc, Arcedit, Arcplot, Grid, and some others are instructed through text commands that are typed by the user after a prompt (such as Arc:). Contrast this with most familiar programs in which you "point and click." Typed input from a command line directed the most powerful ESRI software prior to ArcGIS version 8. The Graphic User Interfaces (GUIs) of ArcCatalog and ArcMap are relatively efficient and elegant ways to invoke Arc and do serious GIS work.

7. Leave the Arc program by typing Quit at the Arc: prompt.

Exercise 1-4 (Project)

Looking at the ArcCatalog Program

ESRI has developed an entire product that is basically designed to help you find, select, understand, and manage geographic data files. This product is ArcCatalog. You'll begin by starting this software and exploring it. Then, considering the definition of GIS—that it is an information system (IS) whose database is a marriage of a geographic database (GDB) and a relational database (RDB)—you'll use ArcCatalog to look at a ridiculously simple GIS data set. A village has developed a system to help with town planning and maintenance. One part of the GIS is a set of features consisting of fire hydrants. You will find the Village fire hydrants theme, named HYDRANTS and explore that data set.

Anatomy of the ArcCatalog Window

Assuming your computer is on and some version of Microsoft® Windows® (or UNIX or LINUX) is running:

1. Start the ArcCatalog component of ArcGIS. Make it occupy the full screen by double-clicking the ArcCatalog title bar. Click the word Catalog at the top of the left-hand subwindow (pane). If you see any negative (minus) signs in the left pane, click them so they become positive (plus) signs. You should now see a window similar to Figure 1-16. That left pane is called the Catalog Tree. At the top level, it is a view of the slow-speed storage devices (hard drives, network drives,[5] CD-ROMs, and so on) of the computer, plus some other entries to be explained later.

---

[4]You dismissed ArcCatalog in a different way. When there is more than one way to accomplish something (almost always the case), I will usually show you alternatives, just by having you use them. Note them in your Fast Facts File.
[5] The network drives may or may not show up, depending on how your system is configured. Even if they don't appear here, ArcCatalog has a way to make them accessible to you, which you will learn shortly.

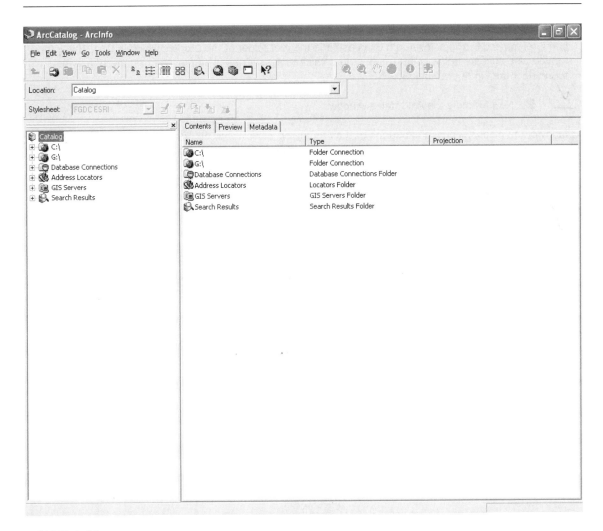

FIGURE 1-16

Near the top of the Catalog Tree you will find the designation for the root folder of the C drive, designated as C:\. Click the folder icon to select it. At the very top left, in the title bar of the overall window, will be some text. What does it say? (Fill in the blank.)

ArcCatalog - _Arcview_ - C:\

Examining the upper left corner of an Arc software window is a way to know (1) which software component you are using (ArcCatalog, ArcMap, and others), and (2) the license your computer is operating under: ArcView (the least expensive with the least capability), ArcEditor, or ArcInfo. Many of the exercises in this text require ArcInfo to run through completely. However, for those of you who have only ArcView or ArcEditor, files that ArcInfo would have developed have been included on the CD-ROM so you can continue past the spots that require ArcInfo.

ArcView is what we are running

35

The title bar of the ArcCatalog window also shows you the path (disk drive and folder) that is currently being referenced by the software (in this case it is the root folder of the C drive, which is C:\). This information also appears in the Location text box, which, unless it has been moved or removed, may be found near the top of the window.

✓ ___ **2.** Now examine the ArcCatalog window.

Many of the menus, icons, and buttons of the ArcCatalog window may be familiar to you from your work with other software. Assuming that your ArcCatalog window has not been customized, you will find the menu headings File, Edit, View, Go, Tools, Window, and Help.

Setting Some Options[6]

✓ ___ **3.** Be sure that the "C" drive icon is selected. Under File, click Properties, then click the General tab. (From now on I may abbreviate an operation like this by saying File > Properties > General.) Since C:\ is selected, you will see some of the properties of the C drive including the amount of disk space—both used and free. See Figure 1-17. Dismiss the Properties window by clicking the "X" in its upper right corner.

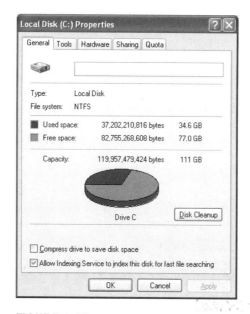

FIGURE 1-17

---

[6]In general, even though you set options, you cannot count on those remaining set if you terminate ArcCatalog and restart it later. You may have to return to this procedure to reset options.

4. Click Catalog at the top of the left pane. Again click File > Properties > General. A bewildering display of possible check boxes and text will show up in an Options window with two scrolling panes. These relate to the types of top-level entries and the types of data that the Catalog displays. Since you are at an early-learning stage, you want to have ArcCatalog display everything possible. By clicking the boxes next to the options in the upper and lower panes, you can toggle a check mark on or off. Make sure each option has a checkmark in the box to its left. Below the lower pane, make sure each of the three options, including Hide File Extensions, is unchecked. Click Apply. (If you made no changes in the pane, Apply will be "grayed out"—a standard feature of ArcGIS software if an action is not possible, not needed, or not appropriate.) The resultant window should look approximately like Figure 1-18.

5. Click the Contents tab in the Options window. Here you may specify the information you want to see about each spatial data set in the catalog when you ask for Details. Read the title over each pane. In the upper pane, put check marks by Size and by Modified. In the lower pane put a check by Projection. The window should look like Figure 1-19. Click Apply. Click OK.

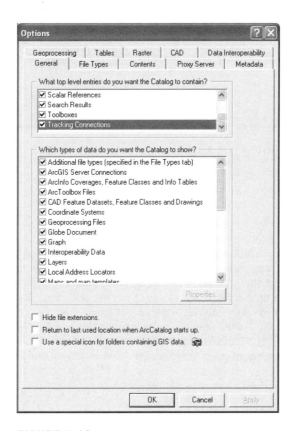

FIGURE 1-18

uncheck "Hide File Extensions"
check "size, modified, projections"

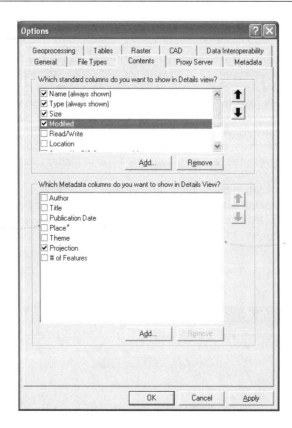

FIGURE 1-19

The Catalog Tree

The main portion of the default ArcCatalog window is usually divided into two panes. On the left you find the Catalog Tree (which you can make disappear and reappear with the Window menu, should you want more visual space for the right-hand pane). The Catalog Tree, when expanded, gives an overall view of the names of your data sets. In this way it resembles what you get when you click My Computer in the Windows Desktop. But you can also look with varying levels of detail at all the data on all the disk drives of the computer by expanding the entries (items)[7] in the tree. (In this way ArcCatalog's presentation is like Windows Explorer.[8] A plus sign (+) indicates that (usually) a given disk drive or folder contains additional folders or files that are hidden from view. A minus sign (–) preceding an entry indicates that any additional

---

[7]If you are a user of ArcInfo Workstation, or of earlier versions of ArcInfo, you know that the term "item" refers to a column heading indicating a coverage's attributes. Although Arc ESRI manuals refer to a line of text in the Catalog Tree as an "item," we will refer to it as an "entry."

[8]Or like My Computer in Windows XP, when you have checked "Folders" under the Explorer bar in the View menu. That's My Computer > View > Explorer Bar > Folders.

folders or files that are directly contained within an entry are displayed below that entry. A click on a plus sign changes it to a minus and expands the entry; a click on a minus sign collapses the list and shows a plus again.

6. Press the Contents tab in the right pane, to make it active. In the Catalog Tree, click the hard drive designator of the path [__].[9] The folders in that will appear in the right pane. Click the plus sign next to the entry for that drive. The folders that are in that are displayed, approximately mirroring the contents of the right pane. Click the plus sign in front of the IGIS-Arc entry in the Catalog Tree and observe the results. Click the entry name IGIS-Arc and observe the results.

You may select an entry name in the Catalog Tree by clicking the name. When you do so, the list of contents of the selected entry shows up in the right-hand pane (provided the Contents tab is selected). (That is, the contents of the right-hand pane is what you would get if you expanded the selected entry in the left-hand pane.) This conveniently lets you see your data names at two levels in different ways. You can select a disk drive, folder, or file from the left pane, and the selection is reflected in the right pane.

In the *left-hand pane* (with the Contents tab of the right-hand pane active):

❑ If you click a folder or file, it becomes selected. If it is a folder, its contents are revealed in the right-hand pane. If it is a file, it may be described in the right-hand pane.

❑ If you double-click a folder, it is selected. It is also expanded or collapsed in the left-hand pane, depending on its previous state.

❑ If you double-click a file, the operating system does something with it—executes it, opens it with an appropriate program, or opens a window that displays its properties. For example, if the file is a text file, it may be displayed in the Notepad text editor.

❑ If you right-click an entry, you get a menu of commands that let you do things like copy, paste, delete, rename, make something new (such as a folder, geodatabase, shapefile, coverage,[10] and so on), search, or reveal the properties of the entry.

In the right-hand pane, with the Contents tab active:

❑ If you click a folder or file, it becomes selected. (The parent folder name is semi-highlighted, instead of blue, in the left-hand pane, just for reference.)

❑ If you double-click a folder, its contents are revealed. To return to the parent folder, click the bent up-arrow icon at the extreme left of the Standard toolbar.[11]

❑ If you double-click a file, the operating system does something with it—it executes it, opens it with an appropriate program, or opens a window that displays its properties.

---

[9]Recall that you recorded the drive and path (if any) of the [__] designator in Step 2 of Exercise 1-2 above. In the event the path to IGIS-Arc does not appear, click File > Connect Folder and navigate to the IGIS-Arc Folder. Select it with a single click. Then OK the Connect to Folder window.

[10]Geodatabases, shapefiles, and coverages are types of GIS data sets, which will be covered extensively shortly.

[11]In the default ArcCatalog window (Figure 1-16), this icon will appear on the line below the Main menu, just under the File keyword, but it may have been moved in your ArcCatalog window.

❏   If you right-click an entry, you get a menu of commands that may let you do things like copy, paste, delete, rename, make a new something (folder, shapefile, coverage, and so on), search, or reveal the properties of the entry.

√ ___ **7.** Experiment with selecting entries in each pane. Also drag the dividing line separating the panes to the left and right to change the space devoted to each. Notice that if there is not sufficient room for the length of the longest entry's text string, a scroll bar appears at the bottom of the window. Further, if you pause the cursor over a text string that is not completely visible, the entire contents of the string will show up in a box that overlays the incomplete string.

The pane on the left side contains, at the least, references to all hard disk drives on the computer,[12] as well as other connections to data. However, you can generate additional references to data sets that you are particularly interested in or want to work with in the future, so you have them right at hand. This is called "connecting to a folder" and is a special feature of ArcCatalog that provides a shortcut to data that you are interested in using. In the next steps, you will make a connection to a folder containing the HYDRANTS data for the village you will explore.

√ ___ **8.** Collapse all the entries in the Catalog Tree completely by clicking each "-". Click Catalog in the Catalog Tree.

Connecting to a Folder

√ ___ **9.** From the File menu, select Connect Folder. A Connect To Folder window appears. Using that window, expanding the entries in its tree as necessary, navigate to a folder named

[___] IGIS-Arc\Village_Data\[13]

Click that folder name so that its little folder icon opens and its name is highlighted. See Figure 1-20. Press the OK button. Observe that the Catalog Tree (in the left pane) now contains an entry (currently selected) that gives you access directly to the desired folder. (Also, the path to that folder is shown in the Location text box and in the title bar of the ArcCatalog window.) Now whenever you want to get to that folder, you merely need to find this direct shortcut in the Catalog Tree, rather than having to navigate to it through levels of hierarchy.

Basically, connecting to a folder provides a shortcut that you should always set up when you anticipate needing a data set more than once or twice.

√ ___ **10.** Using File > Disconnect Folder, remove the entry you just made. Then restore it with the Connect To Folder icon, which you will find on the Standard toolbar that is located under the Main menu. The icon resembles the one associated with File > Connect Folder. Finally, make a folder connection with [___] IGIS-Arc folder itself. Your Catalog Tree should look something like Figure 1-21; it may be different depending on the [___] path to IGIS-Arc.

---

[12]If your computer is linked to a network and that network has drive designations, those designations may not show up in the Catalog Tree. A way of forcing the designations to appear will be covered shortly.

[13]Recall that [___] is the path designation of the data for this text. You determined it in Step 2 of Exercise 1-2.

Connect to Folder

FIGURE 1-20

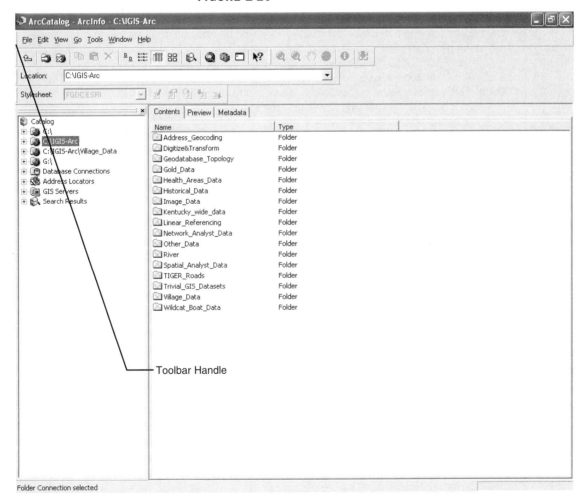

Toolbar Handle

FIGURE 1-21

Chapter 1

The Toolbars and the Status Bar

11. Under View, turn off the Status Bar. The information bar at the bottom of the screen disappears.

12. Turn the Status Bar back on so that messages are displayed at the bottom of the window. To see how the status bar shows information, find a red icon that looks like a tool chest on the Standard toolbar. Move the mouse cursor pointer over to it. The Status Bar should read Show/Hide the ArcToolbox Window. A yellow label (called a ToolTip) with the same message will appear next to the cursor.[14] Also experiment by clicking files and folders in the Catalog Tree and observing the status bar. In particular, look at the status bar as you select [__] IGIS-Arc. Expand that entry. Select Village_Data. Expand. Select Hydrants. Expand. Select Point. Press the Preview tab at the top of the window. You will see the points that represent the hydrants in Village Data.

13. Under View > Toolbars, make sure the Standard, Geography, Location, and Metadata toolbars are all turned on. Active toolbars have a check to the left of their names.

14. Locate an almost-invisible vertical line at the extreme left of the Main menu bar. Consult Figure 1-21 to see where this "toolbar handle" is. Using that handle, drag the menu to the upper left of the right pane. Also use this technique to drag each of the other toolbars so the window resembles Figure 1-22. By "undocking" the toolbars in this manner, you can see the name of each. (When the toolbars are docked, the names are hidden to save space.) Also move the Catalog Tree by dragging it by its top. Then turn it off by clicking the "X" in its upper right corner. Turn it on again by checking Catalog Tree in the Window menu. Re-dock it in its original position by dragging its title bar.

15. Run the mouse pointer over each of the buttons on the Standard toolbar. Notice that, if you pause for a second or so over an option, a small box containing a ToolTip appears. You can, however, get an instantaneous and more complete description of the tool by looking at the status bar at the bottom of the window.

The Standard toolbar provides icons that let you do the frequently-used actions explained immediately following. Of course, as with many software packages, there are frequently several ways to perform an action. Some examples of alternative procedures are given.

Go up a level from whatever selection you have made in a pane. (You can get to the top of the Catalog Tree by repeatedly clicking this Up One Level icon on the toolbar.)
Make a folder connection (substitutes for File > Connect Folder).
Remove a folder connection (instead of File > Disconnect Folder).
Copy, paste, or delete a selection (or use Ctrl-C, Ctrl-V, or Delete keys).

---

[14]Provided ToolTips is set on. To control this: Choose Tools > Customize > Options > Show ToolTips On Toolbars.

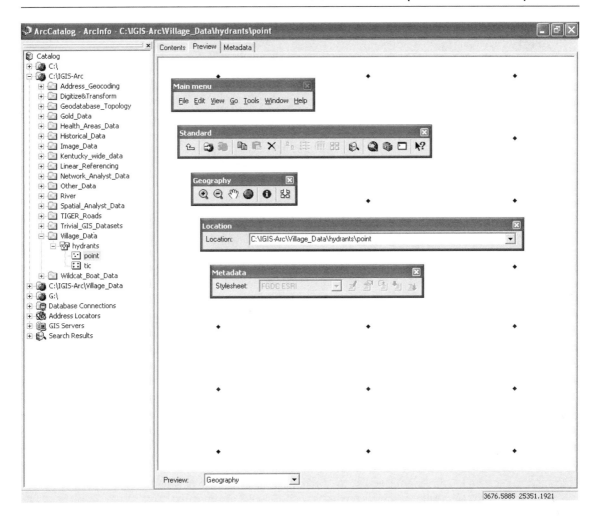

FIGURE 1-22

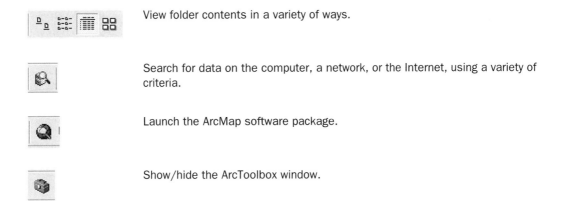

View folder contents in a variety of ways.

Search for data on the computer, a network, or the Internet, using a variety of criteria.

Launch the ArcMap software package.

Show/hide the ArcToolbox window.

Show/hide the Command Line window.

Get specific help on a tool or command.

_____ **16.** Explore the Geography toolbar. This toolbar lets you look at (geo)graphic[15] data in a number of ways.

See features in more or less detail by zooming in and out.

Pan around the area of interest.

Look at the entire geographic region of the selection.

Identify, with text, various features indicated by the pointing cursor.

Make thumbnail sketches of the geographic file of interest so you can recognize it when searching for particular data sets.

_____ **17.** Explore the Location toolbar. The Location toolbar displays a text string showing the current selection from the Catalog Tree and the complete path to that selection. Also, you can type in a different path and selection. If you type in a path and selection (or simply select the text string that is shown) and hit Enter, the string will be placed in the drop-down text box menu below the Location toolbar, making it easy for you to select this location next time. This provides another way to quickly get to data you are using. ArcGIS is full of shortcuts!

_____ **18.** Explore the Metadata toolbar. Metadata is "data about data." It is usually in text form. There are different standards and styles.

The icons of the Metadata toolbar are grayed out because the toolbar is not active. If you click the Metadata tab (next to the Contents and Preview tabs), the toolbar will become active. This is a characteristic of the ArcGIS desktop products. A tool is available to the user when, and only when, it can be applied.

Click the Metadata tab of the ArcCatalog window so that the icons of the Metadata toolbar become active.

This toolbar lets you choose among the metadata stylesheets (the FGDC ESRI stylesheet is the default). It also allows you to do the following:

---

[15]What you are looking at is, of course, graphics. But since the image is tied to positions in the real world, I use the nonstandard term (geo)graphic.

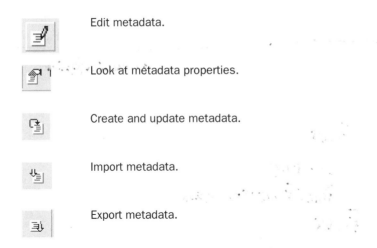

Edit metadata.

Look at metadata properties.

Create and update metadata.

Import metadata.

Export metadata.

When you are finished exploring, you will want the toolbars put back to the positions shown in Figure 1-16. This can happen automatically. If you double-click the title of a toolbar, it will align itself horizontally, in the position it was in before you moved it, and its title will be hidden.

___ **19.** Restore the toolbars to their "home" positions.

Exploring Basic GIS Data Storage Models

Large amounts of computer data are usually stored on mechanical devices called disk drives. Disk drives constitute the "slow memory" of a computer. It takes a computer thousands of times longer to retrieve data from slow memory than from fast memory. An analogy with human processes might be that your brain—where you remember your friend's names, your home telephone number, or the route you drive to work—is your "fast memory" or your "electronic memory." On the other hand, the name of your uncle's second cousin, the phone number of the dry cleaner, or how to get to Punxsutawney, Pennsylvania, are things that you probably have to look up; they reside in slow memory—family documents, telephone book, and road map (or a GIS).

Most of the data that exists in the world on computers is, of course, not in use (not being printed, analyzed, or otherwise processed) at any given moment, and is therefore not contained in the "fast" electronic memory of a computer. Data sets in the slow memory of a computer are, in their most basic form, just sequences of 0s and 1s. But beyond that, well beyond that, such data sets are organized by storage paradigms. You are familiar with the idea of folders on a disk drive. Each folder can contain files and other folders. The operating system of the computer is responsible for keeping files and folders straight. Files may consist of binary sequences that result in typed text, music, photographs, and other products when processed by a computer. For a number of reasons, GIS data sets are composed of fairly complex combinations of folders and files. Also, the techniques for storing spatial data sets have evolved over a number of years, so that, even within the single company ESRI, you will find several storage paradigms or formats. Some of these formats have been devised because of the types of data being represented. Other formats come from different inventions based on progress in hardware and software development. In what follows you use the `status bar` to look at the terminology associated with several of them: coverages, shapefiles, geodatabases, rasters, and TINs.

20. Check that the Options you set in the previous Steps 4, and 5 are still as you left them. If not, reset them. Make sure the Contents tab is active. Collapse the entries in the Catalog Tree as much as possible; you should see no little minus signs. In the Catalog Tree, click the [___] IGIS-Arc designation. Write here what the status bar indicates: _Folder Connection Selected_ Expand the entry. Find the entries indicated in the following, expanding them when possible. Write the status bar text string for each entry. (Those who are artistically inclined may draw the icon they see next to the entry.)

Village Data _____ _Folder_

    HYDRANTS _____ _Coverage_

      Point _____ _Point Feature Class_

  River _____ _Folder_

    Boat_SP83.shp _____ _Shapefile_

    COLE_DRG.tif _____ _Raster dataset_ _____ (say "No" if asked)

    COLE_DOQ64.jpg _____ _Raster data set_ _____ (say "No" if asked)

    COLE_TIN _____ _TIN dataset_

Wildcat_Boat_Data _____ _Folder_

  Wildcat_Boat.mdb _____ _Personal Geodatabase_

    Area_Features _____ _Pers. Geodatabase Feature Dataset_

      Soils _____ _Pers. Geodatabase Feature Dataset_

Close Arc Catalog.

Exercise 1-5 (Major Project)

Exploring Data with ArcCatalog—Fire Hydrants in a Village

Copying Data over to Your Personal Folder

As mentioned earlier, ArcCatalog serves as a sort of operating system for GIS data. One function of an operating system is to make copies of folders and files. When working with GIS data in ESRI formats, you should *always* copy data sets using ArcCatalog—*never* use Windows, UNIX, Linux, or any other primary operating system to copy data sets.

In using this text, you will use your working folder

___IGIS-Arc_*YourInitialsHere*

to do most of the exercises. The idea is to leave the [___] IGIS-Arc folder contents in pristine condition. So you will usually copy over the data sets to your personal folder: ___IGIS-Arc_*YourInitialsHere*. We'll start with the Village Data.

1. Start ArcCatalog. Expand [__]. Expand IGIS-Arc. Select Village_Data. Right-click the selection and pick Copy. Select

___IGIS-Arc_*YourInitialsHere*.

Right-click the selection and pick Paste.

2. Make a folder connection with ___IGIS-Arc_*YourInitialsHere*.

3. Collapse the Catalog Tree as much as possible by clicking all the minus boxes. Then select the folder connection you just made, and expand it as much as possible select HYDRANTS. Click the Contents tab in the right pane. You now see references to both the point and tic components in both panes. Click Point. The Catalog Tree should look something like Figure 1-23. If you see serious discrepancies, use the Delete Key to remove all folders from

___IGIS-Arc_*YourInitialsHere*

and try again. (Be careful not to remove the entire folder IGIS-Arc_*YourInitialsHere* because, if you do, your Fast Facts File will go away.)

FIGURE 1-23

As you determined previously, HYDRANTS is an ArcInfo coverage—coverages are discussed in detail in the next chapter—consisting of 21 points that represent the locations of the fire hydrants in the village. Explore HYDRANTS in the following steps.

4. Select Point in the left pane. In the right pane you see the name and type of that component (feature class). Because of settings you made earlier,[16] you also see its size (in megabytes) and the date it was last modified; you would also see its geographic projection, if it had one. You should also see a blank rectangular image. This will later become a miniature image of the coverage called a thumbnail; we will create it shortly. See Figure 1-24. If you now select the Preview tab *at the top* of the pane, and click Geography in the Preview drop-down list *at the bottom* of the pane, you will see an image showing the positions of these hydrants. See Figure 1-25. This is hardly an exciting picture, but there is a lesson here, so please be patient. If you now select Table from the Preview drop-down list *at the bottom* of the right-hand pane, you will see a table of 21 rows and several columns. See Figure 1-26.

Each *row* (record) of the table represents one hydrant. Each *column* (field) of the table is an attribute (that is, a property or characteristic) of the data set HYDRANTS. Each *cell* in the table (where a given row and a given column intersect) indicates the particular attribute value for the particular hydrant.

| Contents | Preview | Metadata |

Name: **point**
Type: **Point Feature Class**
Size: **0.0772**
Modified: **9/6/2001 5:56:36 AM**
Projection:

point

FIGURE 1-24

---

[16]Repeat Step 5 in Exercise 1-4 if the Options have become unglued.

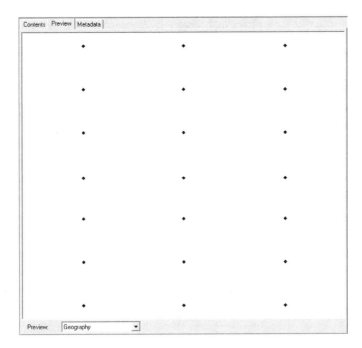

FIGURE 1-25

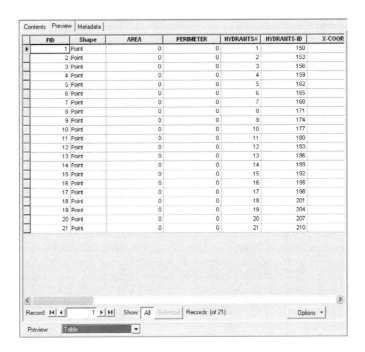

FIGURE 1-26

___ **5.** Locate the record of the hydrant that has a FID (Feature Identifier) of 4. Click the HYDRANTS-ID cell (it has value 159) and then scroll the table horizontally, using the scroll bar at the bottom of the window. You will see that the hydrant COLOR is Red.

Illustrated here is the essence of GIS: the marriage of a geographic database and an attribute database. Each feature in the spatial field has associated with it a row in a relational database table that provides information about the feature. This idea, with variations, is the fundamental underpinning of most of what you will learn about GIS.

___ **6.** Re-read the preceding paragraph.

Examining the Table

___ **7.** The table has a "current record" that, when you first view the table, is the first record. You can change that. Because you clicked on a cell in the fourth row, the current record is now 4. It is marked by a triangle in the box to the left of the record. The number of the current record is also shown in the Record text box at the bottom of the window. You also see the total number of records. You can change the current record by

❑ Clicking in a box to the left of the FID field,

❑ Clicking a cell of a record,

❑ Typing the desired record number in the Record text box (press Enter), or

❑ Clicking the buttons on either side of the Record text box.

You can scroll the viewable area of the table with the horizontal scroll bar and, if the table has a lot of records, the vertical scroll bar. You can also change the widths of the columns for better viewing by dragging the dividing lines between the column headings. (To see more of the table, you can dismiss the Catalog Tree, but the Tree is probably something you will be using frequently, so that should come into consideration.)

___ **8.** Both the AREA and PERIMETER columns are dull and excessively wide. Place the mouse pointer on the column heading text line and drag the column separators to the left to reduce the widths of these columns. On the other hand, the complete titles of the next two columns—HYDRANTS# and HYDRANTS-ID—may not be visible because the columns are too narrow. Fix that. If you double-click a column separator, the column to the left immediately tailors itself to the size that will show the entire column head with no space wasted.[17] However, the column may not be wide enough to completely show the values in the cells of that column.

A word about identification numbers: In coverages there is a record number (FID) as well as a HYDRANTS# number; they are assigned by the software and are the same. HYDRANTS# is sometimes called the Internal-ID. It is there so the software can keep track of which record goes with which hydrant, regardless of the actions of the user. The HYDRANTS-ID number, on the other hand, is a user-assigned integer; it may be called the User-ID or the External ID.

---

[17]If an entire column becomes selected while you are doing this, don't worry about it.

You noticed that an attribute for AREA and another for PERIMITER are shown, and that they are both zero. Since it is obvious that the area and perimeter of a point (a zero-dimensional entity) would be zero, it seems a little strange that the software would bother to store or report these values. The reason is that this same table format is used to represent polygons (such as ownership parcels or school districts) that have nonzero areas and parameters. The reason for storing points and polygons with the same format is historical and will be discussed later.

9. Make record 12 the current record. With the keyboard arrow keys, locate the HYDRANTS-ID value of the record. What is it? __183__ What is the FLOW_RATE of the Hydrant? __365__ What are the geographic coordinates of the hydrant? __4500__ __24733__

Should you wish to do so, you can change the cosmetic appearance of a table through Tools > Options > Tables.

Deriving Information from the Table

Having attribute information in table form allows you to obtain statistics, to search for specific text strings, and to sort the information.

10. You can find out the average of the values of a column along with the count of the number of values, minimum, maximum, sum, and standard deviation.[18] To do so, you right-click the column heading to bring up a menu of choices from which you choose Statistics. For example, select the column FLOW_RATE by placing the mouse pointer over the heading of the column (the pointer becomes a down-pointing arrow) and right-clicking. (You may have to use the scroll bar at the bottom of the table to find the FLOW_RATE column.) Choose Statistics from the drop-down menu. You get information about the values in the column from the `Statistics of point` window. There are 21 values. The minimum flow rate is 250 and the maximum is 400. You also get a frequency diagram that shows the *number* of hydrants that have each given value (e.g., there are four hydrants with a low flow rate, nine with a moderately high rate, and eight with a high rate). If all the hydrants were turned on at once, how much water would flow? __7485__. The mean (i.e., average) and standard deviation, although they probably don't mean much in this situation, are about 356 and 54, respectively.

11. Once you have the `Statistics of point` window up, you can determine the statistics of other columns by selecting the column name in the window's `Field` drop-down menu. What is the average y-coordinate? __24981__ Dismiss the `Statistics of point` window.

Sorting the Records

12. Suppose now that you wanted to have the list ordered so that all the western most hydrants (those with the smallest x-coordinates) appeared at the top of the table. Click the X-COORD column heading to make it active. The column becomes highlighted in blue. Now right-click the column head. Then sort the values in the column by picking `Sort Ascending` from the drop-down menu. Notice that the values of the column are now in order from smallest to largest.[19]

---

[18]You are probably familiar with these concepts. A discussion of basic statistical measures may be found in the Overview of Chapter 6.

[19]A warning: ArcCatalog will sort the selected column, regardless. You can initiate the drop-down menu by placing the pointer over another column, but this column will not be the one that is sorted.

Note that when a value in a given record is moved, say, during a sort, the entire record moves with it. That is, all the values in a given row stay together (including the FID), *regardless*. (If you are familiar with spreadsheets, you should note that this is different from results you may get with that software.)

If you wanted the Y-COORD column values to be a secondary sort (sorted "within" the X-COORD values), you could select both columns (by holding down the Ctrl key while clicking). (An example to clarify the terminology: In a telephone directory the first names are sorted "within" the last names.) When two or more columns are selected, the leftmost one will contain the primary sort, the next selected on the right will contain the secondary sort, and so on.

____ **13.** Sort ascending the X-COORD and, within it, COLOR. Examine the results.

If having the leftmost column as the primary sort order doesn't suit you, you must rearrange the columns, demonstrated next.

____ **14.** Select the COLOR column with a click. Hold down the left mouse button. The cursor changes—a little box is attached to it. You can now drag the column. The column will be placed to the left of whatever column the cursor is in when you release the mouse button. Drag the COLOR column to the left of the X-COORD column. Select both columns. Select Sort Descending (from the drop-down menu as before) and again note the results.

Finding Values in a Table

____ **15.** Make the COLOR column active by itself. Locate and press the Options button at the lower right of the window. (The button may be hidden or partially hidden; if so, increase the width of the window by dragging the separator between the table and the Catalog Tree.) Find all instances of the text string "Red" using the binoculars (Find) icon.[20] Each instance, when found, will have a darker border around it. (You may have to move the Find window to see the COLOR column.) How many records are there with the text string "Red"? _____.

____ **16.** Notice that the Find window gives you considerable flexibility:

- ❏ You can search all the values in a table, or just those in selected columns (fields).

- ❏ The text you type may match only a part of the value or be required to match the whole string.

- ❏ The searched-for string may be required to match the upper- and lowercase of each letter or not, as you choose

- ❏ You may search the entire table or restrict the search to records at or below the current record, or search those at or above the current record.

____ **17.** In the Search drop-down window, pick All. With Match Case checked, search for "yellow." No records will be found, because of the lowercase "y." Now click Match Case off. How many instances do you find? ____. Dismiss the Find window.

---

[20]Do not type the quotes in the Find What text box.

18. Move the COLOR column to the right of the CONDITION column. You have only a limited view of the table. You may not be able to tell which hydrant number (HYDRANT-ID) you are looking at when a given cell in the COLOR column is selected. You can set things up so you can always see a given column or columns. To Freeze the HYDRANTS-ID column, select it, right-click, and pick `Freeze/Unfreeze Columns`. It will now appear always as the leftmost column of the table. Scroll the table horizontally to see the effect. (You can thaw a frozen column in much the same way.)

When you terminate the ArcCatalog program, the results of some actions you took are retained, while others are discarded. The folder connections you have added to the Catalog Tree are remembered by the software. So are the positions of toolbars and some other settings. But if you have sorted or rearranged a table using ArcCatalog, these changes will not be retained. Operations in ArcCatalog such as sorting do not change the attribute table that is stored with the data set. You are merely changing the table's appearance during the time ArcCatalog is being used to look at a particular geographic data set. In fact, to restore the view of a table to its original form, you need only select some other folder or feature class and then return to the table.

19. Click Catalog at the top of the Catalog Tree. Click Contents. Now go back and again preview the HYDRANTS point coverage table. Notice the table is back to its former configuration—no frozen or rearranged columns and no sorted records.

20. Click the Preview tab at the top right-hand pane. Select Geography from the Preview drop-down menu at the bottom of the pane.

Identifying Geographic Features and Coordinates

Coordinates are a big deal, and frequently a big headache, for those doing GIS work. HYDRANTS is referenced by a local (and, in this case, fictional) coordinate system devised by the village or perhaps by the county. The units are survey feet. Suppose the village enters into an agreement of some sort with a regional water system agency to maintain the pipes and hydrants. If the agency uses a different coordinate system, the question might arise as to what the latitude and longitude values, rather than the local coordinate system values, were for these hydrants. As mentioned previously, the latitude and longitude coordinate system is the primary basis for the most accurate geographic information.

21. Select the `Identify` tool from the Geography toolbar. Write here the description of the tool that you see on the `status bar`. _Identifies features you click on._ Notice the appearance of the cursor as you move it around the geographic area. Click in a blank area of the window that shows the hydrants. An Identify Results window appears. (You can move the window around by dragging its title bar and resize it by dragging a side or corner to better see the geographic image.)

22. Click one of the points representing a hydrant. As you click the feature, it is momentarily highlighted. The relational database information for that hydrant appears in the `Identify Results` window. Notice that the `Identify Results` window, in an information box labeled Location, gives you a precise value of the Cartesian (x and y) coordinates of the location of the tip of the cursor pointer. These coordinates should agree closely, but probably not exactly, with the X-COORD and Y-COORD values shown as attribute values. (It is not always the case with point data that the coordinates are part of the attribute database; you usually have to take action to

Freeze Column.   _Identify_

put in X-COORD and Y-COORD; how this is done is described later. However, the information in the Location box is always calculated. *It comes from the geographic location specified by the cursor, not the relational database.*)

____ **23.** With the Identify tool, click again in a place where no fire hydrant exists. Attribute information disappears, but coordinate values of the tip of the pointer are still revealed.

____ **24.** Roughly (to the nearest tenth of a foot), what are the coordinates of the lower left-hand corner of the window? _2680_ , _2412b_ . How about the upperright-hand corner? _5129_ , _4580n_

____ **25.** Dismiss the Identify Results window. If you right-click `Point` in the left pane, you get a menu of options. Of interest to us now is `Properties`. Click that. A `Coverage Feature Class Properties` window appears. Under the `General` tab, you see that the number of features is 21. Of more interest is the Items tab, which tells you the names of all the attributes and the types of fields that are used to contain the data. For example, click the row that says `COLOR` to highlight it. Notice that `COLOR` is a Character field of width 9, meaning that it uses nine bytes of memory. Now look at `Y-COORD`—a numeric field taking up 4 bytes of storage and allowing the storage of a floating-point number, with three decimal places. `HYDRANTS-ID` is also a numeric field, but it is an integer and may therefore be stored in a simple binary format. Dismiss the window.

Looking at GeoGraphics

____ **26.** On the Geography toolbar, click the magnifying glass icon that has a + in it. Copy the text in the Status Bar here. _Zooms in by clicking a point or a box_ click the upper left hydrant. Note that it is moved to the center of the window and the distance between it and its neighbors is increased. You have zoomed in on the layer—not that you will see any more detail in this particular image, but you get the idea.

____ **27.** Notice that although you have zoomed up on the image, the symbols representing the features did not get any bigger. So this zooming action is somewhat different than looking at a paper map with a magnifying glass, which would increase both the distance between the points *and* the symbol size.

____ **28.** Click again on the northwest hydrant and observe the results. Click between that hydrant and its neighbor to the south. Click the Full Extent icon (on the Geography toolbar) to restore the view of the entire layer and bring all hydrants back into view. With the Zoom In tool active, drag a box around the middle three hydrants in the middle column and observe. The lessons: You can zoom in by clicking a point and also by dragging a box. The image is always recentered, either at the point clicked or the center of the box.

____ **29.** Click the "hand" icon on the Geography toolbar—this is the Pan tool. Check out its function on the Status Bar. Move the cursor over the middle point and drag that point to the left side of the pane. When you release the mouse button, you will see that the focus on the image has been shifted ("panned") to the right—in between two hydrants. Click the Full Extent icon.

Identify - Properties   Zooming

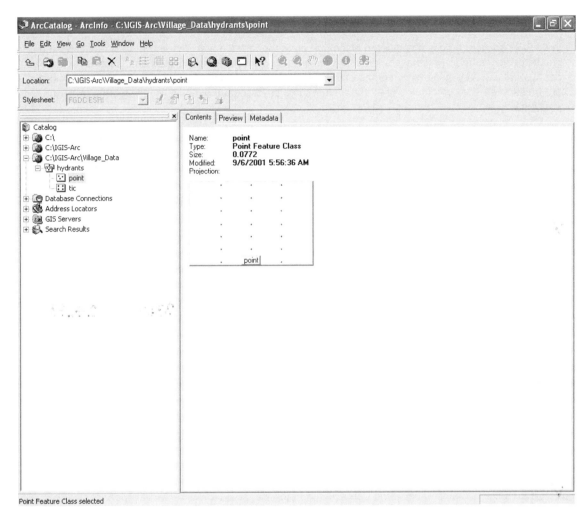

FIGURE 1-27

A "thumbnail" is a small sketch that can aid you in recognizing the contents of a data set. You make a thumbnail sketch in the step that follows.

_____ 30. **Create a thumbnail of the Hydrants layer:**[21] On the Geography toolbar, find the Create Thumbnail icon. Click it, then click the Contents tab and observe. Where there was before a blank rectangle, there is now an image of the layer. See Figure 1-27. Suppose you want only a portion of the coverage to be represented in the thumbnail. Go back to the Preview tab. Use the zoom tool to focus on the six hydrants in the northeast corner. Make a thumbnail sketch of these. Check that it worked by clicking the Contents tab. Now make the thumbnail a third time with all the points shown.

---

[21]A note on using this text: When **bold italic** is used following a step number, it constitutes a general direction. You should read what follows the bold italics to see what specific actions you are to take.

Tics and Ticks:[22] Tying Coverage Geographic Data to the Real World

In ArcGIS parlance, a *tic* is a numbered place that corresponds to a precisely known geographic position. The tic positions tie the coverage's features to accurately known, real-world positions, such as U.S. Coast and Geodetic Survey markers, or perhaps the centers of intersections of major streets whose positions are well known. The coordinates of all the features in a coverage are directly, linearly related to the tic positions.

In the HYDRANTS instance, there are four tics (the minimum for a coverage), and they are at the locations of the four corner hydrants. A given coverage may have a large number of tics. Coverages that describe the same geographic area should have the same tics, if possible. That is, the tic number (and associated location) should be the same, regardless of which data set you are using if the data sets cover the same area. Tics are the key to transforming an ESRI geographic coverage from one coordinate system or projection to another.

31. Where you had selected Point in the Catalog Tree, now select Tic. The Contents tab reveals that these are features of the Tic Feature class. (However, the thumbnail will continue to be that of the features of the coverage.) The Preview tab lets you look at the four tics graphically and examine the attribute table. The attributes of the tic component of a coverage are simply the tic number (attribute name: IDTIC) and the x- and y-coordinates of each tic. Using the attribute table, determine the x- and y-coordinates of tic number 3. **3300** , **24233** . Now use the Identify tool to verify the correctness of what you found. Dismiss the Identify Results window.

A First Look at Metadata

As mentioned, "metadata" means data about data. There are so many things to know about geographic data sets that many people who work with them have recognized the need for established standards for describing such data sets. We will look at metadata in more detail later, using some real data sets. The lesson from what follows is just that metadata exists, that it consists of many elements, and that it can be presented in many formats.

32. Click Hydrants again in the Catalog Tree. Click the Metadata tab. The Metadata toolbar becomes active. On the Metadata toolbar, pick the Stylesheet FGDC ESRI, which is the ESRI presentation of the standard metadata format.[23] See Figure 1-28.

Under Description, you see a thumbnail of the data set (if a thumbnail has been created) and textual information that the creator of the data set provided about the general nature of the data. Some of the

---

[22]"Ticks" (spelled like the name of the blood-sucking creatures) are, among other definitions, small marks along the border of a map indicating the positions of grid lines of the map. A given tick mark indicates either a latitude (or northing) or a longitude (or easting), but not both. "Tics" (spelled like the facial spasms) are geographic control points (consisting of both latitude and longitude [northing and easting]). They tie a location in an ArcGIS coverage to a real-world position.

[23]FGDC stands for the Federal Geographic Data Committee of the United States government. The Web site Internet address is www.fgdc.gov. The committee coordinates the development of the National Spatial Data Infrastructure (NSDI).

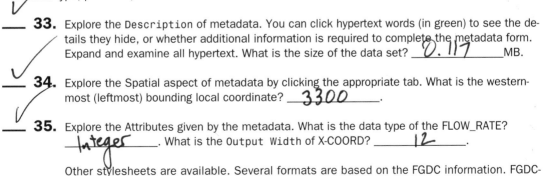

FIGURE 1-28

metadata comes about because ArcCatalog has inspected the data set. For example, under the Spatial tab, you could find the bounding coordinates of the features of the data set. These were derived automatically. In the HYDRANTS case, only the local coordinates were available from the data set. Also, generated automatically is most of the information found under Attributes: the name of each attribute, along with the data type, precision, and so on.

____ **33.** Explore the Description of metadata. You can click hypertext words (in green) to see the details they hide, or whether additional information is required to complete the metadata form. Expand and examine all hypertext. What is the size of the data set? _0.117_ MB.

____ **34.** Explore the Spatial aspect of metadata by clicking the appropriate tab. What is the westernmost (leftmost) bounding local coordinate? _3300_.

____ **35.** Explore the Attributes given by the metadata. What is the data type of the FLOW_RATE? _Integer_. What is the Output Width of X-COORD? _12_.

Other stylesheets are available. Several formats are based on the FGDC information. FGDC-FAQ is a "conversational" format in which frequently asked questions are answered. There are other formats, which display varying amounts of metadata with varying degrees of user friendliness.

36. Using the Stylesheet drop-down menu, look briefly at each of the other stylesheet types to get an idea of its characteristics. For FGDC Classic, describe the general form of the data set. ___outline - hierarchical text___. When you are through exploring, click the Contents tab.

Using ArcCatalog to Place Data in ArcMap

You have explored ArcCatalog and used it to find and connect to data sets you want and to explore them. As previously described, the other major component of ArcGIS Desktop is ArcMap. ArcMap lets you see, create, examine, query, edit, and develop geographic data and maps. Your first step in using ArcMap is to get the data you have found with ArcCatalog into ArcMap. There are several ways to do this. Three are described in the following. First you need to perform a couple of setup steps.

37. From the Standard toolbar (or the Tools menu), launch ArcMap and press OK to start with a new, empty map. Make ArcMap occupy the full area of the monitor screen.

38. Now do either or both Step 39 or Step 40, which follow, to put ArcCatalog and ArcMap on the screen at the same time.

39. Click the middle icon (named Restore Down) of the three in the far upper right of the ArcMap window. That has the effect of reducing the size of the window. By dragging sides and corners of the ArcMap window, make it occupy approximately the right half of the monitor screen. Make the ArcCatalog window occupy the left half of the monitor screen.

40. Right-click an empty area of the Windows taskbar to bring up a menu. (The Windows taskbar will be located on one of the four edges of your monitor screen; it contains the Start button.) From this menu, select Tile Windows Vertically. (If necessary, minimize any windows besides ArcCatalog or ArcMap that appear, and "re-tile.")

41. *Method 1 of inserting data from ArcCatalog into ArcMap:* Drag and drop the point component of the HYDRANTS coverage from the Catalog Tree of ArcCatalog to the pane in ArcMap that says Layers. The "map" will show up in the right pane of ArcMap. (Disregard any warning message about missing spatial reference information.)

42. Dismiss ArcCatalog. Maximize the ArcMap window. Right-click the string "Hydrants point." From the menu that appears, select Open Attribute Table. You can see that the coverage HYDRANTS— the combination of the geographical database and the attribute database—is now in ArcMap. (Of course, you have been writing information into your Fast Facts File all along. Record particularly how to open an attribute table in ArcMap.)

43. Right-click again the reference to the HYDRANTS data. Select Remove to take the data set out of ArcMap.

44. *Method 2 of inserting data from ArcCatalog into ArcMap:* In the ArcMap File menu, click Add Data . . . . In the Add Data window that appears, use the Up One Level icon (repeatedly, if necessary) so that the Look In box reads Catalog. Notice that the contents of the Catalog

putting a data set into ArcMap

Attribute Table

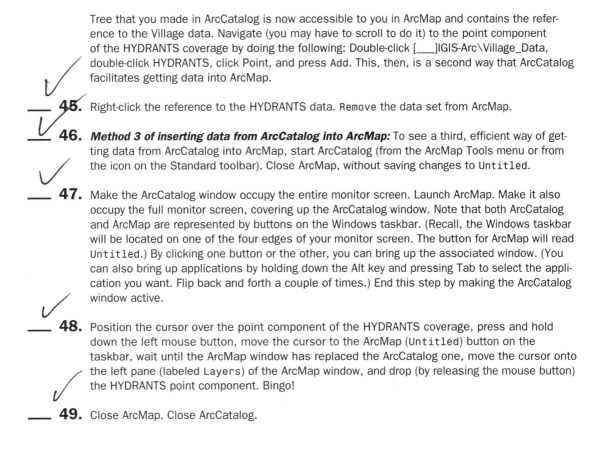

Tree that you made in ArcCatalog is now accessible to you in ArcMap and contains the reference to the Village data. Navigate (you may have to scroll to do it) to the point component of the HYDRANTS coverage by doing the following: Double-click [___]IGIS-Arc\Village_Data, double-click HYDRANTS, click Point, and press Add. This, then, is a second way that ArcCatalog facilitates getting data into ArcMap.

_____ **45.** Right-click the reference to the HYDRANTS data. Remove the data set from ArcMap.

_____ **46.** ***Method 3 of inserting data from ArcCatalog into ArcMap:*** To see a third, efficient way of getting data from ArcCatalog into ArcMap, start ArcCatalog (from the ArcMap Tools menu or from the icon on the Standard toolbar). Close ArcMap, without saving changes to Untitled.

_____ **47.** Make the ArcCatalog window occupy the entire monitor screen. Launch ArcMap. Make it also occupy the full monitor screen, covering up the ArcCatalog window. Note that both ArcCatalog and ArcMap are represented by buttons on the Windows taskbar. (Recall, the Windows taskbar will be located on one of the four edges of your monitor screen. The button for ArcMap will read Untitled.) By clicking one button or the other, you can bring up the associated window. (You can also bring up applications by holding down the Alt key and pressing Tab to select the application you want. Flip back and forth a couple of times.) End this step by making the ArcCatalog window active.

_____ **48.** Position the cursor over the point component of the HYDRANTS coverage, press and hold down the left mouse button, move the cursor to the ArcMap (Untitled) button on the taskbar, wait until the ArcMap window has replaced the ArcCatalog one, move the cursor onto the left pane (labeled Layers) of the ArcMap window, and drop (by releasing the mouse button) the HYDRANTS point component. Bingo!

_____ **49.** Close ArcMap. Close ArcCatalog.

Exercise 1-6 (Project)

A Look at Some Spatial Data for Finding a Site for the Wildcat Boat Facility

You usually don't get the luxury of working on one GIS project from beginning to end. Life in the GIS world frequently isn't like that. We'll be working on one rather large project (the search for sites for the Wildcat Boat facility that you did manually in Exercise 1-1), but along the way, we'll do several smaller ones.

This is the beginning of the GIS part of the Wildcat Boat project. Using ArcCatalog, you will find data for the site. Then, using the techniques you learned in Projects 1-4 and 1-5, you will use ArcCatalog to explore data sets associated with this project.

Most of the Wildcat Boat data is contained in a *personal geodatabase*. This is the newest data structure developed by ESRI, and it has a number of advantages over the coverage data structure. While most of the data you will encounter in the next few years will probably be in the coverage format (or another format called "shapefile"), the move to geodatabases is proceeding at a fast pace. So most of the work we do in this text will be with geodatabases.

Using the Area on the Disk for Your Own Work

Again, we want to keep the [__] IGIS-Arc data sets intact and unchanged, so you can go back to them at any time. But I want you to be able to modify the Wildcat Boat data. To do that, you need to use your own folder, which you made earlier and in which you are keeping your Fast Facts File and the Village_Data. The purpose of

___IGIS-Arc_*YourInitialsHere*

is to let you store files in a personalized folder, without compromising the information in IGIS-Arc.

Copying Data over to Your New Folder

1. Start ArcCatalog; make it occupy the full monitor screen. Expand [__]. Expand IGIS-Arc. Click Wildcat_Boat_Data. Right-click the selection and pick Copy. Select

___IGIS-Arc_*YourInitialsHere*.

Right-click the selection and pick Paste.

2. Using almost the same technique in the paragraph above, copy and paste

[___] IGIS-Arc\Other_Data

into your folder, except this time use Ctrl-C and Ctrl-V for copy and paste.

3. Collapse the Catalog Tree as much as possible. Then expand the entries (except for Village Data) in

___IGIS-Arc_*YourInitialsHere*

as much as possible. The Catalog Tree should look something like Figure 1-29. If you see serious discrepancies, remove the *folders* from

___IGIS-Arc_*YourInitialsHere*

and try again. (Again, be careful not to remove the folder

IGIS-Arc_*YourInitialsHere*

because, if you do, your Fast Facts File will depart with it.) Come to think of it, now would be a good time to back up your Fast Facts File. Put it on a floppy disk, a Zip drive, a thumb drive, a flash drive, a network drive, or e-mail it to yourself. Maybe you should make a couple of back-ups, using different methods. *It is rare that a person has too many backups; it is quite common to not have enough.*

FIGURE 1-29

Searching for GIS Data

In the 1980s and 1990s, the major issue related to spatial data and GIS was how to create the data sets you wanted—either directly from the environment or by converting existing map data. Now the initial emphasis has turned, in many instances, to finding already-existing data—on the Internet and elsewhere. Tremendous stores of spatial data exist, and more comes in every day from satellites and ongoing projects. But you may discover that the data set you want is buried with a lot of other data; it may have an obscure name; the data sets you find may not meet your standards, even if they cover the correct subject and correct geographical area. A major feature of ArcCatalog is the capability to help you discover spatial data sets and, through inspection of their metadata, determine if they meet your needs. Assume that a client or your supervisor asks you to find all the geographic data that could apply to the Wildcat Boat project. You have been told that the data set exists in either a personal geodatabase or coverage. Further, you have been given the geographic coordinates that bound the area of interest.

___ **4.** In ArcCatalog use the Up One Level arrow to get to the top of the Catalog Tree—that's where it says `Catalog`. Click the Metadata tab. Read the paragraphs about the catalog. In addition to the capabilities described there, the catalog can provide the basis for a search for geographic data, which you will now illustrate to yourself. Click the `Contents` tab.

___ **5.** Use Search on the Edit menu (or the Search icon on the Standard toolbar) to bring up the Search–My Search window. Under the Name And Location tab,[24] leave the asterisk (*) in the Name field—since you don't know the name of the data set(s) you are looking for, and the star is a "wildcard" that can stand for any text string. Let's assume you know that the data type will be either a personal geodatabase feature class[25] or an ArcInfo coverage, so highlight those[26] on the menu—after scrolling down to look at the large number of possible data set types that you could search for. Look now at the Search field: In this case you want to search the Catalog, but you see that you could search just the file system or use the ArcIMS (Internet Map Server) Metadata Service. Under Look In, single-click Browse (the yellow folder) to look for

___ IGIS-Arc_*YourInitialsHere*

select it, and click Add. Don't start the search yet! In the next step, you will specify the geographic area to search.

___ **6.** Press the Geography tab. Make sure the box next to Use Geographic Location In Search is checked. For Choose A Location, pick <None>. For Map, select US Counties from the drop-down menu. Suppose you happen to know that the site that you want lies in the United States, south of latitude 41.1° and north of latitude 41.0°. Also, it lies east of longitude –74.1° and west of longitude –73.9°. (In what follows, *do not* hit Enter after typing into a box; just move to a new box with the mouse cursor and double-click. Also be sure not to forget the minus signs that are part of the longitude coordinates.) Specify the bounding coordinates by typing in the North, South, East, and West text boxes (e.g., place –74.1 in the West box; if you double-click the text in a text box, it becomes highlighted and you can type new digits to replace those already there.) As you enter information into the text boxes, note the red crosshatched area becomes smaller and smaller. After entering all four coordinates, click Find Data Entirely Within The Location. Use the Magnify icon in this window to zoom up on the red crosshatched area, which shows the approximate area for verification purposes. The window should look something like Figure 1-30***.[27]

Under the Date tab, make sure All is selected. Ignore the Advanced tab for now. Return to the Name & Location tab. Click Find Now. Wait until the magnifying glass stops moving around the Earth Symbol in the Search–My Search window. The result should look something like Figure 1-31. Dismiss the Search–My Search window.

___ **7.** The ArcCatalog search feature found four personal geodatabase feature classes that fell within the geographic bounds specified: Soils, Streams, Roads, and Sewers, all in the

___IGIS-Arc_*YourInitialsHere*\Wildcat_Boat_Data

folder. It also found a coverage named SOILS in the

---

[24]By "location" here it is meant "location" on a hard drive of the computer (i.e., path to the data), not spatial location. That comes later.

[25]A feature class in a personal geodatabase is somewhat like a coverage, in that it contains geographic features and an associated table. A detailed explanation will be found in future chapters.

[26]You can, by holding down the Ctrl key and clicking, highlight any number of data types.

[27]Figures indicated in the text with three asterisks (***) following are on the CD-ROM that accompanies the book. They are in the folder IGIS_with_ArcGIS_Selected_Figures. These figures are in color, which may provide you with additional information.

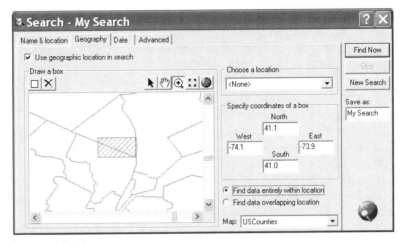

FIGURE 1-30

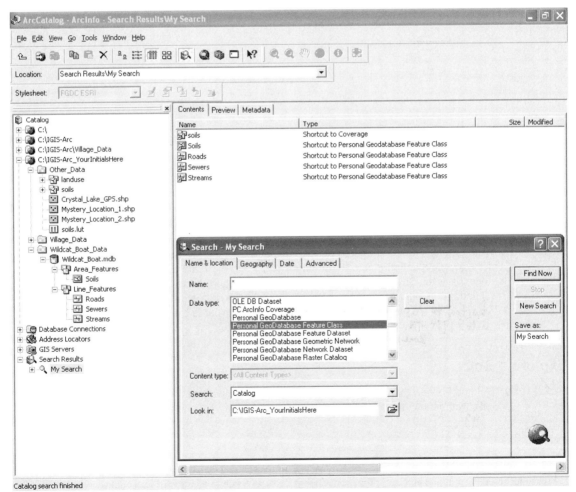

FIGURE 1-31

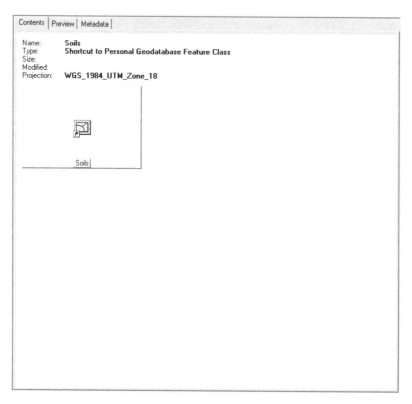

FIGURE 1-32

___IGIS-Arc *YourInitialsHere*\Other_Data folder. (Use the bottom slider bar to see all of the contents.)

___ **8.** In the Catalog Tree, find Search Results. Expand it to find My Search. Expand that. Click each of the five data set shortcuts to determine which are the feature classes and which is the coverage. Click the Soils feature class (not the coverage). With the Contents tab selected, you will find, in the right-hand pane, information about the data set. What is the type of the entry? *Shortcut to Pers Geodatabas Feature Class*. What is the projection[28] of the data? *WGS 1984 UTM Zone 18*. See Figure 1-32.

Exploring Soils

___ **9.** Click the Preview tab. You see a map of the Soils you worked with in Assignment 1-1. In the pull-down menu in the Preview text box, select Table. How many Soils polygons are there? *43*.

---

[28]If Projection doesn't show up, it is probably because the options you set earlier in the chapter are no longer in effect. To set (or reset) the option that shows projection: Choose Tools > Options > Contents > turn on the Projection box.

10. Examine the table. Notice the fields Shape_Length and Shape_Area. These attributes are the perimeters and areas of polygons. This is information you get "for free" by using a GIS. Sort the Shape_Area column into ascending order. To the nearest tenth of a meter, what is the perimeter of the polygon with the smallest area? _673.9_ meters.

11. Click the Metadata tab. Make sure that the stylesheet is FGDC ESRI.[29] Click the Spatial tab. Under `Horizontal Coordinate System`, you find

`Projected coordinate system name` _WGS 1984 UTM Zone 18_

`Geographic coordinate system name` _GS WGS 1984_

Under `Bounding Coordinates`, you find

`Westmost bound: Geographic` _-74.0_ `Projected:` _4008.9_

`Eastmost bound: Geographic` _-73.9_ `Projected:` _6141.9_

`Northmost bound: Geographic` _41.1_ `Projected:` _7376.9_

`Southmost bound: Geographic` _41.0_ `Projected:` _4577.1_

Someone suggests to you that these bounding projected coordinates look very small, given the usual size of UTM coordinates. That is, since the northing is supposed to be the number of meters from the equator (that would be in the millions for 40 degrees latitude),[30] a number like 4500 (that's four and a half kilometers) obviously isn't right. So it is necessary to look more closely at the metadata. Click the green `Details` text and record the following:

`False easting:` _-80,000_

`False northing:` _-4 540,000_

The question of the small number in the northing is now explained. Each UTM northing coordinate has been adjusted by subtracting 4,540,000 meters from the true northing coordinate to get the local coordinates, which are around 4000 to 6000. Why did the developers of the information decide to do this? To keep the size of the numbers relevant to the problem small. Perhaps, at the time the data sets were developed, the software operated in single precision, using only 4 bytes of storage for each coordinate. Numbers as large as 4 and a half million would contain too many digits to provide sufficient precision.

12. As you recorded earlier, the type shown in My Search is a Shortcut. While this has let us explore the data somewhat, we would really like to get closer to the real data set. If you right-click Soils in the Catalog Tree (Search Results\My Search), you will get a menu that contains `Go To Target`. Click that, and `Soils` will be selected in a different place in the Catalog Tree. Scroll to find it. It exists within a personal geodatabase feature data set named Area Features. Area Features, along with another data set named Line Features, is located in a personal geodatabase named Wildcat_Boat.mdb, where "mdb" is the extension for a *Microsoft DataBase*. The

---

[29]If the Metadata toolbar isn't showing: Choose View > Toolbars > Metadata.
[30]It is about 10 million meters from the equator to the North Pole.

Go to Target

FIGURE 1-33

database is contained in a folder named Wildcat_Boat_Data, in the ___IGIS-Arc_*YourInitialsHere* folder. So you can see that we have quite a hierarchy going here: A folder contains the geographic database; the geographic database contains two geographic data sets; each geographic data set contains one or more geographic feature classes. Visually the hierarchy looks like Figure 1-33. So in the Wildcat_Boat geodatabase we have the Soils, Roads, Sewers, and Streams that we need for the Wildcat Boat facility problem.

But Something Is Missing

Assume at this point that you show these results to your client or supervisor, who insists that there is some land use or land cover data somewhere for this region.[31] Perhaps the name is simply LANDCOVER, so you will try searching for a data set with that name.

___ **13.** Initiate another search. This time, under the Name And Location tab, type LANDCOVER in the Name text box to replace the asterisk. You will search in the Catalog. Look In should again contain just ___IGIS-Arc_*YourInitialsHere*. And, again, you are looking for a coverage or a personal geodatabase feature class. Under the Geography tab, remove any check mark from Use Geographic Location In Search, since you already have all the data sets that have metadata indicating the geographic area. Nothing new can show up as long as the area is restricted.

Click Find Now. My Search now shows a data set named LANDCOVER. To find out more about it, dismiss the Search–My Search window, look at My Search in the Catalog Tree, click the Landcover icon, and click the Contents tab. You find that the data set is a coverage located in

___IGIS-Arc_*YourInitialsHere*\Other_Data.

---

[31]The concepts of land use and land cover have a lot in common—and significant differences. Both describe "what's going on" with a particular piece of real estate. A land cover classification basically tells what sort of surface component is present, such as forest, water, urban, wetland, and so on. In a land use classification, the surface component is considered but so also is the human use of the landscape, so one might find more detailed information such as residential, commercial, and agricultural. Land use might also include information related to prescribed zoning, such as single-family residential or industrial.

✓

Right-click the Landcover shortcut icon in the Catalog Tree and Go To Target. Expand LANDCOVER and click Polygon.

14. Click the Preview tab. In the Preview drop-down menu, make sure Geography is selected. Look at the map. Now select Table in the Preview drop-down menu. What is the FID number of the first record? ____2____. What is the number of the last record? ___77___. How many records (and, hence, how many land cover polygons) are there?[32] ___76___ Sort the records. What is the area of the smallest polygon (record all digits)? _927.8128_____ square meters.

927.8128

✓

15. Look at the Metadata for the polygon component of LANDCOVER as you did for the Soils geodatabase feature class before. Use the FGDC ESRI stylesheet. Using the information under the Spatial tab, record all of the following that you can. *The information that isn't available is shown by Xs in the following:*

Projected coordinate system name _____XXX_____

Geographic coordinate system name _____XXX_____

Westmost bound: Geographic _____XXX_____ Projected: ____4008.9_____

Eastmost bound: Geographic _____XXX_____ Projected: ____6141.9_____

Northmost bound: Geographic _____XXX_____ Projected: ____7376.9_____

Southmost bound: Geographic _____XXX_____ Projected: ____4577.1_____

✓

16. So there is a little problem here. No coordinate system is listed. No geographic coordinates are shown. The bounding coordinates are present in "projected or local coordinates" but not in decimal degrees.

Although the lack of a published coordinate system and the absence of geographic bounding coordinates is an issue we will have to deal with, one mystery has been solved. It was the lack of latitude-longitude coordinates in the metadata that made it so the LANDCOVER coverage could not be found by Edit > Search Using Geography. This is why the ArcCatalog search was able to find the coverage SOILS but not the coverage LANDCOVER. The specific lesson: *The way ArcCatalog finds appropriate data sets by Geography is by investigating the metadata.* The general lesson: There are lots of ways to look for data—such as over the Internet (e.g., www.geographynetwork.com), through agencies (e.g., www.census.gov), through personal contacts, and so on. *When one method doesn't give you the results you want, try something else.*

Is the Newly Found Data Applicable?

Although the LANDCOVER coverage and the SOILS personal geodatabase probably occupy the same piece of real estate (note that the projected coordinates you wrote down previously are approximately the same for both), you have to make sure. Suppose that, by looking at some paper documentation, you are able to determine that the projected coordinates shown are in fact in meters and the projection of the coverage is

---

[32]The first record is numbered "2" because an entire polygon ArcInfo coverage an additional, hidden polygon, called the "outside polygon," which is the rest of the Earth—the area not included by the other polygons. It can be confusing. Perhaps for this reason the developers of ArcGIS have decided to leave it out of the table. But it remains polygon "1," so, to maintain consistency with the coverage records, the table starts with FID number "2."

WGS 84 UTM Zone 18, just like your other data. So you should fix up the coverage so it has the same spatial metadata as the other data sets. You can do this with ArcCatalog. Recall that in the original search you also found a *coverage* named SOILS; it was also in the

___:\IGIS-Arc_*YourInitialsHere*\Other_Data folder.

Maybe this SOILS coverage contains spatial metadata that you could assign to LANDCOVER.

____ **17.** In the __IGIS-Arc_*YourInitialsHere*\Other_Data folder, check the spatial metadata for the SOILS coverage against the spatial metadata for the Soils personal geodatabase feature class in the

__IGIS-Arc_*YourInitialsHere*\Wildcat_Boat_Data

folder. Do they have the same coordinates systems? __Yes__

Making a Personal Geodatabase Feature Class from a Coverage

What you want to do is convert the LANDCOVER coverage to a Personal Geodatabase Feature Class so you will have all the data in the same format when you apply analysis tools to it in the future. Also, you need to fix the problem of not having the proper coordinate system associated with the landcover data.

____ **18.** Navigate to

___IGIS-Arc_*YourInitialsHere*\Other_Data\LANDCOVER\polygon

and right-click the icon. Choose Export > To Geodatabase (single). In the Feature Class To Feature Class window that appears, the Input Features should be set. You want to put the new feature class in ___IGIS-Arc_*YourInitialsHere*, so browse to the Output Location of Area_Features so that the window looks like Figure 1-34, making sure Area_Features appears in the Name text box. Click Add. For the Output Feature Class Name, type Landcover. Change the TRUE designations in the Visible column to FALSE for AREA, PERIMETER, LANDCOVER#, and LANDCOVER-ID, so that these columns will not show up in the table of the feature class, where they would be useless. See Figure 1-35. Press OK. When you see that the conversion has completed, close the Feature Class To Feature Class window.

Looking at the Landcover Personal Geodatabase

____ **19.** Navigate to the Area_Features personal geodatabase feature data set. Click it. Within it you should see Landcover feature class that you created from the LANDCOVER coverage. Select it. Click the Contents tab. Verify that the projection is WGS_1984_UTM_Zone_18.[33] Click the

---

[33]If projection is not shown, you will have to go back and reset options.

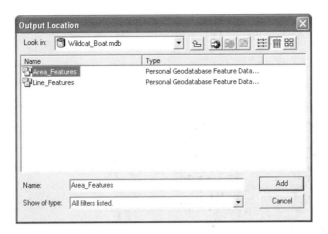

FIGURE 1-34

Preview tab. You see a map of the landcover you worked with in Assignment 1-1. In the drop-down menu in the Preview text box, select Table. How many land cover polygons are there? ____76____ Check this against the number of polygons in the LANDCOVER coverage that you wrote down previously.

20. Examine the Landcover table. Notice the fields Shape_Length and Shape_Area. These attributes are the perimeters and areas of the polygons. Sort the Shape_Area column into descending order. The largest polygon is a portion of the lake, with a land cover code of 500. What is the perimeter of the polygon with the largest *land* area (to the nearest meter)? __1762__ meters. What is the area of the smallest polygon (record all digits)? __927.812932-949213__

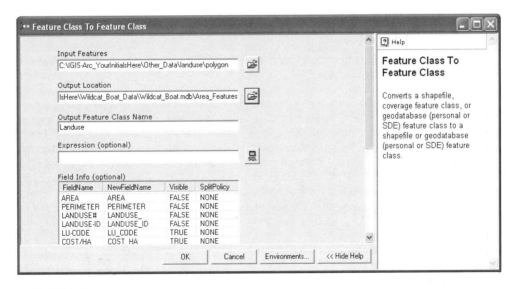

FIGURE 1-35

square meters. Contrast this "double precision" number with the one you previously recorded for the same polygon in the LANDCOVER coverage.

21. Find the Landcover feature class attributes and their properties with a right-click the Landcover icon. Then select Properties > Fields. Fill out the following table, except for Width.

Field Name	Data Type	Width
OBJECT D	Object D	4
Shape	Geometry	0
COST HA	Long integer	4
LC CODE	Short integer	2
Shape length	Double	8
Shape Area	Double	8

22. Cancel the personal geodatabase `Feature Class Properties` window. Find the names of the attributes by using the metadata, check them against what you wrote above, and add the Width. (*Hint:* When you click text that is green, you see more text. Clicking again contracts the text.)

Further Examining the Wildcat Boat Facility Area Data Sets

Do the following steps using the data sets in

_____ \GIS-Arc_*YourInitialsHere.*

23. In the Catalog Tree, click `Soils` feature class, within the `Area_Features` feature data set. Click the Preview tab.

This feature class uses polylines to enclose areas (polygons). Each of the polygons is considered to be homogeneous in a particular type of soil.

24. Using the Identify tool, click a polygon near the center of the map. The polygon becomes momentarily highlighted and the Identify Results window appears. You get the plane area of the feature in square meters in the field Shape_Area. You also get, in the field Shape_Length, the sum of lengths of the polylines (in meters) that enclose the polygon. In the attribute field named SOIL_CODE, you are shown a code that identifies the soil type, such as Tn4. There is also an attribute (SUIT) that indicates whether the soil is suitable for the construction of a building: 1 for unsuitable, 2 for moderately suitable, or 3 for suitable. Click some other polygons. Click the large polygon in the north-east (That's water—note the SUITability and the SOIL_CODE: _____, _____). Dismiss the Identify Results table.

25. Look at the attribute table for the Soils personal geodatabase feature class. Obtain statistics for the Shape_Area column of the table. What is the smallest area? _299_ The largest? _1529796_ Sort the table from smallest to largest and verify the minimum and maximum numbers. What is the area of the largest polygon that is not water? _937263_

_____ **26.** Find the Soils attributes and their properties with a right-click Soils. Then select Properties > Fields. Fill out the following table, except for Width.

Field Name	Data Type	Width
OBJECT ID	Object ID	4
Shape	Geometry	10
SOILCODE	Text	3
SUIT	Short Integer	2
Shape length	Double	8
Shape Area	Double	8
Suit	Integer	1

_____ **27.** Cancel the personal geodatabase feature class `Feature Class Properties` window. Find the names of the attributes by using the metadata, check them against what you wrote above, and add the Width.

_____ **28.** Click the Contents tab. Select the Line_Features personal geodatabase feature data set. Check out the presentation using the four display options on the Standard toolbar: Large Icons, List, Details (scroll to see them), and Thumbnails.

_____ **29.** In the personal geodatabase feature data set Line_Features, select Roads. Make a thumbnail of the northeastern portion of it. (You may need to check your Fast Facts File to recall how to make a thumbnail.) Click the Contents tab and note the thumbnail and some general information about the feature class. What is its projection?[34] WGS 1984 UTM Zone 18

_____ **30.** Click the Metadata tab and, with FGDC ESRI, look at the `description` of the data. Note the thumbnail again. Like much metadata you will find, there is a lot of missing information. Under `Details About This Document`, note the date and time the contents were last updated: Date _20080121_ Time _13392000_ (Cryptically, the date is given as YYYYMMDD and the time as HHMMSS00.)

_____ **31.** Look at the Spatial characteristics of the data. What is the geographic coordinate system name? _GCS WGS 1984_ What is the projected coordinate system name? _WGS 1984 UTM_ To the nearest hundredth of a degree, what are the geographic coordinates of the northwest corner? _-74.0_ , _41.1_ The southeast corner? _-73.9_ , _41.0_ .

_____ **32.** In the projected coordinates, to the nearest meter, what are the coordinates of the northwest corner? _4012_ , _7195_ The southeast corner? _6138_ , _4595_ What are the units of Roads? _Meters_ (Hint: Check under Details.)

[34]If the projection is not revealed, it may be that ArcCatalog reverted to its default options. Reset the options. (How to do this should definitely be in your Fast Facts File.)

_____ **33.** Look again at the Preview of the Roads geography. Zoom to full extent. Then look at the table.

Here again, each row in the table corresponds to a geographic feature. In this case, each feature is a polyline, which is a series of straight-line segments that run between two points. Among the fields of data you see is Shape_Length, which is the sum of the segments that make up each polyline. This is also information that you "get for free" when you use a GIS. (You will see a much more comprehensive discussion of polylines and segments in the Overview of Chapter 4.)

_____ **34.** Again look at the geography of Roads. Zoom up on the northernmost road. Notice that it is not a smooth curve but a series of straight lines. These lines begin and end at Cartesian points, which have *x-y* coordinates. Use the Identify tool to see the particular attribute values for the polyline that represents this road. Be sure that as you click the polyline, the feature is briefly highlighted. What is the length of the road to the nearest meter? ___564___. Each polyline is numbered with an OBJECTID. What is the OBJECTID of this road? ___1___. The road code for this polyline is ___1___. Dismiss the Identify Results window. Zoom to the full extent of the Roads personal geodatabase feature class.

_____ **35.** Find the Roads attributes and their properties. Fill out the following table, except for Width.

Field Name	Data Type	Width
OBJECT ID	Object ID	M
Shape	Geometry	0
RDCOh	ShortInteger	2
Shapelength	Double	8

_____ **36.** Dismiss the personal geodatabase feature class Feature Class Properties window. Find the names of the attributes by using the metadata, check them against what you wrote above, and add the Width.

_____ **37.** Look at the Sewers personal geodatabase feature class. How many polylines are there? ___60___ What are the diameters of the various sewer pipes (inches)? ___45 80___, ___60 in___ Make a thumbnail of Sewers and examine it with the Contents tab.

_____ **38.** Find the Sewers attributes and their properties. Fill out the table below except for Width.

Field Name	Data Type	Width
OBJECTID	Object ID	4
Shape	Geometry	0
Stream Code	ShortInt	2
Shapelength	Double	8

_____ **39.** Cancel the personal geodatabase feature class Feature Class Properties window. Find the names of the attributes by using the metadata, check them against what you wrote above, and add the Width.

40. Look at the Streams personal geodatabase feature class. Make a thumbnail. What is the total length in *kilometers* of all the Streams in the study area? _21588_ What is the longest stream polyline in *meters*? _708_

41. Fill out the following table, except for Width.

Field Name	Data Type	Width
OBJECTCODE	Object ID	4
Shape	Geom	O
Strm Code	Short In	2
Shape length	Double	8

42. Cancel the personal geodatabase feature class `Feature Class Properties` window. Find the names of the attributes by using the metadata, check them against what you wrote above, and add the Width.

Looking at Wildcat Boat Data with ArcMap

43. Launch ArcMap. Tile the ArcMap and ArcCatalog windows vertically. Drag the Sewers personal geodatabase feature class from ArcCatalog into ArcMap.[35] Now drag the Streams personal geodatabase feature class in. Put in the Landcover personal geodatabase feature class. Dismiss ArcCatalog. Make ArcMap occupy the full monitor screen.

Here you see one of the major bonuses of a GIS: the ability to see data of different types and from different sources easily represented on the same map. An additional advantage is that you can immediately choose what is shown and what is not. Notice that the Landcover data set is displayed "under" the Sewers and Streams—both in the Layers pane and, in a different sense, in the map window.

44. Experiment with displaying and un-displaying the different personal geodatabase feature classes by clicking the boxes next to the data set names.

45. Remove both Landcover and Streams by right-clicking the data set name and choosing Remove.

46. Display Sewers. Open the attribute table for Sewers. (Right-click Sewers. From the menu that appears, select Open Attribute Table. Is the way to open an attribute table in your Fast Facts File? If you didn't remember how, it should be.) Shorten the vertical dimension of the attribute table and position it so that you can see it and the graphic image of the Sewers as well. On the Tools toolbar (which resembles the Tools toolbar in ArcCatalog—zoom, pan, etc.), find the Select Features icon (not the Select Elements icon) and press it. Now click one of the sewer lines. It should turn cyan *and the equivalent row of the attribute table should be highlighted.* Pick another arc and observe. Sort the records according to Shape_Length. Click the shortest arc you can find on the map and verify its length in the table.

---

[35]If you run into "schema lock" problems, dismiss ArcCatalog and load the data set into ArcMap directly, using Add Data.

47. In the table, click in the small box to the far left of the last record. The selected record becomes highlighted *and so will the equivalent arc.*

We went through the last few steps to reinforce the point (that maybe you now understand so well you don't want to hear it anymore): A GIS is a special case of an information system whose database is the marriage of a (geo)graphic database and an attribute database.

Next you look very briefly at the capacity of ArcGIS to let you add a completely separate database table to the Table of Contents, view such a table, and then join it together with a feature class attribute table so that the information in the separate table becomes part of the attribute table.

48. Click the New Map File icon on the Standard toolbar. Don't save changes. Add the Landcover feature class from

___IGIS-Arc_*YourInitialsHere*\Wildcat_Boat_Data\Wildcat_Boat.mdb\Area_Features.

49. Open the Landcover attribute table. Notice that the table contains the column LC_CODE but no indication of what the codes mean. Close the table.

50. Add, as data, the database table

___IGIS-Arc_*YourInitialsHere*\Other_Data\LC_Code&Type.dbf.

Right-click LC_Code&Type and Open the table. What are the columns in Attributes Of LC_Code&Type?

OID , _LCCODE_ , _LCTYPE_

Dismiss the table window.

51. Right-click Landcover. Choose Joins and Relates > Join to bring up a Join Data window. You want to join attributes from a table. In box number one, the field in the layer Landcover that the join will be based on should be LC_Code. Such a field is called a *key field*. In box number two, the table you want to join to the layer is, of course, LC_Code&Type. In box number three, the field in the LC_Code&Type table also is named LC_Code.[36] Click OK. (You will be asked if your want to create an index for the join field. When you are processing thousands of records, this can ultimately save time. For our minor demonstration, we can ignore the offer. Select No.)

Seeing the Results of the Join

52. Using the Identify cursor, click a polygon in the map display. Notice that the amount of information you get is substantially greater than before. Specifically, you see not only the land cover code (LC_CODE) but also the type (LC_TYPE) that is associated with the code. The field names look a little strange, but we'll deal with that issue later. The important thing is that we have added the contents of one table to another. Dismiss the Identify Results window.

---

[36]The key fields in the two tables do not have to have the same name. In this case they happen to.

53. Open the attribute table of the Landcover layer. Notice that all the information has been put together. Sort the Landcover type. How may polygons consist of "Barren" land? _____

A note of warning: The product of the preceding process—the table with information gleaned from two tables—will persist only as long as ArcMap is running and Landcover remains as a layer. If you want to save the new table together with the geography of Landcover, you have to take additional steps, which I discuss later.

54. Dismiss ArcMap without saving Changes To Untitled.

Exercise 1-7 (Project)

Understanding the ArcGIS Help System

The Help system in ArcGIS has many facets. There are several ways to access different kinds of Help. Knowing how to access the various pieces of documentation will make your use of ArcGIS more efficient and less frustrating. You will start with a nifty mechanism for getting help on tools and buttons.

A Button for Instant Help: What's This?

1. Start ArcCatalog. Click the ^? tool. Then click the Connect To Folder icon on the Standard toolbar and read all about it. Since you know most of this information, I am mainly trying to show you what you can expect from the ^? tool. Click any blank area (or press Esc on the keyboard) to dismiss the information box.

Press Shift-F1. What happens? _____. Click the Search tool on the Standard toolbar. Dismiss the text explanation.

Click again on the ^? tool. Now click it again and read about the tool itself.[37] Try using Shift-F1 on a context menu item by first moving the cursor down to View > Refresh, then pressing Shift-F1. Read the box.

In the Catalog Tree, click Catalog. On the Main menu, click View. Notice that Identify Results, while shown as one of the selections, is disabled. To get an explanation of the option, and to find out why it is disabled, move your cursor over it and press Shift-F1. An explanatory note is presented. Frequently at the end of such notes is an explanation of why the tool or option is disabled. This can be extremely useful when you are presented with a grayed-out selection.

The notes obtained with the What's This tool and with Shift-F1 are completely separate from, and frequently more useful than, the regular Help system, which is described next.

---

[37]The reference to the Table of Contents has to do with ArcMap, which we explore in the next chapter.

Chapter 1

The Help System and Documentation

Not only is it unlikely that you will ever learn all of ArcGIS, you may not even learn all there is to know about the Help system. But you should make a major effort, because whatever question you may have about the system probably has an answer somewhere in Help. There is an immense amount of information available. But finding it can seem somewhat like a scavenger hunt.

2. On the ArcCatalog Main menu, click `Help`. Pick `About ArcCatalog`. From the first line, you can tell what version of ArcCatalog you are running: ___*9.1*___. From the second line, you can determine the License Type: ___*ArcView*___. The ESRI Web site is: ___*esri.com*___. Click OK.

3. On the ArcCatalog Main menu click Help. At the top of the drop-down menu, you see ArcGIS Desktop Help. We'll come back to that. Next is the help tool you just explored: `What's This?` Depending on which version of ArcGIS you are running, you will find the next menu items in different orders. One is called GIS Dictionary; the symbol next to it indicates that it is a Web page that you get to through the Internet, if you are connected.

4. Click `GIS Dictionary`. If your computer is connected to the Web and your browser cooperates, a page with the banner ESRI Support Center will appear. In a line that says "You are here:," you will be told that you are at the level of

Home > Knowledge Base > GIS Dictionary

Look up "geodatabase". Look up "shapefile". Look up "coverage". Look up "relational database".

This is but one page of a very rich site. Before leaving it, list the green tabs in the ESRI Support Center banner: ___*Home*___, ___*Software Knowledge Base Downloads*___ ___*User Forums*___, _____

Briefly explore each tab—particularly Home, Knowledge Base, and User Forums. You will get at least two impressions: (1) ESRI has a lot of products and services, and (2) there is a lot of help available. Dismiss the page by clicking the "X" in the upper right corner.

5. Click Help again on the Main menu. Click the first item in the menu: ArcGIS Desktop Help (which in the future you can invoke just by pressing the F1 key on the keyboard).[38] Make the window occupy the entire monitor screen. Click Contents. Click GIS Glossary and you will be served up with definitions of GIS terms that reside locally with your software. It is also extensive. To access its search capability, press Ctrl-F to bring up a Find window.

In this text I will not define many terms. Nor will I include a glossary. With all the capabilities you have to get definitions on demand, I decided it would be better to save a few trees. If you want a printed dictionary, you can obtain *The ESRI Press Dictionary of GIS Terminology*.[39]

---

[38]In version 9.1 (check to see which version number you wrote down above), you can also get to Help through the Internet by clicking ArcGIS Desktop Help Online. You can choose whether you want your help to come from your local machine or the Web.

[39]ESRI Press, Redlands, CA. Edited by Heather Kennedy. 112 pages.

Find out your version of the software

6. Dismiss the GIS glossary. The remainder of the Help system under Contents is a hierarchical, nested set of "books." When you get down to the bottom of the hierarchy, you will find a help document. Using the "+" icon in front of the ArcCatalog book, open it. Within that, open Working with metadata. Click the document icon About metadata. A document of that name should appear. Read it. Then click the little triangle in front of How metadata is organized to expand that topic. Click it again to collapse the material. At the top of the pane click Expand All. Note the effect. Click Collapse All. Note that there are other links on this pane. Click Related Topics. Click Learn About Creating Metadata. To get back to the About Metadata pane click the Back arrow below the Main menu.

7. Check out What's New In ArcGIS Desktop 9.1, 9.2, or whatever version you are using. Look at the various book icons. Check out What's New In ArcMap.

8. Right-click any entry in the left pane. Click Open All. Scroll down through the pane. It may be daunting to see how much information there is, because of its implication for the size of the software package. That's why they will pay you the big bucks when you know how to use it. Right-click again and Close all.

9. Click the topmost entry: Welcome To ArcGIS Desktop Help. Read the Getting Help section.

10. Press the Index tab. Type ADD as the keyword. As you can see from the list that appears immediately below, there are lots of topics that relate to adding something. Double-click the Add Rule To Topology tool/command. An Add Rule To Topology (Data Management) document, with an ArcToolbox heading, opens. Notice the Open Tool button. Press it and wait. A window called Add Rule To Topology opens (you may have to display it by clicking its button on Windows the taskbar). From here you could run that tool (if you had any idea what it did—all things in good time). Cancel the Add Rule To Topology window. Now click Expand All. The topics—Usage Tips, Command Line Syntax, and Scripting Syntax—will display details about the command. Now click Collapse All. Click the triangle in front of Usage tips and note the result.

11. Look at the Main menu of the help window (File, Edit, View, Go). Check each drop-down menu, looking at the items on those menus. In the View menu, make sure checks are on next to Search and Highlights.[40] Try out Back and Forward under the Main menu bar.

12. Press the Search tab. Type

catalog tree thumbnail

Click List Topics. About how many topics are displayed? _18_. Add the phrase large icons and press List Topics again. Select An Overview Of ArcCatalog and press Display. Also note the three option boxes at the bottom of the left pane that give you options in searching. Close the ArcGIS Desktop Help window.

There are other parts of the help documentation that you will encounter later, when you use ArcToolbox. And finally, ESRI has a number of tutorials on various parts of the software and on the software extensions. The documentation and tutorials consist of PDF[41] files located, probably, in C:\ESRI_Library. Data for the tutorials may, likely, be found at C:\arcgis\ArcTutor.

---

[40]The terms you search for using the search procedure that follows will be highlighted in the document.
[41]Portable Document Format files, readable by Adobe Reader (available free), a basic part of Adobe Acrobat, at www.adobe.com.

Exercise 1-8 (Dull Stuff)

Using ArcCatalog for Mundane Operations

ArcCatalog serves as an operating system (like UNIX or Windows) for geographic data sets. In this brief exercise, you will copy, paste, rename, and delete a coverage. If you tried to do this with the computer's operating system, you would cause errors.

_____ **1.** With ArcCatalog running, highlight

___IGIS-Arc_*YourInitialsHere*

in the Catalog Tree. Select File > New > Folder and name the folder Housekeeping_Stuff. (Never, ever, accept the proffered name New Folder. Because it has a blank in it, ArcGIS may complain (and fail) later if it is used in a path name. Never use a folder name with a blank in it!

_____ **2.** Navigate to

___IGIS-Arc_*YourInitialsHere*\Village_Data\HYDRANTS

and highlight it. Select Edit > Copy to place the HYDRANTS onto the ArcCatalog clipboard. (In place of Edit > Copy, you could type Ctrl-C, or you could press the Copy button on the Standard toolbar.)

_____ **3.** Highlight the word Catalog in the Catalog Tree, and select View > Refresh (or just press F5). This ensures that the Catalog Tree properly reflects all data sets and puts them in order in the Tree. It is not always necessary to do this, but it doesn't cost much (a bit of time and computer processing). And it will save you on occasion from wondering where something went that was supposed to be there.

_____ **4.** Navigate back to Housekeeping_Stuff and highlight it. Select Edit > Paste (or press Ctrl-V, or press the Paste button on the Standard toolbar). Refresh the Catalog Tree again, expand Housekeeping_Stuff, and display the Geography of the point component of this copy of the HYDRANTS coverage.

_____ **5.** With HYDRANTS highlighted, right-click the name. Choose Rename. (Alternative ways to start the renaming process: press F2, or left-click the highlighted name). Type NEW_HYDRANTS, to change the name of the coverage.

_____ **6.** Attempt to change the name again, this time to NEWER_HYDRANTS. This fails because the number of characters exceeds that allowed for coverages. Experiment to determine the number of characters that may exist in a coverage name. What is that number? _____[42]

_____ **7.** Delete the coverage using File > Delete, or the Delete button the Standard toolbar, or via Delete on the drop-down context menu that you get by right-clicking the highlighted name. Also delete the folder Housekeeping_Stuff. Close ArcCatalog.

---

[42]Most geodatasets may have long names. Coverages, and some others, are exceptions.

ArcCatalog operates pretty much like Windows Explorer or the regular operating system windows when it comes to copying, moving, renaming, deleting, and so on. *Always use ArcCatalog when dealing with geodata sets, whether they be coverages, personal geodatabase data sets, or shapefiles.*

——— **8.** To prove the validity of the preceding admonition, use Windows to perform the same operations. Make a folder named Housekeeping_Stuff in the same place. Copy the HYDRANTS folder onto the Windows clipboard and paste it into the Housekeeping_Stuff folder. Then restart ArcCatalog and try to display HYDRANTS. Nothing doing. I'll explain why later, but, again, know this: You can't properly or effectively copy, paste, rename, or delete geodata sets outside of ArcCatalog.

——— **9.** Use ArcCatalog to delete the Housekeeping_Stuff folder. Close ArcCatalog.

Exercise 1-9 (Review)

Checking, Updating, and Organizing Your Fast Facts File

The Fast Facts File that you are developing should contain references to items in the following checklist. The checklist represents the abilities you should have upon completing Chapter 1.

Important note: This checklist is on the CD-ROM that accompanies the book. It is available in Microsoft Word format. Rather than typing or writing by hand the text that follows, you can copy and paste it into your Fast Facts File from the CD-ROM file.

Name: _____

My User or Logon Identifier: _____

My password is written down in this secure location: _____

The hard drive location of this FastFactsFile is _____

The FastFactsFile is backed up on _____

The date of the last update of the FastFactsFile is _____

The date of the last reorganization of the FastFactsFile is _____

The path to IGIS-Arc, associated with [__], is _____

The path to IGIS-Arc_*YourInitialsHere*, associated with _____, is _____

Operations using ArcGIS Desktop software:

___ To initiate ArcCatalog

___ To determine the level of ArcGIS (ArcView, ArcEditor, or ArcInfo)

___ To see properties of an entry in the Catalog Tree

____ To set options for ArcCatalog

____ To expand or contract entries in the Catalog Tree

____ To copy, paste, delete, rename, make a new something (folder, shapefile, personal geodatabase feature class, and so on), search, or reveal the properties of the entry in the Catalog Tree

____ To see the entire path of an entry in the Catalog Tree

____ To connect to a folder so that it appears as a single entry in the Catalog Tree

____ To disconnect from a folder

____ To turn toolbars off and on

____ To move toolbars around

____ To turn on ToolTips

____ To get specific help on a tool or command

____ To launch ArcMap or ArcToolbox from ArcCatalog

____ To examine and modify metadata

____ To see the components of a coverage (point, line, polygon)

____ To see the attribute table of a geographic data set

____ Ways of selecting a row in a relational database table as the current record are

____ An Internal-ID is different from a User-ID in that

____ To change the appearance of a table cosmetically

____ To get statistics on the numeric values in a table column

____ To sort the values in a column

____ To arrange the records order using both a primary and secondary sort

____ To move a table column

____ To search for values in a table

____ To make a column always visible

____ To look at the graphics of a geographic data set

____ To look at the items in a table that relate to a graphically represented feature

____ To see the coordinates that the cursor is pointed at

____ To see the characteristics and parameters of the attributes of items in a table

____ To magnify the graphics of the area being examined

____ To move around on a map that is being examined

____ To make thumbnails of geographic data sets

___ To determine the locations of tics in a geographic data set

___ To explore the three areas of metadata: description, spatial, and attributes

___ To select among various metadata stylesheets

___ Three ways to get data sets into ArcMap are

___ Three ways of initiating the data set copying process in ArcCatalog are

___ Three ways of initiating the data set renaming process in ArcCatalog are

___ Three ways of initiating the data set deleting process in ArcCatalog are

___ Ways of searching for geographic data sets with ArcCatalog are

___ Find the properties of a geographic data set

___ In ArcMap, to highlight both a selected feature and the corresponding row in the table

What's Next?

In this chapter, you used ArcCatalog to find and explore geographic data sets and to install those data sets into ArcMap. In the next chapter, you begin working with ArcMap extensively—looking at examples of the wide variety of GIS data available.

Quiz 2 on Ch.1 . (handwritten)

Characteristics and Examples of Spatial Data

OVERVIEW

IN WHICH you look at a variety of geographic data sets involving vector, raster, and triangulated irregular networks. Data sets of shapefiles, coverages, and personal geodatabases are considered. And you are introduced to ESRI's ArcMap.

The Original Form of Spatial Data: Maps

Twenty-five years ago spatial data meant maps. A single map is a spatial database and, for many purposes, a very good one. For hundreds, perhaps thousands, of years almost all of the information used to support land-related planning and management, and a myriad of other activities such as navigation, has come from maps. Mapmaking became a well-developed activity. The piece of paper on which the map is drawn is a continuum that can represent the quasi-two-dimensional surface of the Earth in an obvious way. ("A picture is worth a thousand words.") Many times when people are asked "How do you get to . . . ?" they respond, "Let me draw you a map."

Why, then, should we spend millions of dollars on databases composed of discrete symbols when maps are available? Among the answers are that maps alone are extremely hard to use for many of the analyses that human activity requires. There are precious few ways to combine graphic information with other graphic information. Decisions involving the space we live in are becoming more difficult all the time because of the larger number of factors that must be considered. Physical techniques have been evolved for combining maps, such as overlaying one transparent map with another and looking through the composite, but these methods are tremendously time-consuming and have clear limitations in term of useful output.

The reason spatial databases composed of discrete symbols (numbers, letters, and special characters) are overtaking map use is that, in the last 25 years, we have learned a great deal about handling discrete symbols and we have developed both techniques and equipment (primarily digital computers and software) that can manipulate the symbols efficiently and quickly. Thus, an approach that seems basically less appropriate to the task does in fact serve us well—especially since high-quality maps, that is, analogs of the landscape with their innate advantages in conveying information—can now be produced by computers.

Moving Spatial Data from Maps to Computers: Forces for Change

Force #1: There Are Difficulties and Limitations Using Maps for Decision Making.

Maps can depict things beautifully and usefully, so for many applications, a paper map is exactly what is needed. But for many purposes maps are hard to use, for these reasons:

❏ A map is a compromise between a storage function and a display function. As more and more information is stored on a map, it becomes more cluttered. At some point, it becomes unreadable. A map that stored every theme of interest to everyone would be black. Aeronautical charts are a good example of this problem. The aeronautical charts of the 1950s were pretty simple affairs, showing terrain, prominent features, and some airport information. As new regulations came into effect, and new communication facilities were established (whose radio frequencies were placed on the map), as new types of airspace were defined, as new military training grounds were proscribed, the map had to depict more and more. As a consequence, without careful study (not an activity that can easily take place in the cockpit of an airplane), it is easy to misread such a cluttered map. In a GIS the storage function and the display function are separated. When display is required, a map can be constructed of only those themes wanted by the user.

❏ It is difficult to analyze a map. Consider a map that shows highways. Suppose you are interested in knowing the distance from city A to city B. What's meant by "distance"? How about straight-line distance from city center to city center? Obtaining an approximation of a straight-line distance isn't too hard. The map has a scale indicating that a certain linear distance on the map is a certain number of miles or kilometers along an idealized "Earth" with no bumps. You only need a way to measure and compare to the map's scale. Perform a little arithmetic and you're done. But since the map is a projection of the spherical Earth the distance won't be exact. The scale varies over the map's surface, so the scale applies exactly only in very few places on the map. Another reason the straight-line distance is not exact that, even if you could follow the straight-line distance over the surface of the Earth, you would probably encounter hills, which add to the distance.

❏ Let's make the problem harder: You want distance from "A" to "B" along a highway route. Depending on the type of map, you may get some help in this analysis. If you are looking at an oil company map or an automobile association map, it might have one of those triangular matrices that indicates distances between selected cities. Here part of the analysis has been done for you—provided your origin and destination are in the chart. Another approach could be used if the map has numbers printed beside segments of the road that you can add up to get the total distance. Again, some of the analysis has been done for you, but you still have a bit of work to do. If the map doesn't have these features, then you could use the scale of the map to approximate distances along the route. Based on what the graphics of the map show, you could determine the sum of all those curvy road segments. The more the road curves, the more arduous the task. And even when you can sum the segments up, you have to remember that the road curves in three dimensions. Going up and down hills adds miles to the distance a car travels, over and above the distance that would be traveled by simply following the two dimensional line. Once data are in a GIS, such distance calculations are trivial—though they are still are an approximation. So the process of finding the distance—one of the simplest answers you might want from the map—is not simple, and is guaranteed to be imprecise.

❑ If you are unimpressed with the difficulty of analyzing distances on a map, let's move to a harder problem. You have a map of a county that has several parks. Suppose your map shows parkland as green areas. You would like to know the area—in acres, square miles, or square kilometers—that the parks occupy. How do you determine that? You might use the linear scale of the map to make a two-dimensional grid of squares on some transparent material, lay this over the parkland polygons, count squares, and do some arithmetic to estimate the area. Or you could divide each polygon up into triangles and calculate the area of each triangle.[1] Or you might obtain a remarkable device called a planimeter—a gadget that is made to measure the area of a graphically represented planar region—and run its stylus around the boundary to get an approximate value. (You'll find the planimeter in a museum, next to the slide rule, which is next to the abacus.) Or you could paste the map down on a thin sheet of aluminum, use tin snips to cut out the green areas, and compare the weight of all the cutouts to that of a known area of the aluminum sheet. Again, none of these processes is easy, or particularly accurate. The point is this: Maps are hard to analyze. In a GIS, once the data are in, an excellent approximation of the area is a quantity that you get for free.

❑ It is difficult to compare maps. If I didn't convince you how difficult it is to analyze a single map, look at the issue of comparing maps, or analyzing multiple maps to get information from the combination of them. Suppose a municipality wants to build an airport to serve its region and you are to advise the government on how to find adequate locations. What factors must be considered about the several hundred square kilometers around the municipality? Here is a partial list:

❑ Topography (of the site and of the surrounding area)

❑ Geology

❑ Existing houses and other structures

❑ Environmentally sensitive areas

❑ Soil characteristics

❑ Land cost and land availability

❑ Access to ground transportation facilities

❑ Weather patterns (e.g., tendency for fog to occur)

❑ Obstructions in the airspace (e.g., towers, wires)

❑ Existing land use

❑ Surrounding structures and their heights

❑ Habitat of endangered species

❑ Frequent unusual weather conditions

❑ Proximity to populated areas

---

[1]Knowing the three sides of a triangle, you can calculate the area as the square root of $s(s-a)(s-b)(s-c)$, where a, b, and c are the lengths of the sides and s is the semi-perimeter: $(a+b+c)/2$. I mention this because it seems now to be lost knowledge as far as high-school geometry teaching is concerned.

and several more.

❑ You are not likely to find all these features on a single map. If you have multiple maps you certainly have two major problems, and you probably have three.

    ❑ First, just inspecting several maps for the right combination of factors is quite difficult. How would you determine where to put the airport? You could pin the maps up on a wall and look at them. "Here's a location with nice large, flat area on Map A." "Whoops, no. Map B shows the soil wouldn't support a runway." Trying to look at several maps in a serial fashion is not easy.

    ❑ The maps you will be able to get will be different shapes, at different scales, using different projections and different units, and will cover different areas.

    ❑ Maps get out-of-date. A U.S. Geological Survey topographic map covering part of Lexington, Kentucky, was produced in 1965. It was updated in 1993. During that time interval, the city grew by 100,000 people. It takes a lot of time to produce a good map. Even as a map is published, it is out-of-date. Of course, part of this problem occurs because it is impractical to survey and record all significant changes just as they occur. But another major problem is that once a map is printed, the publisher cannot change it. If the accuracy of a map is critical, as with aeronautical maps, an updated version can be reissued periodically. With GIS, changes can be made to maps as soon as data about changes becomes available.

On the other side of this coin is the considerable value that maps, as they have been made for the last few decades, have for many sorts of activities:

❑ Maps provide an intuitive reference to the features and activities of an area of interest. They connect us with our environment with a level of directness and lack of ambiguity that one may not get with the "black box" of the computer.

❑ Maps are easily portable. They don't weigh you down. They display information you can read out of doors—something your laptop computer or palmtop may not do. They don't need batteries or charging; thus, they don't expire in the field at inopportune times.

❑ Maps usually give you large display areas. Yes, you can zoom to great levels of detail and pan on a computer screen or a palmtop screen. But there are times when you need both a reasonable level of detail and a large view. It's hard to beat six square feet of map in those instances.

❑ Many maps are basically honest. Those produced by the U.S. National Mapping Program adhere to rigid standards. (Look at http://mapping.usgs.gov/standards.) Frankly, the ability to provide users with measures of accuracy and quality control is something that all GIS programs fail at because vendors haven't gotten to yet. The move to associate metadata with GIS data sets goes in the right direction, but when you combine GIS data sets, there is no hint as to how good the resulting data sets are.

Force #2: The Need for Better Resource Allocation and Environmental Protection Became Evident

Before the 1960s, it was the view that the source of things humans wanted, such as land, resources, energy, air, and water, was independent of the sink where we put things such as garbage, heat, sewage, and combustion products. (See Figure 2-1).

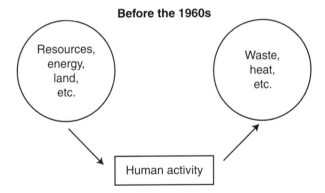

FIGURE 2-1

In the 1960s, perhaps beginning with the publication of *Silent Spring* by Rachel Carson, we began to understand the implications of our freewheeling use of resources and our disposal habits. At the end of the decade, the National Environmental Policy Act (NEPA) was passed and the Environmental Protection Agency (EPA) was established. Some, at least, began to understand our situation as depicted in Figure 2-2: The source and the sink are connected, and in a way that has serious implications for our future. Stuff moves back from the sink to the source. If, following the path of the dashed arrow, that movement occurs naturally, we call it pollution. If the transfer, according to the solid arrows, is back through human activity, we call it recycling. In any event, we have to deal with the fact that our source and our sink are connected. GIS can help.

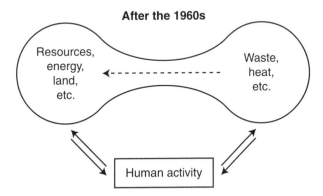

FIGURE 2-2

Force #3: The Evolution of the Garden-Variety Computer System Has Been Amazing

To compare computers, we might use as a gross measure this formula:

(S * M) / $

to be read as "computing speed" times "primary memory size" divided by "machine cost." Let's look at the change of computing power in the past 40-odd years (since the second-generation machines came out, using transistors instead of electron tubes).

❏ A popular machine in 1960 (the IBM 1620) had a memory of 20,000 bytes. A popular machine of today could have 1 billion bytes (1 gigabyte) So today's machine has on the order of 50,000 times as much memory.

❏ The IBM 1620 had a clock speed of 50 kilohertz. Today's machine might run at 2.5 gigahertz. That's another factor of 50,000.

❏ The cost of the IBM 1620 in 1960 was about $100,000 or, considering inflation, $500,000 in today's dollars. Today a popular machine might cost $2,000. This gives us a factor of 250.

So one could say that, considering cost, today's machine was 50,000 x 50,000 × 250 times as powerful as one in 1960. That's a factor of 625 *billion*. To put the idea of a "factor" in perspective, consider a factor of "2." Let's say your income was doubled (or halved). That's a factor of 2. Quite an effect, no? Or consider a factor of "10"—say, the speed of a car compared to a human running, or the speed of an airliner over a car. A factor of 10 changes the nature of whatever is being considered. What does a factor of 625 billion do? Boggles the mind, that's what. Of course, much of that increase is used up with graphics, poor programming, and nonuse (the laptop I'm typing on is loafing along at about 1 percent of its capacity, no matter how fast I type). Still, when the chips are down—so to speak—today's machine is quite unbelievable.

In addition to the sheer increase in power of today's computers, the hardware has become much more reliable, so that failures are quite rare, even over years. However, this is much more than offset by the fact that software has become less reliable. Software vendors tend to want to get their products on the market quickly, whether they are buggy or not.[2]

Spatial Data

When location or position is used as a primary referencing basis for data, the data involved are known as *spatial data*. For example, the elevations, in feet, of Clingman's Dome and Newfound Gap in the Great Smoky Mountains National Park are data. If the primary referencing basis for these data is the "Great Smoky Mountains National Park," or x miles south of Gatlinburg, Tennessee, on U.S. Highway 441, or p degrees latitude and q degrees longitude, then the elevations could be referred to as spatial data.

Spatial data, then, are discrete symbols (numbers, letters, or special characters) used to describe some entity; these data are organized according to the location of that entity in the three-dimensional world. It is data that pertains to the space occupied by objects. Physically it includes cities, rivers, roads, states, crop coverages, mountain ranges, and so on.

---

[2] It is the author's opinion that 30 percent of those who do or supervise computer programming should have to be examined, remedially trained where necessary, and then certified. The other 70 percent should be taken out and shot.

Normally, when it is desirable to describe things in the real world by spatial data, the objects are abstracted into some geometrical or mathematical form, as discussed in Chapter 1. For example, a fire tower might be represented by a point, a stream by a set of connected straight lines, and a lake by a polygon boundary.

Limiting the Scope

Spatial data (again, facts about the real world organized by locational coordinates) can be used to describe molecular structures, a human central nervous system, positions of books in a library, or stars in the universe. Since this book relates to geography, we now exclude several categories of spatial data. Specifically not considered are data that relate to

❑ Conditions that change quickly in time—in a matter of hours, days, or even weeks. Current pollution levels, weather, and tides will not be included, although average pollution at a point, climate, and ranges of tides could be included. Exceptions to this rule include using sensors to collect immediate data about conditions and put those data out to Internet sites for display on the Web.

❑ Objects that move about in space—such as automobiles, animals, or people. However, data about flows of these objects past a certain point at a certain time might well be included. Exceptions to this rule include using GPS to keep up with trucks and cars, or to track animals in the wild.

❑ Circumstances in which the locational identifier must be more precise than 1 decimeter (a tenth of a meter) to ensure that the related data are useful or valid. The smallest distance separating two adjacent entities that can be distinguished from one another is called the *resolution distance,* or, simply, resolution. If the separation in distance is less than the resolution, the data cannot be used to resolve any difference in the condition or situation. Again, there are exceptions: Surveyors are making increasing use of GIS, and they make measurement to within a centimeter.

Databases—What's Meant by "Relational"

Chapter I included a discussion of databases—relational databases in particular. What you have seen so far is one aspect of the RDB: a two-dimensional table in which you store entities as rows and attributes as columns. What is it that makes the relational database such a powerful approach to storing information? One answer lies in the fact that a RDB can be much more than a single table. Usually it is a number of tables that are related to one another, as previously discussed, that provide for efficiency, flexibility, and ease of updating.

Table 2-1 illustrates a database that may be created by a motor vehicles licensing department. Parts of the database connect to other databases formed by other government departments. Shown are nine relational database tables; the RDB table name is on the top line; the attribute names are on the second line. None of the tens of thousands of cells is shown.

Here, mainly for purposes of illustration, the MASTER table is miniscule, consisting only of a license plate number and an owner identification. The key field, shown in bold font, consists of a unique character string (no two license plates are the same) and an owner identification number. The contents of this column may not be unique, since one person may own several vehicles.

Both fields in MASTER refer to—link to—the other eight tables, either directly or indirectly. Plate_# allows the user access to both VEHICLE and ACCOUNTING.

The VEHICLE table describes some of the attributes of the car or truck in question. Since some aspects of all vehicles of a certain year, model, and manufacturer are identical, it would be a waste of space and an updating nightmare to place this information in a record describing a particular vehicle. So each record carries a Type_Code that refers to a key field in the table VEH_TYPE. There you will find the vehicle weight, length, width, horsepower, and fuel consumption data.

The ACCOUNTING table indicates whether taxes and registration have been paid. Also here is a reference to the INSURERS table, which carries information about the companies that insure vehicles in the state.

The other column in the MASTER table contains an Owner_ID code that matches up with the key in the table OWNER. That table could contain a host of information about the owner of the vehicle. Shown in the table are name and address. Also there is the owner's social security number (SSN). Perhaps there is a "motor-voter" effort to register all vehicle owners. With the SSN, which is the key field in the VOTERS table, a user could determine which owners were already registered to vote. Further, with the Precinct_Code, information about the location of the voting precinct could be obtained.

Finally, the SSN also allows a link to the table ARRESTED so that drivers with moving violations or driving while intoxicated could be identified. Both the VOTERS and ARRESTED tables probably would reside in some other department, so maintenance of those tables would not fall to the Department of Motor Vehicles, yet the DMV would have access to the information.

TABLE 2-1

MASTER						
	Plate_#	Owner_ID				

OWNER						
	Owner_ID	Name	Address	SSN		

VEHICLE						
	Plate_#	Color	Model	Year	Type_code	VIN

VEH_TYPE						
	Type_Code	Weight	Length	Width	H_Power	Fuel_Cnsup

VOTERS				
	SSN	Name	Precinct_Code	Regis_Party

PRECINCT				
	Prec_Name	Location	Supervisor	**Precinct_Code**

ARRESTED					
	Name	Alias	SSN	Offence	Convicted	**Court_Doc_#**

ACCOUNTING			
	Plate_#	Date_Paid	Insurance_Co_Code

INSURERS				
	Insurance_Co_Code	Co_Name	Location	Phone

Most governments and organizations have extensive and perhaps sophisticated techniques or systems for storing and manipulating data that can be referenced by these and other schemes. One quite useful referencing basis has not been developed as extensively, however. It may be known by several names: geographic, land, locational, geodetic, or spatial position.

Spatial Data for Decision Making

I have said spatial data relate to conditions, facts, and objects in three-dimensional space. Most spatial data that now exist use a two-dimensional referencing scheme such as latitude/longitude, a projection thereof, or street addresses. Unless the data set is specifically one that addresses the matter of altitude, this third coordinate is either included as part of the attribute data (rather than part of the locational identifier) or is implied by the nature of the data. For example, if the data describe soil characteristics, one understands that the top few feet of the Earth's crust, regardless of altitude, are being described.

Data types that might be part of spatial data sets are exemplified in Table 2-2.

TABLE 2-2   Example Data Types Included in Spatial Data

Soils	Types, physical and chemical properties
Vegetation	Species composition, age
Wildlife habitat	Types, carrying capacity
Hydrology	Ground and surface water, volume, flows
Geology	Rock types, minerals and ores, physical and chemical properties
Physiography	Elevation, slope, aspect
Land use	Activity types, structure type
Land cover	Types
Transportation facilities	Types, capacity, schedules, condition, age
Utility distribution systems	Service areas, capacity historical features and landmarks—importance, condition, ownership, use
Census districts	Population, housing, other demographic information
Fire districts	Equipment rating, insurance rating
Zip code zones	Delineation of zones
Centers of employment	Types, work hours, number of employees, industrial classification
Locations of police stations	Area of jurisdiction, facilities
Pollution sources	Types, duration, occurrence
Land parcel information	Owner name, address, value of land, value of structures, tax information

Sets of Spatial Data: The Database

I have discussed databases in general (a medium containing numbers, symbols, or graphics organized according to some scheme). And I have commented on the idea of spatial data (data describing entities in the three-dimensional world where the location of the thing being described is an integral part of the description). A spatial database, then, is a collection of spatial data, organized in such a way that the data can be retrieved according to their locational identifiers and in other ways as well.

This presentation of a list of data types or variables whose data might be stored in a spatial database does not imply that satisfactory storage schemes are easy to determine, nor that each variable will be stored in the same manner. Whether the geometric abstraction for a data type is best selected as a point, line, area, or volume can be an important consideration in a particular storage scheme.

FIGURE 2-3   An orthophotoquad image of a portion of northwestern Michigan

The development of spatial databases to be used for analysis and decision making is both an art and a science. Thus, when any data-handling program is being developed, a key point to remember is that there is no single best way.

Spatial Databases: Inherent Difficulties

In addition to all the problems one has in building, maintaining, and operating any large database, spatial databases have their own peculiarities and challenges. Some are in the following.

Size

An airplane pilot (in pre-radar days) once said that the thing that kept Air Traffic Control from folding up completely in its attempt to keep airplanes from colliding was that "God packs a lot of airspace in three dimensions." Anyone who has worked planning or management related to the land knows that God also puts a lot of surface area in two dimensions. Thus, any spatial database used for land and resource considerations will either (a) not cover much area, (b) not include much detail, or (c) be very big. Very big databases, regardless of their simplicity, are expensive to build and maintain.

Spatial data in general use up a lot of computer memory and disk space. For example, Figure 2-3***[3] is an orthophoto image of a part of northern Michigan (around Frankfort and Pilgrim) and Lake Michigan. It represents an area of about 30 square miles.

The image consists of squares (picture elements (pixels)) that are 1 meter on a side and can display white, black, and 254 shades of gray. The file underlying the image, represented in the most basic form, binary, looks like this:

```
11111111110110001111111111000000000000000010000010010100100 0110
01001001010001100000000000000001000000010000000000000000000000001
00000000000000010000000000000000011111111110110110000000001000011
00000000000001000000001100000011000000111000001100000010100001000
00000111000001110000011100001001000010010000100000000101000001100
00010100000011010000110000001011000010110000110000011001000 10010
00010011000011110001010000011101000110100001111110001111000011101
00011010000111000001110000100000001001000010111000100111001 00000
00100010001011000010001100011100000111000010100000110111001 01001
00101100001100000011000100110100001101000011010000011111100100111
00111001001111010011100000110010001111000010111000110011001 10100
00110010111111111011011000000000100001100000001000010010000 1001
00001001000011000000101100001100000110000000110100001101010001 1000
00110010001000010001110000100001001100100011001000110010001 10010
00110010001100100011001000110010001100100011001000110010001 10010
00110010001100100011001000110010001100100011001000110010001 10010
00110010001100100011001000110010001100100011001000110010001 10010
00110010001100100011001000110010001100100011001000110010001 10010
00110010001100100011001000110010001100100011001000110010001 10010
00110010001100100011001000110010001100100011001011111111 1000000
00000000000100010001000000001101100110100000101011001010000 0011
00000001001000100000000000000010000100010000000100000011000 10001
00000001111111111000100000000000011111000000000000000000000001
00000101000000010000000100000001000000010000000100000001000 00000
00000000000000000000000000000000000000000000000000000000000000001
00000010000000110000010000000101000001100000111000010000000 1001
00001010000010111111111110001000000000010110101000100000000 0000
00000010000000010000001100000011000001000001000000011000001 01
00000101000001000000010000000000000000000000101111101000000 01
00000010000001100000000000001000010001000010100010010001000 1
00110001010000010000011000010011010100010110000100001110010 0010
01110001000101000011001010000001100100011010000100001000001 00011
01000010101100011100000100010101010100101101000111110000001 00100
00110011011000100111001010000010000010010000101000010110000 10111
00011000000110010001101000100101001001100010011100101000001 01001
00101010001101000011010100110110001101110011000001110010011 1010
01000011010001000100010101000110010001110100100001001001010 01010
01010011010101000101010101011001010111010110000101100101011 010
01100011011001000110010101100110011001110110100001101001011 01010
```

---

[3]Figures indicated in the text with three asterisks (***) following are on the CD-ROM that accompanies the book. They are in the folder IGIS_with_ArcGIS_Selected_Figures. These figures are in color, which may provide you with additional information.

```
0111001101110100011101010111011001110111011110000111100101111010
1000001110000100100001011000011010000111100010001000100110001010
1001001010010011100101001001010110010110100101111001100010011001
1001101010100010101000111010010010100101101001101010011110101000
1010100110101010101100101011001110110100101101011011011010110111
1011100010111001101110101100001011000011110001001100010111000110
1100011111001000110010011100101011010010110100111101010011010101
1101011011010111101100011011001110110101110000111100010111000111
1110010011100101111001101110011111101000111010011110101011110001
1111001011110011111101001110101111101101111011111111100011111001
1111101011111111110001000000000000011110000000100000000000000011
0000000100000001000000010000000100000001000000010000000100000001
0000000100000000000000000000000000000000000000000000000000000001
0000001000000011000001000000010100000011000000111000010000000101001
0000101000001011111111111100010000000000010110101000100010000000
0000001000000001000000010000001000000010000000011000001000000111
0000010100000100000001000000000000000001000000100111011100000000
0000000100000010000000110001000100000100000000101001000010011001001
0000011000010010010000010101000100000111011000010111000100010011
0010001000110010100000010000100000010100010000101001000110100001
1011000111000001000010010010001100110011010100101110000000010101
0110001001110010110100010000101000010110001001000011010011100001
0010010111110001000101110001100000011001000110100010011000100111
0010100000010100100101010001101010011011100011011110011100001110011
0011101001000011010001000100010101000110010001110100100001001001
0100101001010011010101000101010101010110010101110101100001011001
0101101001100011011001000110010101100110011001110110100001101001
0110101001110011011110100011101010111011001110111011110000111001
0111101010000010100000111000010010000101100001101000011110001000
1000100110001010100100101001001110010100100100101100101101001011 1
1001100010011001100110101010001010100011101001001010010110100110
1010011110101000101010011010101010110010101100111011010010110101
1011011010110111101110001011100110111010110000101100001111000100
1100010111000110110001111100100010011001001100101101001011010011
1101010011010101110101101101010111110110001011001110110101011100010
1110001111100100110010111100101101110011111101000011101001111010
1111001011110011111101001110101111101101111011111111100011111001
1111101011111111110110100000000000011000000000011000000100000000
0000001000010001000000011000100010000000001111110000000011110011
1001010000100011000111010000111001001111011010100001110101001011
0111001001111010011101100000001010010001101101111010101110000011
1000111000101001001101110000010001110100000111111000010100000000
0001111110000101000010111101001100111110100001101010001111000001
0011110000010011001110001110110110001010011010010110000000000000
0110001110000001010011111110010110010111101111100010100000000001
1010100111010111000000111010110100111000001110011100100001011010
0100110110000111011101100101011101101011100000011101011010000000 0
1111110011011100111111100001010000000000111000101111110101110010
1001101001000000010011000110001110111111010111000101001000001111
0011100100111000001111010111101011010010110001101010110001111001
0001110100111101011010000000001001000101010111001001000110100011 1
```

(This goes on for about another 630 pages, which the publisher has, understandably, declined to include.)

Continuous Nature of the Referencing Basis

In most databases, a particular unique key points to a unique thing. For example, a given auto license number identifies a particular car; a name or social security number tags an individual person; a house number and street constitute a pointer to a residence. But spatial phenomena do not enjoy any such autonomy: They are a mixture of discrete and continuous. That is, there is no natural and completely satisfying one-for-one correspondence between spatial locators and the related data. There is a virtually infinite amount of data potentially available about even the smallest part of the real world; we can store only a small part. Thus, by choosing a particular technique for organizing the continuous into the discrete, we are screening out or "throwing away" an infinite amount of potential information. Clearly, it takes some sophistication and forethought to select a technique to represent the continuous real world and have a database that will be useful in solving problems.

Continuous Nature of the Data

In addition to the continuum of two- and three-dimensional space just mentioned (i.e., the fact that our basic referencing scheme potentially has infinitely many points in it), there are also problems with the continuous nature of the data themselves. Soil type is probably a good example. Just as no two snowflakes are exactly alike, no two soils are exactly alike. Soils must be categorized into groups and a judgment made about which group a particular soil belongs to. In naturally continuous variables, such as elevation, the parallel issue of precision comes in: Do we measure (vertically) to the nearest meter? To the nearest millimeter?

Abstraction of Entities

The simplest reference that can be made in a spatial database is to a point. But no material entity is ever just a point. Many of the things we deal with are either linear features, areas, or volumes, so the referencing scheme becomes more complicated. Where is a house? Well, it's lots of places when you get right down to it. Do you define it by its corners in plan view? Do you select a single point, a "centroid," and define the house to exist at that point? Do you simply say it exists in acre "X," perhaps with many other houses? There are many fundamental variations in the ways the "real world" is and can be referenced. These varying methods can be incompatible, precluding any easy transfer of data or techniques for manipulating data.

Multitude of Existing Spatial Coordinate Systems

There are many spatial coordinate systems. Most of those in use for planning and resource management rely on the use of flat projections of curved surfaces. Many of the data sets that will be used to build a multivariable spatial database will come from data recorded with distorted and dissimilar methods of representation. Matters of units, datum, spheroid, and projection must be addressed. A single state may use many coordinate systems in its various agencies. Examples are latitude and longitude (both NAD27 datum and NAD83 datum), UTM (both NAD27 datum and NAD83 datum), a state plane coordinate system (one or more zones), road miles, river miles, a special coordinate system for particular features (e.g., oil and gas wells), and so on.

Existing but Inappropriate Data

While it is true that considerable data of the types important to this discussion have been collected, many of them are not directly usable in a spatial database. This occurs principally because these data, collected by groups or agencies with specific missions to serve, have been assembled in nonuniform categories or have been interpreted in a specific manner for a particular purpose. For example, early soils data categories may not contain the necessary information that will enable measurement of some environmental effects of land use activities.

Effort Required for Development

The data in our base won't develop as a natural consequence of some already ongoing process. Other database developers are more fortunate. As a clerk processes applications for auto license tags, he or she may type the pertinent information about the car, owner, and tag directly into a database. Thus, the database develops as a result of the tag-selling process that must occur anyway. Spatial databases about the environment have not evolved as consequences of other processes; the work starts almost from scratch in most cases.

The Changing Environment

One cannot get an entire spatial database of any size developed before part of it is incorrect because some of the values in the real world will have changed over time. Land use is an example of a variable whose data values are changing in many places on a daily basis. Houses are built. Roads are paved. Even such stable phenomena as topography change drastically over time. For example, the Mississippi River was about 1300 miles long when LaSalle floated down in his canoes. When Mark Twain wrote about it 200 years later, it was less than 1000 miles in length. Not only that, very little of what was wet in LaSalle's day was still river in Twain's. And to further illustrate the futility of any attempt at a "permanent" spatial database, Ole Man River has moved at least two towns from one state to another by its meanderings. The moral is that some data values of all variables in a spatial database are going to change over time. Some procedure for updating the base must be developed or the value of the base will be degraded by time. Further, different variables are of different value to the analysis and decision-making process and, of course, change at different rates. In some cases, the efficient thing to do is to note changes as they occur; in other cases, replacement of all the data related to a particular variable is in order. Either way, there are difficulties and costs.

Multiple Paradigms for Storing Geographic Data

Ingenious ways of taking the continuous, virtually infinite environment and storing its important facets in a discrete computer have been developed. Spatial data stored in one scheme are not easily converted to another, and one almost always loses information is such a transfer.

For example, take the matter of representing the elevations above sea level of a geographic area. To begin with, one is dealing, in theory, with an infinite number of values. If you establish the elevation of a given point, then, depending on the precision of your measurements, the point 1 meter to the north will have a different elevation. Elevations of an area may be thought of as a continuous surface, potentially different at every latitude and longitude position. Unless we can model this surface with a mathematical equation (and, with the average mountain or cow pasture, this is usually out of the question), we are stuck with having to select a set of specific points, determine their elevation, and make assumptions about the elevations between those points.

Several ways of representing elevation in GISs have been used. Three are as follows:

Contours

Digital elevation models

Triangulated irregular networks

These will get more detailed treatment later, but let's look at their essential characteristics.

Contours are familiar to you from your experience with topographic maps. Each contour line represents a given elevation. That is, if you walked along the path depicted by the contour line, your elevation would not change. Recalling that a GIS is a marriage of a geographic database and an attribute database, you see that the geographic points along the path form the geographic part of this partnership, while the elevation of the line is the attribute datum. To obtain an estimate of the elevation at a point between two contour lines interpolation might be used.

Digital elevation models (DEMs) rely on the idea of a raster. A *raster* is a set of equal-size squares, arranged in rows and columns, which cover (tessellate, if you want a highbrow word) the plane. Think of a chess-board. Or square tiles on a kitchen floor. The geographic position of each square (e.g., its center) can be calculated by its row and column number. Each square has an attribute that might be its average eleva-tion. (Or, if the DEM were being constructed for aircraft pilots, it would be better if the attribute were the maximum elevation!) The reported elevation is therefore constant within a given square and usually changes at each edge of each square, making for a rather lumpy representation. Obviously, DEMs with more, and smaller, squares potentially represent the surface better. Of course, the value obtained from a DEM for a given position is almost guaranteed to be somewhat in error.

Triangular irregular networks (TINs) represent the surface of a geographic area by a set of triangles whose vertices are points of known elevation. Since three points determine a plane, the computer can come up with an estimated value of elevation for any requested point in between. Of course, the Earth's surface is not made up of triangles, and poor selection of the points of known elevation (e.g., midway up a hill rather than at its top) could dramatically, negatively affect the accuracy of the TIN. Again, the more known points, the more triangles, the (potentially) better representation of the true environment.

Besides acquainting you with three different data structures, my point here is to show the diversity of ways that GIS inventers have chosen to represent a part of the environment.

Information Systems

This issue of "processing" data in map (analog) versus symbol (computer form) brings us to the matter of processing or handling data in general. The conceptual model we will use is that data are processed to produce information. Actually, the terms are not absolute, because what is information to a person filling one role may be data to someone filling another. But the idea of a before-after concept, distinguishing the two states as data and information, turns out to be useful, so we employ it.[4]

An *information system*, in the context of this material, is a set of steps, or processes, that is executed by a "device" to produce information. We choose to call the symbols that are input to the process by two

---

[4]The idea that data precedes information might occur in the following context: Existence, Awareness, Observation, Measurement, **Data**, **Information**, Knowledge, Understanding, Wisdom.

names: data and parameters. "Data" we have discussed along with its formulation into bases. Parameters we consider to be information, which the user of an information system supplies at the time of use of the system. Such parameters might specify which data in the base are to be used, how they are to be combined, what the format of the resulting information—output—is to be, and other specifications and/or constraints.

For an example, a professor may assign a student the task of compiling a word-processed bibliography of the works of Shakespeare. The database might be a library catalog; the parameters are such descriptive terms as "bibliography," "Shakespeare," "word-processed"; the device is the student, who, with his eyes, pencil, word processor, and so on produces the information.

Figure 2-4 is a diagram of a generic information system.

In Summary

Based on the preceding discussion of spatial data, databases, and information systems, I can offer yet another definition of GIS: A geographic information system is an information system that has as its primary source of input a base composed of data referenced by spatial (or land, or geographic) coordinates. The system accepts parameters, examines its database, and provides information for decision making and resource management. In an automated spatial information system, a major part of the device that does the processing is an electronic digital computer. Much of the database it uses is stored on computer hard drives in servers or PCs.

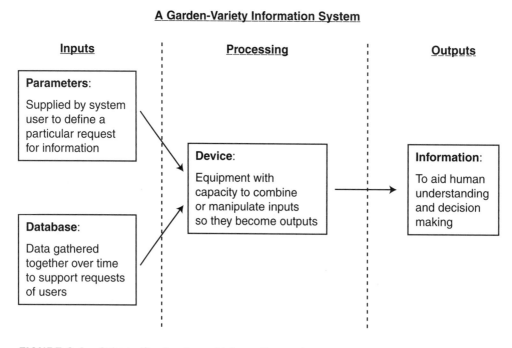

A Garden-Variety Information System

| Inputs | Processing | Outputs |

Parameters:
Supplied by system user to define a particular request for information

Device:
Equipment with capacity to combine or manipulate inputs so they become outputs

Information:
To aid human understanding and decision making

Database:
Data gathered together over time to support requests of users

FIGURE 2-4 Schematic of a general information system

Uses for a Geographic Information System

Land and Its Use

Discounting the possibility of sudden catastrophe, the strongest factor in how things will be tomorrow is how they are today. A planner or manager who fails to provide him- or herself with information about the current state and characteristics of the environment will probably misplan and mismanage.

Perhaps the most important variables in a geographic database are as follows:

1. What now exists on the land (land cover and resources)

2. How the land is employed (land use and human oriented activities)

3. What is legally permitted to happen to the land (zoning and legal control)

Once the present state of the environment, or the portions of it with which we are concerned, is recorded in a form amenable to processing, we can begin to make decisions about its conversions to some other use. A geographic information system can be useful in dealing with at least three general categories of issues:

1. Determining the effect a particular activity or land use will have at a particular location (sometimes called environmental impact analysis).

2. Given a particular activity, with its characteristics known, determining a set of locations where it might be placed (sometimes called locational analysis).

3. Given a particular location or site, determining a set of land use activities that might well be placed there (sometimes called site analysis).

Let us now briefly look at a variety of specific areas in which geographic information systems could have an impact. The thrust of presentation is mostly by example and is far from comprehensive. The format I will use, for the most part, is as follows:

1. Examples of types of spatial (and other) data that might be stored

2. Advantages that might accrue by careful use of the data in (1)

An important factor to notice in the following is the degree of overlap among variables of different areas of concern.

The Natural Environment

Knowledge of the natural environmental state of the land is central to a determination of what should be preserved, what should be enhanced, what activities could be supported, what impacts are likely to occur from given uses, and a host of other questions.

This change—the realization of "spaceship Earth"—has been occurring since the 1960s and has had many profound and far-reaching effects. As a partial result, many of the first attempts to use spatial information systems to support decision making have had the storage of natural science information as their basis.

Storage of geographic (and other) data about the following:

Climate

Bedrock

Surficial geology

Physiography

Hydrology

Soils

Vegetative cover

Wildlife habitats

help us

Identify, delineate, and manage areas of environmental concern

Analyze land-carrying capacity

Write environmental impact statements

Energy

Energy potentials begin their service for humankind in many forms—oil, gas, coal, hydraulic head, wind, tide, sun, fission, fusion—and always wind up in the same way: *heat*.

The problems associated with the efficient and useful transfer from energy potential to heat are myriad; much of the time we throw away large amounts of energy because it does not serve a particular process— for instance, nuclear power plants discharge vast quantities of heat into rivers to the detriment of the fish and the impoverishment of humans paying for waning gas supplies to heat their homes. A geographic information system is not the answer to the sensible use of energy, but it is a tool that can help reduce energy waste. Spatially distributed data on energy sources, energy movements, and energy use of all kinds could lead to a greater understanding of our wastefulness and how to prevent it. These data sets could lead to discovery of new energy sources and how to tap them.

Storage of geographic (and other) data about the following:

Potential energy sources

Location

Size

Cost of extraction or tapping

Surrounding environment

Access

Processing capability

Energy distribution systems

Location

Paths

Capacities

Intermediate storage facilities

Types of energy conveyed

Degree of hazard

Energy use patterns

Industrial

Residential

Peak usage

Distribution among users by user characteristics

may lead to information allowing analysis of

Costs of moving energy

Remaining available energy reserves

Efficiency of different allocation schemes

Waste

Heat pollution

may lead to information for delineating

Areas of danger to humans

Environmental impact

may lead to information for developing

New distribution lines

Resource allocation schemes

Human Resources

It is for people that we operate our governments. It is primarily people who use the land, the energy, and the resources, and, in part, it is people who feel the effects of its ill use.

The vast amount of data about people is not stored in spatial form for at least two reasons:

1. They move around—day to day and year to year.

2. We protect their privacy to a considerable extent.

But the storage of information about human resources and conditions in a spatial context offers two major advantages:

1. It allows us to deal in a very direct manner with our primary concern: humankind.

2. Many sets of data have been developed, largely by the Bureau of the Census in such a way as to permit relatively easy loading into a spatial database, even though—for reasons of privacy and reasons related to the mission of the census—the "grain" of such two dimensional information storage is very coarse for most applications.

Storage of geographic (and other) data about humans:

> Where they live
>
> How much they consume
>
> How much they earn
>
> How old they are
>
> What they discard
>
> Where they play
>
> What crimes they suffer
>
> What mishaps befall them
>
> What facilities are available for their employment, shopping, learning

may lead to information:

> To plan for
>> Mass transit
>>
>> Recreation areas
>>
>> Police unit allocation
>>
>> Pupil assignment
>
> To analyze
>> Migration patterns
>>
>> Population growth
>>
>> Crime patterns
>>
>> Welfare needs
>
> To manage
>> Public and government services

Areas of Critical Environmental Concern

Areas of critical environmental concern are those geographic areas that are important to the needs of humans. Not only do they perform functions related to the health, safety, and welfare of the general public, but they may also serve economic and educational needs as well. Areas become of critical environmental concern when natural resources become scarce or are threatened through the actions of humans, or when the areas themselves present a threat to the human population.

Storage of geographic (and other) data about the following:

Agricultural lands

Natural and scenic resources soils

Aquifers

Geology and geologic hazards

Wildlife habitats

Vegetation

Floodplains

Wetlands

Scientific areas

Wild and scenic rivers

Cultural activities

Transportation networks

may lead to information

to facilitate

Identification of unique resources

Management of designated areas

Determining relative importance of kinds of resources

Water

Water is the most important resource to the functions of natural environmental processes and human activities. It is a dynamic resource—its movement, both as surface water and groundwater, creates a very broad management problem. Through information in the spatial context we can better analyze and manage our water resources.

Storage of geographic (and other) data about the following:

Natural bodies of water

Supplies

Use patterns

Recreation needs

Climate

Water sheds

Elevations

Industrial locations

Settlement locations

may lead to information about

Floodplains

Availability of clean water

Irrigation

Pollution (potential and existing)

Natural Resources

Natural resources are both finite and necessary for the survival of humans and the maintenance of quality of life. Some natural resources are renewable with proper management, while others will simply run out. A continuing supply of information is necessary for a proper evaluation of how we should use our resources wisely.

Storage of geographic (and other) data about the following:

Forests

Mineral sources

Energy sources

Rivers, streams, lakes

Wildlife and fish

Agriculture

Harbors

Geology

may lead to information

to facilitate

Timber management

Preservation of agricultural land

Conservation of energy resources

Wildlife management

Market analysis

Resource allocation

Resource extraction

Resource policy

Recycling

Resource utilization

Agriculture

The production of food has received increasing attention with a growing world population and shrinking agriculturally productive lands. Demands for grain and other crops has increased tremendously as the United States has drawn closer to the world marketplace. The need for good agricultural management becomes more obvious as we try to meet the needs of others. Data demands will also increase as we seek solutions to this growing problem.

Storage of geographic (and other) data about the following:

Land conversion

Soils

Geology

Crop productivity

Climate

Hydrology, water supply

Irrigation

Erosion

Crop disease, blight

Insect control

Pesticides

Fertilizers

may lead to information

to facilitate

Crop management

Protection of agricultural lands

Conservation practices

Prime agricultural land policy and management

Crime Prevention; Law Enforcement; Criminal Justice

The Criminal Justice System has many potential applications for geographic data: to assist the system in predicting likely points of criminal activity and to enable efficient allocation of resources through systematic identification of locations warranting increased manpower, analysis, or resource allocation. Although spatial data are routinely collected by law enforcement agencies, frequently it is done in a nonuniform manner that lacks sufficient precision to be tactically useful or to enable ready comparisons of location information from occurrence to occurrence.

Storage of geographic (and other) data about crime:

Where (specifically) crimes occur

Where stolen property is recovered

Where arrests are made

Where high risk businesses are located

Where arrestees live, were schooled

may lead to information

to plan for

Selection of sites or premises for target-hardening attention

Procedures for establishing risk ratings for particular locations

Tactical (as opposed to strategic) patrol allocation

Selection of particular locations for detailed crime prevention analysis

Crime pattern recognition

Selection of areas or schools for delinquency prevention attention

Homeland Security and Civil Defense

Homeland Security and Civil Defense agencies have been established to respond to natural or human-caused disasters or those caused by war or terrorism. As such, we need plans for both short-term and long-term aid to communities across the nation.

Storage of geographic (and other) data about the following:

Population distribution

Sources of food

Geologic activity (earthquakes)

Transportation

Military installations

Public facilities

Medical facilities

Rescue equipment

may lead to information

to facilitate

Alternative disaster relief plans

Need for stockpiling of foods and medical supplies

Evacuation plans

Proper designation of disaster relief areas

Communications

Communications represents the way in which humans stay in touch with occurrences around them and through which they transmit information. The physical requirements of communication systems have considerable impact on the natural environment. In order to maintain harmony between the two, information on geographic data must be kept before the decision makers.

Storage of geographic (and other) data related to the following:

Communication stations and antennas

Population

Terrain

Power sources

Current events

Technical information

may lead to information

to facilitate

Siting of transmission lines

Location of cellular equipment

Education

Transportation

The movement of people and materials for economic, social, and recreational reasons requires consideration of spatial data. Analysis is required on levels ranging from small-scale local transport to the national and global scale.

Storage of geographic (and other) data about the following:

> Highways, roads, interchanges, and so on
>
> Rapid transit
>
> Airports
>
> Seaports
>
> Railroads
>
> Origins and destinations of travelers
>
> Population shifts
>
> Centers of employment
>
> Commercial traffic

may lead to information

to facilitate

> Alternative transportation plans
>
> Locational analysis
>
> Mass transit
>
> Energy conservation

Characteristics and Examples of Spatial Data

Appreciating Geographic Space and Spatial Data

For this exercise you will need a notebook and may need a calculator and a long tape measure.

___ **1.** Carefully measure the length of your stride. (A procedure that might help with this is to take several steps across a floor with fixed-width tiles, or under a ceiling with uniform length panels.) To the *nearest tenth of a foot*, your stride is _____ feet.[1] How many of your paces would constitute 100 feet? _____

___ **2.** Find an area of landscape that is a square, roughly 210 feet on each side, and walk its perimeter, examining its interior as you go. Aside from giving you an idea of what an acre[2] is, this activity will probably let you view the complexity that can be contained in a small bit of ground—perhaps the land use, soil, rocks, pavement, vegetation, crops, buildings, fire hydrants, parking meters, street lights, pipes, wires, and other features.

___ **3.** How many acres constitute a square mile? (Use the exact definition of acre in the earlier footnote; the area in square feet of a square mile is 5280 times 5280.) _____. The land area of the United States is about 3,500,000 square miles. How many acres would that be? _____. An average-sized state would be about one-fiftieth of that. You can see that recording the data for a state for just one simple theme could use up a lot of computer storage.

___ Open up your Fast Facts text or document file.

---

[1]To convert inches to feet (with tenths) divide by 12. E.g., 30 inches is 2.5 feet.
[2]Defined approximately as the area enclosed by a square that measures 208.7 feet on a side, or defined exactly by a rectangle 1 foot wide by 43,560 feet (about 8 miles) long. In other words, an area of 43,560 square feet.

Exercise 2-2 (Setup)

ArcMap Toolbar Examination and Review

1. Start ArcMap. Select a new empty map and click OK. Enlarge the ArcMap window so it occupies the full monitor screen. The window should look something like Figure 2-5.

2. Using View > Toolbars, turn on the following:

❑ Main menu

❑ 3-D Analyst

❑ Draw

❑ Editor

❑ Layout

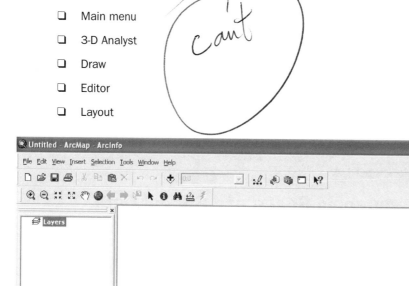

FIGURE 2-5

❏ Spatial Analyst

❏ Standard

❏ Tools

For each docked toolbar, double-click the toolbar handle (a short, vertical bar at the left end of docked each toolbar—shown in Figure 2-6. Also double-click the horizontal bars at the top of the pane on the left of the window—just above where the word Layers appears. Drag the elements around until they resemble Figure 2-7.

When you use ArcMap to display data sets, the Table of Contents (T/C) will contain the names of those data sets and information about the symbology used to display them.

As you saw when you selected the toolbars to turn on, ArcMap has a lot of toolbars, which suggests a lot of capability (and complexity). The preceding toolbars are those you will probably use most often. I will describe each tool later as we come to it, but you can get an idea by running the cursor over them and looking at the status bar and tool tips.

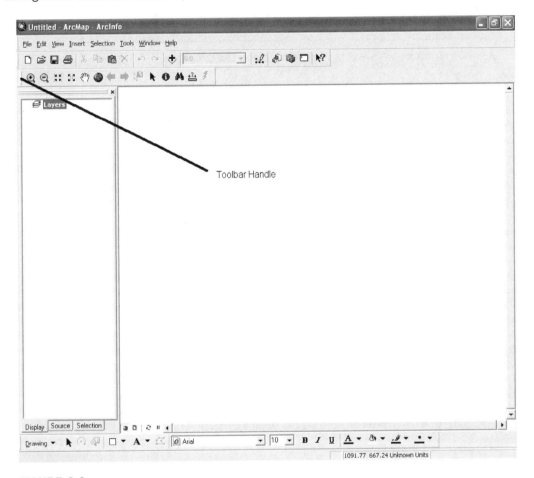

FIGURE 2-6

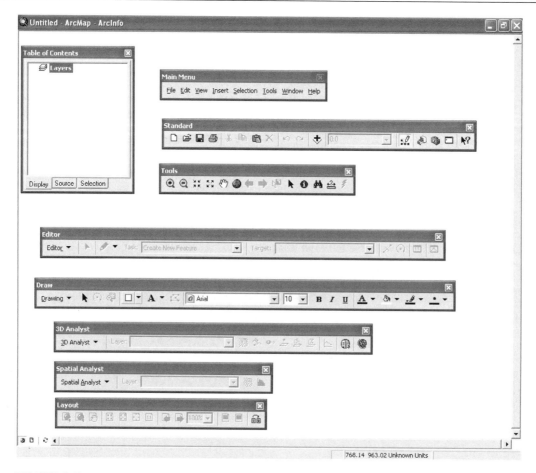

FIGURE 2-7

3. When you have finished exploring, dismiss the 3D Analyst, Spatial Analyst, Editor, and Layout toolbars. Double-click the title bars of the remaining elements to restore them to their original positions. If the result doesn't look like Figure 2-5, drag the toolbars around by their handles until it does. Dismiss ArcMap

Exercise 2-3 (Major Project)

Exploring Different Types of Geographic Data

The Basic Difference Between ArcCatalog and ArcMap

The most general statement that can be made about ArcCatalog and ArcMap is this: ArcCatalog deals with the exploration, examination, and finding of geographic data sets; ArcMap uses those data sets to form layers that display maps and allows analysis of the underlying spatial data.

Exploring Data from the NAVSTAR Global Positioning System (GPS)

A GPS receiver, utilizing the U.S. Department of Defense NAVSTAR system of about 30 satellites, collects positional information—in the form of latitude, longitude, altitude, and time fixes—and stores these coordinates in its memory. Computer files of these points can then be made into ESRI data sets.[3]

In November of 1994, the students and faculty of the Department of Geography at the University of Kentucky participated in a cleanup of the Kentucky River. They took a GPS receiver on their trek; the antenna was mounted on the roof of a garbage scow (originally built as a houseboat).

One file the students collected, along the river from a marina to an "island" in the river, was C111315A.SSF (designating the 11th month, 13th day, 15th hour). Using post-processing differential correction, the fixes in the file were adjusted to yield greater accuracy. Then the file was converted to an ESRI shapefile and renamed Boat_SP83.shp. The file is called Boat_SP83, since the data set was projected to **State Plane** (Kentucky North Zone [1601]) coordinates, using the North American Datum of 1983 (NAD**83** datum).

Preliminary

_____ **1.** Start ArcCatalog. Use Connect To Folder to place the

[____] IGIS-Arc\RIVER

folder in the Catalog Tree.

_____ **2.** Launch ArcMap from ArcCatalog. (Start ArcMap with a new empty map.) Dismiss ArcCatalog.

"Extensions" are additional software packages that extend the capabilities of the main software. If you have the right to use an extension, then it was probably loaded on the hard disk with the rest of the ArcGIS Desktop software. To conserve computing resources (time, memory space), extensions are not loaded into the fast memory of the computer unless the user takes specific action.

_____ **3.** Make the ArcMap window occupy the full extent of the screen. Enable the 3D Analyst extension and the Spatial Analyst extension with Tools > Extensions. See Figure 2-8. Close the Extensions window.

_____ **4.** In the steps that follow, you will add a layer from [____] IGIS-Arc\RIVER. Make sure that the Display tab (rather than Source tab or Selection tab) is active at the bottom of the Table Of Contents pane.

Seeing the GPS File in ArcMap

A GPS file is basically a file of individual three-dimensional spatial points, or fixes. The fundamental form of each point is a latitude value, a longitude value, and an altitude value. When this file was converted to a point shapefile, the latitudes and longitudes were converted to Cartesian northings and eastings. Each altitude became represented as an attribute associated with the appropriate point.

---

[3]See the author's textbook: *The Global Positioning System and GIS*, 2nd Edition, Taylor & Francis, 2002.

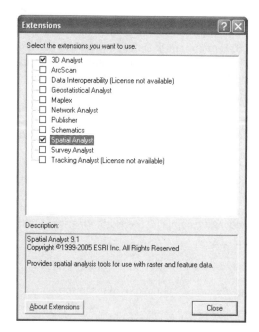

FIGURE 2-8

___ **5.** *Add the layer Boat_SP83.shp:*[4] To add the file, select File > Add Data. In the Add Data window, click the Details icon. Repeatedly press the Up One Level button until it is no longer active. The Look In: box should say Catalog. (See Figure 2-9.) Navigate to the [___] IGIS-Arc\RIVER entry and pounce.[5] Find and select Boat_SP83.shp. Press the Add button. (Ignore any warnings.) In the left-hand pane (the Table of Contents, as you saw previously), you will see the Boat_SP83 entry. In the right-hand pane, which is formally called a *data frame*, you should see a GPS track composed of a plotting symbol that looks like a little diamond or dot. The track resembles a fishhook. Compare with Figure 2-10***.

___ **6.** *Make the color of the plotting symbol bright red:* The plotting symbol color has been chosen randomly by the software, but you can change it. To do so, right-click the *symbol* itself in the Table of Contents to bring up a window-ette containing a hundred or so colors. If you pause the cursor over a color, its name is revealed as a ToolTip. You might use Mars Red here—click it. The process for changing the color of a plotting symbol is something you might want to record as a Fast Fact. Find a tiny textbox next to the plotting symbol and click it. Type GPS point and hit Enter.

___ **7.** *Set the map units and display units:* If you move the cursor around the map, you will see numbers appear in the status bar at the bottom of the screen. These are display units, and are presently shown as Feet. Right-click the data frame to bring up a menu. Click Properties to

---

[4]When the first sentence of a step is in bold italics, like this one, it means that you should read over all of the text of the step before you attempt to complete the step.

[5]A "pounce" is either a double click or a single click (to select) followed by pressing Enter on the keyboard.

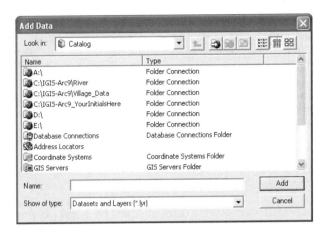

FIGURE 2-9

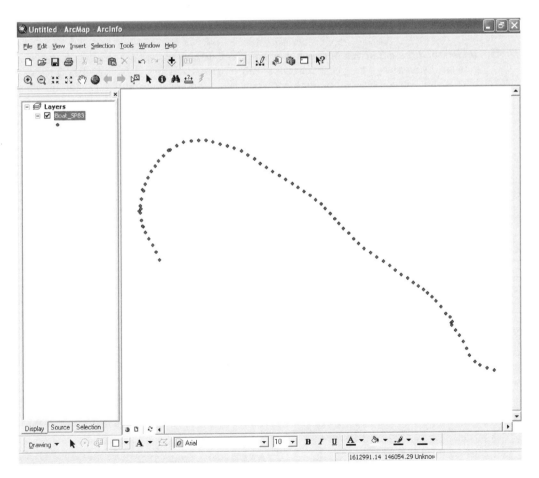

FIGURE 2-10

see the Data Frame Properties window. Click the General tab. In the Units area, you will see Feet in the Map text box. The box is grayed out, so you can't change it. The reason is that the data set has its projection defined already (Kentucky North Zone [1601]), which has survey feet as its unit of measurement.

8. Notice that you can change the display units in Data Frame Properties. They are set as feet, by default—the same as the map units. Place the cursor in the Display text box and press Shift-F1 to bring up context-sensitive help; read the information, then click elsewhere in the window. Select Miles in the Display text box. See Figure 2-11. Click Apply. Click OK. You may note, as you slide the cursor around the map, that now the status bar shows the display units to be Miles. These are the number of miles, in the X and Y directions, respectively, from the origin of the coordinate system.

The map units of all layers shown later in this data frame will be in feet. In order for maps in the data frame to be shown properly any subsequent layers must also appear to have feet as their map units. In the event you add a layer to the data frame that has different units or a different projection, the appearance of that layer will be converted automatically—"on the fly"—so as to be displayed in the form of Kentucky North Zone (1601) and to have feet as the map units. The display units, on the other hand, are up to the user and are changeable at any time.

9. ***Understand the difference between map units and display units:*** Click the data frame. Use the glossary by pressing F1 > the Contents tab > GIS Glossary. Click the letter D to get the entries that will contain display units and scroll down. You might want to highlight the definition (drag the cursor across it), copy it (Ctrl-C) to the clipboard, and paste it (Ctrl-V) into your Fast

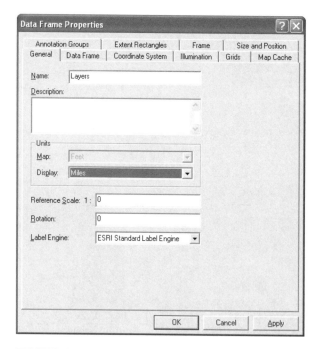

FIGURE 2-11

Facts File. Now look up Map Units. You need to know the difference between map units and display units. Dismiss the Glossary. Dismiss ArcGIS Desktop Help.

____ **10.** Change the display units back to Feet. As you now move the cursor around the data frame, you can see the coordinates of the *tip* of the cursor. Notice that the "easting" (leftmost number, the x-coordinate) increases as you move the cursor directly to the right. The "northing" number (y-coordinate) increases as you move the cursor directly up. What are the approximate coordinates of the east-most GPS fix? Easting (shown first) _1613420_ feet. Northing _138,854_ feet.

____ **11.** ***Bring up the attribute table of Boat_SP83.shp:*** Right-click the text Boat_SP83 and pick Open Attribute Table from the drop-down menu. Note that there are 83 records in the table, numbered with feature identifiers (FIDs) of 0 through 82. See Figure 2-12. You should see columns containing the local time each fix was recorded and its height above mean sea level in feet. By scrolling (vertically) the records in the table, you can note the beginning and ending time for the data collection run. How many minutes elapsed between the start and end time of data collection? _44_ minutes.

Right-click the gray box at the left end of Record 0. Click Identify. The resulting Identify Results window is a quick way to see all the values of a record in a column. What is the GPS Height shown in this record? _579_. In the same way, identify the last record in the table. What is the GPS Height shown in this record? _613_. Dismiss the Identity Results window.

Note the wide variation of altitudes; since the river is very placid and almost level, this gives you an idea of what you can expect in the way of vertical accuracy for individual GPS fixes—even those that have been adjusted through a process called differential correction. On the

FID	Shape	GPS_Time	GPS_Height	Northing	Easting
0	Point	10:03:58am	579.342	138860.827963	1613397.433640
1	Point	10:04:37am	544.422	138928.736470	1613163.723442
2	Point	10:05:05am	548.87	139025.537290	1612924.302071
3	Point	10:05:30am	552.361	139141.636962	1612756.824324
4	Point	10:05:59am	589.331	139328.012071	1612580.537709
5	Point	10:06:29am	565.776	139554.665198	1612480.780640
6	Point	10:07:00am	582.041	139780.049038	1612367.089276
7	Point	10:07:42am	575.831	139980.423435	1612251.479138
8	Point	10:08:32am	539.803	140137.706891	1612120.453564
9	Point	10:09:26am	557.629	140309.367572	1612001.040180
10	Point	10:15:37am	599.512	140382.301147	1611978.604481
11	Point	10:16:44am	547.104	140401.944114	1612016.863947
12	Point	10:17:13am	537.721	140546.847300	1611930.981345
13	Point	10:17:44am	524.355	140668.676295	1611786.301642
14	Point	10:18:32am	539.772	140888.863859	1611628.963336
15	Point	10:19:08am	558.31	141057.253192	1611462.175543
16	Point	10:19:31am	557.138	141189.171565	1611332.880613
17	Point	10:19:53am	558.796	141312.150728	1611198.582286
18	Point	10:20:15am	570.315	141424.165543	1611056.714729
19	Point	10:20:38am	561.976	141532.809672	1610900.640657
20	Point	10:21:04am	572.492	141653.856129	1610728.177759
21	Point	10:21:31am	575.109	141785.622021	1610549.121161
22	Point	10:22:03am	570.81	141901.188797	1610340.729238

Attributes of Boat_SP83

Record: 1   Show: All Selected   Records (0 out of 83 Selected.)   Options ▾

FIGURE 2-12

vertical accuracy
of GPS fixes

117

other hand, this data set was taken a long time ago—things have improved somewhat. But GPS vertical measurements will always be worse than horizontal ones. If any records have been selected click the Options button and pick Clear Selection. Engage the capabilities that let you obtain statistics and see a graph (the same ones that you used in ArcCatalog) to determine the average elevation. _552_ feet. Look at the graph to see where most elevation values are clustered. What is the difference between the lowest elevation recorded and the second lowest? _6_ feet. Dismiss the Statistics of Boat_SP83 window. Dismiss the table Attributes of Boat_SP83.

Looking at the GPS Track in the Context of a Variety of GIS Data

A GPS receiver gives position, not location, information. To recognize a location, you need contextual information. You have available several digital maps and images of a portion of the Kentucky River plus a couple of vector data sets of a few arcs from two USGS quadrangle maps. The county to the north of the river is Fayette; that to the south is Madison. Of the two USGS 7.5-minute quadrangles that cover the area, the westernmost is COLETOWN; the other is FORD.

A Potpourri of Types of Geographic Data

The data exist in two folders:

[___] IGIS-Arc\RIVER

[___] IGIS-Arc\Kentucky_wide_data

You will explore a dozen or so data sets.[6] They are as follows:

❑ A personal geodatabase (PGDB) named Kentucky_River_Area_Data containing two PGDB feature data sets: Quadrangle_Data and County_Streams, described below.

   ❑ Quadrangle_Data contains three vector-based PGDB feature classes:

      ❑ The soil types in the part of Fayette County that is covered by the Coletown quadrangle: cole_soil_polygon.

      ❑ Geologic (surface rock) data in the Coletown quadrangle: cole_rock_polygon.

      ❑ A vector (line) PGDB feature class of the elevation contour lines for the COLE quadrangle: cole_contours_arc

   ❑ County_Streams contains two vector (line) PGDB feature classes that have been derived from TIGER/Line files:

      ❑ The streams of Fayette County, Kentucky: Fay_Tiger.

      ❑ The streams of Madison County, Kentucky: Mad_Tiger.

---

[6]None of the data names in ESRI software are case-sensitive. That is, FORD_VCTR is the same as ford_vctr is the same as Ford_Vctr. However, blanks are not permitted in data set names, nor in the folders in the paths to data set names. Not knowing this causes trouble at times.

6 ☐ A line *shapefile* containing a few features digitized from the Coletown, Kentucky quadrangle: cole_vctr.shp.

7 ☐ A line *coverage* containing a few features digitized from the Ford, Kentucky quadrangle: FORD_VCTR

8 ☐ A *coverage* with both an arc attribute table (AAT) and a polygon attribute table (PAT), showing the Kentucky counties and county boundaries—that is, both the areas of the counties and the lines that separate them are depicted: CNTY_BND_SPN.

9 ☐ A digital raster graphics file scanned from the USGS Coletown topographic quadrangle: COLE_DRG.TIF

10 ☐ A small orthopohoto GeoTIFF[7] image (it is a digital orthophoto quadrangle that is 1/64th of a regular 7.5 minute USGS quad): COLE_DOQ64.JPG

11 ☐ A personal geodatabase named Lexington that contains a single, free-standing PGDB feature class:

PGDB   A line vector data set showing Lexington-area vehicle transportation system, derived from the 2002 TIGER/Line files for Fayette County: Roads.

12 ☐ A digital elevation model (DEM) in the form of an ArcInfo GRID consisting of square pillars or posts of elevation that are approximately 30 meters on a side: COLE_DEM

13 ☐ A triangulated irregular network (TIN) showing elevations derived from the DEM, in the form of an ArcInfo TIN: COLE_TIN

14 ☐ A GRID showing Kentucky landuse: MSU_SPNORTH

In the steps that follow, you will add these feature-based, grid-based, image-based, and TIN-based layers.[8] The photograph in Figure 2-13*** is a picture of a part of the area that you will be looking at.

___ **12.** Enlarge the map window a bit so it occupies almost the entire ArcGIS window, which should itself be set to occupy the full monitor screen. Under `Tools > Extensions`, make sure `3-D Analyst` and `Spatial Analyst` have checks beside them.

[Can't]

Displaying Layers from Vector-Based Data Sets

The *coverage* FORD_VCTR and the *shapefile* cole_vctr.shp contain a few arcs digitized from the USGS 7.5-minute topographic quadrangles Ford and Coletown. In particular, the arcs trace the banks of the Kentucky River and some highways and Interstates from the quad sheets. The data sets have been converted to Kentucky State Plane coordinates (survey feet) in the NAD 1983 datum.

Since we will be adding several layers to the data frame, you should be aware of the different ways to accomplish this. First, you know of the `File > Add Data` sequence. Also, there is an `Add Data button` on the `Standard toolbar`. Finally, if you right-click the data frame itself, you can select `Add Data`. In each case, an `Add Data` window appears. Note these three ways down in your Fast Facts File.

---

[7] A regular TIFF file that is used to portray geographic areas requires a separate world file that provides the geographic coordinates of the TIFF. GeoTIFFs, on the other hand, contain the relevant world file; it is embedded. ArcGIS may use either setup, but not all software does. You may convert a GeoTIFF to a TIFF and a world file with the ArcInfo command CONVERTIMAGE.

[8] GRID and image files may be contained in ArcSDE (Arc Spatial Data Engine) geodatabases, but not in personal geodatabases. TIN files stand on their own at all times.

How to add layers to data frame

FIGURE 2-13   Courtesy of the Lexington Herald-Leader, Lexington, Kentucky

_____ **13.** Add the cole_vctr.shp shapefile from [___] IGIS-Arc\RIVER.[9] Lines appear in the data frame, some of which parallel the set of points of BOAT_SP83 (which will also be referred to as the GPS track). Also, a small island is depicted.

_____ **14.** _Add the arc component of the vector coverage FORD_VCTR from [___]IGIS-Arc \ RIVER:_ To add the arc component of the coverage, bring up the Add Data window, find the coverage name (FORD_VCTR), _double-click the name,_ click Arc, then click Add.

You should see the GPS track, a few arcs from the Coletown topographic quadrangle (topo sheet) depicting the river's banks, and a few arcs from the Ford topo sheet.[10] Both (the shapefile and the coverage) are vector data sets, but they have different structures in the memory of the computer and on the disk drives,

---

[9]Note that the entries are not in strict alphabetical order. Here geodatabases are listed first, coverages are second, and other data sets follow.

[10]Yes, this short trip crossed the boundary between two quad sheets. Not only are you learning about GIS, you are also confirming the First Law of Geography: any area of interest, of almost any size, will require multiple map sheets to represent.

as you will see in Chapter 4. The starting point of the trip is in the southeast; at the other end of the track, you can see the polygon outlining a tiny island.

15. ***Make the FORD_VCTR coverage line symbol a green line of width 2 picas:*[11]** Earlier you changed just the color of a symbol, using a right-click the symbol. Since you now want to change the color *and* the size, you have to open the symbol selector window, which gives you a great deal of control over how features are displayed. Click (left-click, that is) on the *line symbol* that is *under* the text ford_vctr arc to bring up the window. All you need do here is to enter a 2 in the Width box (by clicking or typing) and select a color (say, Medium Apple) by clicking the little Color patch to bring up a pallet of colors. While you are looking at the Symbol Selector window, check out the possibilities for line symbols: everything from highways to narrow gauge railways to aqueducts. Not only is there the set of symbols you see in the window, but under More Symbols you find symbol sets specific to various fields. And, if you want, you can edit symbols with the Properties button. The possibilities are limitless. Now back to the main issue at hand: Click OK.

16. Now change the symbology of the cole_vctr shapefile. Make the line symbol bright red, width 2.

17. Layers may be displayed or not (turned on or off) by clicking the box next to the layer name. Experiment by turning the three layers off and on; leave them all on when you are done.

18. ***Make a group layer:*** Be sure the Display tab at the bottom of the Table of Contents is pressed. Click any of the three feature class names in the Table of Contents. Then, with the Ctrl key held down, click each of the other two layer names to highlight them. Then right-click any one of them and click Group from the resulting menu. All three are now subsumed under the name New Group Layer. In a group layer, the entire set of layers may be turned off all at once. If the group is turned on, whether a particular layer is displayed depends on whether its box is checked or not. In other words, you can turn these layers on and off together. Or you can turn individual layers on and off. Again, experiment, leaving everything on at the end.

19. ***Change the name of the group layer:*** Right-click the text New Group Layer. Select Properties > General. In the Layer Name text box replace New Group Layer with GPS_and_Vectors. Click Apply, and note the name has changed in the Table of Contents. Click OK. Write in your Fast Facts File how to change the name of a group layer. This name-changing technique also works for a plain layer, though it's usually a good idea to keep the names of layers the same, or close to the same, as the names of the data files that created them.

20. Experiment with collapsing and expanding the entries in the Table of Contents by clicking all the boxes containing minus (−) and plus (+). You can reduce the Table of Contents here to a single entry (Layers). Or you can see each constituent layer, and each of those with or without its legend. At the end, expand everything.

21. ***Zoom to the full extent of all data:*** Use View > Zoom Data > Full Extent or find the equivalent button on the Tools toolbar. Examine the map. Those lines that don't represent riverbanks do represent highways in the vicinity of Lexington and the Kentucky River.

22. ***Zoom back to the GPS layer:*** Right-click Boat_SP83, then click Zoom To Layer.

---

[11]There are 72 picas in an inch.

Housekeeping: Saving a Map

As you work, it is a good idea to save your files. While you could easily reconstruct what you have done so far, you have some time invested that you would rather not lose.

___ 23. On the File menu find and click Save. A Save As window appears. Navigate to the ___IGIS-Arc_*YourInitialsHere* folder so that it appears in the Save In box. For a filename use River_Map. The Save As Type field should read Arc Map Documents (*.mxd). The file type MXD is the file extension name for ArcGIS maps. See Figure 2-14. Click Save. Note that the title bar of the ArcMap window has changed to River_Map.mxd and that the button on the Windows taskbar also reflects the change.

___ 24. Minimize ArcMap. Using the Windows operating system (My Computer or Windows Explorer), look (using Details under the View menu or icon) at the contents of the folder

___IGIS-Arc_*YourInitialsHere*.

Notice it contains the file River_Map.mxd. How large, in kilobytes, is the file? **64** KB. Maximize ArcMap.

It is vital to understand what an MXD file is and what it is not. While it contains a great deal of information about *how* the data it refers to is represented (symbols, colors, line widths, and so on), *the MXD file does not itself contain data*. The MXD file contains pointers to the data—that is, the map file tells ArcMap the names of the folders where the data sets are located. You could not, for example, send only an MXD file to a colleague and expect her or him to be able to see the map with only ArcGIS desktop. You would also have to send the geodatabases, coverages, and shapefiles.

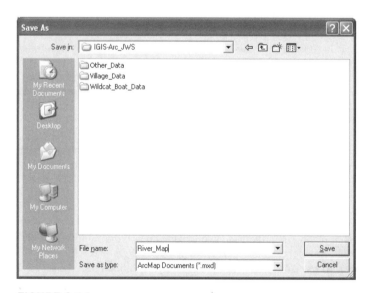

FIGURE 2-14

What a MXD file contains

Selecting: Both Map Data and Attribute Data

In a geographic data set file (cole_vctr.shp, for example), each feature in the map corresponds to a row in the attribute database. Here the features (lines) have lengths, in the unit of measurement of the shapefile (feet). Also, in this case, a description has been added to each arc.

arc

25. See Figure 2-15*** as you work this step. Make sure you are zoomed to the extent of the GPS layer (Boat_SP83). Open the attribute table of cole_vctr. Vertically shorten and move the table down so that you can see the records and the map at the same time. Find the WHAT column. Find the attribute value Island. At the left end of the record you will see a gray box; click inside the box. The _record_ becomes selected—shown by the fact that it is highlighted. At the same time, the _line_ that delineates the small island at the end of the GPS track will also become selected—again shown by the fact that it is highlighted. Again, see Figure 2-15.***

Whenever a record in a layer's attribute table is selected, the corresponding feature in the data frame will be selected. If the layer is being displayed, the feature will be highlighted.

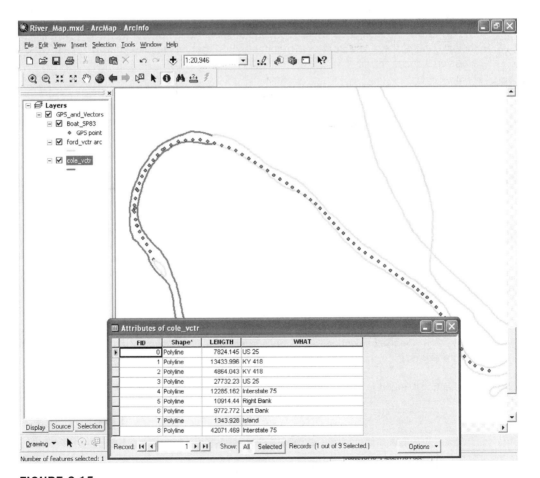

FIGURE 2-15

26. Click the box of the record that has the WHAT attribute value of Right Bank. Notice that an arc bounding the river is highlighted. Hold down the Ctrl key and click the box related to Left Bank. Now both records are highlighted, as are both bounding arcs. At the bottom of the table, press the Selected Records button. The two records highlighted in cyan will appear at the top of the table by themselves. Now click the gray box to the left of the Right Bank record. The record turns yellow, as does the associated line. The lesson: If only selected records are shown, you can highlight records and features *within* selected records and features, using a different color. Show all records. Find the Options button at the bottom of the table. (You might have to expand the table horizontally to see the button.) With the Options button, select Clear Selection.

27. Zoom to the extent of the cole_vctr layer. Again select the records representing the banks of the river. (Move the table window as necessary.) Click the Options button. Choose Switch Selection. Now the records and lines (of cole_vctr) that were selected are not, and those records and lines that were not selected now are. Now press Options, pick Select All, and observe the results. From Options choose Clear Selection. Zoom back to the previous level of magnification with View > Zoom Data > Go Back To Previous Extent (or find this button on the Tools toolbar). The GPS track should again dominate the window. Keep the Attributes of cole_vctr table open.

The ArcMap Tools toolbar contains some additional buttons not found on the ArcCatalog Tools toolbar. One of these is a Select Features button that lets you graphically select features or areas on a map.

28. On the Tools toolbar, find and press the Select Features icon. (In the status bar you will find Select Features By Clicking Or Dragging A Box.) Use the cursor to point at the line that delineates of the island and click. The line again becomes highlighted as does its associated record.

So here is the main message: *You can select either records or features, and the selection is carried through to the associated features or records.*

29. Hold down the Shift key and click again on the arc of the island. Note that it becomes unselected. Select it again. Now hold down the Shift key and click one of the other red arcs; it is now selected as well.

Select a third red arc. (The strange fact that you hold down Ctrl to select multiple records in a table but use the Shift key to graphically select multiple features on a map is something that you might want to write down in your Fast Facts File.)[12] To clear all selections: use Options > Clear Selection on the attribute table window.

30. Drag a small box across the river just north of the island. Notice that all lines and points that have any part within the box are selected. This is one case of Select By Location; there are many more. Clear all selections—this time by simply clicking somewhere on the map where no feature exists.

The Iterative Selection Method in the Selection menu allows you to set up the operation of the Select Features tool.

---

[12]Okay, for the most part, I'll quit reminding you about the Fast Facts File now. Either you are in the habit by this time or you aren't.

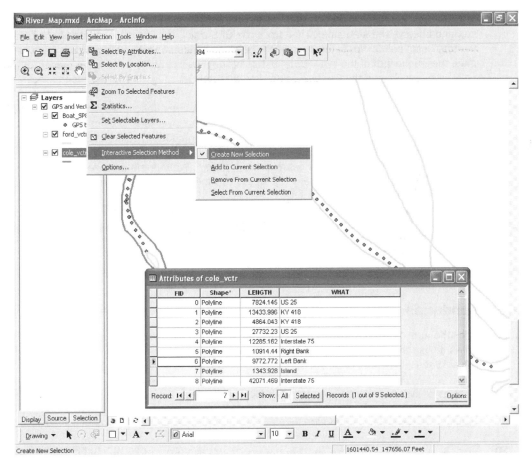

FIGURE 2-16

___ **31.** Zoom to the extent of cole_vctr. In the attribute table select the Island, Left Bank, and Right Bank records. Now select Selection > Interactive Selection Method. Experiment with adding and removing selected graphic features with the Select Features tool. Finish by choosing Create New Selection. See Figure 2-16. Close the Attributes of cole_vctr window. Zoom to the GPS track.

Using the Measure Tool

Also available from the ArcMap Tools toolbar is the capability for measuring distances (using an icon that looks like a ruler), according to how you have some display parameters set. Earlier you set the display units to feet. So the results from the measure tool will be given in feet.

___ **32.** *Measure the length (the distance along the GPS track) of the trip using Measure from the Tools toolbar:* Activate the measure tool by pressing the ruler icon. Using the plus (+) sign as

Selecting

the reference point of the cursor, click the easternmost point of the GPS track, move the cursor, and observe the status bar. Then trace the GPS track, clicking at points along the route of the trip, using points closer together when the route bends so that you can more nearly estimate the true length of the trip. At the last point, double-click (or just move the cursor to the point, click, and press the Esc key) to get the trip total, which is ___17K___ feet.

33. Click the Identify icon on the Tools toolbar. Read the three descriptive statements in the initial Identify Results window that comes up. Set the layer to identify as Boat_SP83.

34. Using the Identify tool, look at the beginning and ending times of the trip and determine the average speed of the boat in feet per minute. ___356___. What is the average rate in miles[13] per hour? ___4___. Dismiss the Identify Results window.

35. *Change the viewing area and measure some other distances:* Zoom up on the lines around the island. What is the length of the island (___612___ feet), and the greatest width of the river near the island? (___500___ feet).

36. Zoom back to the previous extent by pressing the left-arrow icon on the Tools toolbar.

County Boundaries

37. Click the Add Data button. In the Add Data window, find the Connect To Folder icon and press it. Use the Connect To Folder window to select the [___] IGIS-Arc\Kentucky_wide_data folder. See Figure 2-17. Click OK. This action will add the folder to the Catalog Tree.

38. Load as a layer the *arc* component of the CNTY_BND_SPN coverage from [___] IGIS-Arc\Kentucky_wide_data\KY_county_boundaries. Make its symbol a black line of width 2. Compare the boundary between Fayette and Madison counties with the course of the Kentucky River.

39. Load as a layer the *polygon* component of the CNTY_BND_SPN coverage from [___] Kentucky_wide_data\KY_county_boundaries. By right-clicking the color patch of the layer, make its color Yucca Yellow. Use the Identify tool to determine the name of the Area Development District (ADD) that Fayette County is in. ___Bluegrass___ What State Plane zone is it in? ___North___ What was the population of Madison County in the year 1990? ___57500___. What is the name of the county that is to the northeast of Madison and the east of Fayette? ___Clark___ Dismiss the Identify Results window.

40. Zoom to the full extent of the CNTY_BND_SPN polygon coverage to see all the Kentucky counties. Open the attribute table, sort by NAME, find Fayette County, and select the record. By how much did Fayette County's population grow from 1970 to 1990? ___50K___. Note that a highlight outlines the county on the map. Use the Identify tool to point at the highlighted county on the map. As you noted before, there is exact correspondence between the items in the Identify Results window and the column heading names. Dismiss the Identify Results window and the table.

---

[13]There are 5,280 feet in a mile.

44 min → 17,000 ft

356 ft/min × 60 min/hr

÷ 5280   4 miles/hr

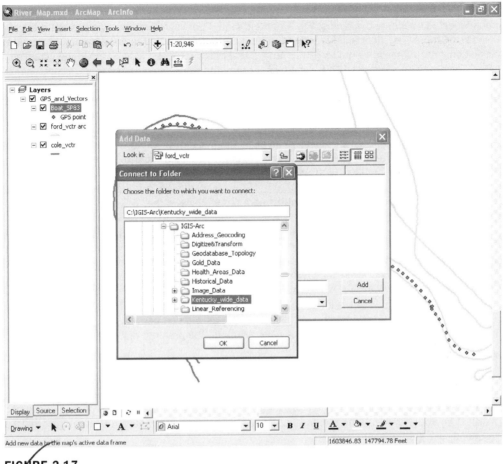

FIGURE 2-17

41. Figure out how to Remove the polygon layer from the Table of Contents and do so. Then put it back. ArcGIS may have picked a different color when the layer is returned. Zoom back to the GPS track. Right-click cnty_bnd_polygon and pick Label Features from the drop-down menu. Labels from the NAME field of the attribute database should appear: FAYETTE, MADISON, and CLARK.

TIGER/Line Files

TIGER/Line files are a product of the U.S. Bureau of the Census. TIGER stands for *Topologically Integrated Geographic Encoding and Referencing*. (You really should write this identifier in your Fast Facts File, unless your memory is better than mine. I can barely parse it, much less remember it. You have to wonder which came first, the name or the acronym.) TIGER/Line files primarily contain data related to streets—names, street numbers, census tracks, zip codes, county and state codes, and the geographic coordinates of these features. They were designed as the framework on which to hang the census data, and for the use of census takers. A bit later in this chapter, you will look briefly at TIGER/Line files that represent streets, roads, and highways.

But TIGER/Line files also include other types of data that can be represented in linear form, such as political boundaries, railroads, and streams. Fay_Tiger and Mad_Tiger represent the streams of Fayette County and Madison County. Fay_Tiger and Mad_Tiger are feature classes in the feature data set County_streams, in the personal geodatabase Kentucky_River_Area_Data, in the folder [___] IGIS-Arc\RIVER. The Fay_Tiger and Mad_Tiger data sets are not TIGER/Line files, per se, but were derived from them.

___ **42.** In the Table of Contents, click Layers to select it. Be sure that you are zoomed to the GPS layer and that Display is active at the bottom of the Table of Contents. Click Add Data and navigate to [___] IGIS-Arc\RIVER. Pounce on it. Pounce on Kentucky_River_Area_Data.mdb. Pounce on County_Streams. Select Fay_Tiger and click Add. Make it a line layer using bright blue as a color and 2 as the size. Pan (using the "hand" icon on the Tools toolbar) so that you can see the two streams that come into the Kentucky River along the GPS track. What are their names? _____ , _____ . Notice also that the Fayette county side of the river is considered a Fayette County stream. Zoom back to the GPS track.

___ **43.** A portion of the dividing line between Clark county and Fayette county is visible in the northeast of the map. Zoom up on it. The line of demarcation between the two counties is given by two different data sets: the county boundaries (cnty_bnd_spn) and the streams (Fay_Tiger), which you will see as separate lines which cross and recross several times. Zoom up again, looking at the place where the two lines seem most divergent. Use the Identify tool to verify the name of the stream: Boone (as in Daniel) Creek. The legal description of the line that separates the two counties is the centerline of that creek. But as you can see, the geographic description of the line is depicted in two different ways. Measure the greatest distance between the two lines. _____ feet. Dismiss the Identify Results window.

The lack of complete agreement between two geographic data sets can pose annoying problems for those trying to do spatial analysis. There exist several ways of coping with this problem—either by eliminating the slivers that are formed between the lines or by using the topological features that come with geodatabases. For now just be aware that data from different sources, which each attempt to represent the "real world," will frequently disagree at some level.

You can see why the quality of GIS data sets must be matched to the needs of the user. These county boundaries might be fine for those delineating watersheds. But surveyors would roll their eyes at discrepancies of 50 feet. They work with inches, and very few of those. Land ownership must be precisely defined. Local wars have been declared over distances of less than a foot.

___ **44.** Zoom back to the Boat_SP83 layer. Display Mad_Tiger in the same fashion as Fay_Tiger, but use a pale blue color of 2 picas. Make a group layer of the two TIGER files. Click in a blank area of the Table of Contents this time change the New Group Layer name by clicking it, waiting a second or two, and clicking again. You can then type the new name and hit Enter. Call it TIGER_Streams. To get an idea of the extent of the total streams data sets, zoom to the extent of the group layer. Then zoom back to the GPS track.

The Table of Contents: Display vs. Source vs. Selection

So far we have looked at shapefiles, coverages, and vector-based personal geodatabases. You will learn the distinctions between them in detail in Chapter 4. The geodatabase will be the data model of choice as this century progresses, but all three are important, mostly because you will find extensive data sets in all three forms. The coverage has been around for decades, the shapefile for years. The geodatabase is the new kid on the block; it and the software to handle it is just being perfected, as are the ways to convert among the different types. One of the major differences between the types is the way in which they are stored on the hard disks of computers. If you look at the Table of Contents of ArcMap, with the Display tab active, the different types are pretty much represented in the same way. If you look at the T/C with the Source tab active, there are some obvious differences.

45. Toggle back and forth between Source and Display, using the buttons at the bottom of the T/C. Widen the Table Of Contents pane horizontally, by dragging its right edge, so you can see all the text. See Figure 2-18. When Source is pressed, the most obvious difference is that you see each data set that underlies each layer and the path all the way to the hard disk. Layers are grouped by their type (shapefile, coverage, or geodatabase) and by their hard drive location. Note, for instance, that while FORD_VCTRarc and cole_vctr are both located in the same place on the [____] path and are also made into a group layer, they show up in different places on the Source Table of Contents, because one is a shapefile and the other is a coverage. Neither of the group layers (TIGER_Streams or GPS_and_Vectors) is anywhere to be found.

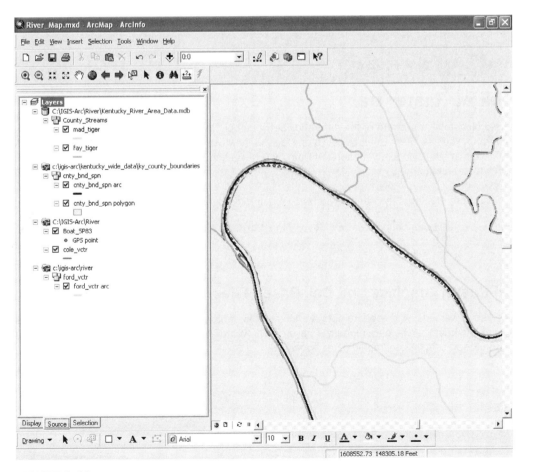

FIGURE 2-18

The Selection tab of the Table of Contents (T/C) lets you determine which layers are "selectable" and which cannot have features selected from them. The T/C also reports on which layers actually have selections made from them.

_____ **46.** Press the Selection tab at the bottom of the Table Of Contents window. Use the Select Features tool and, while holding down the Shift key, select each of the three counties visible in the CNTY_BND_SPN polygon and note the results. All the layers should be listed in the T/C, and if a number of features of a given layer are selected (in this case, counties), that entry is in bold and number is displayed. Experiment by selecting and deselecting other features, such as several of the points along the GPS track, and a stream or two noting the change in the T/C. Then select Selection > Clear Selected Features. Uncheck the box in front of CNTY_BND_SPN polygon and again attempt to select features from this layer. You can't, because it is no longer a selectable layer. Turn the selection box back on. Press Display and narrow the Table Of Contents window.

_____ **47.** Save the map as River_Map_2.mxd in your *YourInitialsHere* folder, where you originally stored River_Map.mxd. Close ArcMap.

Exercise 2-4 (Major Project)

A Look at Raster Data

Here's a dichotomy: Be aware of the difference between opening a map and adding data to a map. To open a map, you use File > Open to get to an Open window, which then allows you to browse for an MXD file. If a map is already open, you use Add Data (which you can get to in a variety of ways) to put an additional data set on the map.

_____ **1.** Start ArcMap. Press the An existing map radio button and select

_____ IGIS-Arc_*YourInitialsHere*\River_Map_2.mxd

and click OK.[14]

Digital Raster Graphics and Cell-Based Files

Shapefiles, coverages, and geodata sets are all capable of storing points, lines, and polygons. These three are illustrative of the data model we have called "vector." Now we turn to a completely different method of representing geographic data: "raster."

---

[14]When ArcMap is already running you can open a map by (a) pressing Ctrl-O, (b) Choosing File > Open, or (c) clicking the Open An Existing Map (see status bar) icon on the Standard toolbar.

Digital raster graphics (DRG) files are images (photographs, pictures) of the 7.5-minute topographic quadrangles produced by the United States Geological Survey (USGS). They have been scanned into a graphic image format (TIFF—Tagged Image File Format) and provided with information that fixes them in geographical space, making them GeoTIFFs.

_____ **2.** ***Display a layer based on the digital raster graphics (DRG) image of the Coletown quadrangle:*** Set the Table of Contents to Display. Remove the layer CNTY_BND_SPN *polygon.* Zoom to the GPS track. In the Table of Contents, right-click Layers. Add COLE_DRG.TIF from [___] IGIS-Arc\River. (If asked, do not build pyramids.) Zoom the view to full extent of the DRG to see what a USGS DRG image looks like. A part of the city of Lexington is in the northwest corner; the river data sets of interest are in the southeast. Then zoom back to the area of the GPS track and notice that only part of the GPS track is in the area covered by the Coletown quadrangle. (See Figure 2-19***.)

When (and only when) the Display tab is selected, you may rearrange the order in which entries appear in the T/C. This is important because the T/C order determines the order in which the layers are drawn on the computer monitor. To change the order, you simply drag a layer's name to where you want it in the T/C. In general, you should put point-based layers at the top of the T/C; line-based layers below them; and polygon, image-based, or grid-based layers at the bottom. ArcMap generally attends to this on its own, but sometimes you have to help out.

_____ **3.** ***Experiment with the drawing order:*** The title COLE_DRG.TIF probably appeared at the bottom of the Table of Contents. Drag it to the top. Notice that it blocks out the display of all the other layers in its vicinity. Drag COLE_DRG.TIF back to the bottom of the T/C.

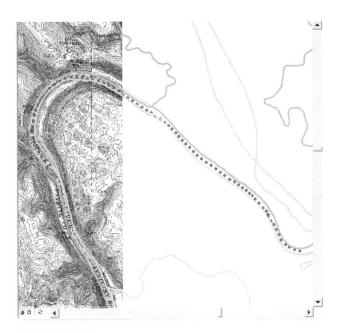

FIGURE 2-19

4. By drawing a box with the Zoom In tool of the Tools toolbar, zoom up to look at the southern-most part of the DRG. Observe the contour lines. Obviously the elevation changes sharply, since the contour lines are close together. Read the map to verify the river name and the names of the counties on each side. Zoom back.

5. Zoom up on the bend in the river—the most northwest quarter circle of it. What is the NORMAL POOL ELEVATION "printed" on the DRG.TIF?_____.

6. *Since you might want to return to this level of zoom, set a spatial bookmark:* Select View > Bookmarks > Create and type River_Bend in the Bookmark Name text box. See Figure 2-20***. Click ok. Check that it works by zooming to the entire DRG, then select View > Bookmarks > River_Bend.

7. Pan so that you can see the human-built artifacts and text just north of the bend of the river. What would you say was the function of this site? _____. Zoom back to the GPS track.

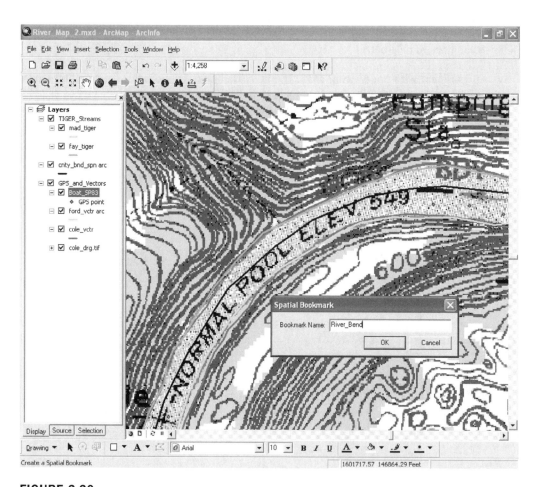

FIGURE 2-20

_____ **8.** Pick a colorful area near the river and zoom way, way up on the DRG, so that you can see individual square pixels. In the T/C, click the plus (+) sign in front of COLE_DRG.TIF. All 256 symbols[15] that could make up the image are now revealed. Actually only the first dozen or so are used.

_____ **9.** Click the Identify tool. The Identify Results window comes up. Look at the possibilities in the drop-down menu labeled Layers. See Figure 2-21. You can ask for results to be shown in:

- ❑   Topmost layer
- ❑   Visible layers
- ❑   Selectable layers
- ❑   All layers
- ❑   Named layer
- ❑   Additional named layer
- ❑   Additional named layer
- ❑   And yet another named layer

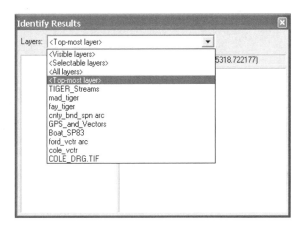

FIGURE 2-21

_____ **10.** Select COLE_DRG.TIF in the Layers text box. Click a pixel and note its Color Index. Match this number with the value in the Table of Contents and note the adjacent color symbol. Select a couple more pixels. Altogether, this should give you some idea of how a DRG TIFF image is put together. Dismiss the Identify Results window.

_____ **11.** Open the attribute table for the DRG. What you see here are attributes of the colors of the pixels (picture elements) on the map rather than information about the geography being represented. See Figure 2-22***.

---

[15]Why 256? Because the index of the colors is stored in a single byte, which as you may know is 8 bits, which allows two to the eighth power combinations, which is 2*2*2*2*2*2*2*2, which is 256.

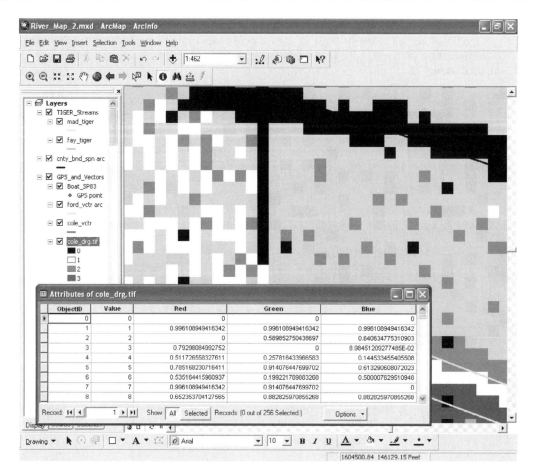

FIGURE 2-22

A DRG is simply a color picture. It may be interesting to you to see how this works. Each color used on the picture is made up of specific quantities of three primary colors: red, green, and blue.[16] Imagine that you have three flashlights focused on a single spot (pixel) and that the amount of light coming from each flashlight can be controlled, with 0.0 indicating none and 1.0 indicating brightest. Of course, when all

---

[16]You may have learned that the primary colors were red, *yellow*, and blue and that you could use these to make other colors. You perhaps verified this by mixing paints that went on paper. Actually, these paints absorbed wavelengths of light for certain color ranges. Paints give off no light on their own. A paint that absorbs red and blue wavelengths from the white light that falls on it and reflects the rest appears as yellow. "Red, yellow, blue" is a subtractive or reflective color model. Light is only reflected, not generated. But when you look at a computer monitor, you are seeing colors from an emissive or additive color model. Light is being generated by the interaction of the particles coming from the electron gun and the special phosphors on the inside of the monitor screen. (Or by some more modern whiz-bang technology that generates colored light.) Here the best combination of pure colors to generate the visible spectrum was found to be red, *green*, and blue. These aren't the only color models. For example, most color printers use magenta, yellow, and cyan (plus black separately) as the primary colors.

three lights are off, you get black. When all three lights are on at their brightest, white light is produced. These two conditions are represented in the table as Value 0 (black) and Value 1 (white). As you can see from the table, when you combine a lot of Red (0.785), no Green, and a small amount of Blue, you get the color with Value 3.

_____ **12.** To get yellow (Value = 7) what amounts of the colors do you combine (to the nearest two decimal places)? Red? _____ Green? _____ Blue? _____ Note that the attribute `Value` in the table corresponds to the legend number in the Table of Contents for COLE_DRG.TIF. Dismiss the attribute table.

_____ **13.** Click the yellow rectangular symbol in the T/C to open the `Color Selector` window. Notice that the Red, Green, and Blue (RGB) values are 254, 233, and 0. These are linearly equivalent values to those in the attribute table; here they are based on 0 to 255, rather than 0 to 1. Move the slider on the Green bar (G) back to 0. Note the resulting color in the lower left-hand corner of the `Color Selector` window: pure Red. Click OK to change the yellow color symbol to that color. Experiment by setting several symbols to different colors. To see the pure primary color Green, set the Green color bar to 255 and the other two to 0. Do the same for the other primary color Blue. Click OK and note the change in the DRG. Change the white pixels to black and the black pixels to white.

_____ **14.** When you have finished experimenting, remove COLE_DRG.TIF from the map. This nullifies any changes you have made to the color symbols. Add COLE_DRG.TIF back to the map; expand its legend to verify that you haven't done any permanent damage to the default color scheme.

_____ **15.** You don't really want every color symbol cluttering up your Table of Contents, so collapse the data set legend if it is showing. Initiate a zoom out with `View > Zoom Data > Fixed Zoom Out`. Find the equivalent icon on the Tools toolbar and click it repeatedly. Observe how the visually meaningless pixels combine to form an image.

Experimenting with Different Ways of Seeing the Data

_____ **16.** Zoom to the GPS track. Click the Window menu and click Magnifier. When the Magnifier window appears, drag it, by its title bar, over the Normal Pool Elevation number and verify its value in the magnified view. _____. Then drag the window over the island and compare the COLE_VCTR arcs with the elevation lines depicted by the topographic map. Notice that you can resize the Magnifier window by dragging its sides and corners. Right-click the title bar of the Magnifier window and click `Snapshot` to lock the view, so you can move the Magnifier window around and not change its contents. Open another Magnifier window and check out its menu options, including those in Properties. Notice that you can "see through" the `Snapshot` window with the Magnifier window. See Figure 2-23. Dismiss the `Magnifier Properties` window, the Snapshot window and the `Magnifier` window.

_____ **17.** Zoom to Boat_SP83. Select `Window > Overview`. In the `Layers Overview` window, right-click the title bar and click `Properties`. In the `Overview Properties` window, find the `Reference Layer` text box, pull down the menu, and click COLE_DRG.TIF. Now click Apply, then OK. Resize and drag the `Layers Overview` window so it fits comfortably in the lower left of your screen.

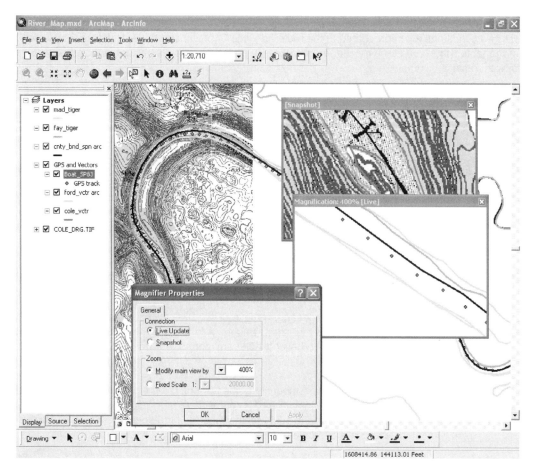

FIGURE 2-23

See Figure 2-24. The Data Frame window shows the GPS track, the DRG, and other data. The Layers Overview window shows the entire DRG, with a red crosshatched box that corresponds to whatever is displayed in the data frame.

____ **18.** Dragging the crosshatched box in the Layers Overview allows you to pan around the COLE_DRG.TIF layer. Move it over to the city of Lexington. Then move it back to the river bend. Also if you use the Zoom control or the Pan control on the map in the data frame, the new view is reflected by changes in the size and location of crosshatched box in the Layers Overview window. Use the Zoom control on the data frame so that the scale box in the Standard toolbar indicates a scale between 1:13,000 and 1:15,000. Drag the crosshatched box to the blue patch at the top of the Layers Overview window, about in the center. What is the number of the Lexington Reservoir that you see? _____.

____ **19.** You may have multiple Magnifier and Overview windows open at a particular time. Before you experiment with these features together, save your map so you can get it back if something

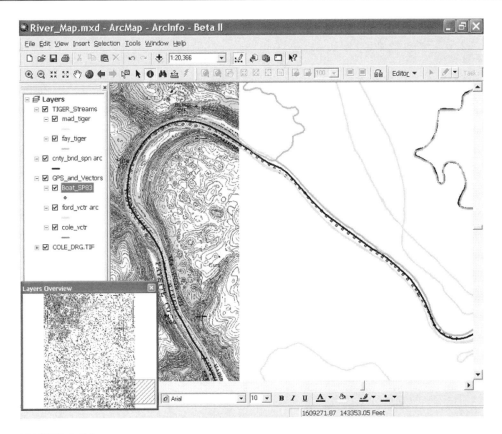

FIGURE 2-24

goes wrong and ArcMap closes. What you are about to do is a sort of high-wire act for the software. And frequent saves are a good idea anyway.

___ **20.** Open another Layers Overview window, resize it, and place it above the other one at the left of the screen—obscuring the Table of Contents. Make its Reference Layer GPS_and_Vectors. Open a Magnifier window and plunk it into the data frame. See Figure 2-25***. Spend some time looking at the capabilities here. Zoom and pan the data frame with the usual tools. Pan with the crosshatched areas in the Layers Overview windows. Move the Magnifier window.

___ **21.** When you are done playing (and being impressed), close all magnifier and overview windows.

___ **22.** Be sure that the scroll bars are turned on by selecting View > Scrollbars. You can zoom to the full extent of all the data sets in the Table of Contents by selecting View > Zoom Data > Full Extent or by pressing the Full Extent icon on the Tools toolbar. Do so now. Notice that the slider bars at the right and bottom of the data frame are fully extended. Zoom to the extent of the DRG. Now you can shift your view of the data frame (a different way of panning) by dragging the scroll bars. Experiment.

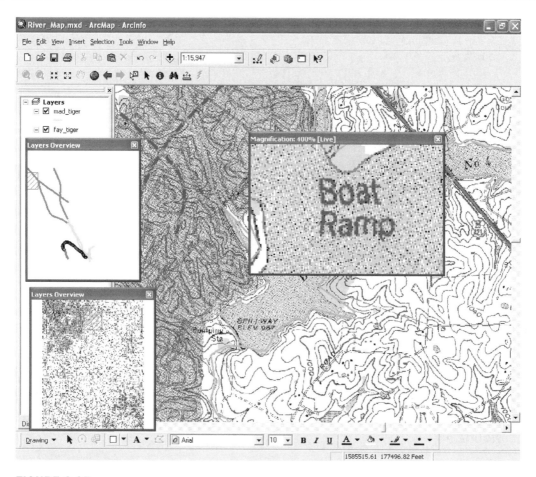

FIGURE 2-25

Digital Orthophotos

A digital orthophoto is an aerial photograph that has been rectified (adjusted) so that it may be used as a map with constant scale throughout. Fixing aerial photos so that they become orthophotos is complicated process, which I describe it later.

_____ **23.** ***Display a layer from an orthophoto "JPG" file:*** Zoom to the GPS track. Navigate to [____] IGIS-Arc\ RIVER and add a layer based on COLE_DOQ64.JPG. (Ignore any warnings about projecting layers. If asked, do not build pyramids.) If necessary, drag the DOQ title next to the bottom of the T/C (with only DRG layer below it). Zoom to the orthophoto image layer. Pan slightly so the island on the DRG is also displayed. See Figure 2-26. You can get an interesting perspective on the images by flipping COLE_DOQ64.JPG off and on. Notice the water filtration plant and other artifacts on both the DRG and the DOQ.

_____ **24.** Experiment with different levels of magnification on the DOQ. Notice that when you are zoomed in very tightly, the "meaning" contained in the image "disappears" in that you see only squares

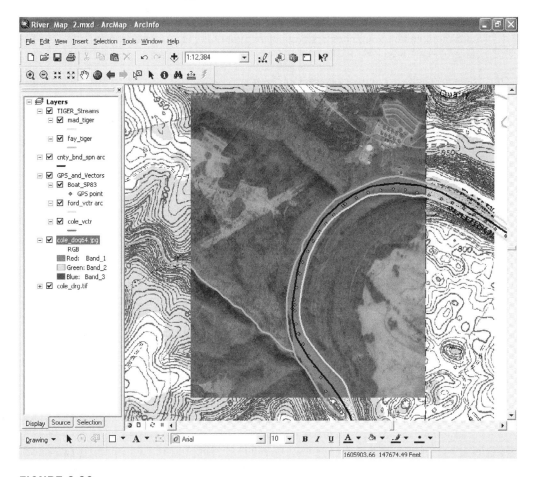

FIGURE 2-26

(pixels) of various shades of gray. How many feet across is each square? _____. What length is this in meters?[17] _____. This is the resolution of the data set. Crudely, the resolution is a measure of the limit of how much detail you can wring out of the data. Zoom out some, so that there are about 10 pixels across the screen.

_____ **25.** Expand the COLE_DOQ64 entry in the Table of Contents so you can see the three bands that make it up: Red, Green, and Blue. Right click on the layer name. Notice that the option to open an attribute table is grayed out. Unlike a TIF dataset, a JPG dataset has no associated table, but you can use the `Identify` tool to query the values of each color. Make the `Identify Results` window show the COLE_DOQ64 layer and then click on various places on the photo. You will notice that the 256 shades of gray are made up of intensity values, which are virtually identical for each color. With the `Identify` cursor, pick a light square on the map and note the values. Do the same for a dark square. Again, the lightest color has the most of red, green, and blue. Click on each of the legend color boxes and uncheck Visible. The image will disappear.

---

[17]There are 39.37 inches in a meter.

Now turn each on separately and note the image. Look at them in pairs to get an idea of how equal amounts of different colors produce different results. Turn all bands back on to restore the gray-scale map image. Close the `Identify Results` window.

____ **26.** Zoom to full extent of the DOQ. Find the northern boundary of the DOQ, where it meets the DRG. Zoom way up on this area so you can see the pixels of each layer. What distance on the ground does the DRG pixel width cover? _____ feet. Zoom back to the extent of the COLE_DRG.

____ **27.** By right-clicking the data frame and choosing `Properties > General`, change the `display units` to `Miles`. Use the `Measure` tool to obtain the height and width of the DRG in miles. _____, _____. How many square miles is this? _____. Zoom to the DOQ. Obtain the height and width of the DOQ. _____, _____.

____ **28.** Under Properties again, notice that you cannot change the map units. They are locked in by virtue of the fact that you originally set up the map with a coordinate system that used survey feet as the basic map unit.

____ **29.** Change the display units back to Feet.

More TIGER/Line Files

____ **30.** Zoom up on the northeast corner of the DOQ, so that the water filtration plant occupies most of the window. You will notice a road that starts in the northwest of the image and traverses the area toward the east-southeast.

By using the Roads personal geodatabase feature class, contained in the PGDB Lexington, you will be able to determine the name of that road. This feature class is somewhat different than the feature classes you saw previously, in that it does not exist within a PGDB feature data set. Rather it is found directly within the Database. This sort of feature class is called "freestanding."

____ **31.** *Form a layer based on the feature class Roads:* Navigate to [___] IGIS-Arc\River\Lexington .mdb\Roads and add it to the data frame. Make it bright green, width 2. You should see a line crossing the DOQ. You will notice that it doesn't fit the road in the image very well—which is sometimes a characteristic of TIGER/Line data. That census application is less concerned with geographic accuracy than with the topology—that is, the main emphasis is on what connects to what. Try to ignore the fact that the TIGER-depicted road takes a little detour through the plant machinery. See Figure 2-27***.

____ **32.** Start the `Identify` tool. Select ROADS from the Layers drop-down menu box and determine the name of the road. _____. What would you say the zip code for the plant was? _____. The TIGER/Line files contain several attribute fields, which we will explore later. Dismiss the `Identify Results` window.

____ **33.** Just to get an idea of the volume of data in Roads, zoom to that layer. Use the `Zoom In` tool to look at the level of detail. Lexington contains a (mostly) limited-access "ring road" that circles the city; its "diameter" is about 5 to 7 miles, depending on where you measure. Can you determine its name? _____.

FIGURE 2-27

_____ **34.** Zoom to the DRG, then to a small square in its northwest corner. Look at the fit, or lack of fit, of the Roads with the DRG. See Figure 2-28***.

_____ **35.** Open the attribute table of Roads. Notice that, at the bottom of the window, where the number of records is shown, the 2000 has a * in front of it. This means that there are more than 2000 records and ArcMap doesn't want to load all of them unless asked. To force ArcMap to display all of the records, find the word "Record" at the bottom of the window. You will see a text box and four buttons. Click the rightmost button. It will take you to the last record in the file. How many records are there in the feature class? _____

_____ **36.** Sort the table into ascending order by feature name (FENAME). Scroll the table. You will notice that many road names are repeated. That's because each road segment, between intersections, has its own record. Since TIGER/Line files exist for every bit of real estate in the United States, you can see that the whole thing is a monster-sized database. It is available on about 50 CD-ROMS. Use My Computer or Windows Explorer to determine the size of the Lexington Personal Geodatabase (called Lexington.mdb) which, recall, contains only Roads. _____ megabytes. Dismiss the attribute table, then remove the Roads layer.

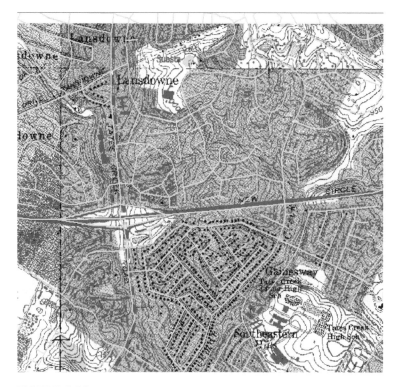

FIGURE 2-28

Another Tie Between Attributes and Geographics

—— **37.** Zoom to the GPS track. Choose `Edit > Find > Features`. Move the `Find` window to the right, past the DRG. Type `Island` in the `Find` text box. In `Layers` verify that the text box contains `<Visible layers>`. Press `Find`. At the bottom of the window you should see that a reference to Island was found in the cole_vctr shapefile table, in the field WHAT. See Figure 2-29.

After you have "found" a feature with Find > Features, you have a remarkable number of options for examining that feature. Here are some:

—— **38.** While looking at the island on the map, double-click the text `Island` in the `Find` window to flash the feature.

   ❏ Right-click the text `Island` and click `Select Feature(s)`.

   ❏ Right-click the text `Island` and click `Identify Feature(s)`. Look at the `Identify Results` window, and then dismiss it.

   ❏ Right-click the text `Island` and click `Set Bookmark`. Select View > Bookmarks > Island. Observe, using View > Bookmarks > Island, temporarily moving the `Find` window as necessary. Then select `Go Back To Previous Extent`.

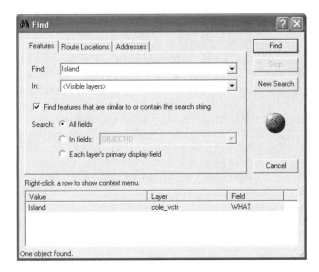

FIGURE 2-29

❑ Right-click the text Island and click Zoom To Feature(s). Zoom back.

❑ Right-click the text Island and click Unselect Feature(s).

Dismiss the Find window. Have you been adding to your Fast Facts File lately?

More Housekeeping: Shutting Down and Restarting ArcMap

____ **39.** ***Shut down ArcMap:*** Use File > Save (or its equivalent: press Ctrl-S) to save your map file. Now click the "X" in the upper right corner of the ArcMap window (or press Alt-F4) to close the ArcMap program.

____ **40.** Restart ArcMap. Unless someone has changed some settings you will be given a choice of a new empty map, or of loading River_Map.mxd, or River_Map_2.mxd. Pick the latter. You should see what you had before you terminated ArcMap.

____ **41.** ***Stop and restart ArcMap using River_Map_2.mxd to invoke the program:*** Again click the "X" in the upper right corner of the ArcMap window. Now using My Computer or Windows Explorer, navigate to

__IGIS-Arc_*YourInitialsHere*\River_Map_2.mxd

Pounce on it and ArcMap should open with the map you saved.

So you see you can easily save your work, and easily restart ArcMap so that it will pick up where it left off.

Digital Elevation Model Files

A digital elevation model (DEM) data file is a digital representation of elevations, using a raster format. A DEM consists of an array of elevations for many ground positions at regularly spaced intervals. In this data set they appear as square cells, or, visualizing them three dimensionally, "square posts" whose flat tops represent the height of the land or water at the particular geographic positions.

____ 42. *Display a layer from the Digital Elevation Model (DEM) that is in the form of an ArcInfo GRID:*[18] From [____] IGIS-Arc\RIVER add a layer based on the Coletown Quadrangle digital elevation model (COLE_DEM). What is the range of elevations shown in the Table of Contents? _____ to _____. Drag its title to the bottom of the T/C. Turn off the two image layers above it. Zoom to the extent of the DEM layer. You can get an idea of the topography by looking at the DEM. White areas are the highest elevations; dark areas the lowest. See Figure 2-30. Select the Identify tool on the Tools toolbar. Set the Layers text box to COLE_DEM (scrolling the drop-down menu as necessary). Shrink the Identify Results window in both directions. If you now use the Identify tool, you can see the elevation where you click, recorded as Pixel value. Click a light area. Click a dark area. Click the river. Dismiss the Identify Results window.

____ 43. Zoom up on an area where there is a large variance in elevation. Zoom up some more. At some point you will see the individual pixels that make up the DEM. Now, with the display units set to Feet, measure the pixel size. _____ feet. This is one of the standard cell sizes for DEMs.

____ 44. Zoom to the GPS track. Pan the image to the east. Yes, the appearance of the DEM is unimpressive—mostly black, due to the way the heights of the grid posts are classified. To fix this, right-click the DEM layer name in the Table of Contents and click Properties. Under the Symbology tab in the Layer Properties window, select Classified in the Show area. Select a color ramp that goes from Yellow to Dark Red. You can do this either by (1) looking at the actual color choices in the drop-down menu or (2) right-clicking the Color Ramp bar and turning off the Graphic View, so that the drop-down menu shows you text descriptions. When all this is set up, click Classify. Look at the Classification Statistics. How many DEM posts are there? (see Count) _____. What is the height of the shortest one? _____. What is the height of the tallest one? _____. What is the height of the average post to the nearest tenth of a foot? _____.

____ 45. For Method specify Natural Breaks (Jenks); also specify 30 classes. Examine the histogram. Specify 40 columns for it. Your Classification window should look like Figure 2-31. Note that the great majority of the elevations lie around the 1000-foot level. But there are some at around the 550-foot level, which you may recall from the DRG, is near the normal pool elevation of the river. Click OK. Click Apply. Dismiss the Layer Properties window. Note the range of elevation values, and the associated color symbols, by widening and scrolling the Table of Contents.

____ 46. Zoom up on the part of the GPS track that is within the Coletown quadrangle. Click again on the Identify tool to bring up the Identify Results window. Notice how quickly the elevation

---

[18]In order to see this GRID, you will need the Spatial Analyst extension or the 3-D Analyst extension. As indicated earlier, you may need to tell ArcGIS to install the extension(s) in the current session. To do this, go to the Tools menu and click Extensions. Put a check mark by Spatial Analyst and 3D Analyst.

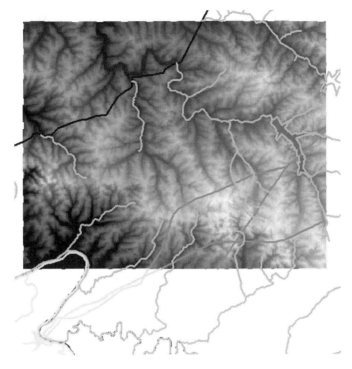

FIGURE 2-30

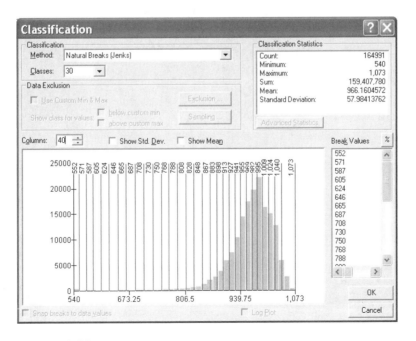

FIGURE 2-31

changes in the vicinity of the river. Getting the results from the COLE_DEM, what is the elevation of the surface of the river? _____. Record an elevation taken on the island? _____. Does the DEM river elevation agree with what you wrote before for the normal pool elevation? How much different? _____. Dismiss the Identify Results window.

___ **47.** A faster way to see the values of the DEM posts is: right-click the DEM layer name, then click `Properties` > `Display` > `Show Map Tips` > `Apply` > `OK`. Now when you move the mouse cursor around the map, even if the Identify tool is not being used, you will see the elevation of the DEM post when you pause over a given geographic location. See Figure 2-32***. Using `Measure` and `Map Tips`, find the steepest average slope you can—where the land drops _____ feet in elevation over a distance of _____ feet.

___ **48.** Examine the COLE_DEM entry in the Table of Contents, after widening the pane enough to see all the text. Collapse the COLE_DEM entry, to hide all the values and their symbols. Open the attribute table for COLE_DEM. Under ObjectID, 529 different heights are represented. The `Value` field contains the height of the grid "post" in feet. The Count field indicates how many posts there are of that height. How many posts are there of height 1000? _____.

___ **49.** Vertically expand and horizontally shrink the `Attributes of cole_dem`. Select the records whose elevations lie between 901 and 1000, by holding down the Ctrl key, clicking the box to the left of the record with elevation 901, and then dragging the cursor toward the bottom of the

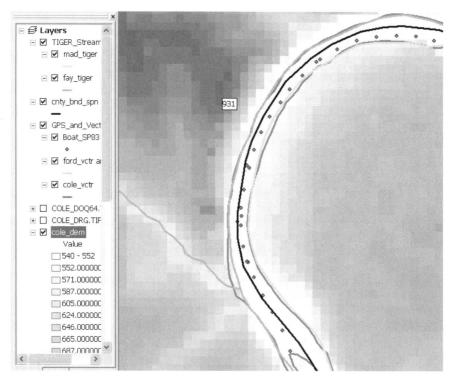

FIGURE 2-32

table, stopping at the record that shows elevation 1000. (If this doesn't work very well because the table scrolls by too quickly, do it in stages, using the vertical scroll bar to scroll the list; hold down the Ctrl key while clicking to select each successive set of records.) Once those 100 records are selected, run statistics on the Count attribute. Recall that you will get results based on the selected records only. Use the Selection Statistics of cole_dem window to determine how many posts (the Sum) there are with elevations between 901 and 1000 _____. (If you find these numbers confusing, read the discussion in the paragraph that follows.) How much surface area in acres would that be? _____ acres. How many square miles?[19] _____.

You may have found the preceding numbers a little confusing. In the attribute table, count refers to the number of posts of a given height. In the selection statistics window, count refers to the number of records being considered, while sum refers to the total number of posts represented by those records.

___ **50.** Dismiss the Statistics window and minimize the DEM attribute table. Zoom to the extent of the DEM. The selected area shows those elevations between 901 and 1000 feet, which, as you will note, is almost all the quadrangle. Restore the attribute table. Under Options, click Clear Selection. Dismiss the DEM attribute table.

___ **51.** Again experiment with various levels of magnification, including full extent of the DEM. Notice that in most places the TIGER streams seem to flow down the valleys shown by the DEM, but sometimes they don't. Again, GIS data is like other data: wide variations in quality—perhaps even within a given data set. Data from different sources are not always consistent. Pick a stream and attempt to determine its direction of flow by looking at elevations. Some stream designations in the southwest area of the map stop abruptly at the black line. Can you recall why? _____.

Comparing the DEM and the DRG

___ **52.** *Compare the elevations "printed" on the DRG with those given as values by the DEM:* Zoom to the bookmark River_Bend. Turn off the DOQ64 layer if it is on. Arrange the DRG and DEM layers so that the DEM is at the bottom and the DRG is just above it. Turn both layers on. The DRG image will obscure the DEM. Turn on the Identify tool and set up the identified layer as COLE_DEM. Use the Identify tool and Map Tips on the topographic quadrangle, pointing at contour lines where the text on the map shows elevations (e.g., 600, 700, 850 and 900 feet, zooming out as necessary) near the bend in the river. You visually see the DRG but the Identify Results and Map Tips come from the DEM. Note the agreement (or, sometimes, lack thereof) between what is printed on the topo map and the values of the underlying grid. Zoom out somewhat and try this exercise again. You may notice better agreement between the two data sets as you get further away from the steep slopes around the river. Dismiss the Identify Results window.

___ **53.** Zoom to the extent of the DRG. Locate Lexington Reservoir No. 4 in the north central part of the quad. Zoom up on it. By turning the DRG on and off, flip viewing between the DEM and the DRG. Determine, from the DEM, the elevation of the surface of the water in the reservoir. _____ feet. According to the topo map, what is the elevation of the spillway? _____ feet.

_____

[19]There are 43,560 square feet in an acre. There are 640 acres in a square mile.

Contour Line Files

Another way of representing elevation on a map, familiar to you from your examination of the topographic map, is through "isolines" or "contour lines." For any given such line, every position on that line is supposed represent the same elevation. The contour lines you have seen so far have been "printed" on the DRG. But a GIS can represent these lines as data as well.

____ **54.** While still zoomed up on the reservoir, turn off all layers except COLE_DEM. Add the feature class cole_contours_arc from

[____] IGIS-Arc\River\Kentucky_River_Area_Data.mdb\Quadrangle_Data

using a Fire Red line of width 1. Start the Identify tool and set its window menu so it shows cole_contours_arc. Adjust the Identify Results window so you can see all the attributes. (Map Tips will still show the DEM values.) Click the map. Investigate the contours around Reservoir 4 by clicking individual contour lines. See Figure 2-33***. What is the name of the

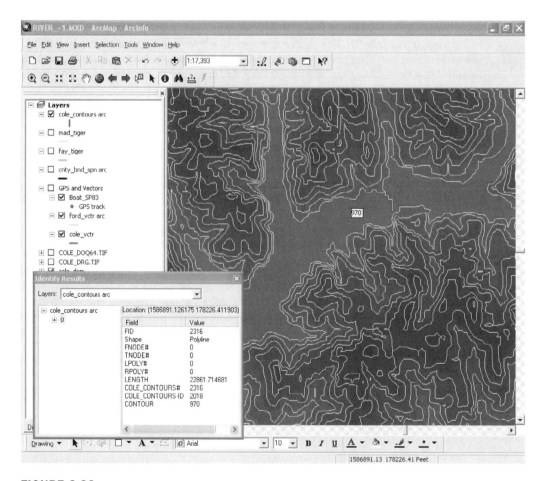

FIGURE 2-33

field in the attribute table that specifies the elevation of the contours? _____. Notice how it is helpful that the feature you click is highlighted briefly. What is the contour interval (the number of feet measured vertically between adjacent contour lines)? _____. Turn off the DEM.

_____ **55.** Turn on the DRG layer. Use the Identify tool to compare the elevation values of the red contour lines with those shown on the DRG. Notice that the contours from cole_contours_arc and the contours from the DRG match pretty closely. Zoom up some more on areas where you see some discrepancies. Using the Measure tool, find a place where the difference between two corresponding contour lines seems excessive. How far apart are they horizontally? _____ feet.

_____ **56.** Zoom up on the GPS track. Look again at the differences between the two corresponding contour lines data representations. The process of comparing contour lines may work better if you pan to a location away from the river. Near the river the contour lines are so dense that it is hard to see what is going on. Pan toward the west. You will notice that the contour lines from the two different data sets are close, but not congruent. Knowing the elevation at every point in an area is a tricky (actually impossible) task. These two data sets were derived in very different ways. They actually show a remarkable degree of agreement.

_____ **57.** Turn off the DRG and turn on the DEM. Again compare DEM elevations with the contour lines. Use Map Tips to see the DEM elevations while using the `Identify` tool to check the contour lines. Dismiss the Identify Results window.

_____ **58.** Save the map as River_Map_3.mxd in your *YourInitialsHere* folder, where you originally stored River_Map.mxd. Close ArcMap.

Exercise 2-5 (Project)

Triangulated Irregular Network Files

_____ **1.** Start ArcMap and Other Data with River_Map_3. Under Tools > Extensions make sure 3D Analyst and Spatial Analyst are active. Open River_Map_3.mxd from your

___IGIS-Arc_*YourInitialsHere* folder.

TINs (triangulated irregular networks), in addition to contour lines and digital elevation models, constitute another way of representing surfaces (particularly topography) in a GIS. A TIN represents elevation by tessellating[20] a surface with triangles. Each triangle is positioned above a horizontal plane such that each of its three vertices is at a known elevation. The result, then, is a surface made up of triangular plates positioned at different angles. Each such plate, therefore, has a lot of information associated with it: elevation of its centroid and vertices, elevation of any point on the plate, slope of the plate, and the compass direction of a line perpendicular to the plate (called aspect).

---

[20]Tessellated is a 50-cent word that means "completely covering an area with geometric figures without gaps or overlaps." Actually, it is the *projection* of the triangles on the plane that tessellate it, since the sum of the areas of the actual triangles is more than the area of the plane—as the actual triangles that make up the surface are not parallel to the plane.

_____ **2.** ***Add a layer based on a Triangulated Irregular Network (ArcInfo TIN):***[21] Navigate to [____] IGIS-Arc\ RIVER, add a layer based on the COLE_TIN data set. Drag its title to make it the second from the bottom entry of the T/C, with the DEM below it, turned on, and with its Map Tips still active. Turn off any layers above it, including cole_contours_arc so you can see the TIN better. Zoom to the TIN layer. Make sure the TIN is expanded in the Table of Contents so that you can see the elevations associated with the different colors. Look at the TIN, comparing the colors on the map with the elevations shown in the Table of Contents. Turn all entries in GPS_and_Vectors on. Now zoom up, using the River_Bend bookmark. Zoom in more—on the southeast quarter— to see the construction of the TIN. (If the surfaces of two or more triangles have the same slope and aspect, you will not see the line separating them.)

_____ **3.** Zoom back to River Bend. To get a better idea of the changes in elevation, pick a monochromatic color ramp—say, Yellow to Green to Dark Blue—by selecting COLE_TIN Properties > Symbology > Elevation. Right-click Color Ramp, unselect Graphic View, scroll to the desired ramp, and click it. Then, using Classify, pick Defined Interval as a method and 25 as the interval size. Press OK. Make sure the box next to Show hillside illumination effect in 2D display is checked. (It produces a neat presentation that lets you see the individual triangles.) Click Apply, then OK. See Figure 2-34***.

_____ **4.** ***Use the Identify tool to notice that, with a TIN, the elevations are calculated rather than stored in an attribute table:*** Zoom up so that you can view a dozen triangles or so in the southeast corner, and launch the Identify tool. In the Identify Results window, set Layers to COLE_TIN. Pick a triangle and get the elevations close to its vertices. Since the triangle is tipped, you will get different values for Elevation when you use Identify. Map Tips will continue to show you the value of the underlying DEM. Now pick a vertex where three or more triangles come together, and get the elevations of the adjacent corners of those neighboring triangles. Those values should be about the same, since the vertex has a single elevation.

_____ **5.** The Identify tool also presents the slope of the triangle (in percent, 45 degrees being 100 percent) and the aspect (the direction in compass degrees that it is inclined towards). See Figure 2-35. As illustrated shortly, you may also view the TIN at oblique angles, rather than simply vertically above. You could spend a full day exploring the different ways to display TINs. Come back to this if you are interested. Get guidance from the ArcGIS Help files.

If you try to open an attributes table of a TIN data set, you will discover that it has none. In most geographic data sets the values produced by the Identify tool come from an attribute table. But in the TIN the values placed in the Identify Results window are calculated on the fly immediately—based on where you have placed the cursor.

_____ **6.** Collapse the legend for the TIN. Zoom to the GPS track and pan it to the east to fill the window with TIN. Again, with the Map Tips of the DEM active, compare some TIN elevations as revealed by the Identify tool. Under the Display tab in the Properties window > Display, set Show Map Tips on, which will take the Map Tips away from the DEM and assign them to the TIN. Slide the cursor around the display looking at elevations.

_____ **7.** Change the Primary Display Field of the TIN (in Layer Properties > Fields) so that Map Tips shows Slope. Click Apply, then OK. Experiment by sliding the Select Elements cursor

_____

[21]In order to see this TIN, you will need the 3-D Analyst extension.

FIGURE 2-34

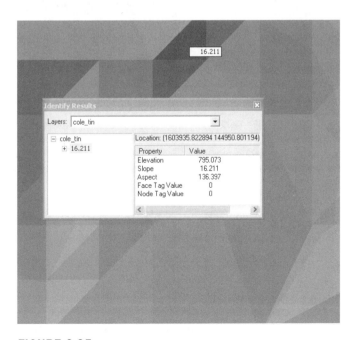

FIGURE 2-35

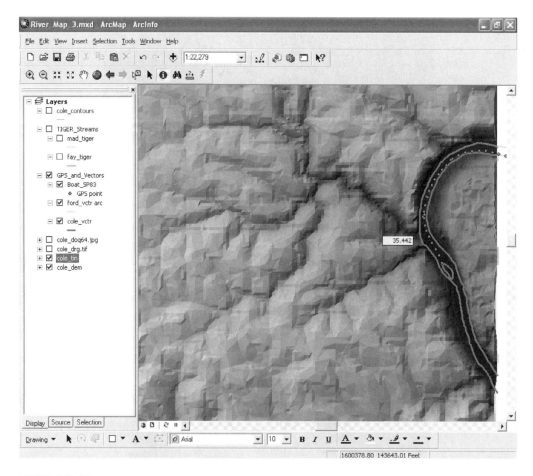

FIGURE 2-36

around on the map. See Figure 2-36***. The slope value is expressed as a percentage: the rise in elevation divided by the run along the surface of the Earth, times 100 to make it a percentage. So the very steep slope of 45 degrees, where the rise is equal to the run, is 100 percent. If you want the *angle* of a given slope, divide the percent slope by 100, and find the angle whose tangent is that value.[22]

___ **8.** Now have TIN Map Tips display aspect. The aspect is the compass direction (zero degrees to 360 degrees) of a vector[23] that is perpendicular to the surface of the triangle. So an east-facing slope would have an aspect of 90 degrees. Experiment again, looking near the river where you know the directions of the slopes.

___ **9.** Save River_Map_4 in ___IGIS-Arc_*YourInitialsHere*. Close the Identify Results window.

---

[22]Such an angle is called an arctangent, defined as "the angle whose tangent is x."

[23]To e precise, it is the direction of the horizontal component of that vector.

TINS Are Three-dimensional Data Sets

_____ **10.** To look at a TIN in another way, and to get a glimpse of the 3-D viewing capability of GIS, we go back to ArcCatalog by selecting `Tools > ArcCatalog` (or find the icon on the `Standard` toolbar). Make ArcCatalog occupy the full monitor screen. Under the `Tools` menu in ArcCatalog, activate the `3D Analyst` extension. Find COLE_TIN in [___] IGIS-Arc\River. Press the `Preview` tab and observe. In the `Preview` text box, select 3D View. A cursor appears. The position of the cursor is initially unimportant, but if you move it around, slowly, holding down the left mouse button, you can dramatically affect the view of the TIN. Play with it. Then do the same with the right mouse button. Finally, if your mouse has a center wheel, press it and move the cursor. (Dragging while holding down the left and right mouse buttons has a similar effect.)

_____ **11.** Now make sure `View > Toolbars > 3D View Tools` is turned on (checked). Put the toolbar somewhere out of the way. Use it to zoom in on the area of the river. Now use the Navigate cursor (left end of the toolbar) to look at the gorge of the river, from the eastern edge of the TIN and somewhat above. You can take your own little helicopter ride around the area. See Figure 2-37***. Play.

_____ **12.** Most of the 3D View Tools icons are familiar to you. Experiment with those that aren't. At the right-end of the toolbar is an icon to launch ArcScene.

ArcScene is like ArcMap, but for mapping in three dimensions. It is a 3-D viewing application with many capabilities, including letting you drape raster and vector data over surfaces. We discuss it in Chapter 9 and show you only enough here to whet your appetite.

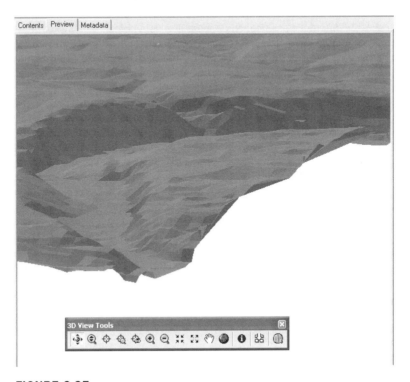

FIGURE 2-37

_____ **13.** Launch ArcScene from the 3D View tools of ArcCatalog. Add Data: Boat_SP83 (use a red dot) and COLE_DRG.TIF. Turn on the Tools toolbar and note the tools it contains. Experiment. Zoom up on the area of the river. (Try out the zoom control that you get with the right mouse button.) Now add the TIN layer.

With the addition of the TIN layer, things became a bit tricky. The GPS path and the DRG appear to be under the TIN—as, upon reflection, you would expect. The GPS track has an assumed elevation of zero—recall that its elevation data was simply recorded as an attribute. Likewise, the DRG is simply two-dimensional. But the TIN is truly 3-D!

_____ **14.** Use the Navigate cursor to view the image in a horizontal state from the eastern edge, so that you see the TIN layer floating above the other two. Dismiss ArcScene without saving changes. Dismiss ArcCatalog. In ArcMap, zoom to the GPS track. Dismiss the Identify Results window if it is showing. Turn off all layers except the GPS and Vectors group layer. Save the map as River_Map_5.

Exercise 2-6 (Project)

Geodatasets of Soils, Rocks, and Land Use

The type of soil that exists at a location is important for agriculture, for assessing the effects of erosion, and for other uses. The rock below the soil is of great interest to those who investigate the resources of the planet. In the database Kentucky_River_Area_Data.mdb we have personal geodatabase feature data set Quadrangle_Data, which contains feature classes cole_soil_polygon and cole_rock_polygon.

In the soils data set, the different classifications appear in a field named MINOR1 of its relational database table. In the geological feature class, the different classifications of surficial rock types are identified in the NAME field of its table. The origin of these databases is not given, nor the meaning of the classifications. While these are real data sets, they are not valid for display or analysis, other than in a tutorial sense.

_____ **1.** *Display a soils personal geodatabase feature class:* Work with River_Map_5 in ArcMap. In [___] IGIS-Arc\RIVER, find the MDB (Microsoft Database) file named Kentucky_River_Area_Data. Pounce. Find the PGDB data set Quadrangle_Data. Pounce. Click cole_soil_polygon and Add. Drag its title to the bottom of the T/C. Make sure all image layers, the DEM, and the TIN above it are off. Turn everything else off as well except the GPS and Vectors group. Zoom to the extent of cole_soil_polygon. You will note two blank areas—one in the southwest and a little one in the crook of the river. These occur because this type of soils feature class is not compiled primarily by quadrangle, but by quadrangle *within* county—in this case the part of Fayette County that is in the Coletown quadrangle.

_____ **2.** Open the attribute table for cole_soil_polygon. How many records do you find? _____. This type of data set is sometimes called "detailed soils"; you can probably see why. Run Statistics on the field MINOR1, which designates the soil type. How many polygons are there? _____.

The Summarizing Procedure

The rest of the statistics doesn't tell you much, because the numbers in MINOR1 are "nominal," meaning that the numbers are used just as names. Under these circumstances, you can't do any mathematical operations other than to compare for "equal" or "not equal." Even the histogram (Frequency Distribution) doesn't give any definitive information because pairs of adjacent numbers (e.g., 34 and 35) are clumped together in the graph. The frequency distribution shows, again, for example, that 34 and 35 together have about 250 polygons associated with them, but this only tells you the total number, not how many polygons are of type 34 and how many of type 35. You can use Summarize to tell you how many polygons of each MINOR1 type there are.

3. Dismiss the statistics window. Click the field MINOR1 to highlight its column. Right-click the field name. Click Summarize. The Summarize window has three text boxes. Number 1 should indicate that MINOR1 is the field to summarize. Do not click in Number 2. Number 3 should indicate that a table named Sum_Output will be placed in the database folder. Click the Browse Folder button and locate

___IGIS-Arc_*YourInitialsHere*

Pounce. Click Save. Without changing the path name, change the output table name (Sum_Output) to Summary_MINOR1. Press OK. When the Summarize Completed window appears, agree to add the table to the map. Close the Attributes of Cole_Soil_Polygon. In the Table of Contents, note the appearance of Summary_MINOR1 and the fact that the Source tab is now active. (Press the Display tab and notice that the reference to the table goes away. Press the Source tab again.) You can only see reference to ancillary tables when looking at the sources of the data. Right-click the Summary_MINOR1 entry and select Open. The Attributes Of Summary_MINOR1 window appears, showing the number of records (and hence polygons) with each different value of MINOR1. What is the count for the MINOR1 value of 34? _____. How about 35? _____. Close the table. Press the Display tab.

4. Zoom up on the part of the GPS track that lies within the soils layer. You see arcs outlining the polygons, but the polygons are all the same color. You can use the Identify tool (be sure to set the proper layer) to see the attributes, including the value of MINOR1. Note that a polygon "flashes" when you click it. What is the "soil type code" for water? _____. What is the "soil type" for an area with no data? _____. Dismiss the Identify Results window.

5. *Label each polygon with the number that indicates soil type:* In general, you can automatically place a label in each polygon. The label, which comes from the attribute table, is a value from the record associated with the feature.[24] Bring up the Layer Properties of cole_soil_polygon. (Recall: To get Layer Properties, you may either double-click the entry name or right-click it and choose Properties.) Click the Labels tab. Turn on Label Features in this layer by clicking it. Apply. A number appears in the polygons, but for each it is 999. That's because these are the values found in the MAJOR1 field of the attribute table. By default, the software picks the leftmost "real" field in the attribute table to use in labeling. Not what we want. Change the Label Field from MAJOR1 to MINOR1 and press Apply again. Now you should see the two-digit value for the soil type in each polygon. In the Layer Properties window, press OK. See Figure 2-38.

6. *Display the feature class with different colors based on the field Minor1:* In order to more easily distinguish between polygons, you can change the symbology so that the polygons are displayed in different colors. Bring up the Layer Properties of cole_soil_polygon. Click the Symbology tab. Notice that you can Show

❏ Features

❏ Categories

---

[24]You saw labels very briefly earlier when we labeled counties.

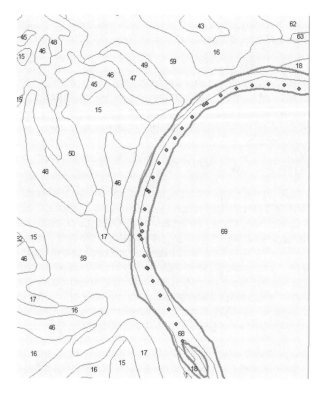

FIGURE 2-38

- ❏ Quantities
- ❏ Charts
- ❏ Multiple attributes.

Pick Categories, and within that, Unique Values. This setting will ensure that when you leave this window the software will assign a color to each polygon depending on the value of a field in the attribute table. For the Value Field select MINOR1. For the Color Ramp choose one that has strongly different colors, say, Basic Random. Click the Add All Values button. Click Apply, then OK. See Figure 2-39***.

_____ **7.** In this step, turn on other layers as needed. Determine the MINOR1 value of the soil type classification (1) of the island, _____, (2) around the stream that flows southeast into the river. _____, (3) that is used to indicate water, _____, and (4) for areas not in Fayette County _____. Use the Identify tool to verify your answers. Also check that the color used for the island polygon corresponds to the color next to its number in the Layer pane—you will have to scroll down to see the values and associated colors. Collapse the list of colors by clicking the minus sign next to cole_soil_polygon. Turn off cole_soil_polygon.

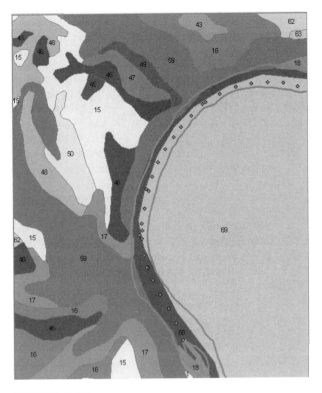

FIGURE 2-39

Some Geological Data

_____ **8.** Also in the Quadrangle_Data data set you will find cole_rock_polygon, which shows the classifi-cations of the surface geology. Add it as a polygon layer, just as you previously added the soils feature class.

_____ **9.** *Run through the same exercises with cole_rock_polygon as you did with cole_soil_polygon, looking at different levels of zoom:*

- ❑ Summarize (on the NAME field) calling the resulting table Summary_NAME.
- ❑ Identify some polygons.
- ❑ Label polygons (use the NAME field).
- ❑ Display the polygons with different colors.

Using the Identify tool, indicate what sort of rock the river bed is made of. _____

___ **10.** In the Layer Properties window for cole_rock_polygon, click the Labels tab and change the size of the labels to 10 picas. Experiment with different scales by typing into the text box on the Standard toolbar. (You only need to type the denominator of the scale fraction—the software will understand the scale you mean.) Notice that with scales smaller than about 1:30,000 (like 1:50,000), the labels are pretty much useless. And annoying! Select Layer Properties > Labels > Scale Range, and click the Don't Show Labels When Zoomed radio button. In the Out Beyond 1 text box, type 30000 in the text box. Click OK, Apply, and OK. Now notice that at a zoom level of 1:30000, you see labels. At 1:30001, you don't.

The concept of showing features only between specified scale levels applies to layers as well as labels. Suppose you didn't want to see the streams when the data frame was zoomed out beyond 1:80,000.

___ **11.** Turn on all TIGER streams. Right-click TIGER_streams and click Properties. Under the General tab, in the Don't Show Layer When Zoomed Out Beyond 1 field, type 80000. Click Apply, then on OK. Using the Magnifying glass icons and/or the Fixed Zoom In And Out tool on the Tools toolbar, zoom in and out on the data frame. Observe the scale text box for the scale and observe the map for the presence (or absence) of streams. Also notice that when the streams aren't being displayed the on-off check box next to TIGER_Streams changes to checked but grayed out.

A GRID of Land Use Data

Of considerable interest to urban and regional planners is what is called "land use" (or the allied but somewhat different concept of "land cover"): the vegetation, water, natural surface, human activity on the land surface, and so on. In the following you will load a data set that describes land cover in the state of Kentucky at a generalized level.

___ **12.** Load as a layer the *GRID* MSU_SPNORTH, which you will find in

[___] IGIS-Arc\Kentucky_wide_data\KY_Landuse_MSU\MSUdata.

If asked, do not build pyramids. Turn all other image and polygon layers off. Zoom to the full extent of the data set to get an idea of the amount of data. Zoom into the GPS track. See Figure 2-40***. Based on the color of the cells and looking at the legend, what would you say the code for water is? _____. What is the code for roads? _____. Verify each of these using the Identify tool. Zoom further in so that you can measure the width of a cell. What are the dimensions of the cells in this data set? _____. How much area in each? _____.

___ **13.** In the Kentucky_wide_data folder you will find the text file KY_Landuse_MSU\MSUdata\msu_lut.txt. Open this file with a text editor or word processor. Verify that your answers about water and roads (transportation) above were correct.

___ **14.** Open the attribute table of MSU_SPNORTH. Based on the value of Count how many cells represent water? _____. Therefore, how many square miles of water would you calculate cover land in Kentucky? _____. How about transportation? _____. Compare. _____.

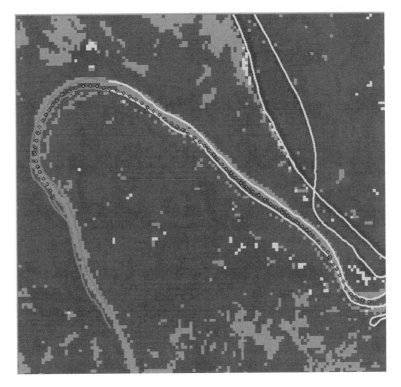

FIGURE 2-40

Next Steps on Your Own

Of course, there is a long story behind each of the types of data that we have looked at briefly here. If you are interested in a particular type or source of data, I suggest that you look first at the Web for sources and then study Web pages or texts for more detail.

Exercise 2-7 (Review)

Checking, Updating, and Organizing Your Fast Facts File

The Fast Facts File that you are developing should contain references to items in the following checklist. The checklist represents the abilities to use the software you should have upon completing Chapter 2.

Important note: This checklist is on the CD-ROM that accompanies the book. It is available in Microsoft Word format. Rather than typing or writing by hand the text that follows, you can copy and paste it into your Fast Facts File from the CD-ROM file.

___ To initiate ArcMap:

___ Square feet in an acre; acres in a square mile

___ To move toolbars by double-clicking

___ To turn toolbars on or off

___ To include extensions in the software

___ The Table of Contents contains _____ when Display is pressed.

___ The Table of Contents contains _____ when Source is pressed.

___ To change just the color of a plotting symbol

___ To change the color and size of a plotting symbol

___ To set the map and display units

___ Map units are

___ Display units are

___ To open an attribute table

___ A personal geodatabase (PGDB) is

___ A PGDB feature data set is

___ A PGDB feature class is

___ A PGDB freestanding feature class is

___ A coverage is

___ A shapefile is

___ A DRG is

___ A DOQ is

___ A GRID is

___ A DEM is

___ A TIN is

___ Vector means

___ Raster means

___ TIGER refers to

___ To group layers together

___ To rename a layer

___ To zoom to the extent of a layer

___ To zoom to the extent of all layers

___ The extension of a map document is

___ A map document contains

___ To save a map document

___ The difference between Add Data and Open is

___ Ways of selecting records are

___ Ways of selecting features are

___ To find a text string in a database

___ To set a bookmark

___ To return to a previous zoom level

___ To measure distance on a map

___ To see the values of all attributes of a given feature

___ To connect to a folder in ArcMap

___ TIFF refers to

___ To change the drawing order in the table of contents

___ The additive colors are

___ To set layers for the Identify Results tool

___ To collapse and expand entries and legends in the table of contents

___ The Magnifier window

___ The Overview window

___ Ways of panning the image are

___ To set map tips for a layer on

___ A TIN does not have an associated attribute table because

___ To see a 3-D data set with ArcCatalog

___ ArcScene is

___ To run statistics on an attribute

___ To summarize an attribute

___ To label features

___ To symbolize features with different colors

___ To bring up a layer's Properties window

___ To change whether on not labels are shown, depending on scale

___ To change whether on not layers are shown, depending on scale

___ To classify features

CHAPTER **3**

Products of a GIS: Maps and Other Information

OVERVIEW

IN WHICH you investigate the relationships between GIS and mapmaking, look at a variety of types of output from a GIS, and learn the rudiments of mapmaking with ArcGIS.

GIS is the worst thing that ever happened to cartography.[1]

Unattributable, circa 1990

GIS and traditional cartography share a fundamental idea: Depict a part of the real world so that humans can understand it and use the knowledge for navigation, planning, resource management, or other forms of decision making. The traditional cartographic product, the map, is a single entity that is usually the result of a major project. For example, a USGS topographic map may take several years to produce. Decisions about the font and placement of textual information, colors used, and so on are carefully considered. It has a fixed size and scale. It is, to many, a work of art.

GIS and Cartography— Compatibility?

Earlier, in Chapter 2, I discussed the differences between maps and GIS. Here I want to illustrate a dramatic, seldom recognized difference between published maps and GIS-developed maps. It turns out that, no matter how good the tools in GIS become in facilitating the production of paper maps, there is a fundamental incompatibility between GIS data and even a fairly large-scale paper map. In addition to the previously discussed advantages of GIS over maps is the fact that cartographers sometimes have to distort the positions of features in order to let the map convey important information. For example, suppose in a mountainous region there is a populated valley. Running through this valley may be several linear features: a stream, a road, a power line, a gas main, and a

---

[1]Probably because the tools weren't there to make high-quality maps and because it let amateurs like the author, with all the artistic ability of a can of spray paint, make maps.

railroad track. These elements may have to exist within a corridor of a few hundred feet. Further, the road may cross the stream and the railroad. A distance of 250 feet translates to only an eighth of an inch on a "2000-scale map" where 1 inch corresponds to 2000 feet. (The scale is 1:24000.)

An eighth of an inch certainly does not supply enough space to show these features and the relationships between them. If the symbols for these features are shown in their correct spatial locations, they will appear on top of each other. What is the cartographer to do? Fudge! Widen the valley and exaggerate the distances between the features, which means putting some of them in places where they, in fact, do not exist. Generalize is a term that is used. The cartographer rightly deems it more important to show the correct relationships among the features than to be precisely spatially accurate.

One can correctly argue that such adjustments should not be part of a GIS database. The different uses, at different scales, of GIS data suggest that everything placed in the database should be recorded as accurately and precisely as reasonably possible. This means that if a map is made or data displayed at the 1:24000 scale from such a database, you will not be able to distinguish among the features. To see the features in their proper relationships, you will need to zoom to a larger scale.

This incompatibility between the paper map and GIS will be exacerbated as more and more detail is introduced into GIS databases. This incompatibility, along with the lack of tools for good mapmaking (a problem that is actually fast disappearing), is why many cartographers convert GIS maps into drawing programs like Freehand and Illustrator. In any event, the virtual map or the Web-based map, will not push the paper map into antiquity, but the trend will be in that direction. After all, what is usually desired is to have spatial data and analysis tools that will create products for decision makers.

Products of a Geographic Information System

Let's shift focus from learning GIS to thinking about the desired end results. By this I mean the decisions taken by humans to change, navigate, or protect the environment.

C. P. Snow wrote, "To be any good, in his youth, at least, a scientist has to think of one thing deeply and obsessively for a long time. An administrator has to think of a great many things, widely, in their interconnections, for a short time."[2] It is to the attention of these administrators that products of a GIS are mostly directed. You, as a producer of information, need to be mindful of limits on their time.

Overall Requirements for Utility

To be useful in decision making, products of a GIS must meet several criteria:

1. The decision makers must know it is available.

2. They must be able to understand it.

3. They must have some reason to believe that it is worth their time to determine how to use it.

4. Assistance to aid the decision makers' understanding of the product must be available.

---

[2]Snow, C.P. 1961. *Science and Government*. London: Oxford University Press.

5. The product must be available at the time it is needed.

6. It must be relevant to the area of concern.

7. It must have considerable accuracy and integrity; if the product lets decision makers down, a long time will elapse before they depend on such information again.

Classification of GIS Products

Products from a GIS might be classified in several ways. I'll use the terms media, format, purpose, and audience.

Media: I use the term "media" to denote the physical carriers of the information presented to the decision maker. Common media are paper, photographic materials (opaque ones like photographs, and translucent ones like slides and films), and electronic visual devices like computer monitors. Three-dimensional electronic displays activated by laser beams—called holograms—may be available in the future, but more conventional products are now available that can meet more important, if less exotic, criteria. Almost all products of existing GISs are designed to respond to the sense of vision in some manner.

Format: While the number of visual media that carry information is limited, the number of forms or formats that information can assume is without limit. An infinite variety can be obtained with characters—the 26 letters of the alphabet, 10 Arabic number symbols, and some special symbols. This type of information is called character-based.

Character-based information can appear in the form of text, tables, lists, formulae, and so on. The way in which information is organized has a tremendous impact on whether or not it will be useful. Character-based information can be processed by an individual in serial fashion (like a reader "processing" this line of text) or in search mode—a procedure in which a person examines unconnected groups of characters in order to find desired information. Looking up a number in a telephone directory exemplifies use of the search mode—followed, of course, by serial mode.

For purposes of mental model building, the best products allow a user of character-based information to quickly grasp two things: the overall scheme of organization of the information (revealed by tables of contents, sections on "how to use this information," executive summaries, etc.) and the subject of the information itself (illustrated by introductions, table titles, lists of parameters relating to the information, etc.). Development of products that can meet these criteria is an art and a science.

Graphic information—pictures, photographs, drawings, maps, displays, graphs, diagrams, and so on—is equally as versatile as character-based information.

Simplistically, character-based information is read, while graphic information is viewed. Both can help a decision maker form a more complete mental model of an issue, but each provides information in different ways.

It is rare that any information is either totally character-based or graphic. Combinations of the two are the most effective (graphs have descriptive headings and designations; reports have diagrams and illustrations), though the process of "marrying" the two is not always straightforward, particularly when computers are used.

Give an example of the media used for GIS products

Purpose: Another classification that might be considered during design of a product from a GIS is the general purpose or purposes of that product or information. Some of the possible purposes include the following:

- ❑ Inventorying

- ❑ Analyzing

- ❑ Explaining

- ❑ Documenting

- ❑ Designating

- ❑ Defending

- ❑ Managing

- ❑ Forecasting

- ❑ Monitoring

The design of the product is frequently more effective if the purpose is kept well in mind. For example, if the major purpose is monitoring, the most appropriate product might be one that reflects a change over time rather than the production of two documents, each of which shows the situation at a given point in time. This sounds elementary, but the amount of effort that has been spent in trying to compare two similar documents, side by side, to ascertain the differences between them is staggering. A GIS has the capability to generate the "difference" and that capability should be used.

Other purposes will be most appropriately met by differing formats. The important point to consider, for each product, is *how that product will be used*.

Audience: A good GIS will be capable of producing many sorts of information products at varying levels of detail and sophistication. An additional classification for these products might be the audience for whom the product is intended.

Attempts to develop a "superproduct" should be avoided. As various products evolve, this becomes a strong temptation. Those responsible for system and product design and evolution keep adding more bells and whistles, which the designers, of course, understand completely. But a person charged with making a decision, who is looking at the product for the first time, may find that an elaborate demonstration interferes with his or her understanding of—and use of—the information.

One approach designed to avoid this problem is the development of an information product series—several forms of information of similar origin, media, format, and purpose. This type of series might show different levels of detail that can be used by anyone exploring all perspectives of an issue. As a product series is used, the first priority should be development of a product appropriate for the needs of the decision maker. That product should be supplemented by others of the same general form that provide more detail or additional information. For example, if information on the limits of the 50-year floodplain for a reach of river is needed, the information product should not show the 20-year plain, the 100-year plain, the normal yearly range of bodies of water, or the expected annual rainfall, and so forth. Instead, it should clearly show the 50-year floodplain, with a notation that more detailed or sophisticated information is available for other floodplain limits in roughly the same format. If the system can produce a complicated map, it should also have the ability of producing less complicated ones.

166

Name 4 possible purposes of GIS products

Documenting Products

Developers of GIS products sometimes become so caught up in the task of providing primary products to decision makers that they neglect a second but vital component of the operation: providing explanatory, documenting, and context setting information. When the information relates to data sets, we call it, as you know, metadata. Products need equivalent attention. Some of the information about products can be obtained from the metadata; some cannot. A partial list of defining information for a report, document, or map might include the following:

- ❑   A title
- ❑   A descriptive paragraph on the content of the document and how it is to be used
- ❑   The geographic area covered
- ❑   The date the information was produced
- ❑   Identifiers that allow the user to determine the defining information about the data which support the information in the report
- ❑   Statements about the precision of the information in the report
- ❑   Estimates of the accuracy of the information in the report
- ❑   The variables that went into the production of the report
- ❑   A name, phone number, e-mail, and postal address of a person (or agency) to contact for information about the report
- ❑   All of the parameters that were supplied by the user in the production of the report
- ❑   Any warnings to users
- ❑   Identification of agencies and individuals responsible for the report

Not all of this information need appear in the same format in the same place. Some of it may be generated by the computer and appear on the printout: date, data identifiers, parameters, precision and accuracy information, and so forth. Various descriptive information might better be printed separately and attached to the computer-produced report. Regardless of the methods of production or dissemination, a product from a GIS should be a complete package. And, of course, pointers to the metadata of the constituent data sources must be provided.

It may be determined that each user of an institution's GIS should be given a user's manual of the system that mainly describes any customization of an off-the-shelf GIS such as ArcGIS. Such a manual could set the context of the entire system and then describe each product series. Such a manual should be loose-leaf and modular; it should also be available online. An updating scheme should be thought out carefully.

When defining information about a report is contained in the user's manual rather than in the product itself, the product must contain references to the user's manual; the loss of the link between the report and its defining information can prevent acceptance and use of the system.

[handwritten: 12 essentials of a GIS product] [handwritten: Like Lost Sand]

Thoughts on Different Types of Products

A major reason to have a GIS is to either:

(a) provide new information or

(b) provide information in new forms.

The obvious sort of information that comes from a GIS is in graphic form: the map. But there are many other ways to convey information about the environment—some more attuned to the style of the decision maker than the casual user.

Don't Ignore Character-Based Information

Character-based information is the sort that many decision makers are most used to using but graphic information is the form that most planning professionals who advise decision makers deal with in formulating their recommendations.

Some humans, perhaps innately, are better at dealing with graphic information and some with character-based information. There is also physiological evidence to suggest that the two different types are processed by different hemispheres of the brain. One might draw a parallel with left-hand and right-handedness.

It is not my purpose to suggest (or deny) that each person has an inherent dominance of ability to process character-based or graphic information but to point out the danger that an individual may well naturally opt for information in a particular form—just as he or she might naturally use a screwdriver with his or her left hand—when another form might be more appropriate in helping the person gain the necessary insights needed.

GISs can provide both character-based and graphic information, and any GIS that is used to provide only one type may be missing a good bet.

Don't Hesitate to Sort Information

For example, suppose some 348 "areas of critical environmental concern" have been nominated and an identification code attached to each; your GIS has the information to calculate a factor between 1 and 100 that suggests the degree of danger to which each is subject. The output you envision is a list, in order of identification code, with the "danger factor" printed out adjacent. As you begin to use the output, you discover yourself looking through the list to find the area with the highest factor, then the next highest, and so on. At this point, while all the information you want is there, it clearly could be in a better form: in order of the danger factor rather than (or in addition to) the identification code order.

Consider Hard Copy

Consider the use of a GIS in some applications to have it print out lists, catalogs, or tables of numbers—of which 99 percent are never viewed. This type of output can be replaced on a periodic or as-needed basis. There is also the option of putting such information on the Internet, but that may make it less accessible in some instances.

While printing a lot of paper sounds wasteful, it may be cost-effective. Consider the example of a telephone book. Despite duplication, paper, and distribution costs—and the fact that an information service is provided over the phone—it is less costly to organize and provide mostly unwanted information to each customer in a region rather than respond dynamically to the customer's needs for information at particular points in time. On the other hand, the phone company does not provide a list of subscribers for the entire nation. The key point is that the issue of product utility and cost must be looked at in a comprehensive way—not just in terms of time, materials, or human effort alone.

Consider Balance in Product Content

In the design of a product, there should be a balance between simplicity and generality. In one sense the best product is one that speaks directly to the decision maker, respecting his or her particular abilities, relating to the issue he or she is dealing with. On the other hand, it is nice to have a product of such good design that it can serve the decision maker, his or her advisor, those in other areas, perhaps the public and the courts. Mapmakers, of course, are well aware of this design problem. A map's usefulness is increased by adding a new type of information and yet decreased at the same time because the map becomes more cluttered. A map that "shows" everything shows nothing, because it is black.

Elements of Product Design

Product design is an art that marries what is possible with what is needed. Many factors go into successful design:

❏ *Examination*—Of products from other systems

❏ *Innovation*—The ability to conceive of more meaningful ways to display information within the constraints of the devices which put the information together

❏ *Refinement*—Of products that the GIS is already producing by obtaining, and heeding feedback from users of the product

❏ *Knowledge*—Of what data are required (and the characteristics of those data) to produce the necessary information

❏ *Lack of bias*—Toward either character-based or graphic information and an ability to provide information in the best format for the given customer

Units, Projection, and Scale

In Kentucky, eight different coordinate systems for spatial information are in use by state agencies alone. No doubt, other states share this problem. How should distances on a map be presented—metric or otherwise? What shows up on the product should take into account the nature of the audience. If you can, avoid cluttering products with multiple coordinate systems. If the user doesn't mind the maps looking weird, you might make a map using latitude and longitude coordinates. Usually though, users like maps whose scales are consistent in all directions.

Thoughts on Resolution and Scale

Resolution is, basically, the smallest length in ground units at which the identity or characteristic of something can be resolved by looking at the product. For example, what is the diameter of the smallest object that can be seen in an aerial photograph? Clearly, the answer depends on several factors besides the diameter of the object. It involves the reflectance of the object compared with the surrounding ground, the quality of the vision of the person looking at the photo, and the scale of the photo. An 8x10 photo of Arizona might allow only an object the size of Phoenix to be visible. Through enlargement of the same photo, however, it might be possible to increase the resolution so that an object 100 feet across could be identified. If further enlargement—no matter how extensive—did not allow identification of objects with diameters smaller than 100 feet, then the resolution of the photo is said to be 100 feet. Thus, it is obvious that resolution and scale are closely linked—the larger the scale, the greater the resolution—up to a certain limit. Beyond that limit, no further information can be obtained from the product simply by enlarging the scale. In previous exercises you have had the experience, looking at raster images, of watching the information disappear as you zoomed in.

If a specific ground area is to be covered, designers must then weigh the relative factors of size, scale, and resolution to determine the most appropriate method of producing an information product. Too often, these determinations are made without sufficient thought, or are influenced by habit and conference room table size. The issue of appropriate scale must be considered early in the process because some GIS products utilize overlays and base maps of other materials to make them meaningful. Obviously, a nearly exact physical match must occur. Further, the size, scale, and resolution of both base maps and overlaid products must be appropriate. It is worth mentioning here that reproductions of paper map output, from any source, are not always the same size as the original—and hence not the same scale as the original. Unfortunately, statements regarding scale printed on the map, such as 1 inch = 1000 feet, are reproduced along with everything else, creating a built-in lie.

Making Sure There Is a Base Map

Your GIS will include, probably, a multitude of layers. The existence of a base map—a cartographically or photographically produced product of, usually, great accuracy, high precision, and great detail to which all the other products of the GIS can be referenced—exists and is prominent. To attempt to build GIS products without first developing (or otherwise obtaining) a geographic base map is folly. But be aware: To use primarily GPS to create GIS data sets to serve the base map function is a monstrous undertaking.

Measures of Quality Assurance

One of the values of a GIS is its capability to combine different types of information that have the same geographic base. For example, a user might need a map displaying a single variable derived from a combination of soil type, bedrock type, and depth to bedrock. An important aspect of the defining information of a report combining these three variables is the confidence one can place in the accuracy of the resultant information. Each of the constituent individual variables is stored in the database, and each has its own characteristics of precision and accuracy—measured in terms of geographical coordinates and the values of the variable. Therefore, just as the values of the three individual variables are combined to produce a single variable, the precision and accuracy attributes of these variables should be combined— according to appropriate numerical techniques—determine the accuracy of the final result. Statements of accuracy should be included as part of the resultant report.

The process of assessing accuracy is not always easy, nor are the results always encouraging. For example, in a study done by the Australian Commonwealth Scientific Industrial Research Organization a number of years ago, involving three variables, an analysis was made of the output. When particular points on the Earth's surface were picked out, it was found that at least one of the constituent variables was incorrect more than 60 percent of the time. The implications for the accuracy of the combined report are obvious.

Frankly, the lack of sufficient quality reporting of derived products is a major shortcoming of all commercial GIS software.

The Decision Maker—Product Interface

The act of a decision maker sitting down at a computer terminal is one that cuts out virtually all the "people buffers," between the decision maker and the computer. Many decision makers will not want to spend the time necessary to either learn or operate the software. There is also a certain element of justified fear involved for even the ablest individual in doing something new with others looking on.

A person familiar with the product (including the assumptions underlying it and other products that might be useful to the decision makers) should be present when the product is used. People charged with making decisions have a way of asking questions no one thought they would ask. Anyone who presumes to provide them with new information in new forms had better be ready.

The person charged with the responsibility of understanding and explaining a document is also in an ideal position to recommend changes in the document's structure or information content based on conversation with the users of the product. The dissemination of a product containing information is very much a two-way street and relies on user feedback for its successful continuation.

Often, user needs are not correctly perceived by product designers. Further, user needs change. These and many other factors suggest that a continuing dialogue between the providers and users of information products must exist.

It may be that, instead of having a person assigned to a particular set of products as the interface between the decision maker and the product, personnel will be assigned as liaisons to various departments using the products. Whichever scheme is chosen, personnel charged with the function of interfacing products with decision makers should meet among themselves regularly to aid in improving the effectiveness of the GIS operation.

In Summary

You can see that the number and diversity of potential information products from a GIS is almost unlimited. These products, however, are useless if they don't fill a need or if they are inappropriate for their intended audience. The song "Alice's Restaurant Massacree" by Arlo Guthrie contains the line: ". . . and the judge [with the seeing-eye dog] wasn't going to look at the twenty-seven 8 by 10 color glossy pictures with the circles and arrows and a paragraph on the back of each one . . . "

Design of an information product must, of course, proceed from an enlightened view of the data used to support it. But equally important is a clear understanding of the needs of decision makers or administrators who will be using it.

Products of a GIS: Maps and Other Information

Up until now the maps you have worked with have fundamentally been portrayals of geographic data. But if you look at any map that is produced commercially or by government sources, you notice that the spatial data reside in a context of other text and graphics. For example, you will probably find a title of the map, an arrow indicating the north direction, a legend showing the scale, and so on. Also, a single map sheet might consist of several maps at different scales or maps showing different data from the same geographic area. While you can print a map directly from the sort of view you have been working with, ArcMap has capabilities that let you produce a map with the additional elements that form sophisticated cartographic products.

While the thrust of this book is to lay the groundwork so that you can use GIS to do geographic data analysis and modeling, I would be less than candid if I didn't let you know that the most popular use of GIS currently is to *display*, in map format, geographic information. This chapter gives you the beginnings of how that is done. And, of course, when you do use GIS to do analysis, you will need to display the results, so what follows is essential. Before we launch into how this works you need to know some terminology:

❏ *Feature*—A representation of a real-world thing, like a house, a city, or a pipe

❏ *Object*—A point, line, network, or area that represents a feature

❏ *Layer*—A set of objects that represent a number of features

❏ *Data frame*—Several layers displayed in a particular way (scale, style, etc.)

❏ Layout—One or more data frames, with optional tables, graphics, and so on—the finished graphic and text product that will become a map sheet

The Data View and the Layout View

To use ArcMap to make maps, you need to be aware of two distinctly different ways to display geographic data: the Data View versus the Layout View. The Data View is the view that you are familiar with. The main ArcMap window shows the data sets you have selected; they fill the window pane. In the Layout View you see an image of a map sheet meant to be printed. In this view you "lay out" the map elements, including the data sets that make the map a true cartographic product. A lot is involved in transforming geographic data into a map worth the name. You discover how to begin the process in the exercises that follow.

⎯⎯⎯ Open up your Fast Facts text or document file.

Exercise 3-1 (Warm-up)

Templates

A template in ArcMap is somewhat like a preprinted sheet of drawing paper. It can be almost blank or can contain graphics and/or data. We start by looking at some examples.

____ **1.** Start ArcCatalog. Copy the folder Trivial_GIS_Datasets from

[__] IGIS-Arc to ___IGIS-*Arc_YourInitialsHere*. Make a connection to this folder.

Under ___IGIS-Arc_*YourInitialsHere*

create a folder named Map_Making. Make a connection to this folder. Launch ArcMap from ArcCatalog

____ **2.** When the initial ArcMap dialog box—Start Using ArcMap With—comes up (assuming it does[1]), pick A template. Click OK. In the New window that pops up, select the My Templates tab. Click the button at the lower left of the New window. One template (probably the only one) will be called Blank Document. It is based on Normal.mxt—the one you get when you start fresh with ArcMap. It is a file, whose location is given in the text box in the New window once you click its name. The file extension of all templates is .mxt. Write down the location of (the path to) the file Normal.mxt.[2] Do *not* click OK.

_____

____ **3.** Select the General tab. Among your choices of templates here are LandscapeModern.mxt. Single-click that to see what it looks like in the preview pane. See Figure 3-1. Write down the location of that file. Do *not* click OK.

_____

____ **4.** In general, all ESRI-supplied templates, except Normal.mxt, are stored together here. By using the up and down keyboard arrow keys, you can see miniatures of several templates. Look at some of the other templates. Do *not* click OK.

____ **5.** Click the button to the immediate left of the text box to see thumbnail presentations of the possible templates. Click LandscapeModern.mxt. Now you may press OK. This makes LandscapeModern.mxt the template for the current ArcMap document, as you can see from the map display area window.

____ **6.** In View > Toolbars make sure that there is a check mark beside Layout. Locate the Layout toolbar. It looks something like this:

---

[1]The dialog box is optional and may have been turned off. Or ArcMap may be already running. If so, choose File > New to get the New window. (To turn on the startup dialog for the future you may choose Tools > Options > General and make a check by Show startup dialog.)

[2]The path name may be too long for the textbox. If so, place the cursor in the text box, click once, and press arrow keys to see the whole thing.

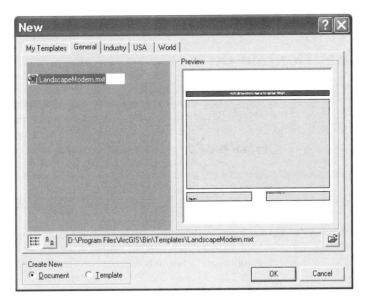

FIGURE 3-1

Some of the icons you will recognize from the Tools toolbar, but note that most include the image of a page as well. These tools are for working with map layouts and they operate somewhat differently, as you will see shortly.

7. *Examine and use the map template:* There is some tiny text in the lower right box of the template. Zoom in on it with the + magnifier on the Layout toolbar by dragging a box around the text. Click the Select Elements tool on the Tools toolbar and double-click the text you zoomed in on. In the Properties window that pops up, select the Text tab. In the Text box, type Just Playing (replacing the text that is already there) and press Apply. Under Change Symbol, which brings up the Symbol Selector window, make the text red, bold, Arial, 36 point. Click OK, then Apply, then OK.

8. Use the Zoom Whole Page button[3] in the Layout toolbar to see the entire layout again. If necessary, click again on the Select Elements tool on the Tools toolbar. Change the text you put in previously to More Play and click OK. When the cursor is inside the selection box, you should see a cursor that has four arrows. Using this cursor, you can now drag the text around. Drag the text to the center of its box.

9. In black, underlined, bold, italic, 40-point Arial, enter a title at the top of the layout that reads I'm Done Here. Center the title.

---

[3]Find it with the status bar and/or ToolTips. If ToolTips is not on, select: Tools > Customize > Options > Show ToolTips On Toolbars.

Chapter 3

Templates That Contain Data

1. Using `File > New.`[4] in ArcMap pick either the USA template category or the World category, depending on where you live. By flipping between the two buttons to the left of the path text box, you can look at the possibilities in different ways. Pick a template that contains your region. Click OK. If asked, don't save your previous layout. What template did you pick? _____ *Southern US* _____.

Notice that the title of the *Layer* in the Table of Contents reflects your choice and that a number of data sets are represented. Notice also that the map window contains a title, a scale bar, and a legend. The *thin black* line represents the boundaries of the page that may be printed. Outside of that are rulers.[5]

2. *Switch between the Layout View and the Data View.* From the View menu, pick Data View. You will see the sort of map display you are used to seeing. Now pick Layout View from that same menu. For a shortcut way of switching between the two views, locate the four icons to the left of the horizontal scroll bar.[6] (See Figure 3-2. The leftmost one, an Earth symbol, is the Data View. The next, a page, gets you the Layout View. The third is a refresh button, should the display not look quite right and you want it "repainted." The fourth pauses the drawing. Flip back and forth between the two sorts of views.

FIGURE 3-2

3. Select File > New again. In the New window, locate the question mark in the upper right corner of the window. Click it. Now click one of the tabs at the top of the window and read about templates and folders.

4. Click the title bar of the New folder and then pick the USA template category. Click SouthwesternUSA.mxt, and then OK (not saving changes from the previous map). The map pane comes up in Layout View. (See Figure 3-3.) Observe, then switch to Data View.

5. Since geographic data sets are being shown, there is likely to be some underlying coordinate system. Determine what it is: Right-click the map, then in the context menu, click > Properties > Coordinate System. _____ *No. Amer. Albers Equal Area Conic*. What is its Central Meridian? _____ *-96* _____ What is the Latitude of Origin? _____ *40°N*. Look under the General tab; what are the Map Units? _____ *Meters* _____ Check that the Display Units are Miles. Click Apply, then OK.

---

[4]File > New is not the same as just pressing the New Map File button on the Standard toolbar, even though they present the same icon. File > New (or Ctrl-N) gets you to the New Template window. The New Map File button starts you off with Normal.mxt.

[5]If the rulers aren't showing, right-click the layout page outside the thin black line, then Rulers > Rulers.

[6]If the scroll bar isn't showing, select: Tools > Options > Layout View > Show Scroll Bars.

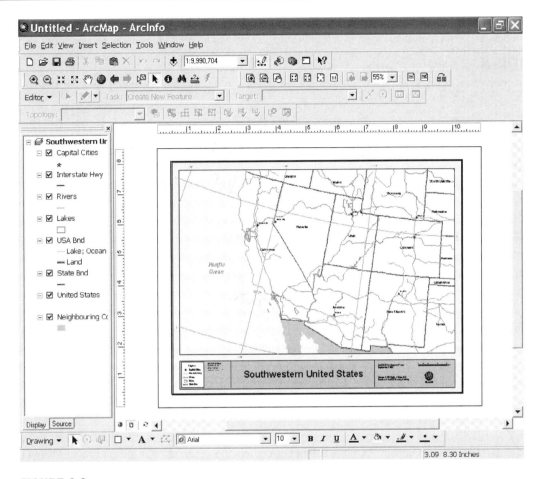

FIGURE 3-3

6. Switch back to the Layout View. When the cursor is on the map proper (where geographic features are being shown), two types of coordinates are displayed: geographic coordinates and page coordinates. Place the cursor at the lower left corner of the map proper. What are the approximate coordinate values?

Map: Easting -1700 Northing -591 in Miles (units)

Page: X-coordinate 0.76 Y-coordinate 1.77 in inches (units)

Do the same for the upper right corner of the map proper:

Map: Easting -237 Northing 358 in Miles (units)

Page: X-coordinate 10.03 Y-coordinate 7.8 in inches (units)

The map coordinates you wrote down in the preceding blanks are the distances in miles from the Central Meridian and the Latitude of Origin of the projection, which you recorded previously as longitude −96° and latitude 40°. By the way, it is easy to get latitude and longitude confused. It is also easy to confuse easting and northing. Usually latitude is given first when coordinates are in degrees. And usually the x-coordinate (easting) is given first when you are operating in a Cartesian coordinate system. Note that these orders are the reverse of each other.

Now check out the lower left and upper right coordinates of the entire rectangular map page.

Page: X-coordinate __0.50__ Y-coordinate __0.50__ in __inches__ (units)

Page: X-coordinate __10.3__ Y-coordinate __8.0__ in __inches__ (units)

Controlling Your View of the Map: Zooming

The two sets of zooming and panning controls take some getting used to. Let's look at zooming first. If you are in the Data View, the Layout toolbar is inactive. But in the Layout View, both toolbars may be employed; their tools have different effects.

____ **7.** Go into Data View. Using the familiar Zoom tool, drag a box around Colorado, leaving significant margins around the state boundaries. You will be able to make out the word Colorado and may be able to see the name of the capitol city Denver (where all the interstate highways come together) and the green star next to it. Zoom up on that area, by dragging about a 1 square inch box. Notice that the place names and the symbol get no larger, but you see more detail of the highways. Notice also that the state name Colorado is visible, having moved from where it was before. Zoom in again tightly on Denver and its symbol. Again you see Colorado. And again, the symbols get no bigger.

____ **8.** Click the Go Back to Previous Extent button until the entire state is shown again.[7] Switch to Layout View. Now using the Zoom In tool on the Layout toolbar, zoom in on Denver. Notice that now you don't see Colorado and that Denver and the star have become larger. When you use the magnifier on the Layout toolbar, the Zoom In feature really does act like a magnifying glass; that is, in addition to seeing a smaller area in more detail, the graphics and symbols get bigger. And the symbols do not relocate as they do in the Data View. In the Layout toolbar, click the Go Back To Extent button so that the full state is visible.

In the Layout View, the zoom buttons simply let you view the map—geographic elements, title block, text, and so on—with different levels of magnification. Zooming and panning do not move the labels and symbols of the map.

____ **9.** *Experiment with the buttons on the Layout toolbar:* Turn on the labels for interstate highways. Then zoom up using the Zoom In tool on the Layout toolbar. Notice again how place names, highway route designations, features, and boundaries become larger as you zoom in. Again recall that this is different from your previous experiences with the Data View zoom control.

---

[7]At any time you run into difficulty and feel that you would like to start over, do. Just choose File > New, pick the USA template category, click SouthwesternUSA.mxt, and click OK.

_____ **10.** Use the What's This (^?) button to determine the use of the Zoom To 100% button (labeled 1:1) on the Layout toolbar. Try the button. Notice that the text box on the Layout toolbar reads 100. Then press the Go Back To Extent button.

_____ **11.** Click the Zoom Whole Page button. The entire layout returns, title block and all. (This button will be your good friend in a number of circumstances—use it often.) Try the other Zoom buttons, the Go Back To Extent and Go Forward To Extent buttons, and the Zoom Control text box. Experiment with panning the map sheet at different levels of zoom. Notice that the information provided by the rulers changes as you zoom and pan.

_____ **12.** Observe the scale textbox on the Standard toolbar. Press the Zoom Whole Page button again. Notice that the scale on the Standard toolbar does not change when you change the zoom level with the Layout toolbar. Again, you are merely looking around a paper map, studying different parts of it with a magnifying glass.

_____ **13.** Experiment by flipping back and forth between the two view types at different levels of magnification. Remember, you can alternate views easily by using the two mini-buttons at the left of the horizontal scroll bar. Go into Data View. The Data View of this template includes a lot of territory, as you can see by going to Full Extent. On the Tools toolbar, press the Go Back to Previous Extent button to get the more detailed map of Colorado again. Return to Layout View, and use Zoom To Whole Page. If the map does not show exactly what you want, alternate between Data View and Layout View, panning and zooming, until you get the desired result.

Understanding the Panning and Other Controls

_____ **14.** To eliminate any confusion after all that map manipulation, let's start over: Click File > New, pick the USA template category, click SouthwesternUSA.mxt, and click OK. In Data View, zoom up on Colorado, as before. Change to Layout View. Using the "hand" icon on the Tools toolbar, (Pan the map layout by dragging it—see the status bar) pan over to Utah. Get back with the Go Back To Previous Extent button. Now pan the map by dragging it using Pan on the Layout toolbar. Notice this simply moves the entire layout (and adjusts the rulers). Return to the original image by using the Go Back To Previous Extent button on the Layout toolbar. Pan again. Go back this time with Zoom Whole Page.

_____ **15.** The image of Colorado looks crooked on the page. The Albers projection does this for areas not near its central meridian—in this case, 96 degrees west. Challenging question: On about what meridian does Denver lie? __102__. The boundaries of the state are in fact parallels and meridians of the latitude and longitude coordinate system, so it would be nice if they ran east-west and north-south. Fix the crooked image by clicking View > Toolbars > Data Frame Tools, and then typing the desired number of degrees of *counterclockwise* rotation into the text box. Try 3 degrees and press Enter. Not enough? Clear rotation and try 6 degrees. (If you wanted a clockwise rotation, you would use a negative number.) Dismiss the Data Frame Tools window.

_____ **16.** *Since you have changed the scope of the data, the title on the map is now incorrect, so change it:* In Layout View, press the Select Elements button on the Tools toolbar. Click various places on the map proper and on the other parts of the page. Notice the blue rectangles and "handles" that appear. Note that multiple clicks sometimes result in different selections. You could resize these elements and move them around—but don't. Zoom up on the area of

179

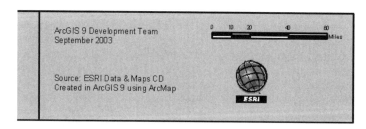

FIGURE 3-4

the title of the map, using the Layout toolbar zoom. With Select Elements active, click the title of the map. A blue dashed-line box should appear around the text of the title; no handles should be seen. (If this doesn't happen the first time you try, keep trying.) Right-click and select Properties. Under the Text tab type `Colorado`. Press Apply, then OK.

____ **17.** Pan and zoom (Layout toolbar) to the left end of the title box. Check out the Legend and the Projection. Verify the Central Meridian and the Latitude of Origin. _____ *40*

____ **18.** Pan to the right end of the title box. Check out the text, the ESRI logo, and the scale bar. See Figure 3-4.

Adding Other Map Elements

The layout has no north arrow. Let's put one in. We'll erase the ESRI logo and put the north arrow there.

____ **19.** With the Select Elements tool, click the logo, so it gets the selected border and also blue handles. Tap Delete on the keyboard.

____ **20.** Hmmm. Maybe erasing the logo isn't such a good idea. An "undo" command doesn't seem to be anywhere around. Try the universal undo: Ctrl-Z. Bingo! (You *will* put that in your Fast Fact File, won't you?) Also check out the `Edit` menu.

____ **21.** Drag the logo to the left to make room for a north arrow.

____ **22.** Click `North Arrow` on the `Insert` menu. Pick ESRI North 6 and click OK. It appears, selected, with handles, on the Layout. If you place the cursor over one of the handles, the cursor changes to a double-headed arrow. If you drag the handle with this cursor, you can resize the selected element. Resize the arrow that appears and drag it into the title bar area. Press the Esc key, or click away from the arrow, to unselect the element and thus turn off the handles. Then move the logo and scale around to make things look neat. Examine the result. See Figure 3-5.

Well, that's annoying. Upon inspection it appears that the North Arrow doesn't seem to be pointing quite north. We went to some trouble to line Colorado up so its meridians were as north-south as possible,[8] and now apparently our actions to rotate the map 6 degrees is reflected in the direction the North Arrow is

---

[8]Of course, no north arrow is exactly right for the whole map. On the eastern edge of the state, the boundary points north, as does the western boundary. And the two lines are not parallel.

Control Z = Undo

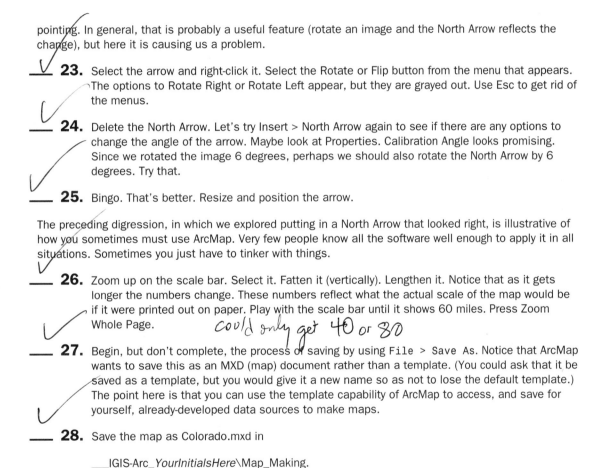

FIGURE 3-5

pointing. In general, that is probably a useful feature (rotate an image and the North Arrow reflects the change), but here it is causing us a problem.

23. Select the arrow and right-click it. Select the Rotate or Flip button from the menu that appears. The options to Rotate Right or Rotate Left appear, but they are grayed out. Use Esc to get rid of the menus.

24. Delete the North Arrow. Let's try Insert > North Arrow again to see if there are any options to change the angle of the arrow. Maybe look at Properties. Calibration Angle looks promising. Since we rotated the image 6 degrees, perhaps we should also rotate the North Arrow by 6 degrees. Try that.

25. Bingo. That's better. Resize and position the arrow.

The preceding digression, in which we explored putting in a North Arrow that looked right, is illustrative of how you sometimes must use ArcMap. Very few people know all the software well enough to apply it in all situations. Sometimes you just have to tinker with things.

26. Zoom up on the scale bar. Select it. Fatten it (vertically). Lengthen it. Notice that as it gets longer the numbers change. These numbers reflect what the actual scale of the map would be if it were printed out on paper. Play with the scale bar until it shows 60 miles. Press Zoom Whole Page. *Could only get 40 or 80*

27. Begin, but don't complete, the process of saving by using File > Save As. Notice that ArcMap wants to save this as an MXD (map) document rather than a template. (You could ask that it be saved as a template, but you would give it a new name so as not to lose the default template.) The point here is that you can use the template capability of ArcMap to access, and save for yourself, already-developed data sources to make maps.

28. Save the map as Colorado.mxd in

___IGIS-Arc_*YourInitialsHere*\Map_Making.

Exercise 3-3 (Major Project)

Data Frames

A *data frame* is a "visual container" for layers in ArcMap. Up to now you have been working with data applied to a single data frame—named "Layers" by default. But Arc lets you work with multiple data frames. Among other advantages, this lets you make a map that shows different data sets and also data at different scales on the same map document. In the Layout View, multiple data frames may be seen. In the Data View, only the data sets in one data frame (the active data frame) are displayed.

1. Use Ctrl-N to bring up the New template window again. Under the General tab, change the template to LetterLandscape.mxt. Click OK.

2. Now make certain that you are looking at a Data View, not a Layout View (View > Data View).

3. **Make two new data frames:** Select Insert > Data Frame, and then Insert > Data Frame again. Note that the name of the last data frame you created appears in bold type. This is the *active* data frame. There can be only one active data frame.

4. Change the name of the active data frame from New Data Frame 2 to LINES by right-clicking the name, selecting Properties, pressing the General tab, and typing in the new name in the Name text box of the Data Frame Properties window. Press Apply, then OK.

5. Make New Data Frame active by right-clicking the name and clicking Activate. Now click the layer name LINES to highlight it. Notice that New Data Frame is still in bold, while LINES is not. *New Data Frame is the active data frame; LINES is the selected data frame.*

6. Change the name of New Data Frame, using a different way than you did previously: Click the on name, wait a full second, and click its name again. You will be able to type the new name— call it ALL. Press Enter.

7. Highlight the data frame that is labeled Layers and change its name to POINTS.

Adding Data to Data Frames

You have to be careful here. If you use the Add Data option from the File menu, or use the Add Data button on the Standard toolbar, a data set is added *to the active data frame*. But if you right-click a data frame name and get the menu that says Add Data (showing the Add Data icon), the data will be added to the *selected* data frame, regardless of whether it is active or not. In other words, Add Data operates differently depending on the source of the command.

8. Making sure that POINTS is the active data frame in the Table of Contents, add the *shapefile* ___IGIS-Arc\Trivial_GIS_Datasets\some_points to the POINTS data frame. Now right-click the name some_points in the Table of Contents. Examine the Properties window of the shapefile named some_points. What is the Projected Coordinate System (under the Source tab)? ___WGS 1984 UTM Z 19N___. Using the Measure tool, determine the approximate distance covered by the data set in an east-west direction. ___3185___ meters.

9. Activate the LINES data frame. Add the geodatabase feature data set some_lines_arc (found in

___IGIS-Arc\Trivial_GIS_Datasets\Carto.mdb)

to the LINES data frame. Change the name to just some_lines.

_____ **10.** From ___IGIS-Arc\Trivial_GIS_Datasets, add the *polygon component* of the *coverage* SOME_POLYGONS to the ALL data frame. Change the name to just some_polygons.

_____ **11.** Switch to the Layout View. Something of a mess appears in the map pane. All three of the data sets, albeit at different sizes, are put in the same place. Make the Select Elements pointer on the Tools toolbar active. Move the pointer to the approximate middle of the pane and click. A box with eight blue handles will appear. Click again, and again. What is happening here is that you are clicking inside all the data frames at the same time, so different data frames are being made active. Keep clicking (slowly—don't double-click), noticing now the data frame names in the Table of Contents. With each click a different data frame becomes active, as you can tell from the bold font.

The data frame boxes will not all be the same size, nor will the images inside be the same scale. We will fix the scale problem now and worry about the aesthetics of making the boxes the same size later.

_____ **12.** Make ALL the active data frame. You should see a cursor that has four arrows. Using this cursor you can now drag the active data frame around. Drag the ALL data frame to the lower right corner of the layout. Drag the LINES data frame to the upper right corner of the layout.

If you place the cursor over one of the handles, the cursor changes to a double-headed arrow. If you drag the handle with this cursor, you can resize the data frame and the image inside it. The image's dimensions remain proportional—good, since it contains geographical features and you wouldn't want to change their proportions.

_____ **13.** Make the POINTS data frame active. Drag the handle in its lower right corner towards the upper left corner, reducing the size of the data frame and making its location the upper left portion of the layout.

_____ **14.** Drag a box around the legend, north arrow, scale bar and text, selecting all of those elements together. Drag the whole works to the lower right bottom corner of the map sheet, pretty much on top of the ALL data frame. Select the ALL data frame and drag it to the lower left corner of the layout.

_____ **15.** Using the tools you just learned about, move and resize the three maps so that the boxes around the data look approximately like Figure 3-6. Don't worry if the geographic elements don't seem to be the right size.

Recap: When you see several data frames in the Layout View, you may click in any of them to make it active. Your choice will be reflected by the bold font in the Table of Contents.

_____ **16.** Click the POINTS data frame, in the map pane, to make it active. Check the Table of Contents for the bold font. Notice also that the Map Scale textbox on the Standard toolbar reflects the scale of the map in the data frame.

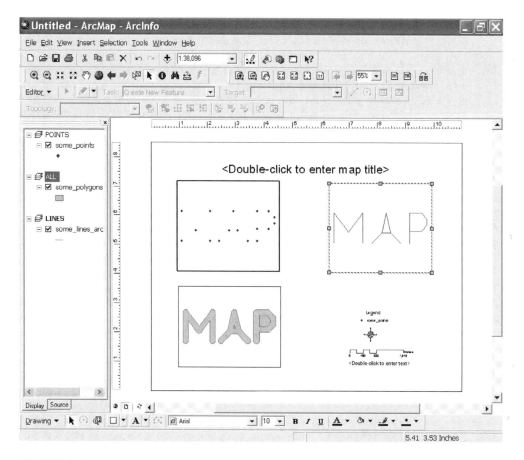

FIGURE 3-6

A Summary of the Graphic Indicators

❑ A blue dashed-line rectangle, with or without handles, indicates a *selected* element or data frame. Multiple elements or data frames can be selected by clicking while holding down the Ctrl key, or by dragging a box.

❑ A line of black dashes around a data frame indicates that it is the *active* data frame.

❑ Just for completeness: A hash-mark border around the active data frame indicates that the frame is "in focus," which means you can edit what is in the data frame in the Layout View. (Normally you can only edit within a data frame when you are in Data View. This is a subject for later.) If you double-click a data frame you put it in focus.

Tinkering with the Map—Scale Bars

___ **17.** Make POINTS the active data frame. Using the Layout toolbar, zoom up on the scale bar. What number appears at its left end? ___0___. Right end? ___1920___. Flip to the Data View.

184

Now select the other Zoom In tool—the one on the Tools toolbar. Drag a box around the points that make the "P." Go back into Layout View and look at the scale bar. Now what are the numbers at the ends of the scale bar? ___◯___ ___9 2∂___. Now select Zoom Whole Page. Notice that you have made the "P" in the POINTS data frame much larger, and that that change was reflected in the legend of the scale bar.

✓

___ **18.** Flip back to the Data View and zoom to the some_points layer. As you return to the Layout View you can see that the scale bar has changed back to about the original number.[9]

✓ **19.** In the Layout View select the POINTS data frame. Drag the box larger. As you do its contents become larger, which is reflected by the scale bar as well.

There are three ideas to be understood here:

❑ The scale bar dynamically keeps up with the true map scale. It may not appear so on the screen, but it will be pretty close to right when the map is printed out. Or when you select the Zoom To 100% (1:1) button.

❑ The *scale* of the geographical elements shown on the map is controlled by the Tools toolbar buttons and by the size of the data frame on the layout, while the *portion* of the map shown on the screen is controlled by the Layout toolbar buttons.

❑ The scale bar is keyed to the POINTS data frame—because POINTS was the original, default data frame (originally named Layers, which you changed to POINTS) that came up with the template. If you wish, you can go through the steps above with one of the other data frames and notice that the scale bar does not change.

✓

___ **20.** Make sure you are in Layout View, whole page. Adjust the three data frames so they are about the same size. Since the scale bar applies only to the POINTS data frame, click the scale bar and drag it inside that data frame.

✓ **21.** Activate the LINES data frame. On the Insert menu, pick Scale Bar. Choose the Stepped Scale Line, since it most resembles the scale bar on the POINTS data frame. OK. Position it in the LINES data frame, shortening it as necessary. Zoom up on the Layout so you can see both scale bars better.

Ahh, heck! The two scale bars don't agree. That would be because the data sets, although they cover about the same territory, are shown at different scales. Let's use a common scale (e.g., 1:50,000) that will let the geographics fit comfortably in each data frame.

✓

___ **22.** Zoom to Whole Page, make Select Elements active, click the POINTS data frame, and type 50000 into the Map Scale text box on the Standard toolbar. (ArcMap knows you mean "1:50000," so that is what appears when you press Enter.) Do the same with the LINES data frame.

✓

___ **23.** The scales of the two data frames are now the same, but the scale bars differ in appearance and length. And since the data really beg to be displayed at the same scale, it seems unnecessary

---

[9]If at any time the geographic elements disappear from a data frame, click on the Refresh View button (next to the Layout View button). If that doesn't restore the image, drag the dividing line between the Table of Contents and the map viewing area well to the right over all the data frames, let it go, and then drag it back to its original position. Another trick to get around this problem is to get to the properties of the data frame and click on a tab or two.

to have multiple scale bars. Click the LINES scale bar, right-click, and from the context menu, select Delete. Drag the POINTS scale bar back where it came from.

_____ **24.** Fix the scale of the ALL data frame to match the other two.

_____ **25.** Add some_points and some_lines (changing the names as before) to the ALL data frame. Save the map as PntLnPlygn_1.mxd in ___ IGIS-Arc_*YourInitialsHere*\Map_Making.

Legends

_____ **26.** Zoom up on the Legend and notice that it applies only to the POINTS data frame. Zoom to whole page and slide the Legend into POINTS' layers lower right corner.

_____ **27.** Make the LINES data frame active. Select Insert > Legend. A Legend Wizard window appears. Arc will let you pretty much click through (Next > Next > Next) and come up with a legend that is right for the data frame. But, as you can see as you do this, the user has lots of control over the appearance of the legend. Slide the legend into the LINES data frame.

_____ **28.** Make the ALL data frame active. Select Insert > Legend. Because ALL contains three different layers you have some options about what the legend will contain. The arrow keys let you move layer names from one pane to another. Arrange it so that the Legend Items pane has only some_lines and some_polygons in it, as in Figure 3-7. Set the number of columns to 2. Finish the wizard. Drag the new legend to the ALL data frame.

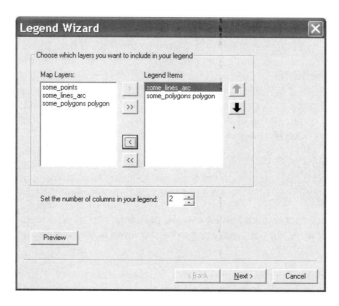

FIGURE 3-7

29. In the Layout View, zoom so you can see the legend of the ALL data frame. In the Data View change some symbology (color, width) of polygon and line features in the ALL data frame. Check the Layout View. Notice that the legend has automatically changed.

30. In Layout View, give the map a title of your choice at the top. Beneath the scale bar put in your name. Save the map as PntLnPlygn_2.mxd in ___IGIS-Arc_*YourInitialsHere*\Map_Making. Keep ArcMap open.

Exercise 3-4 (Mini Project)

Looking at the Plethora of Mapmaking Tools and Options

The ability of GIS systems to make maps has evolved over a period of some four decades. It has become quite sophisticated. With that sophistication has come complexity. I wish I could tell you that there is some underlying grand theory of automated mapmaking, but there isn't. The tools you are about to look at are a monument to ad-hockery.

In this exercise you get a look at only the links to the vast array of options that the serious GIS mapmaker has in ArcMap. What follows will seem something like busy work, but it will serve as a reference for you later, should you become more interested in making maps.

1. With PntLnPlygn_2.mxd in Layout View, click the ALL data frame to select it and make it active. Now right-click the data frame. A menu of 15 possibilities appears, all of which apply to this particular data frame. List them here:

1. Add Data
2. New Group Layer
3. Copy
4. Paste Layer
5. Remove
6. Turn All Layers On
7. Turn All Layers Off
8. Expand All Layers
9. Collapse All Layers
10. Reference Scale
11. Advanced Drawing Options
12. Labeling
13. Convert Features to Graphics
14. Activate
15. Properties

✓ Run the cursor over each possibility that has a triangle mark to note the options within them. We will use a couple of them later.

— **2.** Click Properties to bring up the data frame Properties window, which you have seen before. The window has ten tabs. List them.

1. General
2. _Annotation Groups_
3. _Data Frame_
4. _Extent Rectangles_
5. _Illumination_
6. _Frame_
7. _Grids_
8. _Size + Position_
9. _Coordinate System_
10. _Map Cache_

— **3.** Close the Data Frame Properties window. Click somewhere away from any data frame. Now right-click in that same place. A menu with 13 possibilities appears. These choices relate to the map sheet as a whole. List the choices.

1. Page and Print Setup
2. _Zoom Whole Page_
3. _Go Back to Extent_
4. _Go Forward to Extent_
5. _Toggle Draft Mode_
6. _Cut_
7. _Copy_
8. _Paste_
9. _Delete_
10. _Select All_
11. _Unselect_
12. _Zoom_
13. _Options_   other

Again, run the cursor over the choices that have subchoices and examine them.

____ **4.** Click Options to bring up an Options window. You should see nine tabs. List them.

1. General
2. _____ Data View
3. _____ Layout View
4. _____ Geoprocessing
5. _____ Raster
6. _____ CAD
7. _____ Table of C
8. _____ Data Interoper
9. _____ Tables

You get the idea: myriad possibilities for actions on your part. None of them particularly complicated, as it turns out, but mind-boggling when taken all together. What we have just done is to let you take a first cut at seeing just the names that suggest the possibilities that lie underneath.

In what follows you will experiment with some of the data frame properties and layout options.

____ **5.** Right-click the layout page, somewhere outside all data frames. Then select Options > Layout View. Specify: Don't stretch the contents when a window is resized. Show the scroll bars, horizontal guides, and vertical guides. Show a dashed line around the active data frame. Show the rulers, but not the grid. Make the smallest increment one tenth of an inch. Snap elements to the guides, but nothing else. Click Apply (if you changed anything), then OK.

____ **6.** Right-click the horizontal ruler at the 1-inch mark. Click Set Guide. Notice the little arrow that appears and the light blue vertical guide line. With the menu, clear that guide. Set it again. If the guide isn't exactly on the 1-inch mark, use the double-headed cursor and slide it over.

____ **7.** Actually, if you just click a point on the ruler, a guide will be set. Set additional guides on the horizontal ruler at 5 inches, 6 inches, and 10 inches. Set guides on the vertical ruler at 1 inch, 3.5 inches, 4.5 inches, and 7 inches. You will make the data frames fit into the rectangles created by these guides.

You may have noticed that every time you change the size of a data frame, the scale changes. If you want to keep, say, a constant 1:40,000, you have to keep typing it in the scale text box. There is a way around this.

____ **8.** Pick a data frame and open its Properties window. Click the Data Frame tab. Click the Fixed Scale radio button and type in 40000 in the rightmost text box. Click Apply, then OK. Notice that the Standard toolbar now shows 1:40,000 but the number is grayed out—fixed at that value by the Properties window. Prove this by greatly enlarging the data frame. Its contents will not change size. Return the data frame to a reasonable size. Set the scale on the other two data frames to 1:40,000 as well.

___ **9.** Grab a corner of a data frame and bring it close to an intersection of guidelines. Notice that the boundaries of the frame snap to the guidelines. Fix up all three data frames so they fit the guides.

___ **10.** Use Pan on the Tools toolbar to slide the geographics up toward the top of each data frame— center it as well as you can. (Notice that while Pan is still an active tool, the zoom controls are disabled—because you fixed the scale.)

___ **11.** Use Select Elements to adjust the positions of the legends of each data frame. If you are a perfectionist you can fix up the sizes of the legend texts so they are all the same. To do this: select a legend, right click on the selected legend, choose Properties to bring up a Legend Properties window, and change virtually anything you want about the appearance of the legend. Cancel the Legend Properties window if you brought it up.

In the lower right corner of the layout, you have a north arrow, a scale bar, and a place for your name. In the next two steps, adjust the position of each of these so the layout looks like Figure 3-8.

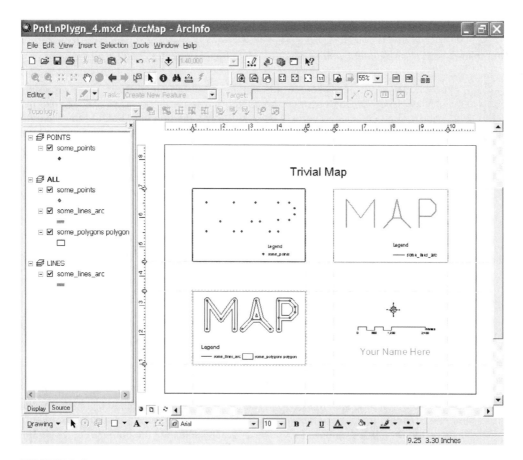

FIGURE 3-8

_____ **12.** Zoom up on the lower right quadrant of the map. Click the `Select Elements` tool. Click the north arrow to select it. Right-click there, then click `Nudge > Nudge Left`. The graphic moves 1 pixel to the left. Actually, if you simply tap the arrow keys on the keyboard, you can move the graphic around by small increments. If you hold an arrow key down, you can move the graphic quickly. Place the north arrow where you want it.

_____ **13.** When you select text, you do not get the sizing handles. Text sizing is done by a `Properties` window, which you get by double-clicking the text (or get to by right-clicking and selecting Properties). Do that, and then click `Change Symbol`. In the `Symbol Selector` window, make your name bold in 24-point type, with the color red. OK everything and observe the results. Move your name around so that it appears properly placed.

_____ **14.** Zoom to the whole page and observe the finished product. Save it as PntLnPlygn_3. Print the map. Close ArcMap.

As you can see, many options for mapmaking exist within ArcMap. We have but scratched the surface.

Exercise 3-5 (Major Project)

Making a Map of the Wildcat Boat Data Sets

We will be using the Wildcat Boat Data a lot in the ensuing exercises in the book. Just to be sure you are working with the proper data, you will delete the entire Wildcat_Boat_Data folder, in which you created the feature dataset LANDCOVER, and replace it with one that is on the CD that comes with the book.

_____ **1.** Start ArcCatalog. Navigate to the folder

[___] IGIS-Arc_AUX\Wildcat_Boat_Data

to be sure that you have it.[10]

_____ **2.** Assuming that you do have [___] IGIS-Arc_AUX, navigate to your folder

___IGIS-Arc_*YourInitialsHere*\Wildcat_Boat_Data and select it. Click on the `Delete` icon on the Standard toolbar or press the delete key. Confirm that you want to delete after making sure that it is only the Wildcat_Boat_Data folder that you are deleting.

_____ **3.** Navigate to the folder

[___] IGIS-Arc_AUX\Wildcat_Boat_Data and select it. Select: Edit> Copy.

Navigate to the folder

___IGIS-Arc_*YourInitialsHere* and select it. Select: Edit> Paste.

---

[10]The folder IGIS-Arc_AUX is on the CD that came with the book. If it is not on the computer's hard drive in the [___] path you should copy it there with the operating system.

——— **4.** Using the techniques and principles you learned previously in Exercises 3-3 and 3-4, make a comprehensive layout, showing the data for the Wildcat Boat project. Use the geodatabase in

___\IGIS-Arc_*YourInitialsHere*\Wildcat_Boat_Data.

The layout you create should contain *three separate data frames;* name them:

1. SOILS

2. LANDCOVER and ROADS

3. SEWERS and STREAMS

Snap the data frame edges to guidelines as you did in the previous exercise. Use a consistent scale throughout. Use distinctive symbols to indicate features. Use labeling as appropriate.

——— **5.** Save the map in ___IGIS-Arc_*YourInitialsHere*\Map_Making. Print the map.

Exercise 3-6 (Major Project)

Enhancing Communication: Styles, Layer Files, Reports, Charts, and Graphics

Somewhere in the conceptual space between raw data and finished maps lie the ideas of styles, layer files, and map templates. We've already looked at templates, which may or may not have data associated with them.

Layer Files

Layer files are based on raw data files. Basically, a layer file tells ArcMap how to draw a data file—what symbols and colors to use. As you know, if you add a raw data file in ArcMap, the software makes random choices as to how features are drawn. Sometimes this is satisfactory; more often it is not, if you have serious intentions of examining and analyzing the data. Let's look at an example.

——— **1.** Use ArcCatalog to copy the shapefile KY_Streams_spf from

[___]IGIS-Arc\Kentucky_wide_data to

___IGIS-Arc_Your*InitialsHere*\Map_Making.

——— **2.** Start ArcMap. Add

___IGISArc_Your*InitialsHere*\Map_Making\KY_Streams_spf

to the map.

This is a fairly large data set (about 60 megabytes) that contains information about the streams of Kentucky, from the largest (Order 8) to the smallest (Order 1). When two streams of the same order (e.g.,

Order 1) come together, they make a stream of the next higher order (i.e., Order 2). But if two streams of different order come together (e.g., Order 6 and Order 5), the output is just a stream of the higher order (i.e., Order 6).[11] Therefore "stream order" cannot be considered true ordinal data, in terms of stream size, volume, rate of flow, and so on, even relative to those streams above it which flow into it. That is, the Order 5 stream mentioned previously might have a greater flow volume than the Order 6 stream it flows into.

____ **3.** Open the KY_Streams_spf attribute table. How many stream segments are there? _____.

____ **4.** All the KY_Streams_spf are shown with a single color. Suppose we want to see the smaller streams in a lighter blue and the larger ones in a darker color. The attribute table has a column labeled ORDER, which has values from 1 to 8. So let's change the way the streams are drawn: Right-click the shapefile name, then click > Properties > Symbology > Categories > Unique Values. In the Value field select ORDER_. Then click Add All Values.

____ **5.** Double-click the symbol for Order 1 streams. Pick the color Sodalite Blue with a width of 1 to symbolize this stream. For Orders 2 through 8, use width values of 1.33, 1.67, 2, 2.33, 2.67, 3, and 3.33. Pick colors of blue that are darker for higher orders, ending with Dark Navy for order 8. The result will look something like Figure 3-9.

You noticed that it took some time and concentration to symbolize the map in this way. Should you remove KY_Streams_spf from the Table of Contents, all that work would be lost. A layer file (extension LYR) will preserve the symbology.

____ **6.** Right-click KY_Streams_spf and select Save As Layer File from the menu. Place KY_Streams_spf.lyr in

____ IGIS-Arc_*YourInitialsHere*\Map_Making.

____ **7.** While we are at it, let's also make a simple layout of the data and save it as a map named KY_Streams.mxd. Go to Layout View, then click File > Save. Put in the name, and click Save again.

____ **8.** Click the on New Map File icon without saving anything. Go to Data View. Add KY_Streams_spf.**shp** from ___IGIS-Arc_*YourInitialsHere*\Map_Making. Notice that the KY_Streams data set is drawn all in a single color, with no size differentiation. Remove the shapefile.

____ **9.** Click the New Map File icon without saving anything. Go to Data View. Add KY_Streams_spf.**lyr** from ___IGIS-Arc_*YourInitialsHere*\Map_Making. Notice that the KY_Streams data set is drawn as you symbolized it.

There is an important caveat to be mentioned here. The layer file does not contain data; it only contains the instruction as to how the data set is to be drawn.

---

[11]This is according to the Strahler method of stream order analysis. In another method, Shreve, headwater arcs are also assigned an order of 1. But when two or more arcs converge, then the arc downstream of the confluence is assigned an order equal to the sum of the orders of the upstream arcs. Stream analysis is discussed in Chapter 8.

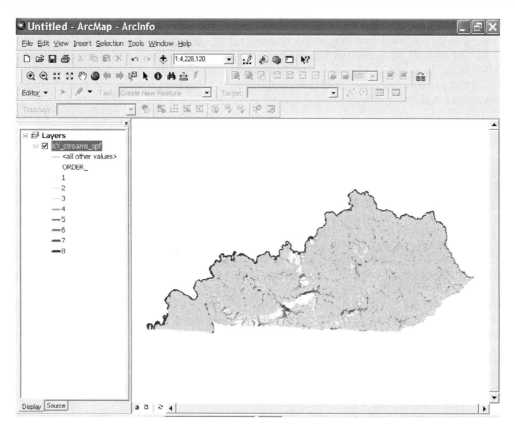

FIGURE 3-9

_____ **10.** Use Windows to navigate to the folder

  ___IGIS-Arc_*YourInitialsHere*\Map_Making.

  Ask for details of the files there. What is the size of:

  KY_Streams_spf.shp _____
  KY_Streams_spf.dbf _____
  KY_Streams_spf.lyr _____
  KY_streams.mxd _____

The SHP file is the geographic feature data. The DBF file holds the attribute data. Together they constitute more than 60 million bytes. The LYR file, in contrast, occupies a mere 14,000 bytes. Obviously, the layer file does not contain the KY_Streams data. Also, the map file is also way too small to hold the actual data.

____ **11.** For a more dramatic illustration that layer files do not contain the data, do the following. (If you should want to preserve the ability to draw the KY_Streams with the layer file, use ArcCatalog to make a new folder and copy both the raw data and the layer file into it.)

In ArcMap, start a new map file. In ArcCatalog, click the word Catalog and press F5 to refresh the catalog tree. In ArcCatalog delete KY_Streams_spf.SHP from

___IGIS-Arc_*YourInitialsHere*\Map_Making.

Now try to Preview the Geography of KY_Streams_spf.LYR by clicking its name. What is the message you get? _____

_____

____ **12.** In ArcMap, click the New Map File icon. Try to add data: KY_Streams_spf.LYR. Interestingly, you get the table of contents and the symbol references. But the raw data set name itself presents a gray check mark and a little red dot, which means the underlying data set is missing. Next, try to open the file KY_streams.mxd. Again, nothing doing.

Note that both the layer file and the map file have been ruined because you made the data on which they depend inaccessible.

In summary, layer files can be useful in several ways. You can make the data available to others—through a network or e-mail—and be sure that the data will be portrayed as you have prescribed. I will warn you, however, that this can be tricky. Obviously you have to send the data along with the layer file. But just as important, the layer file must be able to know precisely where the underlying data set is. Suppose both the data set and the layer file resided on C:\Some_Folder and you sent them to someone who loaded both in D:\Some_Folder; the layer file might not be able to access the data set. There are things you can do (fairly easily—check the help files) to solve this problem, but you have to be careful to preserve or remake the linkages between data sets and layer files.

Styles

Styles basically let you draw maps using the colors, symbols, and patterns developed by other people and organizations. When you have drawn maps before, you have been using a style developed by ESRI. In fact, it is difficult to separate the software, which basically lets you draw points, lines, and polygons—admittedly in a myriad of colors—from the ESRI predeveloped symbols. Let's start your understanding of styles by eliminating all of them.

____ **13.** In ArcMap, click the New Map File icon on the Standard toolbar. Add the *polygon component* of the coverage SOME_POLYGONS, which you find in the ___IGIS-Arc_*YourInitialsHere*\Trivial_GIS_ Datasets folder. Use Data View. Set the Table Of Contents tab to Display.

____ **14.** Set the software so that it uses no styles at all by clicking Tools > Styles > Style References and then clearing all the boxes you can. Which one can't you clear? _____. Click OK.

____ **15.** In the Table of Contents, right-click the polygon symbol. You may recall that usually that brings up an array of distinct colors (e.g., Medium Apple, Sodalite Blue, Mars Red, and so on). But those are part of the ESRI style. Now what you have is the Color Selector window, which you met in Chapter 2, lets you select any color the computer is capable of producing—millions of them[12]—but without benefit of being able to name the color or easily select it again. The Color Selector window lets you define a color in the most basic way. You can move the R, G, and B sliders to determine the amount of red, green, and blue, each on a scale of zero to 255, that go into making up the color that will wind up on the polygons. As you move the sliders, the lower left rectangle in the window shows the new color. Adjacent to it is the current color that is to be changed. In the area of the window just up from the bottom is a box showing, as a continuum, all the colors. Clicking or dragging in this box is also a way to select a color. Try this out, watching the text boxes and the slider bars. Move the cursor both horizontally and vertically. Finally, pick a garish yellow by typing in the boxes—say, R255, G255, B99.

____ **16.** Click the polygon symbol. This brings up a Symbol Selector window. Here you can modify the outline width and the outline color. You could also go to Properties and see another bunch of options. In fact, you could doubtless spend half a workweek exploring the possibilities that ArcMap gives you in the color arena. For now, just press the More Symbols button.

This menu shows you the different symbol sets that come with ArcMap. This list is a subset of the `Style References` boxes that you turned off earlier; this is simply another way to get to that list.

Adding and Using a Style

____ **17.** Clear off the menus and windows. Add the shapefile some_points from the ___IGIS-Arc_ *YourInitialsHere*\Trivial_GIS_Datasets folder. You get the generic dot. Bring up the `Style References` window again. Let's go for something really ridiculous: click 3D Trees and OK. This now gives you the capacity for replacing the generic dot with elements a style sheet called 3D Trees. In the Table of Contents, click the some_points symbol. A Symbol Selector window appears with a plethora of tree images. Slide down through the list, taking a botany lesson as you go. How many types of trees are there?[13]

____ **18.** Pick a Rocky Mountain Maple (from the top row). Click OK. It's pretty hard to see that you have made any difference in looking at the Data View. Return to the Symbol Selector window. Pick the maple again but change the size to 50 points. The symbol has taken on some form. To see it in all its pixilated glory, change to the Layout View and zoom up on a point, using the zoom on the Layout toolbar.

____ **19.** Zoom to the whole page and go back to the Data View.

____ **20.** All the capabilities you had with the software before you have now. Only the symbols you may use have been curtailed. Let's use a different symbol for the points in the M, and the A, and the P.

____ **21.** Bring up the Layer Properties window for some_points. Click Symbology > Categories > Unique Values. Make the value field TYPE_, and add All Values. Double-click the symbol next to the M

---

[12]256 * 256 * 256—you do the math.
[13]Just kidding.

to bring up the Symbol Selector window. At the top of that window you see All in the pull-down Category menu. (This is a different kind of category than on the Layer Properties window.) Pick SUCCULENT to replace All. Select the Century Plant and again make the size 50. Click OK, and OK again. Check out the Data View. Now put the Jade Plant in the A and the Cereus in the P. Observe the results.

_____ **22.** Click Tools > Styles > Style References. Turn off 3D Trees and turn on ESRI.

Reports

As useful as an attribute table is, its format leaves a lot to be desired. It is seldom reasonable to print out a large table in regular form. What is very useful at times is a summary of the information in the table. ArcMap gives you the ability to generate textual reports from an attribute table. You will see this ability demonstrated in a three-step process. First you will create a second table by using the Summarize feature, available by right-clicking a column in a table. Then you will make a report from the second table. Finally you will put that report on a map. You may recall that the Wildcat Boat data contained a personal geodatabase feature data set name Sewers. It consisted of a few linear segments representing lengths of sanitary sewer pipe of two different diameters: 60 inches and 45 inches.

Assume that you need a map of the sewers and want to place a report on that map showing the total lengths of each diameter pipe. You might proceed as follows.

_____ **23.** In ArcMap, start a new map and go to Layout View. Add as data the Sewers feature data set from

___IGIS-Arc_*YourInitialsHere* [continuing on the next line]

\Wildcat_Boat_Data\Wildcat_Boat.mdb\Line_Features

_____ **24.** Open the attribute table of Sewers. Notice that there are four lengths of pipe 60 inches in diameter and two lengths 45 inches in diameter. Just as in Data View, you can graphically select features in the Layout View and see the selections reflected in the table. Try it. Also you can select records in the table and see the results highlighted on the map. Using Ctrl-click, highlight the four pipes of diameter 60 inches. Right-click the Shape_Length column heading and pick Statistics. From the Selection Statistics of Sewers window, determine the total length of 60-inch pipe. _____ meters. Click Options > Switch Selection. What is the total length of 45-inch pipe? _____ meters.

Perhaps you would like to get this information into a report and onto the map, as well.

_____ **25.** Clear all selections. Right-click over the DIAMETER heading and pick Summarize. The field to summarize should read DIAMETER. Skip box 2. Accept the default output table, named Sum_Output. Click OK. When asked if you want to add the results to the Table of Contents of the map, choose Yes. Open the table.

_____ **26.** The new table, Attributes Of Sum(mary)_Output, tells you the numbers of segments but little else. Let's try again. Close the table and remove it from the Table of Contents.

_____ **27.** Again, right-click over the DIAMETER heading and pick Summarize. This time in box 2 expand Shape_Length and check `Sum`. Continue as before. This time you see that you get an additional field: Sum_Shape_Length. Check to see that the numbers you wrote above are the same as those in the table. Write the name of the table here: _____

Now that you have a table that contains the needed information, you may make a report. ArcMap has considerable report making capability.[14] We will create only the most elementary example, primarily to show you that report generation capability exists.

_____ **28.** Click `Tools > Reports > Create Report` to bring up a Report Properties window. Write the names of the five tabs you see. _____, _____, _____, _____, and _____.

_____ **29.** Click the Fields tab. In the Layer/Table drop-down menu, pick the table name that you wrote previously. Move all the available fields to the Report Fields area by clicking the right-pointing double arrow. Now move the OBJECTID back by highlighting it and using the left-pointing single arrow.

_____ **30.** Click the Sorting tab. In the DIAMETER row, click the Sort column and pick Ascending.

_____ **31.** Click the Display tab. You do want to see the field names (under Elements). Click Generate Report.

_____ **32.** The report that you see in the Report Viewer has the information you required. Perhaps you now decide that you want to know the total length of pipe. That capability is easily at hand. Dismiss the Report Viewer to return to the Report Properties window. Click the Summary tab. You want the sum of the Sum_Shape_Length field. Click Generate Report.

_____ **33.** Notice that Sum_Shape_Length Sum[15] has been added to the report at the end.

As you can see, there are myriad possibilities in report writing. If you want, continue to experiment with the options presented to you under the five tabs. You are not committed to any report you generate in the Report Viewer.

_____ **34.** Once you have the report you want in the Report Viewer window, print a copy. Now add the report to the Layout View by pressing Add. Dismiss or minimize windows that you accumulated along the way until you can see the Layout.

_____ **35.** The report is a full page, so it is as big as the Layout. Using `Select Elements`, select the page (it will have four handles, not six) and slide it down so the text fits above the bottom margins. Size it so it fits between the vertical map outline limits. Use the Pan control on the `Tools` tool-bar to slide the geographics up toward the top of the page. Label each segment with the diameter of the pipe with 10-point type.[16] The result should look something like Figure 3-10.

_____ **36.** Display the map at 100 percent of the size it would appear on a page. Using the scroll bars, look around the map to see that things are about the right size. Print the map. Save it in the Map_Making folder as Sewer_Specs.mxd.

---

[14]Further, as an option, you can use Crystal Reports if they are included in your ArcMap configuration.

[15]A little confusing. One "Sum" is short for summary, while the other is a synonym for total.

[16]If this doesn't label each of the six features, go to the Labels tab in the Layer Properties window. Under Placement Properties, choose Place One Label Per Feature.

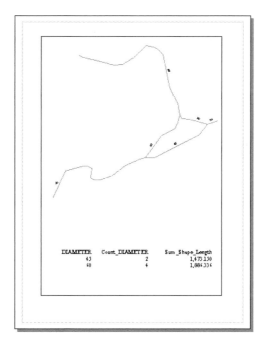

FIGURE 3-10

Charts

Another form of communication—neither text nor map—can also be created by ArcMap: the chart or graph. The software is capable of producing graphs and charts of both two- and three-dimensional appearance.

____ **37.** On a new map in Data View in ArcMap, add the personal geodatabase feature data set named soils from

___IGIS-Arc_*YourInitialsHere*\Wildcat_Boat_Data\ . . . you know the drill.

____ **38.** Open its attribute table and summarize the soil suitability (SUIT), including the sum from the Shape_Area column. Add the table to the Table of Contents of the map. Dismiss the Soils attribute table. Open the Sum_Output table. It should look like Figure 3-11.

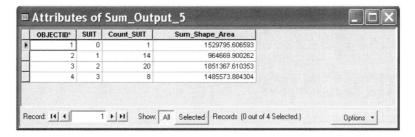

FIGURE 3-11

____ **39.** Select Tools > Graphs > Create to bring up Step 1 of the Graph Wizard. Click Column and look at the many types of column graphs you can create. Click to select the one at the top left. Click Next. In Step 2, in the `Choose The Layer Of Table Containing The Data` drop-down menu, pick Sum_Output. Check only the box next to `Sum_Shape_Area`. Graph the data using `Records`. Click Next. In Step 3 change the title to `Square Meters of Soil Suitabilities`. Uncheck `Show Legend`. Check `Label X Axis With` and get SUIT into the text box. Observing the graph preview, set the bar gap so it suits you. Click Finish. (If the x-axis shows 1, 2, 3, and 4 then a software error has not yet been fixed.)

____ **40.** Minimize, but do not close, Attributes Of Sum_Output.

____ **41.** Assuming that the graph looks as it should (see Figure 3-12), slide your cursor around on it. If you click a column an Identify Results window will appear, showing the attribute values of the table from which the graph was made. Dismiss the Identify Results window.

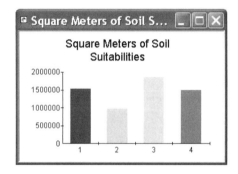

FIGURE 3-12

____ **42.** Right-click the title bar of the graph. Look at the options, then click `Show On Layout`. Dismiss the window that contains the graph.[17] In the Layout View, drag the graph to the bottom of the page. Using the handles on soils_polygon, resize it so that it doesn't conflict with the graph. Size and adjust each so that a reasonable amount of layout space is devoted to the data and the graph.

It would be helpful if the soils map showed the polygons with the same colors as the graph.

____ **43.** Bring up the Soils Properties window. Click the Symbology tab. Show `Unique Values` under `Categories`. Make the `Value Field` SUIT and `Add All Values`. Now, by double-clicking the color patch next to the zero value, make the polygons with SUIT equal zero the same color as the zero column on the graph. Do the same for values one, two, and three. Click Apply, then OK. Double-check that the SUIT code is correct by using Identify Results on some polygons. Save the map in

___IGIS-Arc_Your*InitialsHere*\Map_Making, with the name Soils_with_Graph_1. It should look somewhat like Figure 3-13***.

---

[17]You could save the graph itself if you wanted. Right-click the title bar and proceed.

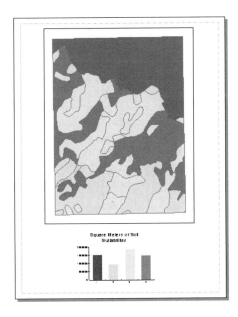

FIGURE 3-13

Suppose you want to see the relative amounts of each soil suitability. A pie chart will be your choice.

___ **44.** Start the Select Elements tool. Click the graph part of the layout. Delete it. Again choose `Tools > Graphs > Create` (or Alt-T,H,C) to bring up Step 1 of the Graph Wizard. Click Pie. Pick the 3-D-looking disk. Click Next. The table again should be to one you created with Summarize. You want to graph Sum_Shape_Area. Click Next. Make the Step 3 window look like Figure 3-14. Click Advanced Options, then click 3-D and play with the up-down slider. Click Cancel or click Apply Now and OK. Click Finish.

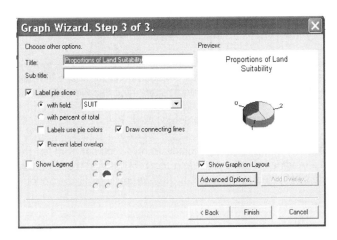

FIGURE 3-14

___ **45.** Arrange the layout so that the graph is again at the bottom, and save the map with the name Soils_with_Graph_2. Print the layout if you want. Delete the graph from the layout.

As you may recall, soil with suitability "zero" is water. Suppose you don't want it included in the pie graph. You can take advantage of the box in Graph Wizard Step 2 that lets you use selected records in the driving table, rather than using all the records.

___ **46.** Restore the Attributes Of Sum_Output table. Select records with SUIT values 1, 2, and 3. Minimize the table. Start the graphing wizard and proceed as before, making sure that the Use Selected Records box is checked. Observe that only the top three soil suitabilities are represented. However, the colors are likely to be wrong, so correct them on the map. The colors of the graph columns cannot be easily changed. Position the graph on the layout and save it as Soils_with_Graph_3.

Graphics

As a last topic, you can put graphic or text information into a data frame directly from a variety of sources. This subject really gets us away from our intended goal—preparing you to do analysis with GIS—but it is a major feature of the software that you should know about, so we will look at it briefly.

Placing ancillary information on a data frame is done primarily with two sets of controls. The first is Insert on the Main menu. The second is the Drawing toolbar.[18]

___ **47.** In ArcMap, Data View, add the shapefile

[___] IGIS-Arc\river\Boat_SP83.shp

to a new map.

___ **48.** Also add the following data sets, in this order:

[___] IGIS-Arc\river\wtp_spn [the point component, water treatment plants]
[___] IGIS-Arc\river\COLE_DRG.TIF
[___] IGIS-Arc\river\COLE_DOQ64.JPG
[___]IGIS-Arc\Kentucky_wide_data\KY_Streams_spf.shp

___ **49.** Make the point symbol for wtp_spn a bright red square of size 10. Make the Boat_SP83 symbols bright green circles of size eight. Pan and Zoom the image until all of the DOQ and the GPS track are in. Your data frame should look something like Figure 3-15. Save this as

___IGIS-Arc_*YourInitialsHere*\Map_Making\LWP1.mxd.

___ **50.** Turn your attention to the Drawing toolbar. Run your cursor from left to right over each of the buttons while reading the status bar and the ToolTips. On those buttons with drop-down menus (little triangle symbol just to the right of the button), look at the options with ToolTips.

---

[18]If the Drawing toolbar is not in the ArcMap window, choose: View > Toolbars > Draw.

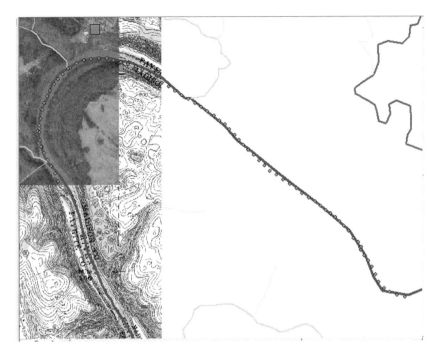

FIGURE 3-15

Suppose you want to put some identifying text on the data frame, pointing to features.

_____ **51.** On the Drawing toolbar, find the "A" (add text to the map by typing it in) and access the menu next to it. Find the callout box, and click. With the Callout cursor click the easternmost point of the GPS track. In the textbox that appears type Beginning of GPS Track and press Enter. Drag the box below the point so that all the text is within the data frame. Click away from the box to deselect it.

_____ **52.** Repeat the procedure of adding a callout box—this time referencing the last point of the GPS track. But instead of typing in the text box, click somewhere off the box, then double-click Text to bring up a Properties window. In the Text area type:

End of
GPS Track

then click Apply and OK. Drag the callout box to the right, off the DRG. Deselect it.

_____ **53.** Make a callout box that says "Filtration Plant" pointed at the facility in the northeast corner of the DOQ.

_____ **54.** Use Save As to save the data frame as a map named LWP2.mxd.

We have available an oblique aerial photo of the water plant. No geographic coordinates come with it. It is just a picture. But we can add it to the data frame.

____ **55.** Choose Insert > Picture and navigate to

[____] IGIS-Arc\river\Lexington_Water_Plant.JPG

Open. Drag the photo to the upper right corner of the data frame.

____ **56.** Bring up the menu next to the callout icon. Click the "A." Click the data frame in the lower right quadrant. Type Lexington Water Plant and press Enter. Start changing the characteristics of the text by double-clicking it. Press the Text tab and then select Change Symbol. Make the text red, 20 points, Arial, bold, and underlined. Click OK, and OK again.

____ **57.** Change to Layout View. You probably would like to rearrange some of the elements, given the change in format. But you notice that you cannot select any element except the entire data frame. ArcMap doesn't let you edit a data frame in a Layout unless you put that data frame "in focus." Do that by double-clicking the data frame (or right-clicking it and selecting Focus Data Frame). Note the hashed border around it.

____ **58.** Select the picture and move it up in the layout. Select the text title and move it down, centered.

____ **59.** Since the title crosses some features, perhaps you want to give it some background. Click the New Rectangle button and drag a rectangle over the title, covering it up. Click the word Drawing on the Drawing toolbar, pick Order, and send the selected rectangle to the back. Pull up the menu for fill color (on the Drawing toolbar) and make the color of the rectangle Lapis Lazuli.[19] Change the color of the text in the box to White. If you have any cartographic design experience, shake your head over what a wretched mapmaker the author is and save the map as LWP3.mxd, after fixing it up to suit yourself.

In what follows you will experiment with some of the drawing tools on a blank data frame. Feel free to vary the process and to experiment.

____ **60.** Click the New Map File icon and go into Data View.

____ **61.** In the New (shape) menu try out the different possibilities. With New Rectangle, New Circle, and New Ellipse, just click and drag. With New Polygon, and New Line, just click to make successive vertices; double-click to end the graphic. New Curve is particularly fun. Again just click to make successive vertices; double-click to end the graphic. With New Freehand simply drag the cursor around.

____ **62.** With any of these graphic elements, you can, after selecting them, right-click and change their positions by rotating 90 degrees or flipping around an axis. You can also bring up a window to get information about and/or change their properties.

____ **63.** Pick a graphic that you have made and select it. Click Zoom to Selected Elements on the Drawing toolbar. Zoom back to full extent. Select another element and experiment with the Rotate (on the Drawing toolbar), by dragging the cursor anywhere in the data frame.

____ **64.** Select another shape. Toward the right end of the Drawing toolbar, change the Fill Color or Line Color to whatever you want.

---

[19]The color of an exotic gem. Also University of Kentucky Wildcat Blue.

_____ **65.** Make some text. Change its font to Courier New, 16 point. Experiment with the New Text options. New Circle Text, for example, will, if possible, place any text you type inside a circle that you initially create. Once the text is no longer selected, the circle's bounding square disappears.

_____ **66.** Click New Map File. Find the button for New Splined Text and press it. Make a spline (like a snake) starting in the northwest corner of the data frame: click, move an inch or so to the right, and click again; continue until you have made a spline that looks somewhat like Figure 3-16. Double-click to end the spline. Type the following into the text box, without using the Enter key except at the very end:

A quick move by the enemy may jeopardize six fine gunboats. Now is the time for all good men and true to come to the aid of their party. The quick brown fox jumped over the lazy dogs.

_____ **67.** When you press Enter you should see the text following along the first part of the spline you made. The text should remain selected. Change the font color. Change the font size to 16. Click away from the text to clear the selection. Observe. See Figure 3-17.

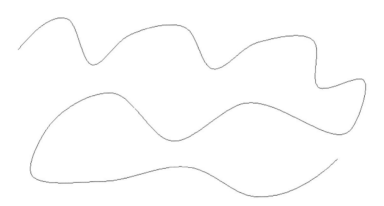

FIGURE 3-16

FIGURE 3-17

___ **68.** Click New Map File. Make a polygon using the New Polygon option. Click Edit Vertices in the Drawing toolbar. Drag the little cyan squares around to reshape the polygon. Use the What's This button on the Standard toolbar to read about deleting and adding vertices with the Edit Vertices tool. Experiment with doing both.

___ **69.** Create a new curve. Edit Vertices again. By dragging the cyan boxes around, you can change the locations of the vertices. By dragging the purple boxes, you can change the shape of the part of the curve that goes into a vertex. By dragging a point on the curve that is not a vertex, you can move the entire curve.

___ **70.** Click New Map File. Add the Layer Sewers from Line_Features of

___IGIS-Arc_*YourInitialsHere*\Wildcat_Boat_Data\Wildcat_Boat.mdb.

Draw a new rectangle so that it covers some of the northern pipe. Just so you can see through it, double-click the rectangle to bring up a Properties window, click the Symbol tab, and make its fill color No Color. Click Apply, then OK. Be sure the rectangle is selected—you will see the eight blue handles. Now from the Selection menu pick Select By Graphics. The northern pipe should become highlighted. If you open the sewers attribute table, you will see its record highlighted as well. In the Selection menu, click Options. Note the various ways you can define the selection process. In the Selection menu, select Clear Selected Features. Dismiss the attribute table.

Making Graphics out of Geographic Features

___ **71.** Right-click Sewers in the Table of Contents. Click Convert Features To Graphics. In the window, specify Convert All Features Of Layer Sewers. Only draw the converted features. Click OK. Click a line of sewer pipe to select it. You will find that you can now move it or change it around like any other graphic. Dismantle the sewer system and pile the pieces up in the southeast corner of the data frame.

Finally, add splined text to the Lexington Water Plant map.

___ **72.** Click New Map File. Open the map file

___IGIS-Arc_*YourInitialsHere*\Map_Making\LWP3. In Layout View, put the data frame in focus. With the Layout toolbar, zoom up on the DOQ. Pick New Splined Text from the Drawing toolbar. Make a spline paralleling the curve of the river, about an inch to the each to the east of the GPS track, starting with the fix at the end of the track and continuing up to the edge of the DOQ. In the text box, type Kentucky River and press Enter. Change the text color to White. Change the text size to 16. Slide and rotate the text until it fits nicely in the bend of the river. If you don't like the result, delete it and try again. Zoom to the whole page, and save the map as LWP4.

As you can see, ArcMap has a remarkable number of tools that aid you in making maps. Admittedly these tools are not as extensive as those in various drawing programs, but don't forget: You retain the advantage of have a dynamic map with all the "intelligence" that GIS gives.

___ **73.** Just to demonstrate that a GIS map is really different, use the Identify tool to click the red square that represents the Lexington Water Plant. As a result, you will see an extensive amount of information on this particular plant. Now right-click wtp_spn point and select Zoom To Layer. These are the water treatment plants in the state. Open the wtp_spn point attribute table and observe the amount of information available. Try that with your drawing program! Close ArcMap.

Exercise 3-7 (Review)

Checking, Updating, and Organizing Your Fast Facts File

The Fast Facts File that you are developing should contain references to items in the following checklist. The checklist represents the abilities to use the software you should have upon completing Chapter 3.

Important note: This checklist is on the CD-ROM that accompanies the book. It is available in Microsoft Word format. Rather than typing or writing by hand the text that follows, you can copy and paste it into your Fast Facts File from the CD-ROM file.

___ The Layout View (contrasted with the Data View) is

___ A map template is

___ The file extension of a map template is

___ The name of the map template that is the basis for a blank document is

___ To get a variety of map templates by

___ A major toolbar used to produce maps is

___ The map templates that contain data are located

___ Ways of changing from Data View to Layout View are

___ To determine the coordinate system of data in the data frame is

___ The types of coordinates available in the Data View are

___ The types of coordinates available in the Layout View are

___ Zoom controls in the Data View and the Layout View

___ To rotate the map display

___ Elements of the map that may be added are

___ To rotate the north arrow

___ Care has to be taken saving a map created from a template with data because

___ A data frame is

___ To make a new data frame

___ To make a data frame active

___ Two ways to change a data frame name are

___ The difference between the active data frame and the selected data frame is

___ Care must be taken when adding data to a data frame because

___ The projected coordinate system could be found under Properties through the table of contents, under this tab:

___ If data frames overlap selecting a particular one of them may be done by

___ To move a data frame within the Layout

___ To select elements on a layout

___ The active data frame appears in the Table of Contents with

___ A blue dashed-line indicates

___ A black dashed line indicates

___ A hash-mark around the active data frame indicates

___ The scale bar will reflect the true map scale when

___ A scale bar is keyed to only one data frame. It is

___ A map scale can be set by typing in the

___ To make different data frames line up one can use

___ A layer file is related to a data file

___ The extension of a layer file is

___ If one erased a data file and then tried to draw the associated layer file

___ You set the software to use particular styles by

___ The Color Selector window

___ To add and use a style

___ The purpose of Summarize is

___ To use the report writing capability of Arc Map, first create

___ A report can be added to a layout by

___ To get to the Graph Wizard

___ Ancillary information can be placed on a data frame in two ways:

___ To add a call-out box

___ To edit data in a data frame while in layout view one must

___ To create text along a spline

___ To make graphics out of geographic features

CHAPTER **4**

Structures for Storing Geographic Data

OVERVIEW

IN WHICH you explore the ways geographic data sets are stored in the memory and on the disk drives of a computer. And you learn the rudiments of using ArcToolbox and ArcInfo Workstation.

Why Is Spatial Data Analysis So Hard?

Spatial (that is, geographic) data sets are notoriously difficult to analyze. In other fields of human endeavor, most of the data sets one wants to analyze are naturally made up of numbers. What is the history of the stock market's ups and downs? Numbers. What are the statistics relating to the grades of students in the sophomore class? Numbers. How many parts per million carbon monoxide molecules may be safely tolerated by different types of breathing animals? Numbers. But the chief way of *storing* spatial data for most of human history has been the map—whether paper, Mylar, or computer image.

Numbers and text are composed of nicely behaved discrete symbols. Each symbol may be represented by a bit of ink, or by a few pixels on a computer screen that fit neatly into a square roughly an eighth of an inch on a side. And there aren't very many different symbols: 10 digits, 26 letters uppercase, another 26 lowercase, and a bunch of special symbols—in total a maximum of 256. Maps use symbols also, but they are not nearly so well behaved. For example, symbolizing a road may result in a wavy line 2 feet long.

As discussed in Chapter 2, maps are difficult to analyze and it is hard to compare maps. Also, the map has been the primary way of *both storing and displaying* spatial data—an idea we discussed earlier. One of the major advantages of a computer-based GIS is that we separate the storage function from the display function.

A physical method of comparing maps involves a set of, initially, clear plastic sheets, one for each theme in the study area. Each map is darkened in certain areas to indicate the lack of suitability of that theme in the location. A completely clear area of the map might mean a completely suitable area on the ground. A totally black area would indicate a total lack of suitability. Other levels of suitability could be indicated by lighter or darker (grayscale) areas. For example, suppose you were searching for a site for an airport. On one sheet, expensive land would be made as dark, less expensive as lighter. On another sheet, areas where structures would have to be demolished might be made black. A third sheet would show a very flat area as clear. Assuming that all these maps were made the same size, shape, scale, projection, and so on (quite a chore in itself), you could then line them up and place them on a light table, making sure that equivalent geographic areas lined up, and look through them. Using this "map overlay" technique,[1] the lighter a resulting area, the more suitable that area would be. You can probably think of several reasons why this method is pretty inexact (relative importance of different factors, for one—are weather patterns as important as topography?), but the overlay method was one way used to analyze spatial data sets that come from several map sources.

How the Computer Aids Analyzing Spatial Data

Computers can aid in spatial data analysis and synthesis in a variety of ways. First off is speed. It helps that computers can add and compare numbers billions of times faster than you can. (Computers, while stupid, are fast and accurate. Human are smart, but slow and sloppy.) Further, a computer is capable of doing repetitive tasks (read: boring) for hours or years on end. You probably would not want to know a person with this capability. A third virtue of computers in GIS is the ability to store very large data sets.

A vital factor in using a computer to analyze spatial data is the paradigm or schema (data model, data structure) that is used to store the data in the memory of the machine. While the issues about the format in which to store data are not unique to GIS, lots of other fields have much less of a problem. Usually when one stores data in a computer, the questions that arise are ones like the following:

❑ Should I use integers or numbers with decimal points?

❑ Is the number likely to be very big or very small?

❑ Would it cause problems if I used a text string to store a numeric value?

Such sets of numbers usually exist in simple lists, in databases, or perhaps in matrices.

Complexity of Spatial Data

With spatial data the problem is much more complex than with numbers or text. The natural and human-made environment we want to work with

❑ Is virtually infinite

❑ Is a mixture of continuous and discrete phenomena

❑ Needs to be considered at different levels of detail

---

[1] A method given prominence by Ian McHarg in his 1969 book *Design with Nature*.

A computer, on the other hand, is finite (small, really) and discrete to a fault (made up, at its most fundamental level, of things, i.e., bits, that either are or aren't, i.e., 1s or 0s—there is no middle ground).

So the question is this: How can we extract significance from the complex, virtually infinite, multidimensional natural and human-made environment and, using only numbers, letters, and patterns of bits, make the computer form a "map" that can be easily analyzed and compared with other maps? A large part of the answer lies in the way we mathematically and logically model the features that make up the environment we are interested in. Put another way, we need to find a way of structuring the geographic data in the computer's memory so that we can derive answers to queries we might make.

Structures for Spatial Data

What are the principles, fields, ideas, tools, and techniques that are in play in the development of a spatial data structure? There are several:

Geometry. A branch of mathematics that deals with the measurement, properties, and relationships of points, lines, angles, surfaces, and solids. With plane geometry we can define a set of polygonal areas with line segments. We can overlay one polygonal set with another, using geometry to calculate where line segments intersect and make new polygons.

Topology. Loosely, a branch of mathematics concerned with the properties of geometric configurations that are *unaltered* when positions of points, lines, and surfaces are altered. (Classic joke: A topologist is a mathematician who can't tell the difference between a coffee mug and a doughnut [since both are solid objects with a single hole].)

Idealization. Easily manipulated symbols are substituted for actual, three-dimensional, real-world objects. All physical objects exist (over time) in three-dimensional space. If the object's measure in one or two dimensions is quite small compared with other dimension(s), we may be able to safely ignore a dimension or two. For example, we tend to think of a single sheet of paper as a two-dimensional object, but of course it has thickness as well. We might think of a fire hydrant (depicted on a map as a dot—just a geometric point) as a zero-dimensional entity, but it is, of course, a three-dimensional artifact. (Ask the engineer who designed it, the workpeople who installed it, the firefighters who use it, or yourself, should you try to lift it.) Just as we idealize objects depicted on maps, we do so in a GIS. We say that the fire hydrant exists at a location specified by a single latitude and longitude pair, when in fact parts of it exist at an infinite number of latitude-longitude pairs—all, admittedly, close together but distinct nonetheless.

Aggregation. Entities having similar characteristics are put together. For example, saying that an area has x acres where corn is grown and y acres where soybeans are grown is a statement of aggregation. Information about where respective acreages of crops are located may or may not be detailed.

Interpolation and extrapolation. We probable-ize. We assume. Data points with a believed high degree of accuracy are interpolated or extrapolated to obtain new information. If we know that the altitude of a certain point on the Earth's surface is 900 feet and that the altitude of another point very close by is at 910 feet, we might interpolate between the two to say that the altitude of a point halfway between them is 905 feet. To get a better estimate, one might also consider the 890-foot contour and the 920-foot contour. In any event, the elevation of such an unknown point is probably known to be not less than 900 feet nor more than 910 feet. Thus, in some cases, there are bounds on the error introduced by the process of probablization.

6 Categorization. We categorize when we break up a continuous set into a number of discrete sets. For example, we might subsume slopes of 0° to 1° in category A, slopes of greater than 1° up to 3° in category B, and so on.

Storage Paradigms for Areal Data

Now we turn to looking at the specifics of the different data structures used by ArcGIS. Representing "almost zero-dimensional objects" (e.g., parking meters) and "essentially one-dimensional objects" (e.g., narrow streams) is relatively simple. If an object is, for our practical purposes, just a point then a simple, single coordinate pair will suffice. If a feature can be represented by a sequence of line segments, then just a sequence of coordinate pairs does the job. Representing areas, however, is a much less straightforward problem.

Representing Areas is Not Easy

Fundamental Bases of Geographic Data Models

Figure 4-1*** is an orthophotoquad showing a picture of a piece of Earth's surface. It shows houses, greenspace, warehouses, roads, trees, railroad, parking lots, a horse race track, and so on. Suppose you have been given the task of determining the area occupied by each of the feature types: x square feet of

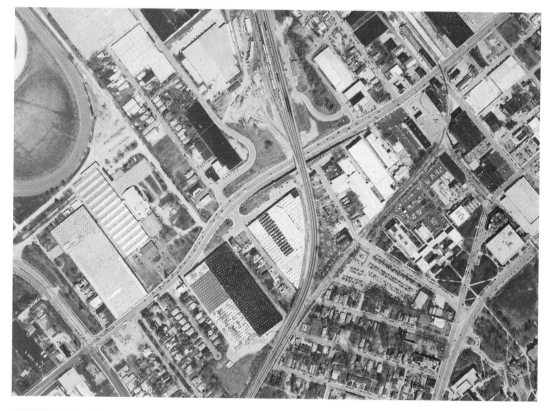

FIGURE 4-1   A color orthophotoquadrangle of part of Lexington, Kentucky

housing, *y* square feet of highway, and so on. Information about where these various land uses exist is also desired. Suppose further that the year is 1960 and you have a computer available to use for the project. If you use the computer, your employer insists that you store the information so that whatever you do can be verified by someone else.

What approach would you take? Basically, to use the computer, you would have to transform the "picture" into numbers and symbols (which the computer will transform into bits). For a given theme (such as land cover) these numbers and symbols must answer two questions at the same time:

❑   WHAT? (entity or quantity)

❑   WHERE?

I don't know how *you* would do this. If you think about it, you may come up with a viable, effective, and efficient scheme that no one else has thought of. If so, head for the patent office.

Here are approaches that others have come up with: *(Raster)*

❑   Systematically divide the overall area up in a regular way into a large number of equally sized subareas (e.g., small squares). Record what is in each subarea. Have a reference scheme so you know where each subarea is. This technique falls under a broad category called *raster* (or *grid* or *cell*). Almost always, a raster may be viewed as a rectangular space composed of rows and columns. A given cell is at the intersection of a given row and a given column.

❑   Completely delineate each of the features—"delineate," *(Vector)* in this case, is a real, physical delineation. It means: In the two-dimensional plane, draw a series of straight-line segments around each area. Develop a method for determining where the lines are and for giving each segment a direction. This is often referred to as a *vector*[2] approach, since a directed straight-line segment is a vector.

❑   Just to exhaust the fundamental types of GIS storage methods, although it doesn't help solve this particular problem: Partition a surface that is above (or below, or both) the area of interest into irregular triangles. Each triangle (except those on the edge of the area) shares a side and two vertices with an adjacent triangle. The triangles approximate the height of the surface (e.g., elevation), the slope, and the direction (e.g., aspect). This sort of data set is known as a triangulated irregular network, or TIN, which you met in Chapter 2.

Raster Data Model

One way of systematically dividing up the area is shown in Figure 4-2. Here, regularly spaced horizontal and vertical lines, like those that generate the squares of a chess board, make a *grid* that creates relatively small areas called *cells*. In the past, and sometimes currently, the practice was to index each cell by a row number and a column number. Generally, the top (north-most) row was numbered one (1) and the left (westmost) column was numbered one (1). More recently the indexing has shifted to strictly geographic coordinates. In this case, the coordinates of the center of the southwestmost (lower left) cell are specified by the easting and northing of (usually) the center of that cell. The horizontal and vertical lines are parallel to the x- and y-axes of the coordinate system. Since the cell size is known, the coordinates of the center of any cell may be easily calculated.

---

[2]A vector is a mathematical or physical entity that has magnitude (in this case, length) and direction (it has a starting point and an ending point, and therefore, points in a geographic direction).

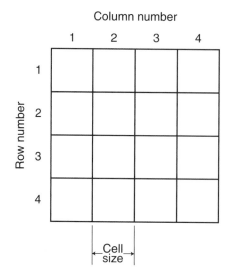

Column number

Row number

FIGURE 4-2 A basic raster that allows storage of categorical data

If the person deciding on the spacing between the grid lines has done a good job, and the overall area being depicted is cooperative, the user will frequently be able to know, for each cell, what feature or condition most occupies the cell, thus answering the "what" question. Actually, several issues, to be addressed later, come into play in determining the "what," when, as will frequently be the case, more than one feature, or condition of the particular theme, appears in the area covered by a cell.

The determination of "where" in the raster case is, on the surface, quite simple. As I indicated before, if the horizontal grid lines run east-west, the location of the upper left (northwestmost) cell is known, and the cell size is known, the geographic location of any cell is a simple calculation based on the row and column number of the cell.

Although a raster of squares (or "almost squares," if the dimensions of a cell are couched in latitude-longitude terms) is a set of discrete areas, the fact that they are regular in nature, and that each one has the same configuration of the four nearest neighbors and the four next-nearest neighbors, makes it a fairly good model for representing continuous surfaces (see Figure 4-3).[3]

A sequence of grids is also an excellent way to represent, analyze, and predict phenomena that change quickly over time, such as the spread of an oil spill or a forest fire.

The raster approach can also represent discrete areas, albeit "lumpily" with straight vertical and horizontal lines separating nonhomogeneous areas (see Figure 4-4).

_____

[3]A raster does not represent a continuous surface as well as a TIN. A surface represented by a raster (such as a DEM) has discontinuous breaks; in a TIN, the surface representation is continuous but not differentiable in places. If you don't know what this means, ignore it.

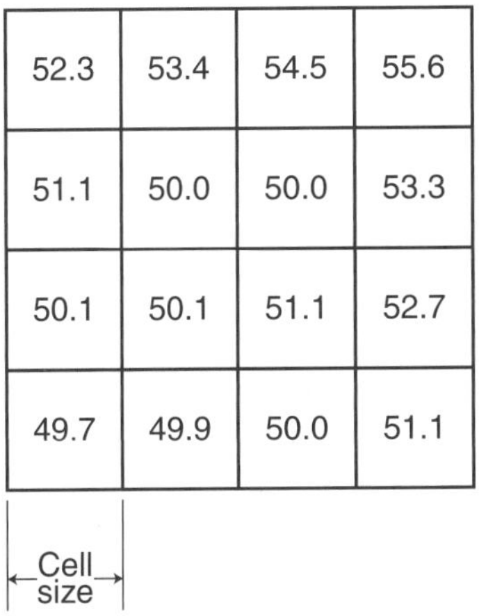

52.3	53.4	54.5	55.6
51.1	50.0	50.0	53.3
50.1	50.1	51.1	52.7
49.7	49.9	50.0	51.1

|←Cell→|
| size |

FIGURE 4-3   A basic raster that allows storage of continuous data

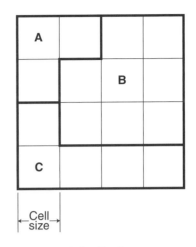

|←Cell→|
| size |

FIGURE 4-4   Raster representations of areas

The computation of "where" becomes more complicated if the matrix of cells is not oriented along Cartesian grid lines or the graticule meridians and parallels, but, after all, computation is something that computers are good at. A more subtle complication occurs if the area covered by the grid is large in a north-south direction because of the issues related to projecting the curved earth onto a flat plane. But for the most part, the matter of location using a raster approach is easily handled.

If a grid cell contains more that one type of area, as many on boundaries between areas do, there is an approach called "quad tree" in which the grid cell is divided into four subcells, as in Figure 4-5. If a subcell is homogeneous in the feature value, then it is left alone. If not, it is redivided into four more subsubcells, and the process repeated. This redivision continues until all the subcells are homogeneous, or they become too small to make further subdivision reasonable.

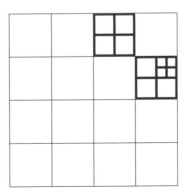

FIGURE 4-5   The quadtree technique for storage of raster data

215

Vector Data Model

Using the vector approach, the level of difficulty of determining "what" and "where" is somewhat reversed. The "what" is relatively simple. Each unique area is enclosed inside a polygon, so the content of the polygon is homogeneous—containing a unique value (or a unique set of values) related to the theme. Contrast this with the raster approach, where several features or conditions may occur in one cell and (usually) one feature or condition is picked for recording in the database.

The "where" with the vector approach is a bit more problematical but usually allows greater precision. Whereas with a raster approach, the location of each area (cell) is but a simple calculation, with the vector approach, each vertex of each polygon has an explicit geographic location. Therefore, the "where" question may have thousands to millions of answers, in terms of coordinate pairs for a large area or one with many polygons that have complex boundaries. And, of course, the issues with respect to projection, datum, and so on are present—for all the coordinate pairs.

A Multiplicity of "Storadigms"

ArcGIS supports, or at least recognizes, several different spatial storage paradigms:

Vector

1. Geodatabases (vector)[4]

2. Coverages

3. Shapefiles

4. Computer-aided design (CAD) drawings

5. Vector Product Format (VPF) data sets

6. Raster (GRID) data sets

7. Triangular irregular network (TIN) data sets

The first five of these are based on the concept of the vector. We will examine and work with the first three extensively. Number 6 is the raster format already discussed. Number 7, as briefly discussed earlier, is a technique for storing data where there is an independent, continuous variable (e.g., elevation) whose values are based on the two dependent variables x and y. That is, a TIN is used to represent a surface.

Ideally, there would be only a single storage paradigm. We would store data in this way, and when we made a query of the database, or asked for a map of a given area and scale, it would be provided. One problem with this approach is that different sorts of data—representing different aspects of the environment—have distinctly different characteristics. Pick a point on Earth's surface. It has elevation. If there is soil there, it has physical characteristics. Someone or some entity probably owns it. At a certain moment it has a certain temperature, and over a year period it has an average temperature. It is probably suitable for growing some kinds of crops, but not others. It has a particular slope and aspect. The vertical distance down to bedrock has a certain value, and that bedrock is of a certain type. There may be a volume of coal or oil under the surface. The atmosphere above it contains some pollutants. We want to represent these facts with data sets.

---

[4]Actually, geodatabases can include, or will in the future include, other paradigms in this list.

Some of this information is more easily and effectively stored in one paradigm or format, other data in a different format. Further, a lot of inventing and creating has gone into the problem of representing the infinite, continuous environment as numbers, letters, and symbols. So we wind up with a lot of different paradigms for storing spatial data, or to coin a term "storadigms." One advantage of using ESRI software is that you may easily convert from one method of storing data to another.

Vector-Based Geographic[5] Data Sets—Logical Construction

Coverages and geodatabases are the two most sophisticated data structures used by ESRI software. They are sophisticated in different ways, as you will see later. The coverage concept dates back many years and was the foundation of ArcInfo. The geodatabase is a more recent development. You will spend a fair amount of time and effort understanding and working with both of these. Another data structure, the middle-aged shapefile, is relatively simple, but because multitudinous data sets exist in this form, nationally and internationally, it is also important. Understanding geographic data structure is vital to being able to doing some forms of analysis with GIS. Please note that you can convert any of the three of these data set forms into any of the others.

The primary elements of vector-based data sets are points (zero-dimensional entities), lines (one-dimensional entities), and polygons (two-dimensional entities). Terms used with all ESRI data models are described in the section that follows, and then again when we look at particular data models. Figures that graphically show the entities follow in the detailed discussions of coverages and geodatabases.

Zero-Dimensional Entities in a Two-Dimensional Field: Points

While vital to the functioning of a GIS, a zero-dimensional thing (generically a point) is pretty dull from a geometric view. It is basically a pair of numbers (x- and y-coordinates, or perhaps, latitude and longitude coordinates) stored as single-or double-precision numbers, or, as with geodatabases, long integers that are converted to floating-point numbers for display.[6]

The concept of a point is used in a variety of ways in GIS to represent features, as end points of lines, as vertices in sequences of line segments, vertices of triangles, as reference points tying the feature set to the real world, as locations to hang labels on, as centroids of areas, as junctions and nodes in geometric networks, as centers or corners of grid cells, and others.

We work mainly with points in a two-dimensional arena, but of course they exist truly in three-dimensional space. ArcGIS will let us add information about this third dimension, sometimes as what amounts to an attribute (a z value) and sometimes (for example, in a TIN) as a measure in the true third dimension. Even when a true 3-D point is represented, the units for measurement of the vertical may well not be the same as those used in the horizontal plane.

---

[5]Or, should you hale from the United Kingdom: Geographical
[6]To really locate a point in real (3-D) space, one would need a trio of numbers—but vector GIS either assumes that the third dimension is the surface of the Earth or is not relevant or is stored as attribute data. An exception to this is a TIN, which stores a coordinate triple.

Points representing features: In these cases a point has associated with it a row in a relational database table that identifies the point and allows the user to add other (attribute) information about the feature that the point represents. You became acquainted with points representing features in coverages in the fire hydrant example of Chapter 1. A point may be used to represent a feature that is too small to have a meaningful area. How can the concept of a point fit into a vector system? A point is simply a vector with zero magnitude.

Multipoints: A multipoint is simply a collection of points that share the same attribute values (e.g., several gas wells which have the same characteristics and the same owner). The collection of points is represented by a single row in a table.

Vertices: Sets of coordinates where two line segments are joined or where sides of a triangle join (in a TIN). Usually, no database table row is associated with a vertex.

Tics: A tic is point in a coverage that is referenced to a point on the ground that is known with a high degree of accuracy and precision; this allows features in the geographic data set to be correctly referenced to the real world.

Nodes: A node is a point in a coverage that begins or ends a sequence of line segments (called an arc). Nodes and arcs are explained in detail in the discussion on coverages.

Labels: A point carrying information about a polygon that the point resides in. Used in coverages.

Junctions Points in geometric networks where ends of lines (edges) are joined. Covered when we discuss geometric networks.

Lattice points: Set of points in a raster, usually defined to be at the centers of cells.

One-Dimensional Entities in a Two-Dimensional Field: Lines

You saw examples of lines representing streams and sewers, again, in Chapter 1. Lines were also used there to delineate boundaries of polygons. A line may be used to represent a linear feature that is too narrow to have a meaningful area. As with zero-dimensional entities, points on a line may have a z (e.g., altitude) value, but it is treated as an attribute.

In GIS terms, a line is a simple geometric entity that consists of a sequence (that is, an ordered set) of vertices, which are simply coordinate pairs. Between each adjacent pair of vertices there is a segment. A segment is frequently simply a straight line, but in vector-based geodatabase feature classes, it can also be a part of a circle or ellipse, or it may be a spline (called a Bézier[7] curve). A line that consists of multiple straight-line segments connected at vertices can approximate a curve. Lines can therefore be used to represent curvilinear features, such as roads and streams. The segments of a line are not allowed to intersect each other.

Paths: A path is a line as described previously, composed of a sequence of connected segments. The term "path" is used in vector-based geodatabase feature classes.

---

[7]Developed by Pierre Bézier in the late 1960s for computer-aided design (CAD) and computer-aided manufacturing (CAM) operations for the Renault automobile company. It involves an anchor point at each end of a segment and two other control points.

Polylines: A polyline is made up of one or more paths. If there are multiple paths, the paths may be connected or disjoint. Even if a polyline representing a feature consists of multiple paths, it has only one row in the attribute table. Polylines are not used in coverages. If they are used in shapefiles, the segments of the path must be straight lines.

Rings: In a geodatabase vector feature class, when a path encloses an area (polygon), the path is called a ring. A ring starts and ends at the same place. A ring is a sequence of nonintersecting segments that form a closed loop. Its primary purpose is to enclose areas. If the segments are all directed straight lines (vectors), then the area enclosed is a polygon, in both the mathematical and GIS sense. If the segments are curvilinear elements (arcs of circles or ellipses, or splines) then the enclosed area is a GIS polygon, but not a geometric one. A ring has an unambiguous inside and outside. The length of a ring is automatically stored in the associated attribute table.

Arcs: Arcs are described in detail in the section on coverages that follows. An arc begins and ends at a node. It is composed of vertices and segments that are vectors. The length of an arc is automatically stored as an attribute in the arc attribute table. Arcs can delineate linear features or serve as partial or complete boundaries for polygons, or both. The arc[8] is the primary element for depicting both curvilinear and plane surface features in coverages, but the term is not used in geodatabases or shapefiles. The terminology that characterizes arcs includes node and vector. These terms also are not used with geodatabases or shapefiles, but many of the same concepts apply.

Routes: Routes exist both in coverages and geodatabases, but in somewhat different ways. Routes are subsets or supersets of arcs or polylines. They allow the user to define collections of linear features (e.g., the parts of a road system that constitute bus route #99), or measured distances along a linear feature (where a stream changes from clear to turbid). Routes make use of an m (measure) number that specifies the distance along a feature that something changes.

Two-Dimensional Entities in a Two-Dimensional Field: Polygons

There is sort of an interesting philosophical question about plane areas and the lines that define them. For example, what is a circle? Is it the curvilinear line? Or is it the plane figure like a coin? That is, is it the locus of points at a distance d from a single point c (the center) or is it the locus of points at a distance d or less than d from the single point c. "Circle" is used in English both ways. What is a triangle? The metal frame of a truss or the sail of a boat? A polygon, however, is almost always considered a plane figure rather than the line elements that bound it.

Polygons: In geometry a polygon is a plane figure with a number of straight-line sides greater than or equal to three. The sides may not cross. Perhaps you think of a polygon as a square or hexagon—and you are correct. But there is no (small) limit to the number of sides a polygon may have, nor do the sides have to be of equal length, as they are in a *regular polygon* where equal length sides and equal angles are the rule. It is not unusual for a GIS polygon to have hundreds of sides.

In GIS we take a lot of liberties in the use of the word "polygon." For one thing, a GIS polygon can contain other polygons (which can contain other polygons, which can contain other polygons, and so on).

---

[8] An historical note: The arc concept is the basis for the designation "Arc" of ArcInfo. "Info" was the name of the original brand of database that the system used.

Further, while polygons in shapefiles and coverages may only have "sides" that are straight lines, geo-database polygon "sides" may be parts of circles or ellipses, or may be Bézier curves. Still further, a GIS polygon may be several polygons, as described later in the chapter.

In geodatabases and shapefiles, a single polygon (an entity with a single row in a table) is formed by a collection of one or more rings—as defined previously. If more than one ring is involved, no rings may touch.

Cells: A (usually) square area that (usually) contains a number related to an entity or condition. Used in raster or grid data models.

Zone: A collection of cells that all have the same value. Used in raster or grid data models.

Triangles: Plane, three-sided polygons used in the TIN data model, where each triangle has a calculated maximum slope and a direction (aspect). Further, each point on or within the triangle has a z value, such as elevation.

Regions: This term is used in different ways depending on which data model is being considered. In a *raster* or *grid system*, a region is a collection of cells, all of which have the same value (i.e., are all of the same zone) and that are connected to at least one other cell in the zone.[9] The term *region* used with the *coverage data model* specifies something philosophically different: A region is a subset of polygons of a single coverage; the polygons may or may not be connected, but they are represented by a single record in the polygon attribute table. A region can represent an entity composed of more than one polygon. An example of a coverage region would be the state of Hawaii, which is composed of several islands, each of which is a polygon. The single record would contain the sum total of all the polygonal areas. It might also contain the name of the governor of the state.

Coverages

The coverage data model was the first one developed by ESRI. It uses arcs, either by themselves to represent linear features or to delineate polygons. Since only a single line (arc) separates each adjacent pair of polygons, there can be no gaps or slivers. Further, polygon borders are not "double digitized," and there-fore, computer storage is conserved. As indicated previously, the "arc" is the key to the coverage and explains the first part of the name ArcInfo. (INFO was the name of the relational database system used by ESRI.) While the geodatabase will ultimately eclipse the coverage, coverages are an excellent way to understand how vector data sets are stored. And there are still a large number of coverage data sets around.

The Coverage Data Model in Detail

I ask you now to take an in-depth look at the coverage data model. In terms of intricacies of linear and areal feature representation, it is the most sophisticated and complex. Once you understand it, shapefiles and geodatabases are a breeze. Geodatabases are more sophisticated in terms of relationships among features and table, and geodatabases have more extensive topological relationships, which we will pursue later in the text.

---

[9]The connection can be specified to be only along edges or to be both edges and corners.

Arc-Node Nomenclature

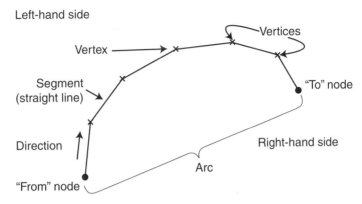

FIGURE 4-6   Defining arcs and nodes (coverages)

Here are some mildly circular definitions relating to coverages; look at Figure 4-6 as you read them.

Node: A geographic or Cartesian point where one or more arcs originate or terminate. That is, a node serves as the beginning or ending point of an arc.

Vertex: A geographic or Cartesian point that is internal to an arc. Vertices allow the arc to change direction.

Segment: A vector that may connect (a) two nodes, (b) a node and a vertex, or (c) two vertices.

Arc: A sequence of segments that originate at a "from node" and terminate at a "to node" (which may be the same as the from-node).

To recap: an *arc* is a sequence of connected straight-line *segments.* A *vertex* is the conjunction of a pair of adjacent vectors, where the tip (ending point) of one vector in the sequence is connected to the tail (beginning point) of the next vector. (The plural of vertex is *vertices.*) Vertices are simply zero-dimensional geometric entities, represented by pairs of coordinates (not unlike point features, but serving a very different function). Each arc begins with a *node* and ends with a node. Usually, the beginning node is different from the ending node, but an arc may begin and end with the same node. Nodes are also simply zero-dimensional geometric gadgets.

Arcs, Direction, and Handedness

By virtue of the fact that an arc is made up of vectors, the arc may also be said to have "direction." The concept of "direction" of an arc is a little tricky formally, since the arc may be made up of several vectors (line segments) that have several different geographic directions, so its "direction" depends on which of its segments is being considered at a given moment. The overall arc's direction is considered to be from the beginning point to the ending point. (Think of the "direction" of a one-way street that has bends and turns in it. Overall it has no single direction in the mathematical or geographic sense, but, as a police officer may explain to you, it has direction in another important sense.)

To determine the direction of an arc, imagine yourself standing on the beginning node and facing in the geographical direction implied by the first vector. Then walk along the vectors, turning, as necessary, at the vertices. You would be walking in the direction of the arc.

When arcs are used to define polygons, you can use the direction of the arc to designate the polygon on the left side and the polygon on the right side. As you walk along the vectors, the space by your left hand is the left side. In a coverage that defines polygons, an arc may not cross itself. Also in a coverage that defines polygons, an arc may not cross any other arc. Arcs that define linear features may cross themselves or other arcs. As an example of arcs that cross when linear features are being depicted, consider a road network where one road goes through a tunnel under another road. A node is basically an intersection. Some elements (like roads) cross but do not intersect, and hence no node is present at the crossing.

Coverage Defined

Basically[10] a *coverage* is

1. A set of points that represent geographic features and the associated attribute tables that describe those point features, or

2. A set of nodes and arcs that represent linear features (such as streams or roads) and the associated attributes of those linear features, or

3. A set of nodes and arcs that enclose geographic areas (polygons) and the associated attributes of those area features, or

4. A combination of (a) and (b), or 1 and 2

5. A combination of (b) and (c). 2 and 3

Coverages and Attributes

We said that a GIS was the marriage of a geographic database with an attribute database. Each row in the attribute database refers to one point, one arc, or one polygon.

Currently, ArcGIS can be connected with a number of database manufacturers' products to store attributes. Among them:

- Access
- ddBASE
- FoxPro
- IBM DB2

- INFO
- Informx
- Microsoft SQL Server
- Oracle

---

[10]I say "basically" because coverages also contain other elements such as tics and annotation, to be discussed later. Here we are concerned with the primary elements of a coverage.

Arcs and Linear Features

Arcs can, by themselves, represent roads, streams, political boundaries, and so on—basically features that are linear in nature.

Arcs Defining Polygons

As indicated previously, arcs have another important function. An arc, or a sequence of arcs, can enclose a plane area, or polygon. Polygons can represent areas such as school districts, ownership parcels, areas of a given soil type, footprints of buildings, and so on. The sides are the segments (vectors) that make up the arcs that define the polygon. The sides of a polygon may not cross, of course.

Polygons of an ESRI coverage partition two-dimensional space into mutually exclusive areas. The assumption is that each attribute value of a given polygon specifies a feature or condition that is the same at every point within the polygon. Put another way, polygons are homogeneous with respect to all attribute values attached to them.

For example, if the features of a coverage are bodies of water and areas of land, each polygon must consist only of water or land. If there is an island within a body of water, then a polygon must be drawn around it.

For another example, if the coverage relates to ownership of land parcels, all the land that is within the boundaries a given polygon (and outside the boundaries of any island polygons—a subject to be addressed shortly) must be owned by a single person or entity.

Polygons divide the area up into mutually exclusive subareas. Any two polygons are either mutually disjoint or one polygon contains the other. In Figure 4-7, in all four instances, polygons A and B are disjoint—although in the first three cases they share an arc. In Figure 4-8, in both instances, polygon D is nested inside polygon C, although in the second case they share a node. By the way, there is no practical limit to nesting. That is D could contain polygons E and F, and F could contain G.

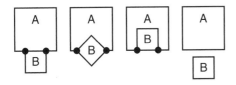

FIGURE 4-7   Disjointed polygons

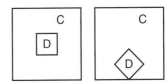

FIGURE 4-8   Nested polygons

Arcs and Polygons Together

A philosophical argument that arises at times between architects is this: "Does the architect design walls, floors, and ceilings, or is the object of the design process the *space* that lies between floors, walls, and ceilings?" With architecture it is obviously simply a matter of emphasis, since to design space, one needs to design enclosing surfaces, and if one designs enclosing surfaces, one automatically designs space. Arcs and polygons interact in a similar way. Consider the set of arcs and the set of polygons in Figure 4-9.

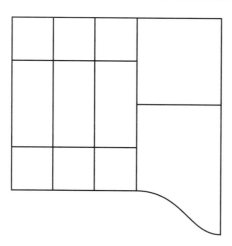

FIGURE 4-9   Arcs and polygons together

Here perhaps the arcs are street centerlines and the polygons are city blocks. Or maybe the arcs are surveyed property lines (one attribute of which might be the name of the surveying firm, another attribute the date the line was surveyed) while the polygons represent ownership parcels (with obvious attributes).

You cannot have polygons without having arcs. You can, however, have arcs without forming polygons, as in a stream network. (See Figure 4-10.)

Even if arcs do form polygons, you needn't pay attention to them. If a coverage consists of arcs representing linear features, then an attribute table (an AAT) provides information about those arcs. If a coverage consists of arcs defining polygons, then an attribute table (a PAT) provides information about the polygons. If a coverage of arcs represents both linear features and areas, then it will have two attribute tables—one for each feature type.

As said previously, nodes form the beginning and ending points of an arc. In that sense, all nodes perform the same function. In another sense, one can consider that there are three kinds of nodes. In the following discussion, I will use the term "arc end" to mean either the beginning or end of an arc.

FIGURE 4-10   Arcs without polygons

"Real" nodes

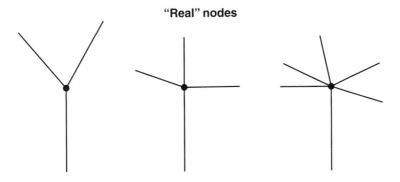

FIGURE 4-11 **Standard nodes**

A "normal" or "real" node is one where three or more arc ends are joined. (See Figure 4-11.)

A "pseudo" node is one where exactly two arc ends come together. (See Figure 4-12; nodes are shown as dots, vertices as x.)

A "dangling node" is where only a single arc end occurs. (See Figure 4-13.)

Pseudo nodes

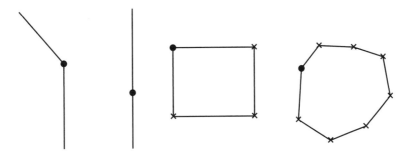

FIGURE 4-12 **Pseudo nodes**

Dangling nodes

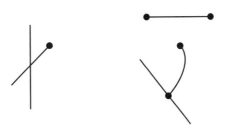

FIGURE 4-13 **Dangling nodes**

Pseudo and dangling nodes are flagged as *potential* errors by the ArcGIS software at the request of the user. If there are two nodes in a geographic location where only one was intended, the set of arcs may not represent what the user wanted. For example, if arcs were to come together at a single node, as in Figure 4-11, but in fact two or three nodes were present, as in Figure 4-14, flagging pseudo and dangling nodes would alert the user to an error.

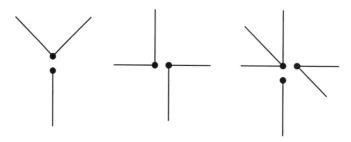

FIGURE 4-14   Potential node errors

Trapping potential errors by the software alerts the user to this possibility by reporting pseudo nodes and dangling nodes.

Can a pseudo node ever be used in representing a real-world situation? Yes. Suppose a single arc represents an oval racetrack. Or encircles a small lake. Or is a contour line. Such an arc must have a beginning node (called a "from" node) and ending node (called the "to" node)—and that node will be a pseudo node.

Can there ever be reasons to use dangling nodes? Again, yes. Consider a stream network. A dangling node might be the location of a spring or the point where the flow of water becomes well enough defined to be considered a stream. Another example would be of a dead-end street or cul-de-sac in a road network.

So while dangling nodes and pseudo nodes might be errors in the structure of arcs in a coverage, they also might not be. Therefore, the user is informed of these "unusual" nodes so a decision can be made about whether they belong there or not.

Feature Attribute Tables

Each feature type in a coverage can have associated with it an attribute table, called a *"feature attribute table"* or *FAT*.[11] You learned about relational database tables in previous chapters. With coverages there are four types of FATs:

❑   Point attribute table (PAT)

❑   Arc attribute table (AAT)

---

[11]Not to be confused with the file allocation table, also FAT, associated with disk drives.

- ❏ Polygon attribute table (PAT)[12]
- ❏ Node attribute table (NAT)

You saw examples of the first three types of tables in the step-by-step part of Chapter 1. The hydrants coverage had a point attribute table, the roads coverage had an arc attribute table, and the land use coverage had a polygon attribute table.

It is also sometimes useful to attach attributes to nodes. For example, nodes could represent road intersections or the confluence of two streams. The road intersection might have a traffic signal requiring description. The confluence of streams might have a name. A node attribute table could contain this information.

Coverage Topology

Look at three plane figures composed of lines connected with nodes (see Figure 4-15). You should understand why configuration A and configuration C are topologically identical and Configuration B is different from both. Nodes are shown by heavy dots. A and C have the same number of arcs and nodes as each other, and you can find equivalences in the connections of the nodes in those two configurations. But you cannot "map" B onto either A or C. If you don't see this, assign letters to the nodes and numbers to the arcs in all three. Make a table for each showing what node connects to what node with what arc.

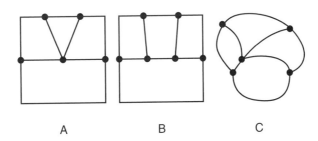

FIGURE 4-15   Topology: What node is connected to what node is all that matters

The reason we say that the coverage is a sophisticated data structure is that it "contains topology." A coverage contains topological information that relates arcs and polygons to other connected and adjacent arcs and polygons. In particular, a polygon "knows" what arcs define it. An arc "is aware" of the polygons on each side of it. Therefore, the computer can easily determine which polygons are adjacent to a given polygon. It can also determine the other arcs that a given arc is connected to.

The ESRI coverage topological data structure is somewhat complex, but makes good sense once you understand it. There is a certain beauty in its integration of polygons and arcs—and in how it allows

---

[12]The use of PAT for both the point and polygon attribute tables dates back to the beginnings of ArcInfo. A coverage may not contain both points and polygons, so it was safe to identify their attribute tables with the same file extension. Older students and TV rerun watchers may recall the Bob Newhart series in which Larry introduces his brother Daryl and his other brother Daryl: same name for two different entities.

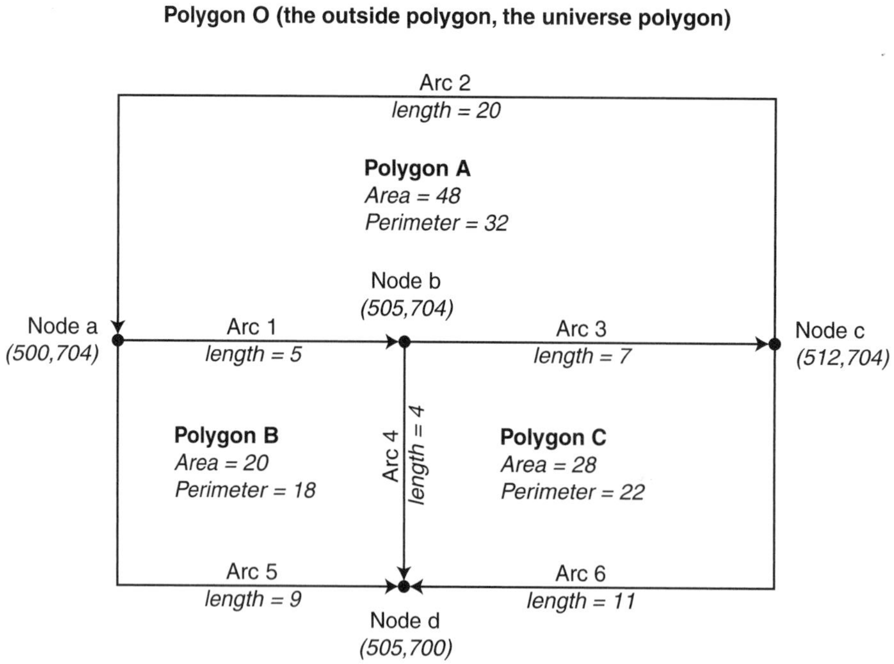

Polygon O (the outside polygon, the universe polygon)

FIGURE 4-16   Simple topological example showing arc lengths and polygons areas

polygons to exist within polygons, which exist within polygons, and so on. Look first at a simple example, but one that demonstrates many of the characteristics of a coverage.

Consider a coverage named ST (for SIMPLE_TOPOLOGICAL). (Coverage names are limited to 13 characters; blanks in names[13] are not allowed.) It looks like Figure 4-16.

Here there are four nodes, six arcs, and three polygons. In the figure, and in the tables that follow, the nodes are labeled as lowercase letters a, b, c, and d. The arcs are shown arcs as Arabic numbers: 1, 2, . . . 6. Capital (uppercase) letters are used to identify the polygons: A, B, and C. This convention is used in order to make it clear which sort of element we are discussing. In an actual attribute table, all of the designators are integer numbers.

Each node, arc, and polygon feature class has its own attribute table, named, respectively, ST.NAT, ST.AAT, and ST.PAT—for the node attribute table, the arc attribute table, and the polygon attribute table.

Most of the field names in the three tables are different. But all tables have the following items:

❑   ST# (the internal identifier, given to the feature by ArcGIS)

❑   ST-ID (the User-ID or Feature-ID, given to the feature by the user)

---

[13]Microsoft's bad decision about allowing blanks in folder names and filenames notwithstanding. It is worth pointing out here that blanks in folder names will create problems for ArcGIS. For example, if you place a coverage in a folder named "New Folder" (the default in some versions of Microsoft's operating systems), you may run into problems.

It is important to understand that the values in these fields (items) in the different tables, despite having the same (column) name, have no relationship to each other. They simply identify a particular node, arc, or polygon, *depending on which table is being considered*.

Look first at the node attribute table ST.NAT (Table 4-1) that follows. The items are as follows:

❏   FID (the Feature Identifier—basically a record number)

❏   ARC# (the identifier of *one* of the arcs that connects to the node)

❏   ST# (the internal identifier, given to the node by ArcGIS)

❏   ST-ID (the User-ID or Feature-ID, given to the node by the user)

❏   X-COORD (the "x" coordinate)[14]

❏   Y-COORD (the "y" coordinate)

TABLE 4-1   The Node Attribute Table ST.NAT

FID	ARC#	ST#	ST-ID	X-COORD	Y-COORD
1	1	a	31	500.00	704.00
2	3	b	32	505.00	704.00
3	2	c	33	512.00	704.00
4	4	d	34	505.00	700.00

The column name in the NAT that heads the identifying numbers is ST# (the coverage name concatenated[15] simply with "#"). This *"Internal ID"* (which is the same as the Record Number) is assigned by the ArcGIS software. If you are editing an attribute table, you should never change a value in this column. This is the number that ArcGIS uses to keep up with the nodes, and you would confuse the software badly if you modified it. Always, in a real attribute table (which this is not), the internal identifier of the node, ST#, is an integer. We are using lowercase letters here for illustrative purposes.

For the *"User-ID"* (or *"Feature-ID"* or *"Cover-ID"*), the column name is made up of the name of the coverage concatenated with the characters "-ID." This field also always contains integers—integers that are of the user's choosing. Here we have chosen 31, 32, and so on. Just how these numbers were assigned we leave for later.

Look at the table ST.NAT together with the graphic representation of the coverage. Verify that node "a" is at location (x = 500, y = 704) and that it does have arc number "1" connected to it. Examine the records of the other three nodes.

---

[14]Usually x- and y-coordinates are not attributes in the NAT. Rather, they are included in the geographic part of the coverage and are hidden from the user. They may be added, however, with a command from ArcToolbox: **ADDXY**. They are included here so you can get a clear picture of the graphic elements of the coverage.
[15]"Concatenated" is a 50-cent word that refers to what you get when you jam a second text string ('#' in this case) up against the first one (the coverage name).

Now examine the arc component of the coverage. Locate the six arcs in Table 4-2, numbered 1 through 6. Then look at the attribute table (ST.AAT) that follows. The diagram numbers on the arcs correspond to the numbers in the column ST#.

TABLE 4-2   The Arc Attribute Table ST.AAT

FID	FNODE#	TNODE#	LPOLY#	RPOLY#	LENGTH	ST#	ST-ID
1	a	b	A	B	5.00	1	20
2	c	a	A	O	20.00	2	40
3	b	c	A	C	7.00	3	50
4	b	d	C	B	4.00	4	30
5	a	d	B	O	9.00	5	10
6	c	d	O	C	11.00	6	60

A value in the column FNODE# in the AAT contains the designation of the "from node"—the departure node for the arc. TNODE# means the "to node." Referring to the ST# item of the node attribute table (ST.NAT), verify that the "from node" and the "to node" for each arc is correct, given the direction indicated by the arrow pointed at the "to node" on the diagram. So, for example, arc "5" starts at node "a" and goes to node "d."

Verify, using the diagram and the table, that the LENGTH of each arc is correct.

There are four polygons, labeled A, B, C, and O in the diagram. O is called the outside polygon. It is composed of all the Earth's area[16] that is not within the areas of the other polygons in the coverage. Let's put polygon O aside for the moment.

Looking at both the graphic representation of the coverage and its polygon attribute table (Table 4-3), verify that the values in the AREA item shown in the following table correspond to the areas of each polygon. Also verify that the *sum* of the arc lengths around each polygon corresponds to the value given for the PERIMETER item for that polygon.

TABLE 4-3   The Polygon Attribute Table ST.PAT

FID	AREA	PERIMETER	ST#	ST-ID
2	48.00	32.00	A	11
3	28.00	22.00	B	12
4	20.00	18.00	C	13

Finally, using ST.AAT together with ST.PAT and the diagram of Figure 4-16, verify that the designations given for LPOLY# (the internal ID of the polygons on the left sides of the arcs) are correct. That is, for example, for the arc designated as "5" in the ST# column of the AAT, the polygon to the left of that arc is "B," and that is also consistent with the ST# column of the PAT. If you are not tired of this exercise, you can also verify that the polygons on the right side of each arc are correct.

---

[16]Earth's area is approximately half a billion square kilometers.

Other Tables Relating to Coverages

Two other tables relate to the coverage ST: ST.TIC and ST.BND. Tics are points on a map or in a coverage that represent points whose coordinates precisely known in the real world. They provide the ability to "nail down" maps and coverages so that they "line up." When you generate an artificial coverage, such as ST, the software places four tics (the minimum number you can have in a coverage) so that all the feature coordinates lie within the rectangle (the *envelope*) implied by the tics. The tic values are stored in one record per tic. For ST the TIC attribute table would look like those shown in Table 4-4.

TABLE 4-4   ST.TIC, the tic attribute table for the coverage ST

FID	IDTIC	XTIC	YTIC
1	1	512.00	700.00
2	4	512.00	708.00
3	2	500.00	700.00
4	3	500.00	708.00

There is also a table called ST.BND, shown in Table 4-5, that contains a single record giving the minima and maxima of the x and y values.

TABLE 4-5   ST.BND, the boundary attribute table for the coverage ST

FID	XMIN	YMIN	XMAX	YMAX
1	500	700	512	708

Polygons Within Polygons—Perimeter and Area Calculations

In representing the natural environment or the human-built world, we frequently want to employ plane areas that are included in other plane areas. Lakes in a county, for example. Islands in a lake. Wetlands that are internal to an island. And so on. As mentioned, many polygons are disjoint (that is, if you look at the area covered by the polygon, you see no other polygons), but others are nested (when you look at a polygon, you see other polygons within it). Nothing said here implies that the areas of polygons are not mutually exclusive. All coverage polygons are. Each has an identification, an area measurement, and a perimeter measurement. Each has its own set of attribute values. Geographically, however, they may be arranged in two different ways. Consider Figure 4-17 with coverage representations A and B. The small squares are 1 unit on a side.

Coverage A consists of two arcs, two nodes, and two polygons (P and Q). Polygon Q is a nested polygon with respect to polygon P.

Coverage B consists of three arcs, two nodes, and two polygons (R and S). Polygon S is not a nested polygon. It is simply an area disjoint from polygon R.

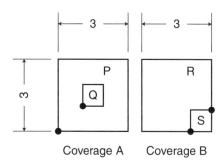

FIGURE 4-17 A nested polygon compared with a disjointed polygon

The area of polygon P is 8 square units, calculated as $((3 \times 3) - (1 \times 1))$; that is also the area of polygon R. The area of polygon Q is 1 square unit; that is also the area of polygon S.

The perimeter of polygon P is 16 linear units. This is the sum of the length of the arc that defines the outside of P (length 12) and the length of the arc that segregates the nested polygon Q (length 4). The perimeter of polygon Q is 4 units.

The perimeter of polygon R is 12 linear units, made up of two arcs. The perimeter of polygon S is 4 units, made up of two arcs.

Again, polygons may be nested to (almost) any depth. Polygon Q could contain three nested polygons, one of which might contain five nested polygons, each of which might contain 33 nested polygons.

In determining areas and perimeters when nested polygons are involved, you might find it useful to invoke a rural analogy. Maybe you are raising llamas and want to know how much fencing you need and how much area the animals would have to graze in. If the animals are to be confined to a given polygon, say, X, the polygon is the area an animal could roam in, which does not include the areas of any nested polygons. The perimeter of X is the lineal units of fencing that would be required the keep the animal in polygon X, including any fencing required to keep the llamas from crossing into any polygon nested within X.

Coverage Topology—More Details

Figure 4-18 is a coverage with both arcs and polygons. The arcs are shown with Arabic numbers, the nodes with lowercase letters, and the polygons with capital letters. All cases of configurations of arcs and polygons you are likely to encounter are illustrated here. You should be able to tell, for each arc, the designation of the polygon on the left and the one on the right. For arc 11 do you get N on the left and P on the right? For arc 12 do you get N on both the left and right? For arc 17 do you get O on the left and S on the right?

Did you ever wonder how the computer "knew" which polygon your mouse cursor was in when you used the Identify tool? You, of course, can look at the cursor and the image and tell which polygon the cursor is in. Your head contains a remarkable spatial data processing system. But how does the computer know? If you are interested in an explanation, locate information (on the Internet or elsewhere) on the

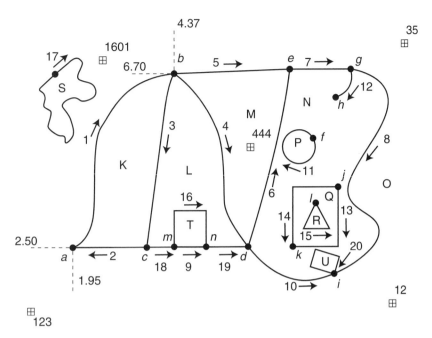

FIGURE 4-18 A coverage illustrating various conditions of arcs and polygons

"point in polygon" problem. To facilitate the solution to this problem, the coverage data model stores, for each polygon, a list of the arcs that compose it. While not directly available to the user, this list is useful to the software when spatial analysis is performed. You could consider that the list of arcs is stored for a given polygon by starting at a "from node" and naming an arc that emanates from the node. Then the next arc is named, and the process continues in this fashion until the first node is encountered again. List elements are separated by commas. For example, the arc list of polygon K could be 1, 3, 2. (or 3, 2, 1. or 2, 1, 3.) Here are a couple wrinkles: If an arc is encountered whose direction is opposite to the one originally established, a negative sign is placed in front of the arc number. Also, if the polygon contains island polygons, their arcs are specified as well, preceded by a zero in the arc list, to indicate the break in the continuity of the arcs. Verify that an arc list of polygon N is 7, 8, –10, 6, 0, 11, 0, 13, –14, 0, 20. If you could add the lengths of these arcs you would get the perimeter of polygon N. Extra credit: Verify that an arc list of the outside polygon is 1, 5, 7, 8, –10, –19, –9, –18, 2, 0, 17.

Regions (in Coverages)

A polygon coverage consists of a set of polygons; each polygon refers to some surface area on the Earth. You may define one or more subsets of those polygons. Each of those subsets is a "region." For example, suppose you wanted a data set that depicted all the states in the United States. You would find yourself delineating two types of areas. First, obviously, there would be those defined by traditional state boundaries, which divided the landscape. But you would also find states like Massachusetts. It consists of a mainland part plus islands. Delineating the surface area of Massachusetts requires several polygons, since the water around those islands cannot obviously be considered land area belonging to the state. So

233

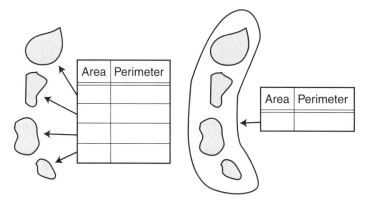

FIGURE 4-19 **Separate polygons and those polygons combined into a region**

you are faced with the prospect of a state consisting of several polygons, which implies that the attribute table would consist of several records for the single state. This is not what you want, since many attributes are statewide descriptors—population, name of the governor, budget figures, and so on. In the case of such a state, counties might consist of several islands. A solution is to define the state to be a region with multiple polygons but only one record in an attribute table. The area of the region would be the sum of the areas of the polygons. The perimeter of the region would be the sums of the perimeters of the polygons. The region could then possess other attributes applicable to the entire set of polygons. See Figure 4-19.

Coverages—Layout in the Computer

Coverages[17] have a somewhat counterintuitive organization in terms of folders and files on the disk drives of a computer. It goes like this: Every coverage resides in a *folder* (or, as it is sometimes referred to, a *directory*) called a *work space*; the choice of the work space folder location and folder name is up to the user. Much, usually most, of the stuff a coverage is made of is in another folder within the work space—a subfolder. This subfolder has the name of the coverage. Inside the subfolder you will find files whose names are arc.dbf, arx.dbf, and potentially many others. These contain, for the most part, the geometric elements of the coverage. There will be such a subfolder, with a coverage name, for every coverage in the work space. Also, there will be a text file named "log" that contains a line for each command executed on the work space.

In addition, the work space folder will contain a subfolder named INFO. This subfolder contains the attribute data files for all the coverages in the work space. The names will be unhelpful designations like arc0005.dat and arc0005.nit. ArcGIS knows which pairs of files go with which attribute tables and will tell you if you ask in the proper way.

To be clear about this: If work space AA contains only coverages QQ, RR, and SS, there will be four subfolders in the folder AA. These will be QQ, RR, SS, and INFO. A most important fact here is that you cannot successfully copy a coverage from one place to another using the operating system of the

---

[17]This discussion relates to Desktop coverages and Workstation coverages. The coverages in ESRI's product "PC ARC/INFO" have a different internal layout.

computer. Every coverage is split into two pieces, and to copy it you must use ArcCatalog. You may, of course, copy entire work spaces from one place to another using the operating system.

Further, you cannot rename a coverage using operating system actions. You may recall that the relational database table for a coverage has column names that include the name of the coverage. For example, coverage MYCOV would have MYCOV# and MYCOV-ID as field names. Changing the name of the folder with the operating system would obviously not work, since it would put the name of the coverage out of sync with the attribute table. You can change the name of a coverage with ArcCatalog.

Geodatabases

Geodatabases are the current "coin of the realm" in ESRI software. All the tools being developed deal with geodatabases. Gone is the integration (and confusion) of lines and polygons. A plethora of topology rules and topology fixes accompany geodatabases. Ideally all coverages (and shapefiles) would be converted to geodatabase form. But that is a monumental undertaking, which, if you continue in the GIS field, you may well be part of.

The Geodatabase Data Model (Personal Geodatabases)

Now please turn your attention to this other fundamental data structure for storing geographic and attribute information: the geodatabase. There are two flavors: the ArcSDE geodatabase—sometimes called the "enterprise database"—and the personal geodatabase. They are distinguished from each other primarily by the class of machine they work on and some slight functional differences.

ESRI developed the geodatabase

❑   To take advantage of increased computing power and data storage

❑   Because ideas of how to store geographic data have become more refined

❑   To "umbrella-ize" the different forms of spatial data storage: vector, raster, and, in the future, TIN

❑   To permit the use of "objects" that depict real-world entities, both in terms of description (which is still done with attributes) and behavior

In describing geodatabases, I will build on what you have learned about coverages. In so doing I will take up the geodatabase topics in somewhat different order than the preceding.

Geodatabases—Layout in the Computer

I said early on that a GIS was the marriage between a (geo)graphical database and an attribute database. The geodatabase still adheres to this in concept. I also said that usually the attribute database was housed in a commercial database management system (DBMS) such as Microsoft Access, dBASE, FoxPro, IBM DB2, INFO, Informix, Microsoft SQL Server, or Oracle. The new wrinkle is that the entire thing—geographic part and attribute part—is housed in a single DBMS file. This means, from the point of view of software, all of the geographic data sets have been rolled up with the attribute data into a single file. For ArcSDE databases this file may be located in one of several commercial relational database systems. Among those used currently are Oracle, Informix, Dynamic Server, Sybase, DB2, and Microsoft's SQL Server.

Personal geodatabases are housed using the Microsoft Access database system. With Arc9, personal geodatabases and ArcSDE geodatabases have virtually the same capability. Differences between personal geodatabases and ArcSDE geodatabases include the lack of ability in a PGDB to automatically keep up with versions of the data and restrictions of who can make changes to the database. Also, personal geodatabases are usually smaller, are run on less powerful machines, and are intended for only a few users in a working group.

With the single file implementation of a GIS in a geodatabase there is, therefore, no temptation to go in with the operating system to move, delete, rename things; the components are somewhat hidden from the user, except through ArcCatalog and ArcMap.

Geodatabases—Logical Construction

Within the single file of a geodatabase there is the framework for quite a complex hierarchy of elements. You have had some experience with this hierarchy earlier, but here is a summary, with a bit of additional information. The description is based on the personal geodatabase, which resides within a folder. ArcSDE geodatabases look somewhat different, but only at the top levels.

The database may consist of the following:

(A) Freestanding, and not necessarily related:

❑ Feature classes, resembling the point, line, and polygon classes you have dealt with

❑ Raster data sets, which may represent surfaces (e.g., elevation), areal phenomena (e.g., land cover), or images (e.g., orthophotoquads, scanned maps)

❑ Triangulated irregular network (TIN) data sets[18]

❑ Tables, which are referred to as *object classes*, and which may be imbued with "behavior," as discussed later in the text.

(B) *Feature data sets*, whose constituents share a common geographic reference (datum, projection, units, and so on) and that are composed of the following:

❑ All the elements cited above in (A)

❑ A relationship class that is a set of relationships between the features of two feature classes

❑ A geometric network that consists of

❑ A junction feature class

❑ An edge feature class.

Geometric networks are useful in a variety of areas, such as routing school buses over a road network and keeping track of electrical or piping systems. While I have avoided trying to divide GIS applications into categories, you could consider that spatial problems that involve flows of entities through conduits to be a major subclass of GIS problems—making GIS of major interest to utility companies. Geometric networks support this sort of activity.

---

[18]As of this writing TINs are not yet included with geodatabases.

Geodatabases—Feature Shape

Though many of the concepts remain the same in the storage of vector-based data, there are differences between coverages and geodatabases. For the most part, those differences that affect the user are mainly in terminology and enhanced abilities of the geodatabase vector model. Almost anything that you can do with coverages you can do with geodatabases in Arc9. And there are several things regarding points, lines, and polygons that you can do easily with geodatabases that are more difficult with coverages.

The concept that a row in a table contains the attribute values of a single feature remains the same for geodatabases as for coverages. However, geodatabases allow considerably more variety in what constitutes a feature. Specifically, geodatabases allow multipoints, multipart lines, and multipart polygons.

Points

In storing point features, geodatabases allow "multipoints." A *multipoint* is a collection of points associated with only a single row in the database, so all the attribute values in that row apply to all points. An ecologist may have mapped gopher holes in an area. The only recorded difference between them is location. So they may be stored together as a single feature. The coverage data model would have required that each be stored as a separate feature. See Figure 4-20, which shows two features—one multipoint feature depicted with dots and another shown with x's.

In geodatabases, the concept of the "tic" is gone. Features are tied to real-world points in a different way, to be described when we discuss editing.

Lines

For linear features, the term *arc* is not used. Linear features are represented by polylines. A *polyline* is composed of one path or several paths. A *path* is composed of sequentially connected segments that may be straight lines, but also geometric curves. You may use a portion of a circle or an ellipse (which are, mathematically, the plots of second-order equations), or you may use a type of spline, called a third-order Bézier curve. The path is a close cousin of the arc in that it is a sequence of segments, and it has a left side and a right side. If a polyline is a multipart polyline, then the paths that it is composed of may be connected, disjoint, or some of each. Look at Figures 4-21, 4-22, and 4-23 for an understanding of segments, paths, and polylines.

Two multipoint features

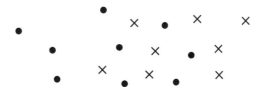

FIGURE 4-20   Two multipoint feature classes: one dots, the other x's

237

Geodatabase segments

A path—composed of segments

Straight line   Circular arc   Elliptical arc   Bézier arc

FIGURE 4-21   **Permitted types of geodatabase segments**

FIGURE 4-22   **A path in a geodatabase is composed of segments**

Single-path polyline

Multipath polyline

FIGURE 4-23   **A polyline may be composed of one or more paths**

Polygons

Polygons in geodatabases are stored quite differently from polygons in coverages, though the quantities of area and perimeter are the same for any given polygon, regardless of the storage paradigm. Whereas polygons in coverages are bounded by one or more arcs, a single-part geodatabase polygon, without any island polygons within it, is simply enclosed by a single ring. A ring might be thought of as a single path (see the previous section, "Lines.") that starts and ends at the same point—that is, a ring is a closed figure. Since it is a closed figure, the software knows whether any given arbitrary point is inside the polygon or outside. One effect of using the complete, single entity (ring) to delineate a polygon is that it divorces a given polygon from its neighbors. (It also means that, for traditional disjoint polygon representation, such as ownership parcels, each vertex is stored twice.) So the traditional coverage topology, which ensured no gaps or overlaps between polygons, is no longer present. This sort of topology has been replaced by a much more general set of topological checks that the user can invoke to assure data integrity.

A geodatabase "polygon" may be what is called a multipart polygon. (A multipart polygon in coverages relied on the concept of "region." previously discussed.) A multipart polygon may be a set of two or more polygons. Either single-part or multipart polygons may have other polygons nested inside them. So the term "polygon" encompasses a multitude of conditions. Please look at the illustrations in Figure 4-24.

Nested Polygons in Geodatabases

The calculations of plane area (called Shape_Area) and perimeter (called Shape_Length) are the same with geodatabase polygons as with coverages. Missing are the complications of nodes and arcs. Look at Figure 4-25, which is the same as Figure 4-16, with the addition of an island polygon D that has dimensions

Polygon defined by a single ring

Polygon defined by multiple disjoint rings

Polygon defined by multiple nested rings

FIGURE 4-24   Different configurations of a "polygon" composed of multiple enclosed areas

of 2 units by 5 units. Be sure you understand the area and perimeter calculations of polygon A, particularly with regard to the Shape_Area and Shape_Length.

Geodatabases and Attributes

I said that a GIS was the marriage of a geographic database to an attribute database. In a geodatabase, each row in the attribute database refers to an "object" that is one point feature, line feature, or polygon feature. When geodatabases are contrasted with coverages, a major difference is in what constitutes a feature—and therefore what (geo)graphic entity it is that deserves a row in a table. The definitions of point, line, and polygon have been well stretched.

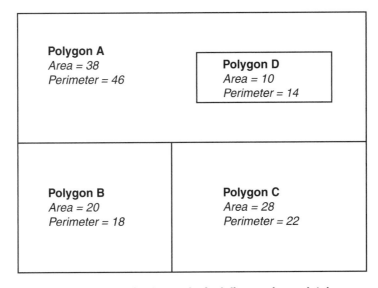

Polygon A
Area = 38
Perimeter = 46

Polygon D
Area = 10
Perimeter = 14

Polygon B
Area = 20
Perimeter = 18

Polygon C
Area = 28
Perimeter = 22

FIGURE 4-25   The simple topological diagram in geodatabase form but with an additional polygon: D

239

One of the indications of maturity of the GIS field is the growing emphasis on attribute data correctness and integrity. All large databases, spatial and otherwise, contain errors. With geodatabases, a number of built-in capabilities promote data quality. For example, suppose you are building a database of the roads in your county. One attribute in the database is the material the road is made of. Perhaps you know that the only allowed materials are concrete, asphalt, macadam, and gravel. With geodatabases you could allow a data entry person to only select among these four. First, this makes data entry faster. Second, it avoids the possibility that someone will type in "asfault" instead of "asphalt." The items concrete, asphalt, macadam, and gravel constitute a domain for the roads feature. You could also set a default value for roads. If no other value is entered, the value of the attribute would automatically be set to "concrete."

Subtypes

Geodatabases go even further in promoting data integrity. Within a feature type, say, roads, you can define subtypes of roads. Perhaps your planning agency has classed all the roads as freeway, major, or minor. You could set up the database so that "concrete" was the default material for freeways, "asphalt" for major roads, and "macadam" for minor roads. And you could set the domain for minor roads so that the number of lanes could only be one or two. A subtype is basically an attribute of the feature that gets special attention from the software.

Geodatabases and Precision

The coordinates of features in a geodatabase are stored, somewhat surprisingly, as long integers. In general, in computing, when one wants to store very big numbers or very small (in absolute value, that is, close to zero) numbers, one uses the floating-point capability of the computer. This stores each number as a mantissa and an exponent. With geodatabases, however, a coordinate (say the x-coordinate) is stored as a 31-bit binary integer. That value is then scaled when display or output is desired, into a floating-point number. The smallest number (of a long integer, in absolute value) is, in binary

    00000000000000000000000000000000

while the largest is

    +1111111111111111111111111111111

The first is, of course, zero. The second is 2 to the 31st power, minus 1, or, in decimal representation:

    4,294,967,295

—that is, a bit over four billion.

Let's suppose that features in an area to be represented span 1000 kilometers and that the coordinate system is couched in meters. This would mean that a million meters (1000 * 1000) must be represented linearly by an integer variable of range zero to roughly 4 billion. How precisely could one represent a given x-coordinate? About 1 million divided by 4 billion, or 0.0025 meters. (The precision reported by the software is actually the inverse of this. That is, it is the number of integer units that "fit into" the real-world unit. So the software would say the "precision" was 1/0.0025, or 400.)

As the size of the area to be represented grows, the precision is reduced. If you wanted the x-coordinate to span the entire United States, say, 15 million feet, then the precision would be about 15 million divided by 4 billion, or 0.00375 feet—roughly a twentieth of an inch. In other words, there is no lack in the software for precise representation, almost regardless of the range of the data. (Accuracy, of course, is a totally different matter.)

Objects—First Acquaintance

A more profound difference between the depiction of features in the coverage data model and that of the geodatabase is that features are not just geometric entities with attributes, but *objects*, in the computer science sense of the word.

The study of electricity and magnetism is customarily divided into two general areas: fields and circuits. The term *fields* refers to the characteristics of the invisible forces that are caused by magnetic material or by current flowing in a wire. The term *circuits* refers to the study of electricity where electrons are confined to wires and other elements. You might think of a loose analogy between fields and areal features, on the one hand, and circuits and linear (network) features, on the other. In networks, entities such as trucks and gas molecules are confined within physical structures, as the electrons are confined within the wires of a circuit. While ESRI products have had network capability for a long time—mainly to deal with transportation systems—the geodatabase takes this capability to new heights. Using the networking features of Arc9 geodatabases, you can represent and simulate complex and extensive linear, human-built infrastructure—loosely: pipes, wires, and roads.

This new networking capability is facilitated by storing geographic features and their attributes in a database system that is "object-oriented." Each row represents an object. Objects are described by attributes. But objects also can have "behavior." For a human example, you might describe a person's characteristics with attributes, such as weight, hair color, and irritability. But if you, in addition, "allow behavior," then the person might be instructed to drive to the store for a jar of pickles. This process could involve other objects: a particular automobile instructed to allow the person to drive it, a cashier who would accept money for the pickles, and so on. To bring this closer to GIS, a road object might be allowed to connect to another road object, but not to a freeway object. For another example, a high-pressure gas line could be connected to a high-pressure valve, but not to a low-pressure valve. Objects in geodatabases bring us one step closer to integrating the various ideas and components of GIS. A detailed discussion of geodatabase objects is beyond the scope of this text.

Shapefiles

Geodatabases and coverages are powerful and sophisticated data structures. In addition to topology, you get for free the area and perimeter of delineated areas and the lengths of linear features. But ESRI also supports a much less complex data structure: the shapefile.

A shapefile combines the same two essential major elements that geodatabases and coverages do: a (geo)graphic component and an attribute database. The database software is a relational database management system named dBASE.[19]

A particular shapefile can represent points, multipoints, polylines, or polygons. With points, each individual point has a record in the relational database. If a number of points are considered the same object,

---

[19]The vendor is dBASE, Inc. The Web site is dBASE.com.

then that object has only one record in the attribute table. As with geodatabases, polylines can be composed of one or more paths, connected or disjoint. However, the paths are composed only of straight-line segments.

A polygon in a shapefile bears similarity to a polygon in a geodatabase, but no topology is present or can be created. Each polygon is a stand-alone affair. It is delineated completely by one linear entity: a sequence of segments that starts in one geographic location and returns to that location. There may be adjacent polygons or not. Other polygons may overlap it. See Figure 4-26.

A problem to which shapefiles are particularly susceptible is that there may be slivers of overlap or slivers of vacancy between intended polygons, and there is nothing you can do about it if the data remain in shapefile format. In Figure 4-26 the sliver "a" is claimed by both W and X, while sliver "b" is in neither.

A geographic area partitioned into mutually exclusive shapefile polygons will have duplicate information. Since the total boundary of each polygon is defined for that polygon, the lines are "double digitized." Within a single coverage, you get the guarantee that only one boundary exists between two adjacent polygons. With shapefiles, there is no such guarantee; there are two independent boundaries, and you have no assurance that they are congruent. With geodatabases, you can make topological rules to ensure that polygons do not overlap or have gaps, but not with shapefiles.

With shapefile polygons, you do not get the area and perimeter as you do when you build polygon topology on coverages. Few of the spatial processing tools in ArcToolbox work on shapefiles.

The advantages of shapefile representation are simplicity, processing speed, drawing speed, and, usually, economy of storage. Shapefiles are useful when you do not need sophisticated geoprocessing. Do be aware that a lot of GIS data sets have been put into shapefile format. There may be considerable conversion to geodatabase format in your future if you want to use those data sets in geoprocessing.

Shapefiles—Layout in the Computer

The format for a shapefile on a disk drive is also much simpler than that of a coverage or a geodatabase. Basically, at least three files in a folder are required for a shapefile. If a folder named AMENITIES contains a shapefile named LAWN_SPRINKLERS, then AMENITIES will contain at least these files:

lawn_sprinklers.shp (contains the geographic information)

lawn_sprinklers.shx (contains a spatial index to the geographic information)

lawn_sprinklers.dbf (contains the dBASE table for the attribute information)

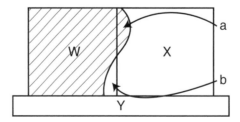

FIGURE 4-26   Shapefile polygons illustrating problems with overlap and gaps

Other files (lawn_sprinklers.sbx, lawn_sprinklers.sbn, and lawn_sprinklers.shp.xml (containing the metadata) may also be present. So the term "shapefile" is somewhat a misnomer, if one expects a single computer file. ArcCatalog, of course, represents a shapefile as a single entity, but to the operating system, it is several files. You could actually move a shapefile from one folder to another using the operating system. You would simply move all the constituent files, but that is not a recommended practice; use ArcCatalog instead. (Recall that you cannot move a coverage by moving its folder with the operating system! With geodatabases there is no issue with moving or renaming because you can't see the constituent feature classes with the operating system; they are locked away in a single database file.)

When you access a folder containing the files that make up a shapefile in ArcMap or ArcCatalog, you see only the designation—continuing our example—lawn_sprinklers.shp. The other files are hidden, so you may think of the shapefile as just one entity.

Shapefiles are important because they have been around for some time and immense numbers of data sets are in shapefile format. Shapefiles were the data format of choice for ESRI's ArcView software (up to version 3).[20]

Summarizing Vector Data Set Features

For various reasons relating to technology and history, ESRI has three major ways of storing vector data sets: geodatabase, coverage, and shapefile. Some of the differences relate to functionality and some mostly to terminology. Somewhat surprisingly, shapefiles and geodatabases are more alike than are coverages to either. Geodatabases and shapefiles allow multipoints. They both delineate polygons with rings. They use polylines, composed of segments, with one or more paths. The similarities end there. Geodatabase segments may be circular arcs, elliptical arcs, or Bézier curves. Geodatabases allow for topological relationships, shapefiles do not. A major feature of geodatabases is the ability to use a geometric network (composed of junctions and edges that possess not only attributes, but behavior). Geodatabases calculate the lengths of lines and rings, and the areas of polygons—and shapefiles do none of that. Geodatabases allow subtypes of features, while shapefiles do not. So the semi-witty comparison that a geodatabase is a "shapefile on steroids" greatly understates the case.

Coverages are derived from the concept of the arc, which serves to delineate both one-dimensional networks and polygons. Topology within a given feature data set is sophisticated, but limited to that data set. Variations on the themes of networks are routes, and polygons may be considered in subsets to form regions. The multipoint feature class is not supported. Neither is the polyline feature class supported, but similar functionality exists through routes and measures, which are described later.

Summary of Logical Structures of Vector-Based GIS Data Sets

In terms of logical structure and layout within the computer, shapefiles are at the surface. The files that constitute a shapefile may reside anywhere on a hard disk, provided that they are all in the same folder. Attribute data is always stored in dBASE tables. Only a single feature type—point, multipoint, line, polygon—is stored in a given shapefile.

---

[20]ESRI has used the term "ArcView" in two fundamentally different ways. ArcView versions 1, 2, and up through 3.3 is a computer program. The ArcView associated with ArcGIS (that is, versions 8 and 9) describe a level of functionality.

A coverage is buried deeper in the file structure of a computer. It exists, in a fragmented way, in a work space that may contain other coverages and must contain a folder named INFO. The work space is itself a UNIX directory or Microsoft folder. The work space may also contain raster and TIN data sets.

A personal geodatabase is a single file in Microsoft Access. An enterprise geodatabase may use other commercially available database management systems.

Raster-Based Geographic Data Sets—Logical Construction

In GIS a raster is a set of equally sized squares[21] that cover a rectangular surface. Rasters are used in two basic ways in GIS:

1. As *grids* that store attribute information about the area covered by the square (such as elevation, or soil type), in which case the individual squares are called cells

2. As *images* such as orthophotoquads, where the squares are picture elements (pixels), containing values that prescribe the intensity of red, green, and blue color.

Grid Rasters

The *grid* data model, which can be very useful in spatial analysis,[22] comes in two flavors: those in which the cells contain integer numbers and those in which the cells contain floating-point numbers. In each case, each grid cell may contain a single number. If that number is an integer, then the grid represents categorical or discrete data. Each different integer represents a type of object or a condition. If the numbers in cells are floating point—that is, contain numbers that may be composed of decimal fractions—then the grid may represent continuous data such as an elevation surface over the area of interest. Integer grids usually[23] have associated with them a value attribute table (VAT). Floating-point grids do not have a VAT.

Each record in the VAT of an integer grid has a minimum of three fields: an ID field, a Value field, and a Count field that indicates the number of cells in the data set with the given value. For example, Table 4-6 shows a part of the Kentucky land use data set you examined earlier.

TABLE 4-6

ObjectID	Value	Count
—		
1	5	2,387,059
—		
5	14	2,606,086
—		

---

[21]Or almost square—some rasters are based on fractions of a geographic degree.

[22]Use of grid rasters will be discussed in detail in Chapter 8. Here I want only to acquaint you with the data structure and how it is stored in the computer.

[23]If there is an immense number of different integers, a VAT might not be created, due to storage limitations.

You may recall that value = "5" indicated water and "14" indicated transportation, communication, and utilities. A set of cells that contain the same number is called a "zone," whether the cells are adjacent to one another or are unconnected. The Count column is useful because, when it is multiplied by the area covered by a single cell, the total area of the zone is obtained.

Each cell in the grid can contain a number (value) or can contain an indicator that says that no number is assigned to the cell: NODATA.

A floating-point grid has no VAT because, usually, almost all the values in the cells are unique. Cell values would be numbers like 45.312 and 46.789. Thus, the number of records would come close to the number of cells and the COUNT for most records would be "1," since it is likely that very few cells would contain identical numbers.

Image Rasters

Raster images come from scanning an area. The two primary types of area scanned are portions of the Earth (scanned by satellites) and maps (scanned by various hardware devices such as flatbed or drum scanners).

You saw the table of an image raster in your examination of the digital raster graphics file COLE_DRG. That table looked partially something like Table 4-7.

TABLE 4-7

ObjectID	Value_	Red	Green	Blue
0	0	0.00	0.00	0.00
—				
3	3	0.79	0.00	0.08
4	4	0.51	0.25	0.14
—				
12	12	0.80	0.64	0.55

The "Value" here is simply a code for a color (or grayscale) with the intensity of red, green, and blue shown by the floating-point numbers in the indicated columns.

Raster-Based Geographic Data sets—Layout in the Computer

The simplest storage of a raster band would be something like this:

```
66666666777777754444449999999
666666777777777777777444444999
6666677777777777777744444445555
```

That is, the integers are stored in sequential locations in the memory of the computer or on a disk. In the preceding case, the number of columns (i.e., the length of a row) is 29 and the number of rows is 3.

The nice thing about storing a raster band is that very little addressing needs to take place to know the location of a grid cell in geographical space. If you know (a) the location of the upper left corner of the upper left cell, (b) the cell size, and (c) the orientation of the grid, you can then easily calculate the location of any cell, given its row number and column number. To know what the value of the cell at column 14 and row 3, you need only multiply the row number minus one by 29 and then add the column number. In the example, this gives $2 \times 29 + 14$ or 72. If you start counting in the upper left corner of the sequence of grid values, wrapping around the end of a row to the beginning of the next row, you will find a "7" when you get to the 72nd value. Though this example is highly simplified, it gives you an idea of an addressing scheme that may be used to store raster bands.

Grids may be very large. A grid composed of hundreds of millions of cells is not unusual. (The Kentucky land use grid that you looked at has about 300 million cells.) The number of cells is the product of the number of rows and the number of columns. If there are 10,000 rows and 10,000 columns, then there 100 million cells. If you make each cell half the size, then the number of rows and columns is doubled (to 20,000 each), so the number of cells goes up to 400 million.

The sheer size of grids produces a problem, which in years past limited either the size of the area covered or limited the level of detail—that is, the cell size. Great advances in computer storage—both electronic (RAM) and mechanical (hard drives, DVDs)—have solved part of the problem. But huge grids are also made possible by advances in data compression techniques. One such approach, which has been around for a long time and is simple to implement, is called run-length encoding. The idea is based on the fact that if you pick any cell in an integer grid, the cell to its right probably has the same value. So we might take advantage of this fact and encode the preceding data as follows:

| Row 1: | 1-8:6 | 9-15:7 | 16-16:5 | 17-22:4 | 23-29:9 |
| Row 2: | 1-6:6 | 7-20:7 | and so on. | | |

This says that in row 1 there are eight sequential values of 6, seven values of 7, one value of 5, and so on.

Whether this approach uses less storage depends on the data. For example, if every cell were different from its neighbor, this scheme would be much more costly, but usually the savings in memory are enormous.

As I have said several times, a computer file is composed of 1s and 0s. The idea of a compression scheme is to store the same information in a new file in many fewer 1s and 0s, and then be able to reconstitute the original file exactly, so that no information is lost. This is vital if the file is, say, a computer program where one wrong bit can sink the whole enterprise. Zipping programs do this sort of compression, called *lossless*. But if one is not picky about being able to exactly reproduce the original file—say, it is a photo in which you will accept a near replication in exchange for a great reduction in file size—then you could use a *lossy* compression method. To know more, type "lossless" and "lossy" into a Web search engine.

Formats accepted by Arc9 for raster data sets are as follows:

ESRI GRID

ERDAS IMAGINE

TIFF

MrSID

JFIF (JPEG)

ESRI BIL

ESRI BIP

ESRI BSQ

Windows Bitmap

GIF

ERDAS 7.5 LAN

ERDAS 7.5 GIS

ER Mapper

ERDAS Raw

ESRI GRID Stack File

DTED Levels 1 and 2

ADRG PNG NITF

NITF

CIB

CADRG

TINs

A surface is a mathematical entity—in particular, it is a function of two variables. That is, if you consider the Cartesian plane, with variables x and y, for every coordinate pair, there is one, and only one, third value z. Visually you can imagine a flat plane with a cloth billowed above it. The distance, measured perpendicularly from the plane to the cloth is the z value. (Actually, some values of z might be negative, which would indicate that part of the cloth was below the z = zero surface.)

An obvious example of such a surface is elevation[24] above sea level. Another is the daily pollution level of a particular contaminant. A third might be wind speed at a given moment at a given altitude. For every geographic point, there is a z value for these themes. Since there are an infinite number of points on a plane, there are an infinite number of z values. Obviously, we could not store an infinite number of values, even if we knew what they were, in a finite computer store. We therefore apply the usual GIS technique, which by this time you are used to, of storing some data and inferring information as we need it. The triangulated irregular network (TIN) is such a device. TINs, which you met in Chapter 1, are described in more detail in Chapter 9, which deals in part with GIS and the Third Spatial Dimension. The goal here is to describe the data model that lets the computer tell you about the surface that the TIN represents.

---

[24]Elevation is an obvious, but not perfect, illustration. A mathematical surface may have only a single value at a given point and almost everywhere on Earth's surface this is the case. However, in a few places, like where there is an overhanging cliff, the surface of the Earth can have three or more values. For GIS to take these into account would create immense complication, so it gets ignored.

The idea of a TIN is to approximate a surface with a set of planes—triangles, to be specific. A triangle has some nice characteristics: If you know the x-, y-, and z-coordinates of each of its three vertices, you can calculate

❏ The z value (e.g., elevation) of any point on the surface of the triangle

❏ The triangle's slope (the maximum quotient of rise over run)

❏ The triangle's aspect (the direction of the projection on the x-y plane of a line perpendicular to the surface of the triangle)

You could also calculate the area and perimeter of the triangle, but it turns out these aren't usually important. Since the triangle is (usually) at an angle to the x-y plane, the area of the triangle will be greater than it's projection onto the plane.

The trick to making a TIN yield useful information is to connect a number of irregularly spaced x-y-z points, of known value, with straight lines so that they form triangles that are not too skinny. The software, based on clever algorithms,[25] takes care of this for you.

TIN-Based Geographic Data Sets—Layout in the Computer

If you look at a TIN in terms of folders and files, it looks a lot like a coverage. There is a single folder, residing within a work space, with a bunch of files within it. As with coverages, if you want to copy it, move it around, or rename it, you must do so with ArcCatalog, not the operating system.

The primary apparent difference between a TIN data set and other GIS data sets is that there is no attribute table. You can, however, use the Identify tool with TINs. It provides the elevation, slope, and aspect, as well as some additional information tags that the user may add. How does it do this? By calculating information on the fly when you click on a point.

For particular problems such as surface analysis, surface display, and hydrological analysis, TINS can be quite useful. You will see more of them in future chapters.

Spatial Reference

Of course, every instance of every spatial data model discussed previously must have as a foundation a spatial reference to the real world. As discussed, this is not simple. If it were, perhaps Microsoft or IBM would be the principle vendor of GIS software. And if the world had been created as a big cube, rather than a big sphere, life would be easier for GIS specialists.[26] But we are stuck with the complexities of

---

[25]Algorithms involving Delaunay triangulation and Thiessen polygons (also know as Voronoi cells and Dirichlet regions).
[26]Except for those living on the edge. (Pun intended.)

spherical trigonometry, geodesy, and myriads of coordinate systems and datums. You could say that the spatial reference consists of three parts:

❑ A coordinate system, with its associated map projection and parameters

❑ A spatial extent and domain, which defines latitude and longitude (or x and y), and in some cases an elevation (denoted by z values), and, in some cases, distances along lines or paths (denoted by m values)

❑ A scale, when display is involved, that relates units of linear measure on the map to those on the ground.

The method of coordinate system application is handled differently with different ESRI products. One way is a projection file, which, in a coverage, might look like this:

```
Projection      STATEPLANE
Zone            3976
Datum           NAD83
Zunits          NO
Units           FEET
Spheroid        GRS1980
Xshift          0.0000000000
Yshift          0.0000000000
Parameters
```

Of course, this is only part of the story. Lines of this file point to additional complexity, which might look like this:

```
PROJCS["NAD_1983_StatePlane_Kentucky_North_FIPS_1601_Feet",
GEOGCS["GCS_North_American_1983",DATUM["D_North_American_1983",
SPHEROID["GRS_1980",6378137.0,298.257222101]],
PRIMEM["Greenwich",0.0],UNIT["Degree",0.0174532925199433]],
PROJECTION["Lambert_Conformal_Conic"],
PARAMETER["False_Easting",1640416.666666667],
PARAMETER["False_Northing",0.0],
PARAMETER["Central_Meridian",-84.25],
PARAMETER["Standard_Parallel_1",37.96666666666667],
PARAMETER["Standard_Parallel_2",38.96666666666667],
PARAMETER["Latitude_Of_Origin",37.5],
UNIT["Foot_US",0.3048006096012192]]
```

And then there is the software that has to interpret all this.

Your job, as a GIS specialist, is not to understand it all, but to be sure that when you combine data sets, there is agreement in all the specifics. If there isn't complete agreement, then you need to consult an expert in geodesy to determine what errors are introduced by mixing data sets with different parameters, and whether those differences are important to your work.

Structures for Storing Geographic Data

The exercises in this chapter will acquaint you intimately with both the logical structure of the ESRI data sets and the methods used for storing those data sets on disk. In the process, you will be introduced to a major component of ArcGIS: ArcToolbox. You will also learn more about ArcCatalog and ArcMap. As before, you should view almost every step as containing a nugget of information that you will find useful in the future. Since the software is quite complex, it is a good idea to ask yourself at each step if the capability it suggests is one you will be able to remember how to tap, or whether it belongs in your Fast Facts File.

In Exercise 4-1 you are introduced to ArcToolbox. ArcToolbox is available, through an icon on the Standard toolbar, from both ArcCatalog and ArcMap. In two exercises that follow, we look at some trivial ESRI data sets, basically to see their underlying structure. Exercise 4-2 looks at point, line, and polygon personal geodatabase feature classes. Exercise 4-3 examines point, arc, and polygon coverage feature classes. In each case you will see the minimum in terms of attribute names. You will look at the geographics, then at the attribute tables. You will augment the attribute tables of some of the data sets by adding the x- and y-coordinates of points, using ArcToolbox.

_____ **Open up your Fast Facts text or document file.**

Exercise 4-1 (Warm-up)

Meet ArcToolbox

_____ **1.** Start Arc Catalog. Verify that the folder Trivial_GIS_Datasets is in

___IGIS-Arc_*YourInitialsHere*

_____ **2.** Start ArcMap. The ArcToolbox icon is a red tool chest on the Standard toolbar.[1] Click it and the ArcToolbox appears. Click it again and the toolbox disappears. ArcToolbox is a window that can be docked or resized.

_____ **3.** Show ArcToolbox and collapse everything that will collapse (i.e., close any little tool chests that might happen to be expanded). The result should look something like Figure 4-27.

---

[1]ArcToolbox is also available as an icon on the Standard toolbar in ArcCatalog.

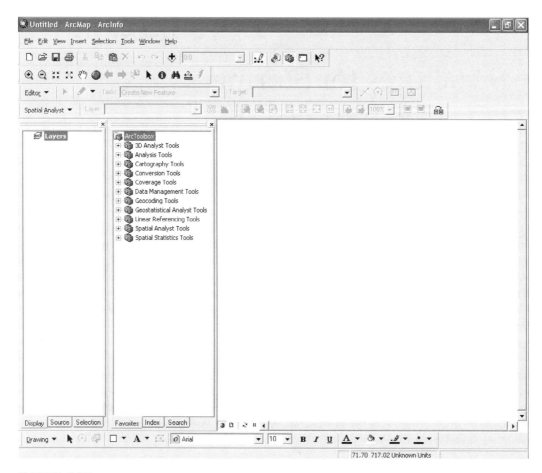

FIGURE 4-27

If you are running ArcGIS 9.0, you will see 11 basic categories of tools available:

❑ 3-D analyst tools

❑ Analysis tools

❑ Cartography tools

Not all of these

❑ Conversion tools

❑ Coverage tools

❑ Data management tools

❑ Geocoding tools

❑ Geostatistical analyst tools

❑ Linear referencing tools

❑   Spatial analyst tools

❑   Spatial statistics tools

In ArcGIS version 9.1, there are three additional toolboxes:

❑   Data interoperability tools

❑   Network analyst tools

❑   Samples

Some of these tool categories we discussed in the Overview of this chapter. But there are hundreds of tools, so the approach you should use is to believe that for whatever you want to do with GIS data sets, there is a tool for it and you just have to find out what it is and how to use it.

_____   **4.** Leave ArcMap running, with the ArcToolbox window open, for the next exercise.

Exercise 4-2 (Warm-up)

A Look at Some Trivial Personal Geodatabase Feature Classes

_____   **1.** *In ArcMap, use a different method to make a connection with a folder:* In ArcMap click the on Add Data icon. In the Add Data window you will see a Connect to Folder icon. Connect to ___IGIS-Arc_*YourInitials_Here*\Trivial_GIS_Datasets.

_____   **2.** From PGDB.mdb\PGDBFD add the personal geodatabase feature class pgdbfc_line_1. Use the fact that the Identify tool highlights a feature to look at the lines. How many lines are there? ___4___

_____   **3.** Open the pgdbfc_line_1 attribute table. Write down the names of the attributes (i.e., columns, fields). ___OBJECT ID___, ___Shape___, ___Shape.Length___

_____   **4.** In ArcMap click the New Map File icon on the Standard toolbar. Decline the offer to save the existing map. Add the pgdbfc_line_2 feature class. Using the Identify tool again. How many lines are there? ___12___

Comparing pdgbfc_line_1 and pdgbfc_line_2: The reason for the difference in the number of lines is that, in pgdbfc_lines_1, the lines cross but do not intersect. An example might be that the lines represent dual-lane highways, where one highway passes over the other. In pgdbfc_lines_2 the lines do intersect like the streets on a city grid.

_____   **5.** Open the pgdbfc_line_2 attribute table. Verify that the columns are the same as above.

_____   **6.** Again initiate a new map and don't save the existing map. Add the pgdbfc_poly feature class. Open the pgdbfc_poly attribute table. Write down the names of the items. ___OBJECT ID___, ___Shape___, ___Shape.Length___, ___Shape.Area___

____ **7.** Finally initiate a new map and add the pgdbfc_point feature class. Open the attribute table. Write down the names of the attributes.

OBJECID, _Shape_

Dismiss the attribute table. Dismiss the Identify Results window.

At this point we are going to use ArcToolbox to perform an almost trivial operation. If you will recall, we included the x- and y-coordinates of the fire hydrants in the Village Data of Chapter 1. You will add x- and y-coordinates here so you get to see a toolbox used in a simple situation.

The tool you want to use is called Add XY Coordinates. How do you find it among all the tool categories and tools available?

____ **8.** At the bottom of the ArcToolbox pane, press the Index tab. In the Type In A Keyword To Find text box, type Add. You will see tools to add all sorts of things. There are two Add XY Coordinates tools. One (arc) works on coverages; the other (management) on geodatabases. You could also search for commands using the Search tab. Try that, typing in Coordinates and pressing Search. Again you see the two tools you are interested in, and the toolbox in which they reside. Highlight the data management one and press Locate. The reference to the tool will be shown in the ArcToolbox window, highlighted. You see that it could have been accessed with ArcToolbox > Data Management Tools > Features > Add XY Coordinates—if you had known it was there.

More Help

____ **9.** Pounce on the tool to start it. In the Add XY Coordinates window, which may need some expanding, notice that at the lower right corner there is a button that controls whether a help pane appears to the right of the main window. Show the help pane. Click anywhere in the expanse of the window and read about the tool. Click the Input Features text box and read about it. If you want more help, you can press the question mark (?) at the top of the pane. That will take you to the appropriate place in the ArcGIS Desktop Help window. Go there, read, expand, peruse, and dismiss.

____ **10.** Back in the Add XY Coordinates window, what you want to do is add the coordinates to the pgdbfc_point feature class attribute table. So browse using the little yellow folder next to the Input Features text box to

___IGIS-Arc_*YourInitialsHere*\Trivial_GIS_Datasets\PGDB.mdb\pgdbfc_point.

Highlight pgdbfc_point to bring it into the Name text box. The Input Features window should look like Figure 4-28.

____ **11.** Press Add. Press OK. The window turns into a dialog box; the Cancel button becomes a Close button when the process is finished. Close the window.

____ **12.** Open the attribute table of pgdbfc_point. List the names on the column heads again.

OBJECTID, Shape, Pt_X, Pt_Y

In the following exercise, you will basically repeat Exercise 4-2 but use coverages instead.

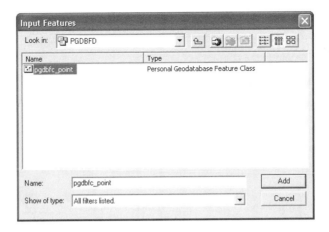

FIGURE 4-28

A Look at Some Trivial Coverage Feature Classes

1. To a new map in ArcMap add the arc, node, and tic components of the ARC_COV_1 coverage from ___IGIS-Arc_*YourInitialsHere*\Trivial_GIS_Datasets.

2. Open the ARC_COV_1 arc attribute table. Write down the names of the items.

_____FID____ , __Shape__ , __FNODE__ , __TNODE__# , _Arc_Cov1 ID_

LPOLY , _RPOLY_ , _length_ , _____ , _____

Use the Identify tool to look at the arcs. How many arcs are there? _____4_____ Notice the ARC_COV_1-ID values, added by the user: 61, 62, 71, and 72.

Open the ARC_COV_1 node attribute table. Write down the names of the items.

FID , _Shape_ , _Arc#_ , _ArcCov1#_

ArcCov 1-ID

How many nodes are there? _____8_____

Open the ARC_COV_1 tic attribute table. Write down the names of the items.

FID , _Shape_ , _IDTIC_ , _XTIC_

YTIC

255

How many tics are there? ___4___. What are the coordinates of the southwest tic? ___2000___, ___1000___. What are the coordinates of the northeast tic? ___2012___, ___1012___. Close all tables.

___✓___ **3.** In ArcMap, click the New Map File icon on the Standard toolbar. Decline the offer to save the existing map. Add the arc and node components of the ARC_COV_2 coverage. Open the attribute tables.

Open the ARC_COV_2 node attribute table. Write down the names of the items. ___FID___, ___Shape___, ___ARC#___, ___ARCCOV2#___, ___ARCCOV2-ID___

How many arcs are there? ___12___ How many nodes? ___12___

Comparing ARC_COV_1 and ARC_COV_2: As before, the reason for the difference in the number of arcs and nodes is that in ARC_COV_1 the arcs cross, but do not intersect. Such arcs could not be used to represent polygons, since arcs in a polygon, coverage must intersect at a node. In ARC_COV_2 the arcs do intersect. Notice in the arc attribute table that the ARC_COV_2-ID values (61, 62, 71, and 72) now appear multiple times. This is because we made the second coverage by forcing the original arcs from the first coverage to intersect (using the CLEAN command, to be discussed later) so that nodes were placed at the intersections. This demonstrates why you can't really count on user IDs to uniquely identify features.

___✓___ **4.** Again initiate a new map and don't save the existing map. Add the arc, polygon, and node components of the POLY_COV coverage. Open the POLY_COV polygon attribute table. What is the FID of the polygon in the first row of the attribute table? ___2___. What is the FID of the polygon in the last row of the attribute table? ___10___. How many polygons are there? ___9___. Write down the names of the items. ___FID___, ___Shape___, ___Area___, ___Perimeter___, ___POLYCOV#___, ___POLYCOV-ID___

Open the POLY_COV arc attribute table. Write down the names of the items. ___FID___, ___Shape___, ___FNODE___, ___TNODE___, ___LPOLY___, ___RPOLY___, ___length___, ___POLYCOV#___, ___POLYCOV-ID___

How many arcs are there? ___20___. Explain why some of the arcs are longer than others. ___They are "corner arcs, doubling in size___

Open the POLY_COV node attribute table. Write down the names of the items. ___FID___, ___Shape___, ___ARC#___, ___POLYCOV#___, ___POLYCOV-ID___

The ARC# attribute is the number of *one* of the arcs that connects to the node. Close all attribute tables.

Placing X- and Y-Coordinates in a Coverage

Again, we are going to place the coordinates in an attribute table. With coverages we can associate coordinates with points or nodes, or both. Let's do nodes, for variety. Recall that we have to use a separate tool. By the way, it is not uncommon to have one set of tools for geodatabases and shapefiles, and a different set for coverages. Also, you must edit coverages with an entirely different package than that used for other geodata sets. Chalk it up to software evolution.

The tool we want to use is the other Add XY Coordinates than the one we used on the geodatabase feature class.

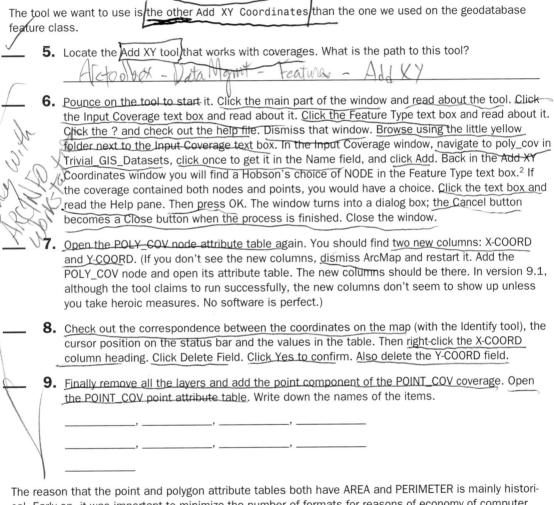

5. Locate the Add XY tool that works with coverages. What is the path to this tool?

Arctoolbox - Data Mgmt - Features - Add XY

6. Pounce on the tool to start it. Click the main part of the window and read about the tool. Click the Input Coverage text box and read about it. Click the Feature Type text box and read about it. Click the ? and check out the help file. Dismiss that window. Browse using the little yellow folder next to the Input Coverage text box. In the Input Coverage window, navigate to poly_cov in Trivial_GIS_Datasets, click once to get it in the Name field, and click Add. Back in the Add XY Coordinates window you will find a Hobson's choice of NODE in the Feature Type text box.[2] If the coverage contained both nodes and points, you would have a choice. Click the text box and read the Help pane. Then press OK. The window turns into a dialog box; the Cancel button becomes a Close button when the process is finished. Close the window.

7. Open the POLY_COV node attribute table again. You should find two new columns: X-COORD and Y-COORD. (If you don't see the new columns, dismiss ArcMap and restart it. Add the POLY_COV node and open its attribute table. The new columns should be there. In version 9.1, although the tool claims to run successfully, the new columns don't seem to show up unless you take heroic measures. No software is perfect.)

8. Check out the correspondence between the coordinates on the map (with the Identify tool), the cursor position on the status bar and the values in the table. Then right-click the X-COORD column heading. Click Delete Field. Click Yes to confirm. Also delete the Y-COORD field.

9. Finally remove all the layers and add the point component of the POINT_COV coverage. Open the POINT_COV point attribute table. Write down the names of the items.

_____, _____, _____, _____

_____, _____, _____, _____

_____

The reason that the point and polygon attribute tables both have AREA and PERIMETER is mainly historical. Early on, it was important to minimize the number of formats for reasons of economy of computer storage. It was decided that geographic points and polygons should not exist in the same coverage. Hence the similar format, and the use of PAT and PAT for polygon attribute table and point attribute table. The last three items relate to the use of labels in polygons. We defer that discussion until later.

10. Dismiss ArcMap.

---

[2]A Hobson's choice is no choice, or a choice between what is offered and nothing at all. To understand the derivation (background/source), check out "Thomas Hobson" and "anecdote" on the Internet.

Exercise 4-4 (Warm-up)

Making an ArcInfo Coverage Named GENWARMUP

1. Use ArcCatalog to make a folder named GenedDatasets in

___IGIS-Arc_*YourInitialsHere*.

The plan here is to make an ESRI coverage of arcs and polygons that corresponds to Figure 4-29. You will do this by the cumbersome method of typing the coordinates of each node and vertex into a text file. After doing this a couple of times you will really appreciate more efficient methods of data entry. The text file you make will become input to the GENERATE tool.

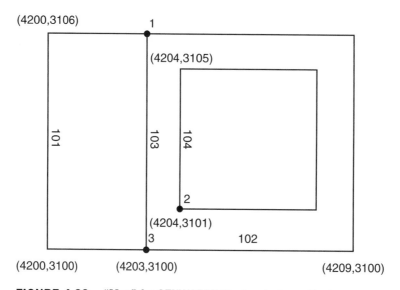

FIGURE 4-29 "Map" for GENWARMUP, showing coordinates

The GENERATE Tool Input Text File Specification

The symbolic format of a text file for the arcs of the coverage is as follows. The symbols < and > are used to enclose characters (letters and numbers) that will appear in the text file. The text file will contain only letters and number; the text file will not contain the symbols < and >.

```
<arc_number>
<x-coordinate of the first node>, <y-coordinate of the first node>
<x-coordinate of the following vertex>, <y-coordinate of the
    following vertex>
<x-coordinate of the following vertex>, <y-coordinate of the
    following vertex>
```

```
and so on until you reach the end of the arc by
<x-coordinate of the last node>, <y-coordinate of the last node>
END
specification of another arc with its nodes and vertices
END
specification of another arc with its nodes and vertices
END
and so on until you reach the end of all specifications and type another
END
```

2. Compare the preceding format with the following actual text file which GENERATE will use to create the coverage. Verify that the drawing in Figure 4-29 has indeed been specified by the text file below. Notice the arc numbers. Notice also that at the bottom of the file there are two END statements.

```
101
4203,3100
4200,3100
4200,3106
4203,3106
END
102
4203,3100
4209,3100
4209,3106
4203,3106
END
103
4203,3100
4203,3106
END
104
4204,3101
4208,3101
4208,3105
4204,3105
4204,3101
END
END
```

In Step 3 that follows you should create exactly the Text File shown immediately above with a text editor,[3] not with a word processor. Suggestion: Notepad.

3. Using a text editor type the file *exactly* as shown immediately above. Save this file (as a text file, with extension txt, and not with rich text format) as GENWARMUP.txt in GenedDatasets in ___IGIS-Arc_*YourInitialsHere*.

---

[3]You could use Notepad in Microsoft Windows. Choose Start > Programs > Accessories > NotePad.

Creating the Coverage from the Text File

_____ **4.** Show ArcToolbox in ArcCatalog. In the ArcToolbox pane expand: Coverage Tools > Conversion > To Coverage. Click once on Generate. The status bar indicates the function of the Generate tool.[4] (You have to be a little careful here. If you click a toolbox instead of a tool, the status bar does not change; it continues to report on the most recently highlighted tool.) Pounce on Generate. For the Input File field, in the Generate window that appears, browse to the text file you just created. In the Output Coverage field, check that the name of the coverage is GENWARMUP, directed to the same folder. The Feature Type should be Lines. See Figure 4-30. Click OK. When the command is finished executing, close the Generate window. Also dismiss ArcToolbox by clicking the "X" in the upper right corner of its pane.

I don't have Coverage

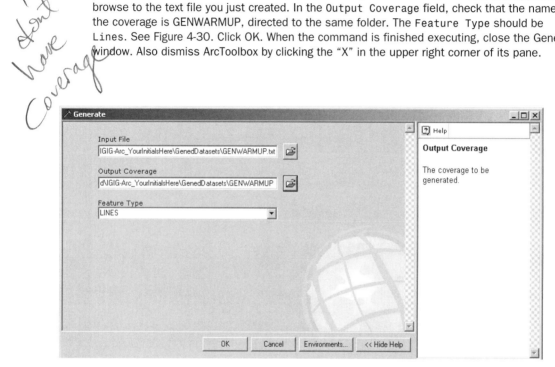

FIGURE 4-30

Making a Folder Connection and Looking at the Coverage with ArcCatalog

_____ **5.** In ArcCatalog, under the File menu, pick Connect Folder (or use the Connect to Folder icon on the Standard toolbar). In the window that appears, navigate (by expansion, not double-clicking) to ___IGIS-Arc_*YourInitialsHere*\GenedDatasets. With GenedDatasets highlighted, click OK.

_____ **6.** Notice that the new entry is selected in the Catalog Tree. As I indicated before, the entry will remain there (on this computer) even when ArcCatalog is stopped and started again. As you remember, this provides a shortcut to this folder to make navigation to it easier.

---

[4]Unfortunately, the description is wrong in version 9.1. It describes what Generate should do, not what it does. Generate doesn't add features—it only makes a new coverage. We will work around this.

_____ **7.** Click the Contents tab and notice your GENWARMUP coverage is named in the right side of the ArcCatalog window. Double click it. You will see that it contains `arc` and `tic` components. Highlight `arc` and click the `Preview` tab. Examine the coverage visually for errors. If it is correct, proceed with the step that follows. If it is incorrect, assess what you did wrong. Then select GENWARMUP in the left pane of ArcCatalog. Right-click in it and select Delete. Fix up the GENWARMUP.txt text file and then return to Step 4 above and perform the steps to make GENWARMUP again.

_____ **8.** Using Preview, look at the attribute table for the arc component of GENWARMUP. The table is unexciting, a condition we will remedy shortly. Look at the attribute table for the tic component. Notice that there are four tics and they correspond to the maximum and minimum x and y extents of the coverage. In other words, the tics are located at the corners of the coverage. This is standard for the Generate tool.

Building Node and Arc Topology

This coverage is only partly done. While there is an arc feature attribute table, and a tic coordinate table, there is only preliminary topology. To create the full topology you have to *build* the topology and the various tables.

_____ **9.** Right-click the coverage name GENWARMUP and click `Properties`. Click the `General` tab in the `Coverage Properties` window. Notice that an arc feature attribute table (FAT) does not exist and the arc feature class has only Preliminary Topology. Click `BUILD` and in the Build window, choose the `Node Feature` class. Click OK. Now build the `Line (arc)` feature class. All three feature classes should now have True in the FAT column. Dismiss the Coverage Properties window.

_____ **10.** Launch ArcMap. Dismiss ArcCatalog. Add the arc and node components of GENWARMUP to the map. Examine the attribute table of GENWARMUP node. Recall, the ARC# number is the number of only one of the arcs connecting to a particular node; it is not of much interest here. However, the internal of each node ID (GENWARMUP#) is important to what follows. This is the number that was automatically assigned to each node when you built the node topology of the GENWARMUP coverage.

Labeling Features

_____ **11.** *Label the nodes on the map with GENWARMUP# from the node attribute table (NAT):* Right-click GENWARMUP node and choose Properties. Pick the Labels tab. As the `Text String Label Field`, choose GENWARMUP#. If necessary, place a check in the box `Label Features In This Layer`. Your Layer Properties window should look like Figure 4-31. Click Apply, then OK. Observe that the three nodes on the map are now numbered. Dismiss the node attribute table.

_____ **12.** Bring up the attribute table of GENWARMUP `arc`. Label the arcs on the map with the GENWARMUP-ID (not GENWARMUP#) column of this arc attribute table (AAT), using the same procedure as described previously for nodes. Your map should now look like Figure 4-32, although the nodes might be numbered differently. The numbering of the nodes is under the control of the software. The arcs will have the identification numbers that you provided in the text file—for instance, 101, 103, and so on.

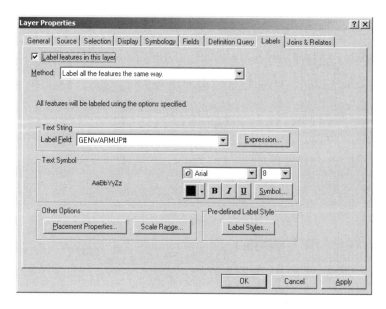

FIGURE 4-31

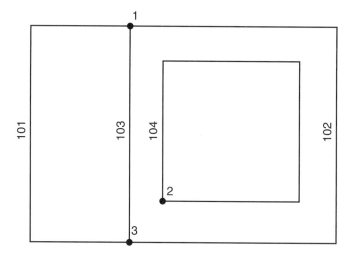

FIGURE 4-32   GENWARMUP with lines and nodes identified

There are a couple of repeat lessons here: (1) you can label a feature with a corresponding value from the attribute table, and (2) you can select the item (column) that you want to use for labeling from the Labels tab of the Layer Properties window.

____ 13. Look carefully at the GENWARMUP arc attribute table together with the map. Use the Select Features tool to select arcs and verify that the LENGTH, FNODE# (From-node number), and TNODE# (To-node number) are correct for each arc.

Building Polygon Topology

At this point, for the coverage GENWARMUP, you have arc and node topology built, but not polygon topology. In what follows you will build polygon topology, after a false start.

_____ **14.** Restart ArcCatalog. Navigate to the GENWARMUP coverage. Dismiss ArcToolbox if it is open. Right-click GENWARMUP and bring up properties. Here is where you would build the polygon topology, but the choice is grayed out. Why? Because building involves modifying the coverage (by adding an attribute table) and the coverage is already in use by ArcMap. So dismiss the Coverage Properties window, restore ArcMap, and save the map as GENWARMUP.mxd in

    ___IGIS-Arc_*YourInitialsHere*\GenedDatasets.

Then free ArcMap of the data sets in its Table of Contents by clicking the New Map File icon and flip back to ArcCatalog. Now again try to build the polygon component of GENWARMUP. You should discover that BUILD is not grayed out, so you can press it. If this doesn't work, try clicking Catalog in the Catalog Tree and then on View > Refresh. Again bring up the Properties window. If BUILD is still grayed out, you will have to dismiss ArcMap, refresh the Catalog Tree, and try once more. Finish building polygon topology.

The need to modify a geographic data set when it is being used by another program crops up frequently and is occasionally frustrating. An error message you will sometimes encounter is Unable to obtain write lock . . . . How you correct this depends on circumstances. If you have an extensive map up and you don't want to remove a data set from it, you can easily save the entire map, with all its settings, pretend you want a new map, make the modifications to the data set, and then reload (open) the original map with the modified data set. ESRI is working on the problem of data sets that are locked out because they were previously used by other programs.

AREAs and PERIMETERs

If you have trouble understanding the areas and perimeters for polygon coverages, recall our rural analogy: Suppose that each polygon represents a field of a farm in which are kept different animals. The questions to be answered are these: How much area does each animal have to graze in and how much fence is required to keep each animal where it belongs? In the steps that follow, you will use the labeling capabilities of ArcMap to display the areas and perimeters of each of the three polygons.

_____ **15.** Go back to ArcMap (restart it if necessary). Open the map GENWARMUP.mxd. Add the polygon component of the GENWARMUP coverage. Right-click GENWARMUP polygon. Click Label Features. What values appear? _____, _____, _____. Verify by calculation of the areas of Figure 4-29 that these are the areas of the three polygons. Write the area numbers on Figure 4-32 in the form: A = xx.

_____ **16.** Right-click again on GENWARMUP polygon. Then click Properties > Labels > Label Field > PERIMETER > Apply > OK. Now you will see the polygons labeled with the total lengths of the arcs that define them (instead of the areas, as before) _____, _____, _____. Inside each polygon on the diagram, write the perimeter numbers: P = yy. Be sure you understand where the 40 came from. Going back to our analogy, it would take 40 units of fencing to keep the animals in that polygon: the sum of arcs 102, 103, and 104. Verify that each number you have written in a polygon represents the lengths of the arcs that bound that polygon. Dismiss ArcMap and ArcCatalog.

Exercise 4-5 (Quick Quiz)

Areas and Perimeters

For each pair of polygons in Figure 4-33, write the Shape_area and Shape_length (perimeter) in the table. Each large square has sides' lengths of 12. Each small square has sides' lengths of 4.

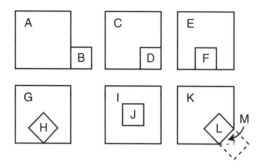

FIGURE 4-33   Area and perimeter quiz

Polygon	Area	Perimeter
A		
B		
C		
D		
E		
F		
G		
H		
I		
J		
K		
L		
M[5]		

---

[5]The dashed lines are shown as a hint regarding the area of polygon M. If a right isosceles triangle has a hypotenuse of length 1, the length of a side is approximately 0.7—or less approximately, 0.70710678118.

Exercise 4-6 (Project)

Making an ArcInfo Coverage for a Foozit_Court

Now you know how to make a coverage with ArcToolbox's Generate tool and how to create topology. In what follows you will make a more complex coverage and examine it. Then you will explore some additional ArcGIS and ArcInfo workstation features.

A Foozit court is a (hypothetical) game surface, somewhat like a tennis court in that it has painted lines. The diagram in Figure 4-34 shows where the lines are. The units are in meters. The lower left corner and upper right corner coordinates are shown in the diagram. Almost all nodes and vertices fall on Cartesian points with *integer* coordinates. If you are familiar with the Cartesian coordinate system, this exercise may seem like so much busy work—but it will go quickly. If not, you will learn a lot from the exercise.

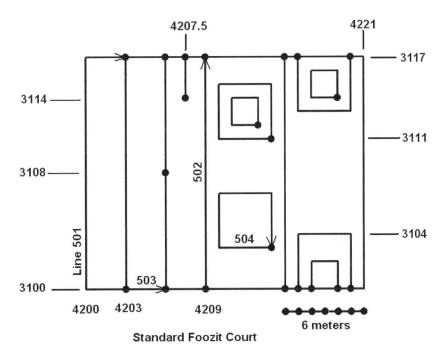

Standard Foozit Court

FIGURE 4-34

1. Examine the Foozit Court diagram in Figure 4-34. The first four arc numbers (501 through 504) are specified for you. On the diagram,[6] fill in your choice of arc numbers for the remaining arcs using any arc numbers you choose and in any order you choose. How many arcs are needed to describe Foozit_Court? _____. Put an arrowhead on each arc. Also, you may find it helpful

---

[6]In the IGIS-Arc_AUX folder of the CD-ROM accompanying this book there is a copy of Figure 4-34. You may print it out so you have a bigger copy that is more convenient to work with.

to write in the Cartesian coordinates of the nodes and some vertices, to help you make the text file. You should be able to easily calculate the position of each node and vertex from the diagram. The nodes on the diagram are shown with black dots. In making the text file, do not place nodes where such dots are not shown—use vertices if necessary to allow a change of direction in the arc. For example, arc 501 should be represented like this:

```
501
4203,3100
4200,3100
4200,3117
4203,3117
END
```

That is, the nodes are located at (4203,3100) and (4203,3117). The vertices are at (4200,3100) and (4200,3117).

_____ **2.** Using the format from the previous project, make a text file named Foozit_Court.txt (in Gened-Datasets in ___IGIS-Arc_*YourInitialsHere*) using the information in Figure 4-34, augmented by your written additions. Be sure to place nodes only where shown.

Be very careful to get everything right. If you make a mistake, you will have to re-create the coverage from the text file—not a big deal, but a nuisance. (There is a way to graphically edit mistakes, but learning a graphic editor is a project in itself, which we undertake later.)

_____ **3.** Start ArcCatalog. Show ArcToolbox if it isn't in the window. You already know how to find the tool that generates a coverage, but let's find it in a different way. Collapse all tool chests. Press the Index tab at the bottom of the ArcToolbox pane. For a keyword, type `Generate`. `Generate (arc)` should be semi-highlighted in the resulting list. Press the Locate button. The ArcToolbox window expands and the Generate tool is now highlighted. The tab that is selected at the bottom of the window should change to Favorites. Press the Enter key to activate the tool.

_____ **4.** In the lower right corner of the Generate window, find the toggle button: Show Help (or Hide Help). Use it to show the help pane. You will see a brief description of the Generate tool. Click on a text box and you will get a brief description of the function of that box. Check out the Feature Type text box. Note that you can create[7] coverages of

```
ANNOTATIONS
CIRCLES
CURVES
FISHNET
LINES
LINKS
POINTS
POLYGONS
TICS
```

_____
[7]In version 9.1 some of the descriptions are not quite accurate. You could not use Generate, for example, to add lines to an existing coverage. Generate would only make a new coverage, and only of one type of feature. Perhaps this will be repaired by the time you need to use it.

_____ **5.** Also recall there is a Help (?) button at the top of this pane. Press it. This action sends you into the ArcGIS Desktop Help system, to the particular help file for Generate. Find expand all and press it. Read over the usage tips. Scroll down and notice that you could get help on the command-line syntax and parameters, should you need them in the future.

_____ **6.** Make the help window occupy the full screen. Scroll up and click Learn more about how Generate works. An extensive discussion of the Generate tool appears. Do not try to comprehend this material, but do absorb (and put in your Fast Facts File) the general ideas related to getting help for ArcToolbox tools. The objective here is to learn about "Help" not about the generate tool. Then click Back and dismiss the help window. Click Hide Help.

_____ **7.** Using what you learned in Project 4-3, use Foozit_Court.txt in the Generate window to create the LINE coverage FOOZIT_COURT. Place the coverage in the

___IGIS-Arc_*YourInitialsHere*\GenedDatasets

folder. When execution has completed close the Generate window and dismiss the ArcToolbox pane.

Looking at the Coverage with ArcCatalog

_____ **8.** In ArcCatalog highlight the FOOZIT_COURT coverage (in

___IGIS-Arc_*YourInitialsHere*\GenedDatasets).

_____ **9.** Pick the arc component of the Foozit_Court coverage. Look at its Geography. Examine the coverage visually for errors. You may see some.

_____ **10.** Correct any errors by changing the text file, save it as Foozit_Court2.txt, and regenerate the coverage (be sure to name the coverage FOOZIT_COURT2).[8] Check it out again with ArcCatalog. If it still isn't right, redo, making the name FOOZIT_COURT3. Continue the process until you have it right—certainly before you get to FOOZIT_COURT9.

_____ **11.** Now that you have FOOZIT_COURT*n* (where *n* is the largest integer you got to, or just a blank if you got the coverage right the first time) looking the way you want it to, preview both the graphics (geography) and the table for tic. How many tics are there? _____ What are the coordinates of the tics?

(_____, _____), (_____, _____), (_____, _____), (_____, _____)

_____ **12.** Preview the associated *arc attribute table (AAT)* for FOOZIT_COURT*n*. Notice that while the arc ID numbers are present (501, 502, and so on), the rest of the table is pretty much zeros. This is because the *topology* for the coverage has not yet been created.

Creating Attribute Tables Using ArcToolbox

In what follows you will create the node attribute table (NAT), arc attribute table (AAT), and polygon attribute table (PAT) for the FOOZIT_COURT*n* coverage using ArcToolbox rather than ArcCatalog.

---

[8]Coverage names are limited to 13 characters. This name is 13 characters long.

___ **13.** In ArcCatalog select the word Catalog at the top of the window. Just to be safe, choose View > Refresh, or press the F5 key. (This may prevent the error of trying to obtain a "write lock" on the coverage while it is in use by ArcCatalog.)

___ **14.** Back in ArcToolbox, collapse all the toolboxes, press the Index tab, type `Build` as a keyword (you should see Build(arc) semiselected), press `Locate` and pounce on the `Build` tool. Browse for the name of the coverage (FOOZIT_COURT*n*) so that it appears in the Name field. Open. Make sure the name and its path appear in the Input coverage field. Before you OK the window, select the Node feature class to make the NAT. Proceed. You may have to wait a bit for the topology to be created.

___ **15.** Use the Build tool to make the AAT and the PAT in the same way as the NAT, using the LINE and POLY feature types. Dismiss ArcToolbox.

___ **16.** With ArcCatalog, examine the arc, polygon, and node tables and the associated geography.

___ **17.** What is the number of arcs? ____ List below each *internal* polygon number (Foozit_Court*n*#) and the associated PERIMETER and AREA.

_____ _____ _____; _____ _____ _____;

_____ _____ _____; _____ _____ _____;

_____ _____ _____; _____ _____ _____;

_____ _____ _____; _____ _____ _____;

_____ _____ _____; _____ _____ _____;

_____ _____ _____; _____ _____ _____

Write each polygon number on the paper diagram. Verify, by looking at the diagram, that the area and perimeter shown for each polygon is correct.

___ **18.** Dismiss ArcCatalog

Exercise 4-7 (Project)

Making an Olympic Foozit Court

The Olympic Foozit Court has additional features, as shown later in Figure 4-35. Also the Olympic Court has one fewer box. Since the Generate tool, as it is presently implemented, will not let us add to an existing coverage, we will work around the problem. In doing so, you will learn the beginnings of using ArcGIS *Workstation*. What you will be working with is the "old" ArcInfo—command-line-driven software (compared with point-and-click, menu-driven software like ArcCatalog and ArcMap). ESRI software has grown and changed over the decades. Arc7 was the last command-line version. Versions 8 and 9 are ArcDesktop, using the point-and-click approach, but, for the most part, the pointing and clicking are a mechanism for constructing commands that go to Arc Workstation. Sometimes it is better or necessary to go straight to

Arc Workstation and use it in command-line form. In any event, any person who professes expertise in GIS should have some familiarity with Arc Workstation.

_____ **1.** Click Start > Programs > ArcGIS > ArcInfo Workstation > Arc. A black window with a blue title bar, labeled Arc, appears.

Command-line software usually shows some sort of prompt that invites you to type a command and parameters. In this case the prompt is Arc:.

_____ **2.** Type the command WORKSPACE.[9] Press Enter. Probably you will see a message stating that the current location is [some drive letter]:\workspace.

A workspace is a folder. If the folder contains ESRI coverages, it is a valid Arc workspace. Using the WORK-SPACE command by itself results in a message from the software telling you the name of the current workspace. The WORKSPACE command also lets you change to a different workspace.

_____ **3.** Type (and, again, case does not matter):

WORKSPACE ___IGIS-Arc_*YourInitialsHere*

Ignore any warning that the location is not a workspace. The Arc software is now pointing to the folder you specified. Type WORKSPACE by itself to verify that the correct folder is the current folder. Now type

WORKSPACE GENEDDATASETS

This should put you in the subfolder containing the coverages you have been working on. To verify, type WORKSPACE by itself.

Once you have established the current workspace, you need only type the name of a data set to access that data set. The path is assumed to be the one to the workspace.

Some economies are available in typing commands. First, you can usually use an abbreviation—W for WORKSPACE, in this case. Second, you can type an entire path at once.

_____ **4.** Type W C:\

This makes the current workspace the root directory of the C drive.

Now type W ___IGIS-Arc_*YourInitialsHere*\GENEDDATASETS

to get back to the folder you want.

Type W. Assuming you are in the right place, type LISTCOVERAGES (or its abbreviation, LC). You should see listed the coverage GENWARMUP and the Foozits.

---

[9]Arc Workstation doesn't care if you use uppercase or lowercase letters in command names. I will use uppercase just for clarity and consistency.

As you can see, command-line software leaves a lot to be desired in the realm of human-computer interaction. You are required to type a lot, and you have to know what to type. There are a few aids. For example, when a command name is typed, usually the software responds with a line telling you the usage of the command—that is, the parameters (sometimes called "arguments") that must or may be included after the command name. For example, let's take the command that deletes a data set. It has the unfortunate name KILL.

—— **5.** Type KILL at the Arc: prompt. The response is

        Usage: KILL <geo_dataset> {ARC | INFO | ALL}

Following are conventions for the usage message:

❑ The angle brackets < and > enclose arguments (text strings) that are required and must be filled in for the command to operate. Here you must give the geo_dataset name—that is to be deleted.

❑ The braces { and } enclose arguments that are optional.

❑ The "OR" signs (|) between brackets or braces indicate a choice is to be made.

❑ If the argument names are uppercase letters, then the argument must be typed exactly as shown. In the case of the KILL command, you are given the option of deleting just the coverage graphics (ARC), just the database table (INFO), or both of them (ALL).

—— **6.** *Being very careful* to *NOT* delete the *latest* FOOZIT_COURTn[10] coverage, delete one you don't need with

        KILL FOOZIT_COURTn ALL

and list coverages again to verify that the coverage you deleted is gone.

Seeing a Coverage with Arc Workstation

ARCPLOT is a module of Arc Workstation. It was the principle way of making maps with "old" ArcInfo.

—— **7.** At the Arc: prompt type ARCPLOT to start the plotting module. ARCPLOT then requires some preliminary commands to get it started. At the ARCPLOT: prompt, to tell the software what sort of display device is in use, type

        DISPLAY 9999

An ARCPLOT window should appear. Move the windows around so you can see the both the Arc window and the ARCPLOT window. Make the Arc window active by clicking its title bar. At the ARCPLOT: prompt, type

        MAPEXTENT FOOZIT_COURTn

This tells ARCPLOT what values of coordinates to expect—here they are those that bound the features of the coverage FOOZIT_COURTn.

---

[10]Perhaps your Foozit Court generation was perfect the first time and you don't have a FOOZIT_COURTn. In that case, just read about the KILL command. Don't delete the coverage.

_____ **8.** Type ARCS FOOZIT_COURTn to draw the arcs of the coverage in the ARCPLOT window. Make that window active. Inspect it to verify that you have the right coverage. Make the Arc window active. Type QUIT to leave ARCPLOT and return to the Arc: prompt.

Using the Workstation GENERATE Command

The Olympic Foozit Court is shown in Figure 4-35. Notice it has a circle at its west end and a grid of boxes at the east end, and it is missing a box from the Standard court. You will make the circle first.

_____ **9.** Type GENERATE

The Usage is shown as

GENERATE <cover>

indicating that you must give the name of the coverage you want to create or add to. So type

GENERATE FOOZIT_COURTn

which will bring up a Generate: prompt.

_____ **10.** Type CIRCLE and follow the instructions. Separate the arguments, which are all numbers in this case, by commas. You want a circle (make the ID 888), centered at x = 4206 and y = 3108.5, with a radius of 4. Press Enter, then type END. Press Enter. At the Generate: prompt, type QUIT (or its abbreviation, Q). Press Enter (after this and every other Workstation command).

_____ **11.** Go back into ARCPLOT and look at the augmented coverage. Type Q to get back to the Arc: prompt.

_____ **12.** Open ArcCatalog. Look at the (geo)graphics of FOOZIT_COURTn in this way also. Close ArcCatalog.

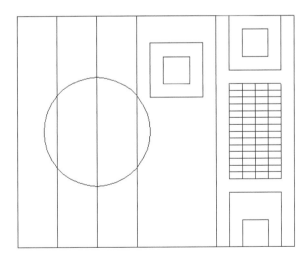

Olympic Foozit Court

FIGURE 4-35

Now you will generate the little grid at the east end of the court by using the GENERATE command with the FISHNET subcommand. First look at that command using the Arc Help system (which is different from the ArcGIS Desktop Help system, and also different from the ArcToolbox help system).

_____ **13.** At the Arc: prompt, type HELP, and wait. In the Help Topics: ARC/INFO Help window, click Contents and look at the dozen or so "books" that appear. Open Command References For ARC/INFO Prompts by pouncing on it. Open the ARC book. Pounce on the Alphabetical list of ARC commands. Scroll through the list and recognize that this is a large software package you are dealing with. And this is only part of it—there are also modules such as ARCPLOT (which you just used), ARCEDIT (which you will use shortly), and Extensions, such as GRID, TIN, NETWORK, COGO, and so on. Click Help Topics to bring that window back.

_____ **14.** Open the "G" book. Pounce on GENERATE and then click CIRCLES. Match up what you read there with the steps you performed previously when you added a circle to FOOZIT_COURTn. Click Back.

_____ **15.** Click FISHNET, and scan the information to get an idea of what the subcommand is capable of. You will use it, but I'll lead you through it. Dismiss the ARC/INFO Help window, after noting in your Fast Facts File how to get help in Workstation ArcInfo.

Using the Generate Module to Add Features to Foozit_Courtn

_____ **16.** In the Arc window, at the Arc: prompt, type GENERATE FOOZIT_COURTn.

_____ **17.** At the Generate: prompt, type FISHNET. For the origin, use x = 4216 and y = 3105. The input Y-Axis coordinates just needs to be a number pair that will make a vertical line. For the X value, use 4216. For the Y value, any number greater than 3105 will do. Use 10000. You want the cells to be one unit wide and half a unit tall: use 1.0 and 0.5. And you want 14 rows and 4 columns. At the Generate: prompt, type QUIT.

_____ **18.** Go back into ArcCatalog. Preview the arc component of FOOZIT_COURTn. You should see a diagram of the Olympic Court, except it still has that extra box, which we will deal with shortly.

_____ **19.** GENERATE made the new arcs, but it didn't build any tables or create topology. As before, use the properties of the coverage to build node and arc (line) component topology.

BUILD? No? CLEAN? Yes.

_____ **20.** Look at the properties of FOOZIT_COURTn. Note that the polygon topology has reverted from Exists to Preliminary. The thing to do, obviously, is to build polygon topology. But when you try to rebuild, you will get a window saying Build Failed Intersection found and, if your computer's speakers are turned on, a rude noise.

The message Intersection found means one arc tried to cross another and there was no node at the intersection. This has happened because you added the circle, which cut across some of the court lines. Arcs can cross (like one highway on a bridge crosses another), but not if those arcs define polygons. The problem now is this: How do we create a node at the intersection? We will use Clean instead of Build. Among other things,[11] Clean adds nodes where they are needed for proper polygon topology.

_____ **21.** Dismiss the Build Failed and Build windows. Under the BUILD button find the CLEAN button. CLEAN will create a node wherever two arcs cross—as in where the circle cuts across vertical arcs. Press CLEAN. Make sure the CLEAN window has Clean Lines Only *unchecked*. (If you are only cleaning lines, the node won't be placed, since you would be indicating that you weren't concerned with polygons.) Click OK. Note in the Properties window that polygon topology again exists. Click OK again.

_____ **22.** Explore FOOZIT_COURTn with the Identify tool. Look at the new nodes, arcs, and polygons that have been created. Also look at the tables for the node, arc, and polygon components.

One lesson here is that you can add to coverages using the Arc Workstation GENERATE command, but that you must then rebuild the topology. If you have put in arcs that create intersections without putting in nodes at those intersections, you must use CLEAN rather than BUILD to get polygon topology back.

The Olympic Foozit Court, in addition to the additions, also has a deletion. Compare Figure 4-35 with Figure 4-34 and determine which box is to be deleted. You can make the deletion with ARCEDIT, the editing module of Workstation ArcInfo. This is, in fact, the program you must use to edit coverages. More and more, GIS users are using geodatabases. The need to edit coverages may be filled by converting the coverage to a geodatabase (or shapefile) and using the powerful point-and-click editing tools in ArcMap. But you should at least know ARCEDIT exists and can be used for editing coverages.

_____ **23.** Invoke ARCEDIT at the Arc: prompt. You have already set the display to 9999, so a graphic ARCEDIT window will appear. ARCEDIT requires that you define several parameters before you can edit. You must set the MAPEXTENT, as in ARCPLOT. You must state the name of the coverage you want to work on. You must select the sort of feature you want to work on. And you have to set up a drawing environment to tell ARCEDIT which features you want drawn in the window when you give the DRAW command. So, after making the Arc window active, type the following at the Arcedit: prompt:

```
MAPEXTENT FOOZIT_COURTn
EDITCOVER FOOZIT_COURTn
EDITFEATURE ARCS
DRAWENVIRONMENT ARCS ARROWS NODES
DRAW
```

Notice that you get the drawing you saw before, and you also see the arrows that indicate the directions of the arcs. Interestingly enough, this is the only ESRI software that shows arc directions in coverages.

_____ **24.** *Remove the box from FOOZIT_COURTn:* In the Arc window, type SELECT, which brings up crosshairs in the ARCEDIT window when you move the cursor there. Place the crosshairs over a portion of the arc that defines the box you want removed, and click. The arc turns yellow. In the Arc window, type the word DELETE, and press Enter to delete the selected element. Type QUIT to dismiss the ARCEDIT window. Agree to keep all edit changes. Minimize the Arc window.

So you can see that Workstation Arc is a program that you can get things done with, but one that is sort of clunky and certainly isn't very user-friendly. However, it performs well, is virtually bug free, and allows you to do some things you can't do with its snazzier point-and-click cousins. We'll do a bit more with it so you can at least say you've been exposed to the concept and know a few commands. When it is necessary to use Workstation Arc, you won't be at a complete loss.

---

[11]The CLEAN process also cuts off dangling arcs (arcs shorter than some prescribed length) and snaps nodes together if they are within a certain distance of each other. We won't dwell on CLEAN because we will be working mostly with geodatabases, which handle the issue of "dirty data sets" in a different way.

Examining the ArcGIS Coverage Directory (Folder) Structure

_____ **25.** Under My Computer, pounce on your ___path, then on ___IGIS-Arc_*YourInitialsHere* folder to look at its contents. Inside the GenedDatasets workspace folder, what folders do you find? (Never mind any that start with xx; they are temporary folders.)

_____, _____, _____

Inside the Foozit_Courtn folder, what files do you find? (Never mind about the extensions.)

_____

Inside the INFO folder, note that you find files like arc<a number> with extensions .nit and .dat.

Looking at the Association Between the Files in the Info Folder and the Coverages

_____ **26.** Restore the Arc window. At the `Arc:` prompt type

`DIRECTORY INFO`

to get a list which associates each feature class with the data files in the INFO directory. The feature class names are obvious. The data files have names like ARC0001DAT, which aren't obvious at all. Again, you are reinforced in the view that a coverage is split into parts.

_____ **27.** Okay, enough command line stuff is enough! Type `QUIT` at the `Arc:` prompt.

Exercise 4-8 (Exploration)

Understanding a Couple of Things That Don't Look Right

_____ **1.** Start ArcMap. In `___IGIS-Arc\`*YourInitialsHere*`\Trivial_GIS_Datasets\FEAT_NUM` you will find the following:

1. A personal geodatabase named Feat_Num.mdb containing three feature classes

   ❑ `feat_num_pts`

   ❑ `feat_num_lns`

   ❑ `feat_num_ply`

2. Three coverages: `feat_num_pts`, `feat_num_lns`, `feat_num_ply`

3. Three shapefiles: `feat_num_pts`, `feat_num_lns`, `feat_num_ply`

The data sets are identical in what they represent. Each data set contains three features: one of points, one of lines, and one of polygons. (The personal geodatabase feature classes were converted from coverages, as were the shapefiles.) What we will do is look at the differences in the tables. In particular, we will look for differences in the FEATure NUMbers and at anomalies in calculation.

_____ **2.** Add all three of the coverages to ArcMap. You will see three points, three lines, and three polygons. The map should look something like Figure 4-36.

Open the attribute table of each layer. Resize, or just slide, each so you can see all three at once. The points have user IDs of 501, 502, and 503. The lines have user IDs of 701, 702, and 703. The polygons have user IDs of 901, 902, and 903. The lines were created to have length 100. The polygons were created to have area 2000 and perimeter of 240.

The anomaly of interest is that the FIDs (feature identifiers) of the polygons are 2, 3, and 4, when you might expect 1, 2, and 3, as in the cases of points and lines. The reason for this is that the first record (not shown), with FID 1, is the so-called outside or external polygon—it consists of the entire Earth except for the areas delineated by the other polygons. We will come back to this shortly.

_____ **3.** Start a new map. Add all the shapefiles. Open and arrange their attribute tables. (Some of the fields have been carried over because of the conversion from coverages. Some are useless. Others, like AREA and PERIMETER were just copied, but you should realize that area and perimeter and *not* automatically calculated for shapefiles.) What you should be aware of here is that the FIDs begin with zero instead of one for all features.

_____ **4.** Start a new map. Add all the personal geodatabase feature classes. Open and arrange their attribute tables.

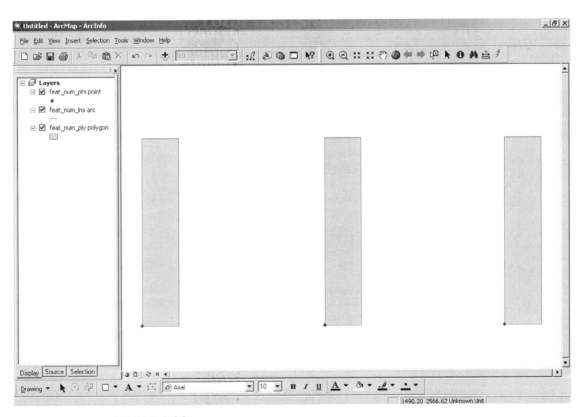

FIGURE 4-36

Here, in keeping with the idea that features are objects, the identifier is called an OBJECTID instead of FID (feature identifier). In each case the records are numbered starting with 1.

Note that the LENGTH, AREA, and PERIMETER field values are carried over from the coverages. They are, respectively, 100, 2000, and 240, exactly. The Shape_Length of each line, in `Attributes of feat_num_lns`, however, is calculated in the conversion. It displays as 99.999999. The Shape_Lengths and the Shape_Areas in the polygon table are likewise slightly inaccurate. This points up a "problem" for those users who want exact answers: Computers cannot be counted upon to give them. The difficulty is that, while computers can usually do exact arithmetic with integers, they cannot be exact with floating-point numbers, which may have fractional parts. The errors occur because humans do arithmetic with decimal (base 10), and computers with binary (base 2). For example, it is not possible to exactly represent the decimal number one-tenth (0.1) in binary. One-tenth in binary is represented (imprecisely) by the sum of some of the fractional powers of two: one-sixteenth, one-thirty-second, one-one-hundred-twenty-eighth, and so on. In binary one-tenth looks something like 0.000110011 . . . To be exact, the number would have to have an infinite number of bits. (Don't look down on binary, however. Every number system has this problem. For example, you cannot represent the number one-third in the base 10 [decimal] system in a finite number of significant figures. An approximation is 0.3333333333, but it is not exact.)

So if you tell the computer to ask a simple question of its data, such as "is A equal to B?" the answer may be "no" even though A and B are meant to be the same The computer will report that they are not the same, but the reason for the difference is that they are calculated in different ways by the computer, so that they don't match, bit for bit. A better question would be something like "is the absolute value[12] of A minus B less than some appropriate very small value (such as 0.00005)?"

Finally, let's look at the infamous "outside polygon" of polygon coverages.

——— **5.** Dismiss ArcMap. Launch the Arc program (short name for ArcInfo Workstation version 9.x). The black window with the prompt `Arc:` will appear. At this prompt type `WORKSPACE`, which displays the current workspace. Change the workspace by typing

　　`W ___IGIS-ARC_YourInitialHere\TRIVIAL_GIS_DATASETS\FEAT_NUM`

By the time you get this typed in correctly, you may really appreciate the point-and-click programs, as differentiated from the command-line program.

——— **6.** Type `WORKSPACE` again (or its abbreviation, `W`) to confirm that the workspace pointer is aimed properly. See Figure 4-37.

——— **7.** Type `LC` to list the coverages in the workspace. Type

　　`LIST FEAT_NUM_PLY.PAT`

to list the contents of the FEAT_NUM_PLY polygon attribute table. It will look like Figure 4-38.

---

[12]The absolute value of a number is, crudely, the number with the sign stripped off (so it is assumed to be positive). Formally, if q is the absolute value of a number p, then q is equal to p if p is positive, and q is equal to "negative p" if p is negative. (Recall that a negative of a negative number is a positive number.)

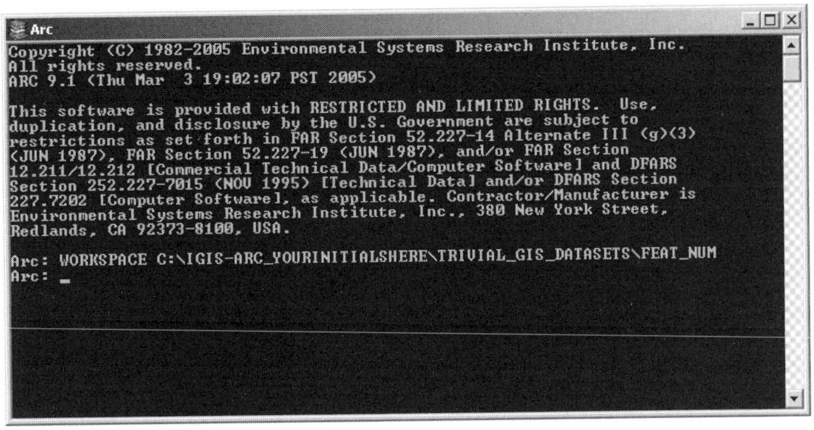

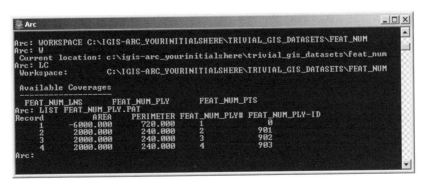

FIGURE 4-37

Here you see Record 1, which relates to the outside polygon. The PERIMETER is the sum of the perimeters of the three polygons—that is, the lengths of the arcs that bound the polygons that are nested (islands) within the outside polygon. The AREA is the sum of the areas of all the polygons, with a minus sign placed in front of it. A "negative area" you wonder? If you need a theoretical basis for this construction think of it as the last part of this expression:

```
[the area of the entire earth] minus 6000.000
```

which is shown only as [-6000.000].

The outside polygon does not appear in the table shown by ArcMap or ArcCatalog. The developers of ArcGIS wisely decided not to burden the user by displaying its record. But it does leave an ArcMap attribute table looking a little strange—starting with FID = "2".)

___ **8.** Type Q at the Arc: prompt.

FIGURE 4-38

Geodatabase Topology

The concept of topology takes a much different form in geodatabases from coverages. In coverages, topological concerns were limited to a single data set. In that data set, as far as the user was concerned, the topology was used for:

❏ Making sure arcs were connected properly

❏ Making sure polygons had no overlaps

❏ Flagging pseudo nodes and dangling arcs

In geodatabases all sorts of conditions of data misbehavior can be tested for, and topology can be constructed between different data sets, *as long as they are in the same personal geodatabase feature data set.* (Recall that a geodatabase can contain a feature data set, which can contain feature classes. Those feature classes all have the same extent, projection, and so on. Topological relationships can be formed among those feature classes. A geodatabase can also have freestanding feature classes, which may bear no relationship with each other. Topological relationships may not be formed within or among those feature classes.)

Let's look at a simple example. Recall that the first data set you looked at was a coverage consisting of fire hydrants in a village. Here you look again at that same data, but it has been converted to the form of a personal geodatabase data set. The first step is to copy the data from the original into your personal space.

1. Start ArcCatalog. Navigate to [___] IGIS-Arc\Geodatabase_Topology and highlight it. Select Edit > Copy. Navigate to ___IGIS-Arc_*YourInitialsHere* and highlight it. Select Edit > Paste.

2. Expand all the entries of

___IGIS-Arc_*YourInitialsHere*/Geodatabase_Topology

so that you see all the constituents of the Village_Water geodata set, including Fire_Hydrants and Water_Lines.

3. Explore Fire_Hydrants in the Preview pane. Then look at Water_Lines.

The idea is that the Water_Lines (called laterals) are supposed to connect to the Fire_Hydrants. You will use the topology capabilities in geodatabases to see if they do, or, if they don't but are close enough, to move the water lines.

It is rare that two sets of coordinates meant to refer to the same point in space will be identical if the points are created in different ways. So it is wise to ask if two represented points are sufficiently close together to be considered to be in the same place. You, the user, can indicate what is "sufficiently close" by specifying a "cluster distance" when you develop topology for a feature data set.

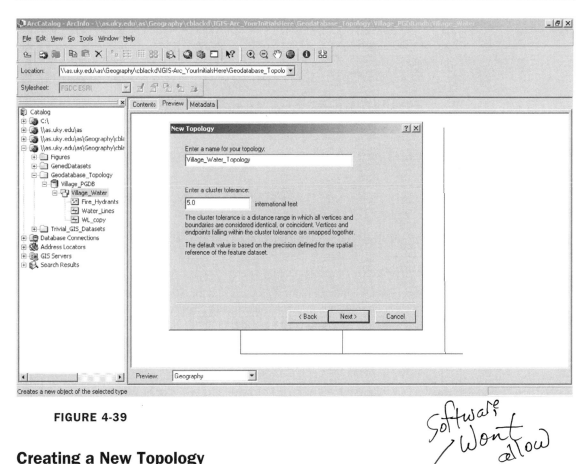

FIGURE 4-39

Software Won't allow

Creating a New Topology

4. Highlight the Village_Water feature data set. Select File > New > Topology. Read the New Topology window. Click Next. Accept the default name Village_Water_Topology. Enter a cluster tolerance of 5.0. See Figure 4-39. Click Next.

5. Put checks in the boxes of Fire_Hydrants and Water_Lines so they will participate in the topology. Click Next.

The process of generating and validating topology in geodatabases may result in moving some features—somewhat like CLEAN does in coverages. With coverages you have no choice as to which features moved or changed. With geodatabases you do. Each feature class is given a rank. The highest rank is 1. When a feature must move to solve a topological error, it is the feature whose feature class has the lower rank that moves. Let's presume that the locations of the fire hydrants are highly accurate, but that the locations of the water lines are less well known. So we will give Fire_Hydrants a higher rank so their coordinate representation will not change.

What Moves When Features Are Adjusted: Rank

6. Leave the number of possible ranks at 5. Change the rank of Water_Lines to 2. Leave the rank of Fire_Hydrants at 1. See Figure 4-40. Click Next.

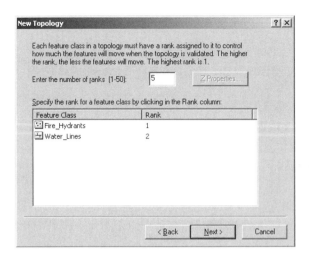

FIGURE 4-40

Geodatabase topology is based on rules that define the required spatial relationship between two feature classes, or sometimes specify a requirement within one feature class. For example, a rule might be that no two polygons may overlap each other. Currently there are 20-some rules defined in the software. In the following you will choose a rule that says that points of one feature class must be covered by (that is, have a point coincident with) a feature in a given line feature class.

Topology Rules

—— **7.** Press Add Rule. In the Add Rule window, select Fire_Hydrants in the Features Of Feature Class drop-down menu. Click the down arrow in the Rule text box to see the rules. There are four rules that relate to point feature classes. List them below:

_____

_____

_____

_____

—— **8.** Select Point Must Be Covered By Line, since we want the water lines to go to the fire hydrants. The third text window (Feature class) should indicate Water_Lines. Now give your attention to the right side of the window, where a cartoonish explanation of the rule is displayed. Read what it says. Then toggle the Show Errors box to see which points satisfy the rule and which don't. Points that turn red indicate points that don't obey the rule. Click OK.

Validating Topology

—— **9.** Since we will only need the one rule, you may press Next after you examine the New Topology window. Examine the summary and, if correct, click Finish. You will get a message that the new topology has been created, and will be asked if you want to validate it. Select Yes.

In this case validation means that the computer looks at every point in the Fire_Hydrants feature class to see if a line from Water_Lines comes to it, crosses it, or comes within 5 feet of it. Those points that do not meet the criteria will be flagged as errors. For those cases where the water line is within 5 feet but does not touch the hydrant, the water line will be moved so that it does touch the hydrant.

____ **10.** Once validation is complete, the Village_Water data set will contain an entry labeled Village_Water_Topology. Click it and preview its geography. You should see two red squares. This means that two fire hydrants were not within the cluster distance of a water line.

Let's use ArcMap so we can look at all three layers at once to see what is going on.

____ **11.** Launch ArcMap. Add Fire_Hydrants to the map. Now add Water_Lines. Finally add Village_Water_Topology. (You will be asked if you want to add all the feature classes that participate in the topology. Since you have already added them, select No.[13]) Pull the Topology layer down below the Fire_Hydrants layer. You can see that the two hydrants shown in the red squares are in fact slightly off the eastmost lateral. See Figure 4-41***. Zoom up and measure the distance from the end of the east lateral to the northeast hydrant. What is it? _____ feet. Zoom back to the full extent.

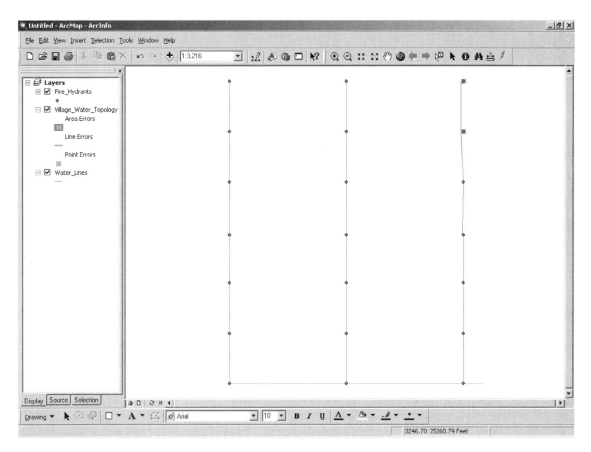

FIGURE 4-41

---

[13]It turns out that if you take advantage of this offer and add everything at once, you cannot rearrange the Table of Contents.

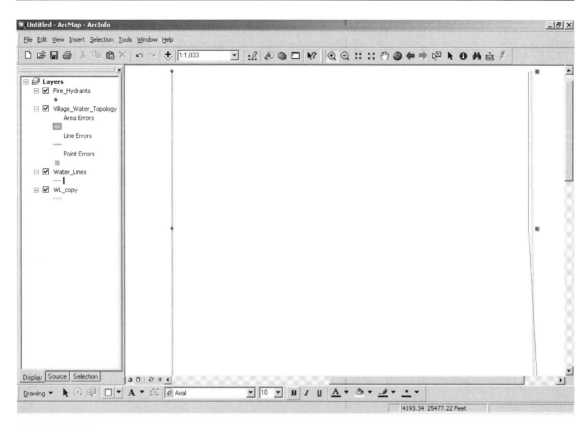

FIGURE 4-42

_____ **12.** Use the Identify tool to determine which water pipes are not lined up properly. Their OBJECTID numbers are _____ and _____.

_____ **13.** Look further into this matter of misaligned water lines. Add as data WL_copy, which is a copy of the original Water_Lines feature class. Pull it to the bottom of the Table of Contents. Make its size 4 and its symbol bright green. Make the symbol for Water_Lines bright red, size 1.

_____ **14.** Using zooming, look at each of the hydrants on the east lateral, starting with the southernmost one. What you will find is that the original water line diverges away from the hydrants as it goes north. However, the water line that participates in the topology covers the first five hydrants, but not the last two. What has happened here is that the water line came within 5 feet of the first five hydrants, so that, during validation, the lines were snapped over to each hydrant. The other two hydrants were further away than 5 feet, so they were reported as errors, and the water lines were not moved to the hydrants. In fact, moving the lines to five of the hydrants actually caused the lines to diverge from their original path, pushing them further away from the last two hydrants, as indicated by the lack of congruity between Water_Lines and WL_copy.

A Warning: Changes Made Through Topology Are Permanent

Please note that the changes made to the Water_Lines feature class during validation cannot be reversed, except by editing. It is important to keep a copy of the original of any feature class involved in a validation, as we did with WL_Copy, in case the results are not what you want.

Checking, Updating, and Organizing Your Fast Facts File

The Fast Facts File that you are developing should contain references to items in the following checklist. The checklist represents the abilities to use the software you should have upon completing Chapter 4.

Important note: This checklist is on the CD-ROM that accompanies the book. It is available in Microsoft Word format. Rather than typing or writing by hand the text that follows, you can copy and paste it into your Fast Facts File from the CD-ROM file.

___ To make ArcToolbox appear

___ The categories of tools available in ArcToolbox are

___ Two lines may cross in two different ways. They are

___ Sometimes you will see two tools that appear to do the same thing. That's because they operate on

___ To find a particular tool

___ Help for ArcToolbox can be found in several ways:

___ For a command to make a coverage with precise coordinates use

___ Topology and complete attribute tables for coverages are made with the BUILD command, which is found

___ To label features in a coverage

___ Sometimes frustrating difficulty may be encountered in trying to modify a geographic data file because

___ "Unable to obtain write lock" means

___ To start ArcGIS Workstation

___ The WORKSPACE command allows the user to

___ If two arcs cross in a coverage and no intersection is present, an error occurs, preventing the use of BUILD. The command that can be used instead is

___ The row identifiers in the attribute tables for polygons in geodatabase feature classes, shapefiles, and coverages are different in these ways:

___ If two numbers are in theory the same the computer may calculate them to be different because

___ The outside polygon in a coverage is

___ To participate in a geodatabase topology the feature classes must be part of a

___ The objects with lower feature class ranks are

___ Topology rules are defined in the software. These rules define

___ Since validation of a topology may move objects, it is a good idea to

Geographic and Attribute Data: Selection, Input, and Editing

OVERVIEW

IN WHICH you explore some ideas about collecting and selecting data, including using the Global Positioning System. Also you digitize and edit map data, transform spatial data, and examine combining attribute data with geographic data.

"Garbage in, garbage out."

An often quoted, but seldom heeded, admonition in the computing world.

In the Step-by-Step section of this chapter, you will gain experience working with mostly small data sets. You will examine the "nitty-gritty" of digitizing and manipulating data. But most GIS projects deal with large amounts of data. Sometimes these data sets are found. Sometimes data are collected from the field. What follows here is advice for getting the right data for the products that might be produced by a medium to large GIS project.

Concerns About Finding and Collecting Data

Data sets form the basis of a GIS. These systems are referred to as being "data driven" to emphasize the importance adequate data play in their operation. The product of a GIS is the most important contributor to its utility; the data are the chief ingredient of that product. The single message emphasized by this section is that any determination of what data are needed and what the characteristics of those data should be, comes after a very careful look at what sorts of information products are required for specific decision making. The products, one would hope, are developed to satisfy the needs discussed in Chapter 2 and the requirements for analysis that we will take up in later chapters.

Determining what data sets are needed is not an easy process. It requires the concentrated effort of experts, from the decision makers who will use the ultimately produced information to the scientists who gather and analyze the data. The ideal order of matching data to needs is from the specification of the information product back to the data gathering—that is, in the direction opposite from the one in which the process of producing the information takes place.

The fact that a lot of already collected data exists will influence not only the process of turning those data into information but also the types of information produced. However, to allow the fact that some inappropriate data are at hand to dictate the output of a GIS reminds me of the man who, one evening, lost a ring on the north side of a street but searched for it on the south side because "the light was better."

Looking for Data on the Internet

The fact that suitable spatial data sets are hard to find, and that a major impediment to sharing data sets is simply knowing of their existence, the U.S. government undertook two distinct but related efforts in the 1990s: the Federal Geographic Data Committee (FGDC), set up by an Executive Order (12906) from the President, and the National Spatial Data Infrastructure (NSDI).

The FGDC (www.fgdc.gov) is a governmental interagency committee composed of representatives from the Executive Office of the President, Cabinet-level, and independent agencies. The FGDC is developing the NSDI (www.fgdc.gov/nsdi). Cooperating in the effort are state, local, and tribal governments, the academic community, and the private sector. The NSDI encompasses policies, standards, and procedures for organizations to cooperatively produce and share geographic data. The FGDC also sponsors the National Geospatial Data Clearinghouse (www.fgdc.gov/clearinghouse) to provide a mechanism for searches for geographic data.

At the same time, various governmental agencies (federal, state, regional, and local) are developing spatial data sets, which may or may not be accessible, and which may or may not be free to the public. For example, the U.S. Geological Survey (www.usgs.gov) has amassed large and varied data sets, dealing not just with geology, but elevation, hydrography, land cover, and other themes. USGS produced the Digital Raster Graphic quadrangles that you saw in Chapter 2. Also, the Census Bureau develops massive spatial data sets.

Commercial data sets are also available—sometimes for considerable amounts of money, sometimes inexpensively, and sometimes for free. The Geography Network, sponsored by ESRI (www.geographynetwork .com) is a place to search for data sets. Finding the data you need on the Internet is a matter of, first, casting a wide net, and, second, drilling down to see what you can find—realizing that you are in an immense interconnected array of information sources. If you type "Geographic Information Systems" into a search engine, you get almost 20 million hits. (You can also get many hits by typing "Geogrpahic Information Systems," which should tell you something about quality and the Internet.)

Steps in Developing the Database

1. Determine what types of data are needed to produce the product(s).

2. Determine what characteristics of those data sets (accuracy, timeliness, coverage, and so on) are demanded by the product; set priorities.

3. Make some preliminary studies to determine that the data and characteristics specified will produce the information wanted.

4. Begin a data acquisition effort—search for already-collected data that meet specifications, or begin a subarea data collection effort.

5. Put data acquired into proper form (reformat, or encode it) for inclusion in the GIS database.

6. Check accuracy of each step of the collection process; also check the first form of the data against the form in the GIS.

7. Repeat Steps 4, 5, and 6 for the complete set of data for the base.

8. Employ techniques for monitoring and updating of the database.

We now look at these steps one at a time:

1. Determine what types of data are needed for the information product(s).

This assumes that you have determined the needs of the decision-making apparatus and, further, that you can identify the sort of information which will satisfy those needs.

The most important component in this step is to have people involved who understand (a) the information required (and how it is to be ultimately used), (b) the characteristics of data that might be used to generate the information, and (c) the manipulation of the data necessary to produce the information.

At this early stage, it is wise to consider alternative and innovative ways of getting to the same information. For example, if the information required is a delineation of areas of potential high soil erosion, then calculating the potential soil loss might be produced using the universal soil loss formula and data including rainfall, soil erodability, slope length and gradient, and vegetative cover.

However, it might be produced by interpreting aerial photography for existing erosion conditions. Many projects that get into trouble at this stage do so probably because communications with the analysis and decision-making group ceased after the initial contact. Problems are constantly changing in both importance and type. Thus, to be most responsive, the data support sector of a decision-making process must have its roots not in the data collection area, but rather in the analysis and decision-making area.

2. Once the basic types of data required have been identified, more thought must be given to the characteristics of the data to be acquired.

In determining the characteristics of the data needed, several fundamental questions should be asked:

What geographic area is involved? What geographic identifiers are necessary for use of the data? With what accuracy must the coordinates be known? Are the values of the data "continuous" (like elevations above sea level) or "discrete" (classifications of land cover)? How frequently do data values change? What causes these changes? Are the most basic data types in use, or can other data be derived from more basic sources? If the latter, what are the advantages and costs in using the most basic information available? What degree of detail is required? When the detail level desired is "multiplied" by the area involved, how large are the data sets? How sensitive to errors in the data is the process that is used to get information products from the data?

3. Make some preliminary studies to determine that the data and characteristics specified will produce the information wanted.

How this task is done depends greatly on what is available. If a GIS exists and new data sets to support a new product are being added, the best course might be to generate some typical subarea data to try out the process. If the GIS does not exist but is installed to produce the product, the issue of preliminary testing of the data-product relationship must be approached differently. Perhaps an analysis of your plans could be contracted to a consulting firm for checking. However it is done, someone other than the originators of the techniques for development of information from data should independently examine the projected course of action.

4. Begin the data acquisition effort.

So far I have described a rather idealized process for the formation of a portion of a database for a GIS. I think the idealized approach is worth sticking to. Too many times the existence of some collected data sets not only dictates the process used to manipulate them but also the kinds of products that get produced. It often turns out that when the cost of conversion of already collected data is counted, the error rate discovered, and the lack of suitability of the data for the task at hand realized, more money and time will have been spent than if an original data collection effort were begun. And yet, it is unreasonable not to make an examination of existing data sources, after you know what you want, to see if, considering the millions spent on data collection in this country, there are some that will meet your needs.

A search for relevant data may not prove easy. Although there are myriad Web sites that allow downloads (paid or free), the questions of appropriateness and quality will keep occurring. It may be the type of frustrating undertaking during which one never really knows when to surrender; how long do you keep looking before you elect another route? A common problem is that very few people seem to have both the depth of understanding required to manipulate data into information in particular areas and the overall view of how to collect the data that could be relevant to the decision-making process.

A real search, therefore, should be undertaken, and it should look widely, not eschewing any possible source of data. Interviewing of individuals in various agencies and companies is probably as profitable as searches through documents on the Web; interviews may produce more up-to-date information about data sources or data collection efforts.

There is no dearth of data, spatial or otherwise. But there are three major problems with existing data:

1. The data files themselves are spatially distributed, hither and yon, in offices and computing centers, in desk drawers and filing cabinets. Perhaps the first task in developing data for a specific geographic area should be a list that includes data sources, characteristics, and owners.

2. The data sets are not in a common format. Granted, different sorts of data should be presented in different form because of their inherent characteristics and uses. But the variability vastly exceeds the requirements for different formats.

3. Already-existing data sets are getting older and less correct every day. The accuracy of the data will decay slowly over time.

A method for assessing the correctness of data after a length of time might be borrowed from the concept in physics of a radioactive half-life period. Such a period, measured in time units (from millionths of a second to years to millennia) and different for each radioactive substance, is the length of time required for half the mass of the substance to have decayed into something else. In general form this idea could have a parallel with "correctness of a set of data" as the variable instead of mass of material. The correctness of spatially distributed data of a given type, say, land use, decays at different rates depending, as you might imagine, on its location. A greater rate of decay would be expected adjacent to urban areas than distant from them. In any event, the age of the data used is one of its most important characteristics.

Assuming a set of already collected data has been found that approximately meets the specifications, the data should be carefully analyzed according to a number of characteristics:

- ❑ Can the data be used directly or must they be manipulated before they can be used in analysis?

- ❑ Are the data the most specific and detailed available? When accuracy is critical, and you have to resort to digitizing maps, can you find a version printed on nonshrinking, nonstretching Mylar?

- ❑ Is the resolution of the data sufficient to fill information needs?

- ❑ Are data mapped at an appropriate scale for the resolution required?

- ❑ In what geographic coordinate system are the data recorded? What complications will occur in converting the data if conversion becomes necessary? Will accuracy or resolution be lost in the process?

- ❑ What process was used to collect the data? Do statements about the precision and accuracy of the data accompany the data set?

- ❑ Are the data uniform? Is the medium on which they are recorded also uniform and free from the kind of distortion found in some aerial photos?

- ❑ Are the data sets truly available? Who owns them? Are they in the public domain? Can "originals" be obtained, or only copies; what information is lost in the copying process?

- ❑ Are there administrative obstacles precluding use of the data? Are they subject to provisions of confidentiality? Are they classified by the military?

- ❑ How much time is required to obtain the data? How much time must be allowed to reformat or encode them?

- ❑ Will information updates be available from the same source? If not, will new updates, possibly from different processes, mesh with existing data?

- ❑ When all considerations are combined, what will the data cost?

If you can't find data sets that meet your specifications—or even if you can—you may embark on a data collection effort. In many ways, if data sets that approximately meet your specifications are available, the issue of whether to use them or collect your own is much like the issue of whether to buy a used or new car. There are advantages, disadvantages, and uncertainties associated with both courses of action. The decision can become very much the classic avoidance-avoidance conflict that college sophomores learn about in psychology courses: the more you look at other people's data for your requirements, the more you want to collect your own; the more you examine what you have to go through to collect your own, the more attractive the existing data seem.

If you decide to collect data anew, many of the concerns for characteristics still apply, but the question changes, from "Do these data have the properties I want?" to "How do I construct a process to produce the data and characteristics I want?"

Probably the best advice to anyone planning a large data collection effort is to start slowly . . . and carefully. In fact, with all operations involving a data-handling program, one should probably use a "10 percent planning rule." This rule says that if x dollars are to be spent over a period of time, then 10 percent x

dollars should be spent over a previous period on the same subject. For example, if 1 million dollars is to be spent on GIS data development in a year period, $100,000 should have been spent in the months before for planning, analysis, and testing. And, by extension, $10,000 should have been spent before that to determine how to spend the $100,000. The 10 percent planning rule suggests, then, that a small but substantial and representative amount of data be collected, encoded, and validated before the major data collection effort gets under way.

The process of data collection must be carefully planned and executed. It is possible to spend a lot of money at it and wind up with nothing very useful. Among the points to consider are the following:

1. Some work may well be contracted out. Be sure that all understandings with the contracting firm are both written down and completely comprehended by both parties.

2. There aren't very many firms that do good intermediate- and high-altitude orthophoto work. Those that exist may be scheduled for months or years in advance.

3. Rigid timetables for collection of data about the environment cannot be followed. Clouds form, trees get leaves, airplanes malfunction, the ground gets wet. Timetables must be based on probabilities.

4. Consider data sets that are collected in an ongoing fashion by satellite or high-altitude aircraft. Data sets showing features in color at resolutions of somewhat more than a meter are now available— available, but not cheap. These data sets primarily depict land cover (from which, in many instances, land use may be determined). The use of data collected in this way has a number of advantages. Among them: (a) the data may be obtained in already digitized form, (b) updating takes place on a periodic basis, (c) sophisticated computer software is available to manipulate these data.

5. Put the data acquired into proper form for inclusion in the GIS database. Basically, the process of encoding the data means transforming it from the basic form in which it is collected (or acquired if already collected data are used) into the symbolic or graphic form required by the GIS. The process depends on the types of data, the precisions required, the equipment available, the scheme used to represent the data in computer memory (the storage paradigm), and other factors, discussed in Chapter 4.

6. Check the accuracy of each step of the collecting or reformatting process; also check the first form of the data against the form in the GIS.

Two elements must be constantly monitored: (a) the process of collecting and reformatting the data, and (b) the quality of the data themselves. Perhaps the most important statement that can be made about this "checking" process is that it be accomplished by someone other than the person or group doing the data collection.

Such independent checking has many virtues: It provides for a more objective view by those doing the checking; it ensures that the checking activity is a project in itself and not just an adjunct to the data collection effort; and it reduces the temptation to use the same techniques to check the data as are used to develop the data.

There must be more to the checking process than simply ascertaining and reporting error rates. An understanding of why errors occur must be developed. If the encoded value for evaluation at a certain point is 1023 feet and a checker with an altimeter set on the spot finds 999 feet, what happened? Was the problem in measuring altitude? Are the positional coordinates off? Is there some systematic or random error in the encoding process or equipment?

It is vital to understand that all large databases contain errors. If a variable in a database is a continuous quantity, such as elevation, there will be values outside the established accuracy standards. If the base is one of classifications, such as land use activity, some uses will be misclassified. If the base is geographically referenced, there will be disagreements of actual locations between points of the Earth's surface and where the system has them located. The purpose of validating data is to develop an understanding of how great these error rates are and, if they are too great, take steps to reduce them.

7. Repeat Steps 4, 5, and 6 for the complete set of data for the base. Realize, however, that the database will be a growing, evolving entity, supplying useful information to decision makers for years to come. These steps are simply the first ones. By the time the pilot data have been satisfactorily collected or acquired, encoded, and validated, the database developers should have a firm grasp on the associated problems, costs, and techniques. There should be some serious thought given to the differences between the pilot project and the major data-gathering effort.

There should be no letup in the testing of data as they come in and are encoded. And if you really want to put your data collection techniques on the line, recollect some data from the pilot area and compare it against that originally collected. The results of that may be very instructive, if disheartening.

One of the worst losses at this point can be of key personnel, now much more valuable than when the pilot data collection began. Their importance to the success of the system should be recognized and appropriately compensated, if possible.

8. Employ techniques for monitoring and updating the database. Even as the major database collection effort is going on, the world will be changing and new problems will be appearing; updating must be a constant activity to keep the database reasonably current.

GPS and GIS

In Chapter 1 you learned a bit about the NAVSTAR Global Positioning System. Let's explore some of the main reasons for making GPS a primary source of data for GIS:

❏ *Availability*—In 1995, the U.S. Department of Defense (DoD) declared NAVSTAR to have "final operational capability." Deciphered, this means that the DoD has committed itself to maintaining NAVSTAR's capability for civilians at a level specified by law, for the foreseeable future, at least in times of peace. Therefore, those with GPS receivers may locate their positions anywhere on the Earth.

❏ *Accuracy*—GPS allows the user to know position information easily and with remarkable accuracy. A receiver operating by itself can let you locate yourself within 3 to 8 meters of your true position. (And using two GPS receivers, when one is positioned over a known [surveyed] point, a user can get accuracies of 1 to 3 meters.)[1] At least two factors promote such accuracy:

First, with GPS, we work with primary data sources. Consider one alternative to using GPS to generate spatial data: the digitizer. A digitizer is essentially an electronic drawing table, wherewith an operator traces lines or enters points by "pointing"—with "crosshairs" embedded in a clear plastic "puck"—at features on a map.

---

[1]These accuracies pertain to "mapping grade" data collection. By spending more money and much more time one can shift to "survey grade" GPS, with accuracies down to a centimeter.

One could consider that the ground-based portion of a GPS system and a digitizer are analogous: The Earth's surface is the digitizing tablet, and the GPS receiver antenna plays the part of the cross-hairs, tracing along, for example, a road. But data generation with GPS takes place by recording the position on the most fundamental entity available: the Earth itself, rather than a map or photograph of a part of the Earth that was derived through a process involving perhaps several transformations.

Second, GPS itself has high inherent accuracy. The precision of a digitizer may be 0.1 millimeters (mm). On a map of scale 1:24,000, this translates into 2.4 meters (m) on the ground. A distance of 2.4 m is comparable to the accuracy one might expect of the properly corrected data from a medium-quality GPS receiver. It would be hard to get this out of the digitizing process. A secondary road on our map might be represented by a line five times as wide as the precision of the digitizer (0.5 mm wide), giving a distance on the ground of 12 m, or about 40 feet.

On larger-scale maps, of course, the precision one might obtain from a digitizer can exceed that obtained from the sort of GPS receiver commonly used to put data into a GIS. On a "200-scale map" (where 1 inch is equivalent to 200 feet on the ground), 0.1 mm would imply a distance of approximately a quarter of a meter, or less than a foot. While this distance is well within the range of GPS capability, the equipment to obtain such accuracy is expensive and is usually used for surveying, rather than for general GIS spatial analysis and mapmaking activities. In summary, if you are willing to pay for it, at the extremes of accuracy, GPS wins over all other methods. Surveyors know that GPS can provide horizontal, real-world accuracies of less than one centimeter.

❑ *Ease of use*—Anyone who can read coordinates and find the corresponding position on a map can use a GPS receiver. A single position so derived is usually accurate to within 10 meters or so. Those who want to collect data accurate enough for a GIS must involve themselves in more complex procedures, but the task is no more difficult than many GIS operations.

❑ *GPS data points are inherently three-dimensional*—In addition to providing latitude-longitude (or other "horizontal" information), a GPS receiver may also provide altitude information. In fact, unless it does provide altitude information itself, it must be told its altitude in order to know where it is in a horizontal plane. The accuracy of the third dimension of GPS data is not as great, usually, as the horizontal accuracies. As a rule of thumb, variances in the horizontal accuracy should be multiplied by 1.5 (and perhaps as much as 3.0) to get an estimate of the vertical accuracy.

Anatomy of the Acronym: "Global Positioning System"

Global:—Anywhere on Earth. Well, almost anywhere, but not (or not as well):

Inside buildings

Underground

In very severe precipitation

Under heavy, wet tree canopy

Around strong radio transmissions

In "urban canyons" amongst tall buildings

Near powerful radio transmitter antennas

or anywhere else not having a direct view of a substantial portion of the sky. The radio waves that GPS satellites transmit have very short lengths—about 20 cm. A wave of this length is good for measuring because it follows a very straight path, unlike its longer cousins, such as AM and FM band radio waves that may bend considerably. Unfortunately, shortwaves also do not penetrate matter very well, so the transmitter and the receiver must not have much solid matter between them, or the waves are blocked, as light waves are easily blocked.

Positioning—Answering brand-new and age-old human questions: Where am I? How fast am I moving and in what direction? What direction should I go to get to some other specific location, and how long would it take at my current speed to get there? *And, most importantly for GIS, where have I been?* To collect GIS data with GPS, one moves the receiver antenna around areas of the Earth, leaving a tracing of points in the memory of the receiver, which you later transfer to GIS software.

System—A collection of components with connections (links) among them. Components and links have characteristics. GPS might be divided up in the following way.[2]

The Earth

The first major component of GPS is Earth itself: its mass and its surface, and the space immediately above. The mass of the Earth holds the satellites in orbit. From the point of view of physics, each satellite is trying to fly by the Earth on a horizontal path at 4 kilometers per second. The Earth's gravity pulls on the satellite, so it falls vertically. The trajectory of its horizontal movement and its vertical fall is a track that parallels the curve of the Earth's surface, so it never crashes. All satellites, including our moon, are subject to the same principle.

The surface of the Earth is studded with little *"monuments"*—carefully positioned metal or stone markers—whose coordinates are known quite accurately. These lie in the "numerical *graticule*," which we all agree forms the basis for geographic position. Measurements in the units of the graticule, and based on the positions of the monuments, allow us, through surveying, to determine the position of any object we choose on the surface of the Earth.

Earth-Circling Satellites

The United States GPS design calls for a total of at least 24 and up to 32 solar-powered radio transmitters, forming a constellation such that several are "visible" from any point on Earth at any given time. The first one was launched on February 22, 1978. In mid-1994 all 24 were broadcasting. The minimum "constellation" of 24 includes three "spares." As many as 30 have been up and working at one time.

The newest GPS satellites (designated as Block IIR) are at a "middle altitude" of about 11,000, nautical miles (nm), or roughly 20,400 kilometers (km) or 12,700 statute miles above the Earth's surface. This puts them above the standard orbital height of the space shuttle, most other satellites, and the enormous amount of space junk that has accumulated. They are also well above Earth's air, where they are safe from the effects of atmospheric drag. When GPS satellites "die," they are sent to orbits about 600 miles further out, where they will remain virtually forever.

GPS satellites are below the geostationary satellites, usually used for communications and sending TV, telephone, and other signals back to Earth-based fixed antennas. These satellites are 35,763 km (or

---

[2]Officially, the GPS system is divided up into a space segment, a control segment, and a user segment. We will look at it a little differently. One of the many places to see official terminology, at the time this book went to press, is http://tycho.usno.navy.mil/gpsinfo.html#seg.

19,299 nm or 22,223 sm) above the Earth, where they hang over the equator, relaying signals from and to ground-based stations.

The NAVSTAR satellites are neither polar nor equatorial, but slice the Earth's latitudes at about 55°, executing a single revolution every 12 hours. Further, although each satellite is in a 12-hour orbit, an observer on Earth will see it rise and set about four minutes earlier each day. There are four or five satellites in slots in each of six distinct orbital planes (labeled A, B, C, D, E, and F) set 60 degrees apart. The orbits are almost exactly circular.

GPS satellites move at a speed of 3.87 kilometers per second (8,653 miles per hour). Different versions of the satellites have evolved over the years. They weigh 1100 to 2200 kilograms (1 or 2 tons) and have a width of about 11.6 meters (about 38 feet) with the solar panels extended. Those panels generate roughly 1000 watts of power. The radio on board broadcasts with about 40 watts of power. (Compare that with your maximum permitted FM station with 50,000 watts.) The radio frequency used for the civilian GPS signal is called "GPS L1" and is at 1575.42 megahertz (MHz). Each satellite has on board four atomic clocks (either cesium or rubidium) that keep time to within 3 billionths of a second or so, allowing users on the ground to determine the current time to within about 40 billionths of a second.

Ground-Based Stations

While the GPS satellites are free from drag by the atmosphere, their tracks are influenced by the gravitational effects of the moon and sun, and by the solar wind. Further, they are crammed with electronics. Thus, both their tracks and their innards require monitoring. This is accomplished by four ground-based stations near the equator, spaced around the world. Each satellite passes over at least one monitoring station twice a day. Information developed by the monitoring station is transmitted back to the satellite, which in turn rebroadcasts it to GPS receivers. Subjects of a satellite's broadcast are the health of the satellite's electronics, how the track of the satellite varies from what is expected, the current *almanac*[3] for all the satellites, and other, more esoteric subjects that need not concern us. Other ground-based stations exist, primarily for uploading information to the satellites. The GPS Master Control Station is at Schriever Air Force Base near Colorado Springs, Colorado.

Receivers

A GPS receiver consists of the following:

❑   An antenna (whose position the receiver reports)

❑   Electronics to receive the satellite signals

❑   A microcomputer to process the data that determines the antenna position and to record position values

❑   Controls to provide user input to the receiver

❑   A screen to display information

More elaborate units have computer memory to store position data points and the velocity of the antenna. This information may be uploaded into a personal computer or workstation, and then installed in GIS software database. Another elaboration on the basic GPS unit is the ability to receive data from and transmit data to other GPS receivers—a technique called "real-time differential GPS" that may be used to considerably increase the accuracy of position finding.

---

[3]An almanac is a description of the predicted positions of heavenly bodies.

Receiver Manufacturers

In addition to being an engineering marvel and of great benefit to many concerned with spatial issues as complex as national defense or as mundane as re-finding a great fishing spot, GPS is also big business. Dozens of GPS receiver builders exist—from those who manufacture just the GPS "engine," to those who provide a complete unit for the end user. Prices range from just over $100 USD to several thousands.

The United States Department of Defense

The U.S. DoD is charged by law with developing and maintaining NAVSTAR. It was, at first, secret. Five years elapsed from the first satellite launch in 1978 until news of GPS came out in 1983. In the two decades since—despite the fact that parts of the system remain highly classified—citizens have been cashing in on "The Next Utility."

There is little question that the design of GPS would have been different had it been a civilian system "from the ground up." But then, GPS might not have been developed at all. Many issues must be resolved in the coming years. A Presidential Directive issued in March of 1996 designated the U.S. Department of Transportation as the lead civilian agency to work with DoD so that nonmilitary uses can bloom. DoD is learning to play nicely with the civilian world. They and we all hope, of course, that the civil uses of GPS will vastly outpace the military need.

Users

Finally, of course, the most important component of the system is *you*. A large and quickly growing population, users come with a wide variety of needs, applications, and ideas. From tracking ice floes near Alaska to digitizing highways in Ohio. From rescuing sailors to pinpointing toxic dump sites. From urban planning to forest management. From improving crop yields to laying pipelines.

What Time Is It?

Although this is a text on GIS—and hence is primarily concerned with positional issues, it would not be complete without mentioning what may, for the average person, be the most important facet of GPS: providing the humans of Earth with a universal, amazingly precise, and accurate time source. In fact, GPS probably should be called a GPTS (Global Positioning and Timing System). Allowing any person or piece of equipment to know the exact time has tremendous implications for things we depend on every day (like getting information across the Internet, like synchronizing the electric power grid and the telephone network). Further, human knowledge is enhanced by research projects that depend on knowing the exact time in different parts of the world. For example, it is now possible to track seismic waves created by earthquakes, on one side of the Earth, through its center, to the other side, since the *exact* time[4] may be known worldwide.[5]

---

[4]Well, okay, there is no such thing as "exact" time. Time is continuous stuff like position and speed and water, not discrete stuff like people, eggs, and integers, so when I say "exact" here I mean within a variation of a few billionths of a second—a few nanoseconds.

[5]The baseball catcher Yogi Berra was once asked "Hey Yogi, what time is it?" to which Berra is said to have replied, "You mean right now?" Yes, Yogi, RIGHT NOW!

Geographic and Attribute Data: Selection, Input, and Editing

——— **Open up your Fast Facts text or document file.**

You will want ArcCatalog to display file extensions and to indicate dataset projections. Relevant information on setting this up should be in your Fast Facts File—recorded from the early steps in Chapter 1. Also, for exercises in this chapter you will need 3D analyst and spatial analyst.

Exercise 5-1 (Warm-up)

Looking at Areal Representations of the Real World

The land of the earth is partitioned into areas. Nature started this. First take a look at a Globe View of the planet.

——— **1.** Start ArcCatalog. Dismiss ArcToolbox if it is present. Click Catalog in the Catalog Tree. Choose View > Refresh. You are going to be looking for ESRI's ArcGlobeData. The first place to look might be

C:\Program Files\ArcGIS\ArcGlobeData.[1]

——— **2.** Expand ArcGlobeData. Click wsiearth.tif. The geography preview doesn't give you any hint that the world is spherical, so click Globe View in the Preview drop-down menu. If necessary, click once on the black background that appears. You get a vision of the Earth that can only be described as neat, especially when you

rotate it by dragging the mouse cursor, and

zoom by holding the right mouse button down and dragging back and forth.[2]

———

[1] If you don't find the folder ArcGIS here, then it will probably be in a root directory of some other local drive or buried in the file structure of a network. Ask your instructor if you have difficulty finding it.
[2] If you have a "wheel mouse," you can zoom with the wheel.

___ **3.** Find a view in which the Earth appears to be almost all water. Now find the continent of Europe. Then Asia.

It's pretty hard to understand, from visual inspection, if the word "continent" has any physical meaning at all, just how Europe and Asia were ever considered separate entities. This may be the first major indication that humans divide land up in strange ways. But read on. It does get better.

___ **4.** Done playing? Click country.shp. And wait. And, depending on the speed of your computer, wait some more. If you get tired of waiting, click the map. When a black screen finally appears, click it once. Look around.

Here you should see the divisions humans have recently imposed on the land masses. They have divided up planet Earth into areas for reference and jurisdiction (a generous way of describing the results of, mainly, wars and political strife over the millennia). Now, perhaps more than ever, it takes documents and time to cross many of these division lines.

___ **5.** Click Countries.lyr. Look at the Globe View of this, then shift to Geography in the Preview drop-down menu. Examining the number in the lower right of the window, what would you say the coordinate system of Countries.lyr is? _GCS WGS 1984_ Zoom in on your country. Start the Identify tool and click on the country to bring up the Identify Results window. Set it so that All Layers will be identified. Determine the populations of your country and two or three around it. Look at some other countries. What is the monetary currency type used in Mexico? _Peso_ Is Bolivia (latitude 18° south, longitude 65° west) landlocked according to Identify? _Y_ How many square miles are there in the United States? _3648943_ What population is listed for Antarctica? ___ _0_

___ **6.** Zoom back to the full extent. Note the distortion in areas away from the equator where a degree of longitude is much less than a degree of latitude. The most dramatic example is Antarctica, where a single point (the South Pole) becomes a line five times the width of the United States, which itself looks pretty squashed.

___ **7.** Look at the table for Countries.lyr. How many countries are there, according to this table? ___ _250_

Begin now to look at more rational ways of dividing up the landscape.

Looking at Reference Systems

___ **8.** In the Catalog Tree contract ArcGlobeData. Expand Reference Systems. Here are ten plus ways of dividing up the planet's surface that are scientifically, if somewhat randomly, devised. Click usgs250q.shp. These are polygons that represent the United States Geological Survey 1-to-250,000-scale quadrangle maps. Zoom in on one quadrangle. Using the coordinate values in the lower right corner of the window determine how many degrees it is from west to east? _2°_. How many degrees from south to north? _1°_. Zoom back to the full extent. Pick a quadrangle at random with the Identify tool and look at the attributes. Why do you think there are four possible state names?

States overlapping w/ grid

Geographic and Attribute Data: Selection, Input, and Editing

_____ 9. Look at the table of usgs250q.shp. Select the QUAD_NAME column. Click Options > Find. Find Boston. Parts of how many states would you think should appear on that quadrangle, based on the ST_NAMEn columns? ___4___. Select the ST_NAME1 column. How many quads cover only Washington state? ___9___. Just for a diversion, click Globe View. Spin it around a little. Also notice that quadrangles in the northern United States cover less land area than those in the southern United States, since the "rectangles" are defined in terms of meridians and parallels. Now bring back the geography.

_____ 10. Look at the USGS 1-to-100,000-scale map divisions. Zoom in on one quadrangle. How many degrees is it from west to east? ___59___. How many degrees from south to north? ___25___. Then look at the 1 to 24,000 scale quadrangles. Zoom in on one quadrangle. How many degrees is it from west to east? ___~59___. How many degrees from south to north? ___25___. Go to the full extent, then zoom up on the southernmost tip of Lake Michigan and determine the name of the quadrangle there. ___Gary___. What is the state? ___Indiana___.

_____ 11. Examine georef15, which divides the Earth up into 15 degree by 15-degree "squares." Also look at georef1, whose table, together with identifiers from georef15, identifies every one degree by one degree "square" on the planet. How many such "squares" are there? ___64800___ (see attr. buter of the shapefile)

_____ 12. Click World Time Zones.shp and display its geography. Use Identify. What does the ZONE attribute refer to? ___hours from Greenwich M.___ How many square kilometers are in the zone that contains Greenwich, England? ___3~ 2 I M___. From the table: how many time zones are there? ___37___ How many would there be if time zones were defined simply by meridians that were one hour apart? ___24___ For extra credit (find the polygon, zoom in, go back to Countries, or use the Internet): what place relates to a time zone that is 5:45 later than (ahead of) GMT? _____.

_____ 13. FYI: JOG stands for Jet Operation Graphic. ONC means Operational Navigation Chart. Take a quick look at each.

The two major reference systems that are left in this list are the United States State Plane coordinate system (usstpln83.shp) and the Universal Transverse Mercator coordinate system (utm.shp). UTM is a very regular, almost worldwide system.[4] The State Plane system is a set of one or more zones defined by each state in the United States, and not intended to be used outside that state. You saw some examples of data sets in these reference systems earlier.

Looking at Coordinate Systems

There is not much to see here in terms of geography. These are the files that store the formulas for converting from one coordinate system to another. What is most impressive is how many different ways people have referenced positions on the Earth.

_____ 14. Collapse Reference Systems. Expand Coordinate Systems > Geographic Coordinate Systems.[5]

---

[3]UTC (Universal Time Coordinated), previously designated GMT (Greenwich Mean Time), is the basis for all time zones, so the difference between UTC and time in the city of Greenwich is zero.
[4]The area around the poles is done with a different projection.
[5]You can also get to Coordinate Systems directly from the top level of the Catalog Tree.

15. Expand Europe and look at the Catalog Tree. The point here is that many countries and some cities each have (or had) its own coordinate system—just as a century or two ago every locale had its own time system. And these are latitude and longitude systems; consider that each of these lat/lon systems may be projected onto the Cartesian plane in dozens of different ways, making for hundreds of different ways of assigning a pair of numbers of a given point on Earth's surface.

16. Collapse Europe (and the subsequent systems after you explore them). Check out North America, where there are fewer geographic coordinate systems, but still a lot. If you double click on a coordinate system you can view its properties. Check out Solar System. Based on year 2000 data, what is the diameter of Venus? __12 M__ meters.[6]

17. Check out World. At the bottom of the list is WGS 1984. This is the most current estimation of where the latitude longitude graticule falls on the Earth's surface. What is the diameter of Earth? __12,600__ kilometers. _(12,600,000 meters)_

18. Collapse Geographic Coordinate Systems and expand Projected Coordinate Systems. Look at World and you will see names of projections that may be familiar from some past geography class. Expand the Sphere-based folder and you will see a lot more. The number of approaches taken to representing the quasi-ellipsoidal Earth on a flat plane is mind-numbing. Collapse World.

19. You see a folder for UTM, with which you have some familiarity. Open the WGS 1984 folder. Widen the Catalog Tree as necessary to see all the text. There you will find, in addition to some complex UTM zones, the pairs of UTM zones, in the northern and southern hemispheres numbered from 1 to 60: 1N (meaning zone 1, northern hemisphere), 1S (meaning zone 1 southern hemisphere), 2N, 2S . . . 60S.[7] To repeat the warning issued earlier: In the United States, the UTM system based on the World Geodetic System of 1984 is just different enough, in terms of geographic space (up to a few hundred meters), from UTM based on the North American Datum of 1927 to cause major problems for those trying to determine accurate locations. Contract UTM > WGS 1984 and UTM.

20. You also see there a folder for State Plane. Expand it. Even if you restrict yourself to NAD 1983 (almost exactly equivalent to WGS 1984 in the United States), you will notice the units of measurement in the state plane system is a hodgepodge. Meters are available for all states (under NAD 1983), and some states use meters as the primary survey unit. But other states use feet (called, variously, feet, survey feet, or US feet), and still others use international feet.[8] Further, it is usual for multiple zones to represent each state. How many zones is Michigan, which uses international feet, divided into? __2__. What are the FIPS numbers of the zones? __2111__, __2113__. Collapse the State Plane folder.

---

[6]The Semimajor Axis is equivalent to the radius at the equator.

[7]UTM zones customarily have a letter suffix, such as P in 18P. These letter designations are not necessary. The N and S suffix in the ESRI designations are not these, but rather reference the northern or southern hemisphere.

[8]The meter is the fundamental unit of length measurement in the world. Both the survey foot and the international foot are based on it. The survey foot is defined by 1 meter = 39.37 inches exactly. The international foot is defined by .0254 meters equals exactly 1 inch. These two definitions differ by about 2 parts in 100,000. So, for example, a distance of 100 miles in one system would be about 11 feet different than in the other system.

Using the Reference System to Discover the Boundary Coordinates of a State Plane Zone

We can take advantage of what we saw earlier in "Looking at Reference Systems" to set the extent of a geodatabase. We begin by setting the data frame to the correct coordinate system.

____ 21. Using ArcCatalog, make a new folder under

___IGIS-Arc_*YourInitialsHere*.

Rename the new folder Digitize&Transform. Make a Folder Connection to that folder.

____ 22. Start ArcMap with a new, empty map. Add data:

C:\Program Files\ArcGIS\Reference Systems\usstpln83.shp.

You will see a somewhat squashed coterminous U.S. map and a really large Alaska. Bring up the Data Frame Properties window (View > Data Frame Properties; or double-click the word Layers in the Table of Contents; or right-click the data frame and choose Properties). Press the General tab. What do you find for display units? *Deg Min Sec* Press the Coordinate System tab. What is the coordinate system? ___GCS North Amer. 1983___

____ 23. Under the Coordinate System tab, navigate to the predefined projected coordinate system NAD 1983 State Plane Kentucky North FIPS 1601 (survey feet). Under the General tab the map units and the display units should now read Feet. Click Apply, then OK.

A map appears—distorted in a different way from the previous one. The reason it looks strange is that the latitude and longitude coordinates that are natural to this data set have been converted "on the fly" (meaning instantaneously , for display purposes) to the Kentucky state plane coordinates. Those coordinates work well in middle America, but not elsewhere, particularly, as you note, in Alaska. In the next step you will find the Kentucky north zone, isolate it, and export it as a shapefile all by itself.

____ 24. Zoom in on Kentucky (KY), using Identify if you need to. Zoom in on its north zone. Use the Identify tool to make sure you are in the right place—the ZONENAME83 should be KY_N. What is its FIPSZONE83 number? _1601_ . To the nearest tenth, how many square miles are in the zone? _10,352.3_

____ 25. From the Tools toolbar, pick Select Features. Click the north zone. It should now have a colored border. Right-click usstpln83 in the Table of Contents. Click Data > Export Data. In the Export Data window, make sure that you are about to export Selected Features. Press the radio button to select Use the same Coordinate System as the data frame so that the north zone will be changed from geographic coordinates to state plane coordinates. For output, browse to

___IGIS-Arc_*YourInitialsHere*\Digitize&Transform

and make the name KY_N_sp83.shp. Click Save, then, after checking that it looks like Figure 5-1, OK the Export Data window. When asked if you want to add the data to the map, select Yes. Remove the layer usstpln83.

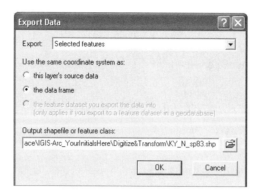

FIGURE 5-1

_____ **26.** Using as a cursor the Select Elements tool, slide the cursor around the map. Verify that the map display is in feet. Imagine a north-south line that is west of the entire zone, and whose easting is a nice round number like 1110000. Write it below, along with three other boundaries (again with round numbers):

Western boundary ___1110000___

Eastern boundary ___2,170,000___

Northern boundary ___610000___

Southern boundary ___740,000___

You will use these numbers later when you make a personal geodatabase feature data set that encompasses the Kentucky north zone.

_____ **27.** Collapse Program Files in ArcCatalog. Dismiss ArcMap without saving changes to the Untitled map.

Primary Lesson

If you take only one point away from this exercise, it should be that there are a multitude of ways to reference points on the surface of the Earth, and that if you combine data sets, either (a) they must all agree in all parameters or (b) you must know exactly what you are doing. For the case of (b), In the absence of having access to a geodesist, I recommend two resources to start with. First is the text *ArcGIS 9—Understanding Map Projections*, by ESRI. The second is to browse to the ESRI Web site, www.esri.com. Then select Support > Knowledge Base > Search Our Site> and type 21327. You should see an article entitled "How To: Select correct datum transformation when projecting between datums." Reference material will be found at the bottom of this Web page.

Exercise 5-2 (Project)

Look at Geographic Data on the Web

Starting with some of the Web addresses in the Overview of this chapter, locate five spatial data sets both in your geographic area and on a topic interesting to you. Make a list of those sources you find.

Wetlands fws. er. usgs.gov/ .

_____ .

_____ .

_____ .

_____ .

Exercise 5-3 (Project)

Digitizing and Transforming

"Digitizing" is a process in which a graphic representation (a drawing, a map) is turned into numbers (digits) and characters that give the computer some of the information that is "in" the drawing. Generally that information makes it possible for the computer to reproduce the drawing on its screen or on paper, and to answer questions about it. We, of course, are most interested in digitizing maps.

You have had experience with one method of digitizing: Generating a coverage by typing in the "digits" of the coordinates of the Foozit Court for the elementary school. You made a coverage named Foozit_Court. Doubtless you have had all of that sort of data entry you want, although sometimes it is necessary when the absolute coordinates of some important feature are known precisely.

Another way of digitizing, requiring much less human effort, at least initially, is _scanning_, where a map is placed on a rotating drum, or laid flat on a table, and through a combination of electronics and mechanics, the map's contents are finely gridded into little squares. The pixels that result are stored in a computer file. At this point all the user has is a picture. But, with some direction from the user, sophisticated computer programs can then interpret the picture according to rules and can create an intelligent GIS map. For example, the program could recognize a contour line—perhaps even to the extent of reading its height off the map and placing it as an attribute in a table. High-precision scanners are expensive, and the work to make programs that interpret map image files is ongoing.

A third technique of digitizing is to mount the paper map[9] on an electronic drawing board, called (surprise) a _digitizer_, and trace over the lines of the map with a stylus or puck; this process transmits x-y coordinates to the computer. The user can add attributes to the lines or areas that are digitized.

---

[9]Or Mylar map—used because it is a "stable" medium, unlike paper, which stretches and shrinks with environmental factors like humidity.

A Plan for Digitizing and Transforming

Here we will look at a fourth way of getting graphic information into the computer; it is called "heads-up" digitizing—so named because you look up at the monitor rather than down at a paper map, as in traditional digitizing. In heads-up digitizing, an electronic image of a paper map is made by *scanning*. In this case the map is the Foozit Court. You will be able to see this image on the screen of your computer. By tracing around the lines of the map with crosshairs (a cursor or pointer, controlled by the mouse), you will be able to automatically supply coordinates to ArcMap.

This process is a little like learning to ride a bicycle: not really hard, but difficult to describe in all its detail (e.g., if the bicycle begins to tip over to the left, turn the handlebars to the left—just a little now—and . . .) So I will simply provide general directions and let you figure out how to digitize mostly on your own.

The product of your digitizing will be an ArcGIS *shapefile*. As you know, since we have discussed the nature and characteristics of these elements in the last chapter, a line shapefile is made up of *beginning points, ending points*, and *vertices*, which define *polylines*.

Our ultimate goal is to produce a personal geodatabase polygon feature class and have it appear at a particular location on a real world map. For purposes of illustration, you will do this in two different ways. Here is an overview of Exercise 5-3 (which breaks into 5-3a and 5-3b, after some preliminary steps).

1. Make a blank shapefile with ArcCatalog.

2. Add the image of a scanned "map" as data in ArcMap.

3. Add the blank shapefile as data in ArcMap.

4. Use the Editor to digitize (trace over) the lines of the map.

5. Save the shapefile with the digitized lines in non-real-world coordinates.

In Exercise 5-3a the features in the shapefile will be converted to a geodatabase feature class, with real-world coordinates. The steps will be as follows:

1. Make a blank personal geodatabase feature class (PGDBFC) in the proper projection and with the proper extent.

2. Convert the shapefile to a geodatabase feature class using ArcToolbox.

3. Move the Foozit_Court feature class into the real world with the Spatial Adjustment tool.

Exercise 5-3b mirrors 5-3a, but you will make the shapefile into a coverage, place that in a location with real-world coordinates, and, finally, convert that to a PGDBFC. The steps will be as follows:

can't do it

1. Convert the shapefile to a coverage.

2. Build topology (the node, arc, and polygon tables) for the coverage.

3. Set tics in the coverage based on the digitized coordinates of the map.

4. Make a new, blank coverage that will ultimately contain real-world coordinates.

5. Change the *tic* coordinates of the new coverage from digitizer coordinates to real-world coordinates.

6. Transform the coordinates of the *features* of the coverage to real-world coordinates.

7. Convert the new coverage into a PGDBFC

Getting Started

_____ **1.** Make sure your Fast Facts File is open; you'll need to enter information into it. Your FFF should contain at least the name of each ArcGIS command, module, feature, tool, or wizard that you use.

_____ **2.** With ArcCatalog running, highlight your Digitize&Transform folder in the Catalog Tree. Choose `File > New > Shapefile` to bring up the `Create New Shapefile` window. Create a shapefile named Dig_Lines_shape. Make the `feature type Polyline`. Click the `Edit` button. Choose `Select > Projected Coordinate Systems > State Plane > NAD 1983(Feet) > NAD 1983 StatePlane Kentucky North FIPS 1601(Feet).prj` and click Add. In the Spatial Reference Properties window, read over the details of the coordinate system you have selected. Then click Apply, then OK, then on OK again.

Loading an Image File as a Layer in ArcMap

_____ **3.** Launch ArcMap using a new map. Dismiss ArcCatalog.[10] From

[_____] IGIS-Arc\Image_Data

add the image file Foozit_Court.TIF as a layer in ArcMap. If asked, do not build "pyramids." An image of the court should appear. See Figure 5-2. No reference numbers are provided, because, unlike when you typed in the values, the node positions and line lengths will be determined by where you click when you trace the image with the cursor.

_____ **4.** Make the ArcMap window occupy the full screen. Zoom in so the line drawing occupies the full map display area. Slide the cursor over the map area and notice that the coordinates are (probably) in degrees, minutes, and seconds. This is because, if the coordinate numbers are small, ArcMap assumes a geographic "projection." Fix this by right-clicking the map, picking Properties > General, and changing the map and display units to Unknown Units. Click Apply and OK. Use Measure to determine the dimensions of the court to two decimal places. The result: ___5.00___ by ___4.00___

_____ **5.** Click the pointer (Select Elements) icon. Slide the cursor around the image, noticing the coordinates. Zoom in on the southwest corner. By placing the cursor on the corner, write the coordinates in the table that follows with precision of two decimal places. Zoom back. Find the coordinates for the northwest corner. Repeat this procedure for each of the other two corners of the overall rectangle; write the coordinates of each corner of the court to two decimal places:

Corner	x-coordinate	y-coordinate
Southwest	0.19	2.32
Northwest	0.19	6.32
Northeast	5.19	6.32
Southeast	5.19	2.32

---

[10]Throughout this project you will be asked to close ArcMap and ArcCatalog and then reopen them. This may not be necessary. The problem that sometimes occurs is that you are "locked out" of performing operations by "program A" on a feature class if "program A" believes (rightly or wrongly) that "program B" is using that feature class. The solution that always works is to close "program B," but other solutions may exist. Symptoms of the problem are messages like "unable to obtain write lock" or simply grayed-out fields or buttons.

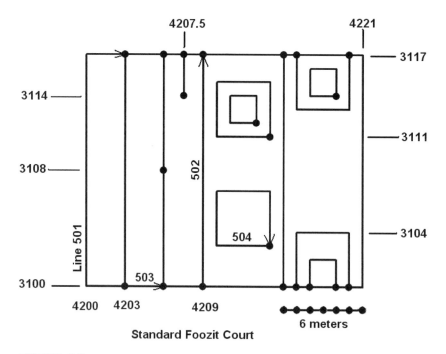

FIGURE 5-2

These local coordinates will be used later to provide the reference locations you will need to move the court to real-world locations. Zoom back to the full extent.

Loading the New, Blank Shapefile into ArcMap

6. Again press Add Data. Navigate to

___IGIS-Arc_*YourInitialsHere*\Digitize&Transform\Dig_Lines_shape.shp

and add it as a layer.

Adding Line Features to a Shape File by Using the Editing Facility in ArcMap

ArcGIS Desktop contains a rather extensive editing facility. To use it, you add the Editor toolbar to ArcMap.

7. Use the Tools menu (or View > Toolbars) to get the Editor toolbar onto the ArcMap window if it isn't there already. Park it where you want it—but be sure it is horizontal.

8. Click Editor > Start editing. The Target of editing, found on the toolbar, should be Dig_Lines_shape.

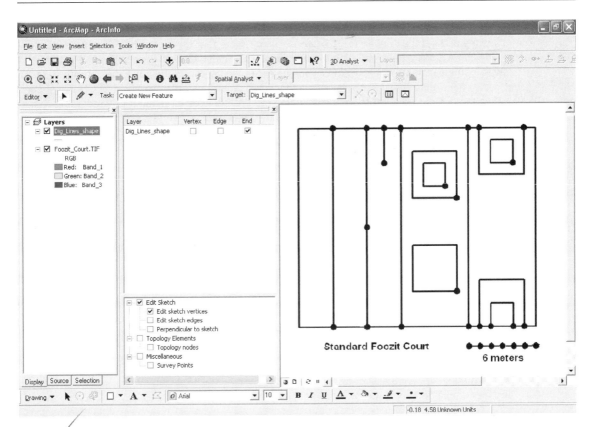

FIGURE 5-3

> **9.** In the Editor toolbar drop-down menu, click Options. Select the General tab. Display measurement using two (2) decimal places. Set the snapping tolerance to 0.1 map units (not pixels). Show Snap Tips. Click Apply, then OK.

Snapping is a process that helps you connect parts of features (e.g., ends of lines) that need to be connected. The word probably comes from the effect caused when the two parts of a snap on a piece of clothing are pressed together. In the case of the ends of lines, it means that when the cursor comes within a preset distance of the end of a line, it moves over exactly to the end of that line. This allows you to start a second line precisely at the end of the first one. That preset distance is specified as the snapping tolerance. You have set it to one-tenth of a unit because that will provide for ease of editing lines in this coordinate system but will prevent ends being snapped together that shouldn't be.

> **10.** In the Editor menu, click Snapping. In the upper pane check the box End on so that the *ends* of the polylines of Dig_Lines_shape will snap together. Also, in the lower pane, set the sketch snapping so that the vertices (only) of the "Edit sketch" will snap together.[11] (This will ensure that, as you draw a "sketch" consisting of lines, you can snap the end of a polyline back to the beginning of that line, as when you have a single polyline around a polygon.) There is no Apply or OK here. (See Figure 5-3.) Just dismiss the window and the snapping environment will be set.

---

[11]The term "Edit sketch" is a noun phrase, not and adjective followed by a noun, as you might suspect. An "Edit sketch" is a graphic entity that you will work with.

___ **11.** In the Editor toolbar, ~~click the~~ Sketch tool (a pencil) to initiate the task. Click Create New Feature and set the target to Dig_Lines_shape. Note that the cursor has become a cross-hairs with a small blue circle. With that cursor you can begin to digitize polylines by digitizing three of the segments of the leftmost rectangle. Put the cursor on the black disk at (4203,3100) as shown in Figure 5-4). Now click once and you will create a starting vertex in Dig_Lines_shape. Move the cursor to the left to the place where the line changes direction (at the corner of the court), and click again to create a second vertex. Create the next vertex. Create the ending vertex, again by clicking once. Press F2 to complete the sketch of the polyline. Notice the bright cyan line that indicates the polyline that you have digitized.

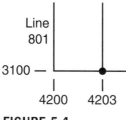

FIGURE 5-4

___ **12.** Begin the second polyline at the southern endpoint of the first polyline. Notice that when you move the cursor near the already existing endpoint, the little circle jumps over to lie precisely on that endpoint. This is the effect of snapping. Notice also the Snap Tip. What does it say? ___Dig_Lines_Shape:a___ Since you set the snapping tolerance to 0.1 map units, this effect occurs when you get within 0.1 units of an existing endpoint. Once the circle is snapped to the end-point (never mind where the cross hairs are), you may click to begin the new polyline. Move the cursor to the northern end of the line. Instead of using a single click followed by F2 to end the sketch, use a double-click on the northern point.

___ **13.** Zoom in on the portion of the drawing as shown in Figure 5-5***. Click the Pencil icon again. You see that you can be as precise as you like in beginning a line on a disk on the graphic image. However, you are stuck (for the moment) with the locations of the endpoints you have already digitized. Digitize the line in the figure. Return to the previous zoom extent.

If you wanted to be really precise, you could zoom in on every vertex. Don't bother on this project, since you are simply learning the digitizing process.

___ **14.** You may have trouble distinguishing the lines of the shapefile from the lines of the image. So change the symbology of the shapefile polylines to a red line with width of 2.

___ **15.** Continue to make polylines until you have created each polyline that corresponds to a line in Foozit_Court.TIF. Be careful to make polylines along the lines between the black disks shown on the image. Where there are no black disks but the line changes direction, make vertices.

If you start to make a line where there shouldn't be one, you can press Ctrl-Delete to do away with it. If you complete making a line where there shouldn't be one, you can click the Edit tool (next to the pencil), select the offending line by clicking it, and press the Delete key.

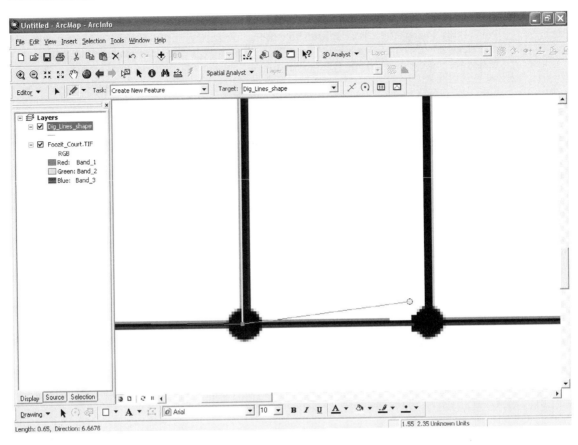

FIGURE 5-5

Draw polylines only to black disks, never through a black disk. The order in which you digitize the polylines is unimportant. The direction in which you digitize the polylines is unimportant. It's a good idea to save edits (in the Editor menu) along the way.

_____ **16.** Check carefully that all the black lines of the image have red polylines over them. Turn off Foozit_Court.TIF. Check again. Everything there? If not, turn Foozit_Court.TIF on again and continue digitizing. Click Editor. Click Save edits. Click Editor. Click Stop editing.

_____ **17.** Close ArcMap without saving the map. The empty shapefile, Dig_Lines_shape, that you started now contains the polylines you digitized, due to you having saved the edits. Its coordinates are not real-world coordinates, but you will be able to move it, intact, to where you want it. Further you will be able to rotate it and change its size.

_____ **18.** Start ArcCatalog. Navigate to

___IGIS-Arc_*YourInitialsHere*\Digitize&Transform\

Dig_Lines_shape.shp

and click it. Press Ctrl-C to copy its contents onto the ArcGIS clipboard. Click Digitize&Transform. Press Ctrl-V to copy Dig_Lines_shape.shp into the folder. The name will appear as Dig_Lines_shapeCopy.shp. You do this in case something goes wrong in the upcoming transformation to real-world coordinates and you need to start again. Backups are almost always a good idea.

Converting a Shapefile to a Geodatabase Feature Class and Giving It Real-World Coordinates

The first step in getting the contents of the Foozit court shapefile into a geodatabase feature class is to make the personal geodatabase, and within it, a feature data set that will contain the spatial parameters of datum, projection, X/Y domain, coordinate system, and units. Finally you create the feature class.

1. Start ArcCatalog if necessary. Click

___IGIS-Arc_*YourInitialsHere*\Digitize&Transform

to highlight it. Choose `File > New > Personal Geodatabase`. In the Catalog Tree, change the name of the PGDB to Recreational_Facilities.mdb. Make sure the new entry is highlighted in the Catalog Tree. Select `File > New > Feature Dataset`. For Name type North_Zone. Click the Edit button. Choose `Coordinate System > Select > Projected Coordinate Systems > State Plane > NAD 1983(Feet) > NAD 1983 StatePlane Kentucky North FIPS 1601(Feet).prj` and then click Add. In the Spatial Reference Properties window, read over the details of the coordinate system you have selected. Then click Apply, but *don't click* OK. Press X/Y Domain. *Read the following discussion carefully.*

For the reasons discussed previously, ArcGIS stores coordinates as long integers. A long integer consists of 32 bits. This allows for more than 4 billion possibilities (all the integers ranging from −2,147,483,648 to 2,147,483,647, to be exact), which provide great precision, even over long distances. For example, if your data set spanned 4000 miles in the east-west direction, and the units of measurement were inches, you could have maximum precision down to about a sixteenth of an inch.[12] But you have to tell ArcGIS at the outset the possible range (X/Y Domain) of your data set, because once the spatial domain of a feature data set is defined *you may not change it*. Also the data set may not contain features whose coordinates fall outside the specified range.

We are going to move the contents of the Foozit_Court shapefile into a feature class in the StatePlane Zone Kentucky North, with its units of (survey) feet, using the North American Datum of 1983. We know the approximate coordinates of the extents of the zone, so we could make a geodatabase with those specifications. You can check the previous exercise to see that the western bound is at about 1,100,000 feet, while the eastern bound is around 2,180,000 feet. The south-to-north range goes from 50,000 feet to 610,000 feet.

So you might think you should just put in these for the minimum and maximum x-y coordinates. But we are also going to install the Foozit Court shapefile in this same data set and then move it to where we want it. That means that the ranges of values for x and y *must include* those digitizer coordinates that you wrote

---

[12]Four billion divided by the number of inches in 4000 miles (4000 * 5280 * 12) gives approximately 16—meaning 16 divisions of an inch.

down previously for the coordinates of the corners of the Foozit_Court. As you can note, these coordinates lie around the origin, where x = 0 and y = 0. So use the following: ← *Problem here. Not allowed*

___?___ **2.** For Min X, type 0. For Max X, type 2180000. For Min Y, type 0. For Max Y, type 610000. Note that the precision (the number of "storage units" per foot) is calculated for you in the Precision text box. It should be somewhere around 1000, meaning that the precision of the data set is one-thousandth of a foot. Double-check your entries, by counting zeros. If you get this wrong, you will have to make another feature data set. Once you OK this window (and the next one), the die is cast. Click OK to get back to the New Feature Dataset window.

?

___?___ **3.** After clicking Show Details and scrolling down to verify that everything, including the X/Y Domain, is the way you want it,[13] press OK.

?

___?___ **4.** Make a feature class by highlighting North_Zone (under Recreational_Facilities.mdb) and then selecting File > New > Feature Class. Name the feature class School_Court_A. Click Next, and Next again. In Field Name, click SHAPE. In Field Properties change the Geometry Type from Polygon to Line. Click Finish. *Not allowed*

Converting the Shapefile to a Geodatabase Feature Class

___✓___ **5.** Open ArcToolbox. Expand: Conversion Tools > To Geodatabase > Feature Class to Feature Class. Pounce. Read the Help in the right pane (Use Show Help if necessary). Click the Input Features text box and read the right pane. Then browse with the Input Features open-folder icon. You want to navigate to the shapefile you created earlier: Dig_Lines_shape.shp. Add the shapefile.

___✓___ **6.** Click the Output Location text box. Browse to the North_Zone data set (make sure that name appears in the Name box of the Output Location window) and click Add. You can use the arrow keys, and Home and End keys to see the hidden contents of the Output Location box in the Feature Class To Feature Class window.

___ **7.** For the Output Feature Class Name, type School_Court_A. Expand the window vertically or use the slider bar to see the rest of the window. Under Field Info do away with the unneeded shapefile identifier ID by making its visible field FALSE. Click OK. Once processing stops, close the Feature Class to Feature Class window. Dismiss ArcToolbox. Preview the geography and the table of *Says it already exists*

School_Court_A.

Moving the Foozit_Court Feature Class into the Real World

___ **8.** Start ArcMap with a new, empty map. Add School_Court_A. Change the lines in School_Court_A to Mars Red, with a width of 2. *Nothing visible*

You will notice, as you slide the cursor around the Foozit court, that the coordinates on the screen show in feet. That's what our unknown digitizer units became when we put the contents of the shapefile into a personal geodatabase feature class that had feet as the units. If you measure the width of the court, you

---

[13]If it isn't right, press Edit. What's important here is that the extent be big enough to cover what you want covered. ArcCatalog may take it on its own to make the extent bigger in one direction or the other.

get something on the order of 5 feet—clearly not correct. It's nothing to worry about. When you transform the court into real-world coordinates, the feature class dimensions will change and it will be rotated to fit the spot. All that is important here is that the right lines are connected and that the image is not distorted.

9. Add the image file six.tif from ___IGIS-Arc\Image_Data. (Do not build pyramids if asked. Ignore the warning about the unknown spatial reference.) Zoom to the layer. In the southwest corner of the image is a school with a running track. Zoom to the school, including the track. Bookmark the zoomed up school image, calling it School. Observe, at the south-central part of this image, a large parking lot.

You will place the Foozit court in the parking lot of the school, at the state plane coordinates shown in Table 5-1, which have been provided to you by a surveyor:

TABLE 5-1

Corner Number	Easting	Northing
1	1573792.8	176684.5
2	1573800.0	176700.0
3	1573819.0	176691.1
4	1573811.8	176675.7

The corner numbers are given by Figure 5-6. Corner 1 is where the original southwest corner of the digitized image will go. Corner 2 is the northwest corner. Corner 3 is the northeast corner.

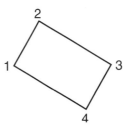

FIGURE 5-6

In order to move the court to the correct coordinates, you will need to create a text file. Each line of the file consists of three elements (a) a point number (the corner number), (b) the digitizer coordinates you wrote down earlier in Step 5, and (c) the real-world coordinates provided by the surveyor—so five numbers will be placed in each row of the text file.

Where the string "corner1x-digcoord" (see the following) appears, you need to substitute the coordinates of the corners from the digitizing process that you wrote down in the table in Step 5 of Exercise 5.3. For example, corner1x-digcoord might be something like 0.21. Be sure to get the right pairs of digitizer coordinates associated with the right pairs of surveyor's coordinates. For example, the southwest corner is represented in the preceding table by line 1. Also the x-coordinate (easting) precedes the y- (northing) coordinate. That is, the text file looks like this:

```
1 corner1x-digcoord corner1y-digcoord    1573792.8    176684.5
2 corner2x-digcoord corner2y-digcoord    1573800.0    176700.0
3 corner3x-digcoord corner3y-digcoord    1573819.0    176691.1
4 corner4x-digcoord corner4y-digcoord    1573811.8    176675.7
```

where you will put in actual numbers for strings like `corner1x-digcoord`.

Use tabs to separate the values! The first line of your file will look something like this:

```
1 0.21    2.28    1573792.8    176684.5
```

except, of course, you would use the numbers you found when digitizing instead of 0.21 and 2.28.

_____ **10.** Make this four-line file with a text editor, calling it Foozit_Parklot.txt and save it in ___IGIS-Arc_*YourInitialsHere*\Digitize&Transform.

_____ **11.** Back in ArcMap, zoom to the full extent of both data sets. You may (or may not) see a tiny red dot on the left of the window at around (0,0), which is School_Court_A, and a little black smudge at the right at around (1575000, 180000), which is six.tif. Start editing from the Editor toolbar. Turn on the Spatial Adjustment toolbar, under More Editing Tools in the Editor toolbar drop-down menu.

_____ **12.** In the Spatial Adjustment drop-down menu, select Set Adjust Data. Make sure School_Court_A is checked as the data to adjust (i.e., transform). Click the radio button to pick All Features In These Layers. Click OK. Again in the drop-down menu, pick Transformation—Affine as the adjustment method. Under Links pick Open Links File, specifying Foozit_Parklot.txt in Digitize&Transform.

If you have done everything right, you will now note lines running from the tiny red dot (the digitized school court) in an east-northeast direction toward the black smudge (the orthophotoquad).

_____ **13.** Zoom in on the digitized court. Then zoom out so the court occupies only about half the window.[14] You will see lines running from each corner, in accordance with the text file table you made. See Figure 5-7.

_____ **14.** For the following operations refer to Figures 5-8 and 5-9*** as necessary; but mostly pay attention to what is happening on the computer monitor. Zoom to the bookmark School. Here you will see the eastmost ends of these "transformation lines." Zoom in on the parking lot where the Foozit Court is to be placed. From the arrow heads note where the corners of the Foozit court will be. From Spatial Adjustment > Links > View Link Table, get the link table up. One at a time, click the ID value in the link table and note the flashing link to the parking lot.

You can also see, as the last column in the link table, the residual error for each link, which is the number of feet by which the corners of the court don't exactly fit. It should be 0.05 or less.

---

[14]In ArcMap 9.1 the lines run in the wrong direction if you zoom to layer. If you zoom out a little bit this problem gets fixed.

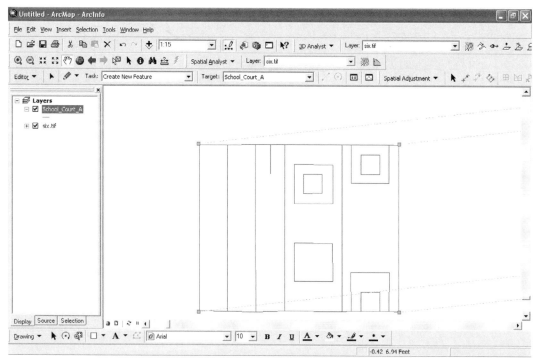

FIGURE 5-7

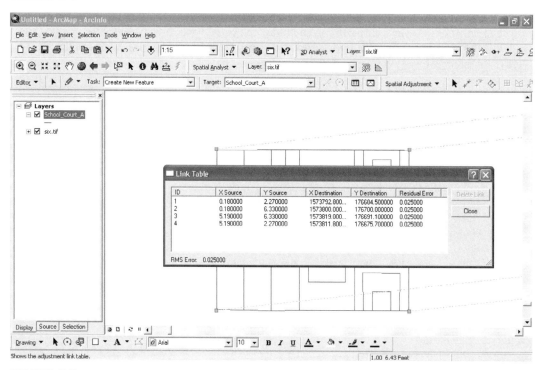

FIGURE 5-8

FIGURE 5-9

That is, the rectangle specified by the corners of the digitized feature cannot be stretched *linearly* to exactly fit the surveyor's corners, but the small numbers tell you that the transformation is going to go very well. If these numbers are large, or if the arrowheads don't appear to look right, something has gone wrong. If there is a problem, Delete Links and go back and fix things—which probably means revising the text file.

_____ **15.** Close the link table. From the Spatial Adjustment toolbar pick Adjust. Bingo. The Foozit Court appears. Zoom in further to check that it arrived intact. Use the measure command to check that the dimensions are correct. What are they? _____ feet by _____ feet. Check the diagram from which you digitized. Are these dimensions correct? _____ (Since this court is for an elementary school we've made it smaller by laying it out in feet instead of meters.) Zoom out to see the court in the parking lot.

_____ **16.** If everything looks okay, from the Editor drop-down menu, click Save Edits. Click Stop Editing. If things do not look right, from the Editor drop-down menu, click Stop Editing. DO NOT SAVE EDITS. Go back to Step 11.

_____ **17.** Save the map as School_map1.mxd in

_____IGIS-Arc_*YourInitialsHere*\Digitize&Transform. Dismiss the Spatial Adjustment toolbar and then dismiss ArcMap.

From Shapefile to Coverage to Real-World to Geodatabase Feature Class

As the title implies, this is an exercise in digitizing, converting, and transforming. Some of GIS is display, some is analysis (more and more, we hope), but a lot is coping with the myriad data types and geographic representations. Here is an example.

Creating and Transforming a Coverage Based on a Shapefile

___ **1.** From ArcCatalog, launch ArcToolbox. Use `Conversion Tools > To Coverage > Feature Class To Coverage` to bring up a `Feature Class To Coverage` window. Make the input feature class Dig_Lines_shape.shp in

___IGIS-Arc_*YourInitialsHere*\Digitize&Transform.

In the Type column, click the blank text box next to the feature path/name, and pick ARC as the type. Append the `output coverage name DIG_COV`, to the path

___IGIS-Arc_*YourInitialsHere*\Digitize&Transform.

Make the cluster tolerance 0.1 Feet. Click OK. When processing stops, close the Feature Class To Coverage window and dismiss ArcToolbox.

___ **2.** In ArcCatalog, navigate to Dig_Lines_shape.shp. Open its table and examine. Dull.

___ **3.** Navigate to the *arc component* of DIG_COV. Open the table of its arc component and examine. More items, but also dull. The software did calculate the length of each arc for you.

___ **4.** Right-click the DIG_COV name. Click Properties.

Building Topology for the Coverage

___ **5.** In the Coverage Properties window, click the General tab. Click Build. In the Build window, choose the Node feature class from the drop-down menu. Click OK.

___ **6.** Now build Line (Arc) topology. Click OK. Now build Polygon topology. Click OK. Dismiss the Coverage Properties window.

If any of the preceding doesn't work, it's probably because the shapefile you constructed has overlapping lines or intersections without a node present. You will probably have to go back and redigitize the image, making sure that you haven't digitized through a node and that all arcs snapped together.

___ **7.** *Refresh the Catalog Tree:* Click Catalog at the top of the tree. Choose View > Refresh (or just press F5). This process ensures that the Catalog Tree is up-to-date—including the building operations you performed previously. Examine the geography and tables of node, arc, and polygon components of DIG_COV.

As you know, coverages also have *tics* (coordinate pairs) that can tie the features of the coverage to particular locations in a Cartesian coordinate system.

___ **8.** Click on the `tic` component in the Catalog Tree. Look at the Geography and note that it is blank. Look at the table: no records.

In order to provide positions for the transformation later, you need to supply the coordinates of the corners of the Foozit_Court, which you determined earlier. You do this in the following steps. For the coordinates of the tics, you will use the ones you wrote down previously in Exercise 5-3, Step 5.

___ **9.** Bring up the Coverage Properties window again. Press the `Tics and Extents` tab. By pressing the Add button you can add a tic.

___ **10.** *Add tics to the DIG_COV coverage:* Click Add to put in a new tic. You will call it ID number 91. It will be the southwest tic. Put in its coordinates from the table in Exercise 5-3, Step 5. Add again. Make the northwest tic 92 and put in its coordinates. Continue, making the northeast tic 93, and the southeast tic 94.

___ **11.** Carefully check the ID, the x-coordinate, and the y-coordinate of each tic. Be sure you included minus signs if appropriate. Click Update. Fix values if necessary—an error here will have serious repercussions later. Click Apply. Click OK. Now look at the table of the tic component of the coverage. Make sure it is okay. Look at the geography.

Verifying the Correctness of the Coverage

___ **12.** Start ArcMap. Add the tic component of DIG_COV. Make its symbol a green circle of size 8. Use Identify to examine each tic. Now look at the tic attribute table. Dismiss it after making sure it is okay.

Add the polygon component of DIG_COV to the map. What is the area of the largest polygon? _____. What is the perimeter of the polygon with the smallest area? _____.

At this point you have made a coverage, with attribute tables, that is equivalent to the map of the Foozit_Court. But all the coordinates, areas, and lengths are in digitizer coordinates. The task now is to make a new coverage in which the digitizer coordinates have been transformed into a real-world coordinate system.

___ **13.** Dismiss ArcMap.

Making a New, Blank Coverage, Called REAL_COV, Using the Tics of DIG_COV

___ **14.** Start ArcCatalog, if it is not running. Highlight the folder name Digitize&Transform in ___IGIS-Arc_*YourInitialsHere*. Choose File > New > Coverage to bring up the New Coverage window.

___ **15.** Make a new coverage named REAL_COV. Check the box that lets you use an existing coverage as a template. Browse to specify DIG_COV as the template coverage. This will make a coverage that has no features but does have the same tics as the DIG_COV. Click Next. Don't be concerned about the Coverage projection. Click Next. The Feature Class should be None. Precision should be Double. Click Finish.

Updating the Coordinates of the Ticks in REAL_COV to "Real-World" Coordinates

_____ **16.** Refresh the Catalog Tree in ArcCatalog by highlighting Catalog and then selecting View > Refresh, or by pressing F5.

_____ **17.** Navigate to REAL_COV in the Catalog Tree. Right-click. Click Properties. Click the Tics And Extents tab.

It has been decided to place a monster Foozit Court inside the track at the school. The surveyor was not available, but the corners for the new court have been determined by tape measure and eyeball to be as follows:

Corner	x-coordinate	y-coordinate
Southwest (91)	1573917	177509
Northwest (92)	1573958	177601
Northeast (93)	1574104	177549
Southeast (94)	1574068	177560

_____ **18.** Modify the x- and y-coordinates of tics 91, 92, 93, and 94 of REAL_COV so that they conform to these real-world coordinates. Click Update to lock in the changes. Check again. Click Apply. Click OK.

_____ **19.** Preview the table of REAL_COV tic. Is it right? Try, by zooming way out and then back in, to preview the geography of the four tics. This may be tricky because the Zoom to Extents tool may not show you the tics—it only zooms to features, not tics.[15] Use the Identify cursor to see the attributes of each tic.

Transforming the Features of DIG_COV to Make REAL_COV

Here you will transform the features (arcs, nodes, and polygons) of DIG_COV, with their arbitrary digitized coordinates, to REAL_COV, with its real-world coordinates. When you run this TRANSFORM tool, all the feature coordinates will be linearly adjusted based on the tic coordinates in REAL_COV.

_____ **20.** Start ArcToolbox.

_____ **21.** To find the TRANSFORM tool in ArcToolbox press the Index tab and type TRANSFORM, then press Locate. Where is the tool? _____. Double-click the highlighted item: Transform.

_____ **22.** The input coverage will be DIG_COV. The output coverage will be REAL_COV. Use an "affine" (that is, linear) transformation. Click OK to complete the process.

_____ **23.** Look at the output messages in the Transform window. Expand it both horizontally and vertically. You won't understand everything that is there, but you should realize that you can get

---

[15]When you do see the tics, you will notice immediately that they are wrong. This is intentional to show you what happens when the numbers aren't right. Continue.

information about the formulas used in the transformation, as well as the accuracy of the process. Actually, the numbers don't look too good—large errors, found under x error and y error. You see errors there of several feet. But let's press on, since we have no obvious alternative at this point.

___ **24.** Close the Transform Messages window. Dismiss ArcToolbox.

___ **25.** Look at the geography of REAL_COV polygon. Whoops. Looks like someone wasn't careful enough with the pencil or the tape measure. Go into ArcMap and add both the arc and tic components. Notice how the arcs miss the tics. Add six.tif from Image_Data and zoom out to see the running track. The court is sort of in the right place but all catawampus.

The lesson here is that a transformation will do strange things if the tics in the output coverage do not bear a strictly linear relationship to the tics in the input coverage. A bit of rechecking shows that the y-coordinate for the southeast tic (94) should have been 177460.

___ **26.** Dismiss ArcMap. Use ArcCatalog to make a blank coverage named REAL_COV_2 under Digitize&Transform, using the tics of REAL_COV. Fix up the incorrect tic in REAL_COV_2 and transform DIG_COV into REAL_COV_2. That should work out a lot better. Notice the errors are on the order of a little more than a foot.

___ **27.** Use ArcMap to look at REAL_COV_2 polygon along with six.tif. Everything should be fine. Close ArcMap.

Converting the Arc Component of a Coverage into a Personal Geodatabase Feature Class

The final task in obtaining a personal geodatabase polygon feature class is accomplished with ArcCatalog: make a personal geodatabase feature class from the coverage.

___ **28.** In ArcCatalog start ArcToolbox. Choose Conversion Tools > To Geodatabase > Feature Class to Feature Class. In the window that appears, make the input features

___IGIS-Arc_*YourInitialsHere*\Digitize&Transform\REAL_COV_2\arc. The Output Location should be Recreational_Facilities.mdb\North_Zone.

The output location can be the North Zone feature data set, because the coordinate systems match. REAL_COV_2 does in fact have a coordinate system associated North_Zone. The coordinate system designation comes from the original shapefile that you made and digitized into in Exercise 5-3. It was carried along in the conversion to the coverage and now to the geodatabase feature class.

Make the Output Feature Class Name School_Court_B. Hide (make nonvisible) all the fields that come from the coverage. Click OK.

___ **29.** When processing stops, close the message window. Examine the geography and the table of School_Court_B.

___ **30.** Start ArcMap. Add School_Court_A, School_Court_B, and the image file SIX.TIF. Zoom in on the school so that both courts are included to make sure everything is placed properly.

_____ **31.** Save the map under the name Final_School_Map in Digitize&Transform. Shift to Layout View. Give the map a title (Foozit Courts A & B) and a north arrow. Print it.

Without looking back in the text (you may use your Fast Facts File) see if you can do the following operation.

_____ **32.** Convert the *polygon* component of REAL_COV_2 into a personal geodatabase feature class called School_Court_Paved. Check to see that it shows up in the right place in six.tif.

Exercise 5-4 (Project)

Digitizing Directly into a Real-World Coordinate System in a Geodatabase

In this exercise you will again do heads-up digitizing, but instead of transforming the product, you will instead move the image file into the appropriate coordinate system and digitize directly. This project focuses on five fictional islands at about 37 degrees north latitude and 171 degrees west longitude, which puts them in UTM Zone 2 North. An ancient, sketchy map of the islands (one of which is artificially square for purposes of illustrating computation) may be found in [___] IGIS-Arc\Image_Data. Some UTM coordinates are written on the sketch for reference. See Figure 5-10.

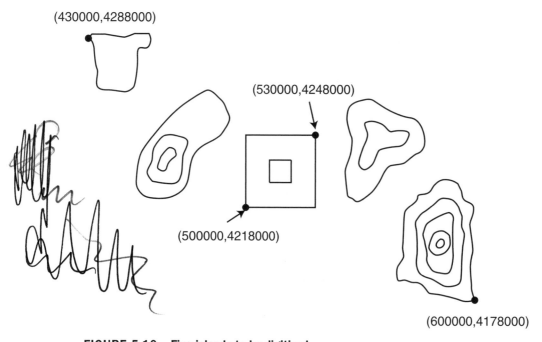

FIGURE 5-10   Five islands to be digitized

Preliminaries

You will use ArcCatalog to copy Five_Islands.TIF from

[____] IGIS-Arc\Image_Data to

____IGIS-Arc_*YourInitialsHere*\Digitize&Transform—actually making a total of three images.

1. In ArcCatalog make a copy of the image data file Five_Islands.TIF by highlighting the name, pressing Ctrl-C, highlighting Digitize&Transform, and pressing Ctrl-V. Again highlight Digitize&Transform and press Ctrl-V. Do this once more. In the Digitize&Transform folder, you should now have Five_Islands.TIF, Five_IslandsCopy.TIF, and Five_IslandsCopy2.TIF.

Making the Feature Class That Will Be Digitized into

2. Using ArcCatalog, in *Teaching*

____IGIS-Arc_*YourInitialsHere*\Digitize&Transform *Not allowed*

make a personal geodatabase named Islands.mdb. Inside that, make a feature data set named UTM_Zone_2 with the following specifications: WGS 1984, UTM Zone 2 (North), with X/Y Domain of –500000 to 500000 in X, and zero to 9000000 in Y. *The X is Easting; the Y is Northing.* Click OK. Show and check details. The extent shown must at least cover the area you specified; it may be larger. Click OK. Within the UTM_Zone_2 feature data set, make a feature class of geometry type Line, named North_Islands_Lines.

3. Start ArcMap. First add North_Islands_Lines to the map (which will, of course, show up as nothing). Then add Five_Islands.TIF to the map. Zoom to the TIF layer.

Georeferencing

4. Choose View > Toolbars > Georeferencing. Place the Georeferencing toolbar where you want it.

Georeferencing is a method of moving images or grids to real-world coordinates. Basically, you identify control points on the source layer and then specify where in a real-world coordinate system these points are to be placed. It is, like spatial adjustment, a procedure that results in transformation. Spatial adjustment moves points, and then features, in feature classes. In contrast, georeferencing moves images and grids.

In what follows you are going to set up the image of the Five_Islands with UTM coordinates so you can digitize directly in the UTM coordinate system. But first you will play with the image and the transformation process, both to see what it can do and how it can get you into trouble.

5. The Georeferencing toolbar should be active. Check out the button with the green and red plus signs, noting both what ToolTips has to say and the status bar, as well. Click the button.

6. Move the cursor on the image to the northeast point on the square island. Observe the y-coordinate of the image. What is it? _____. Click this point. Now move the cursor about 1 inch up. Click again. Notice the image moves up one inch. So far, so good.[16] The entire image maintains its orientation and is adjusted by the amount you specified. After reading about the View Link Table button on the status bar, click the button. A link table window appears, showing the link you just made. The Y source should be about what you wrote down for the position of the point you moved. The Y map location should be some greater. Close the window.

7. Now repeat the operation, but use the southwest corner of the square island instead. If you did this the way I intended, the entire image rotated clockwise. Did you expect that? So the lesson here is that, unless you change some options, each adjustment of the image is based on where it was originally before. It may help to look at the link table to understand this better.

8. On the Georeferencing menu, click Reset Transformation. You see what sort of amounts to the history of the transformations you made. Click `Delete Control Points` to start over.

9. Click the drop-down menu in the Georeferencing toolbar and *remove the check in front of Auto Adjust*. Repeat the procedure in the preceding steps where you made two vertical 1-inch links. *Turn Auto Adjust back on.*

The result probably showed some rotation, because you didn't get the two vertical lines precisely the same length, but nothing like the distortion you saw before. The reason is that the transforming process considered both links at once.

10. Make a random 1-inch line from one place on the image to another. Probably things go all weird. Make another such link. More distortion. Since Auto Adjust is back on the links are again considered sequentially (each individually).

The moral of this story is that if you want to make positional transformations you have to keep Auto Adjust off until you have defined all the links. When you turn Auto Adjust on all the links will be considered at once, moving the figure where you want it.

11. Click the last link (#4) in the Link Table to highlight it. Press the Delete key on the keyboard. Note that the image reverts to the form it had before you added that link. Delete link #3 and note the results. On the Georeferencing menu, click Reset Transformation. Click Delete Control Points.

Moving the Sketch to UTM Zone 2

You have seen how you can change the coordinates of the elements of the sketch. Now you change the coordinates so they conform to locations in UTM Zone 2. These coordinates represent locations much further away so you can't just click-click. Also, you want to place the control points in precise locations.

12. *Make sure Auto Adjust is off* and the Add `Control Points` tool is active. Click the most western point that has UTM coordinates identified. Right-click and select `Input X and Y` from the resulting menu. Put in 430000 and 4288000 for X (Easting) and Y (Northing) , respectively, and click OK. Notice a blue line extending approximately north-northeast by north. Look at the Link

---

[16]If at any time you get in trouble, click the Georeferencing drop-down menu and choose Delete Control Points.

Table to be sure everything is going according to plan. Click each of the other specified points and make links from them as well. The values of the coordinates, from left to right, are as follows:

(430000, 4288000) (extreme northwest)

(500000, 4218000) (southwest of square island)

(530000, 4248000) (northeast of square island)

(600000, 4178000) (extreme southeast)

The result should look something like Figure 5-11.

____ **13.** Dismiss the Link Table. Turn on Auto Adjust. Poof. The image disappears. Select Zoom To Layer of the TIF file. If you have done everything right, you will see the image, undistorted. By sliding the cursor around, you can assure yourself that the new coordinates are now being used. Change the display units to kilometers. Use the Measure tool to determine the dimensions of the square island. _____ by _____. Does this agree with the UTM coordinates? _____ .

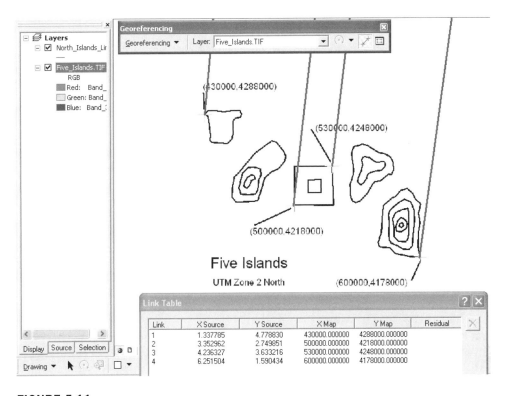

FIGURE 5-11

Digitizing the Line Boundaries of the Islands

____ 14. Note that the Link Table shows residuals—probably of a few hundred meters. Like in other transformations this is a measure of how much "less than perfectly linear" the transformation was. Dismiss the Link Table. Turn off the Georeferencing toolbar. Click Editor. Click Start editing. The `Task` should be set to `Create New Feature`. The `Target` of editing, found on the toolbar, should be North_Islands_Lines. Set `Snapping` to Edit Sketch, and within it, specify `Edit Sketch Vertices Only`. Under `Options`, set the snapping tolerance to 1500 map units (which are meters).

____ 15. Digitize the square island and the square lake inside it. Open the attribute table of North_Islands_Lines. The Shape_Length of the outer line should be about 120,000 meters. What is it? _____. What is the length of the inner line? _____. Dismiss the attribute table.

____ 16. Digitize the other islands. This will be easier and more accurate if you zoom in on each island to do the digitizing. The island in the southeast has a lake within it, an island in the lake, a volcano on that island, and a cauldron defined within the volcano.

____ 17. Click `Save Edits`. Make the feature file line symbol a bright color of width 2, and check that everything has been digitized. If not, complete the editing. Click `Save Edits` again. Stop editing. Turn off the TIFF file; it has served its purpose. Save the map under Digitize&Transform as North_Islands_map1. Dismiss ArcMap.

Making Polygons of the Digitized Lines

The next task is to make polygons based on the lines you have digitized.

____ 18. In ArcCatalog, right-click the entry UTM_Zone_2 in Islands.mdb. Click New > Polygon Feature Class From Lines. In the window that appears, put a check in North_Islands_Lines. Use the name North_Islands_Polygons for the new Feature Class. Click OK. Check that North_Islands_Polygons made it into UTM_Zone_2.

____ 19. Restart ArcMap. Add North_Islands_Polygons. Use the Identify tool to make sure you have proper GIS polygons. Open the attribute table and inspect it. How many entries are there? _____. Minimize the attribute table. Dismiss the Identify window.

Making Multipart Polygons

It turns out that several jurisdictions are involved in the chain of islands. As shown in Figure 5-12, the land area is divided into a West County and an East County, by a north-south line that runs from (515000, 4300000) south to (515000, 4160000). This line splits the square island. However, all inland lakes are controlled by the Region Aqua Board. The volcano is the responsibility of the Safety Board, except for its cauldron, under the jurisdiction of the Seismic Group.

As a first step you will split the square island, except for its lake, into two polygons.

____ 20. Start editing. Set the Edit Task to Cut Polygon Features. Use the Select Features tool on the Tools toolbar to select the land part of the square island. You should see the outer border of the island and its inner border (separating it from the lake) highlighted. Start the `Sketch Tool`. Right-click the data frame and select Absolute X,Y. Type in 515000 for X and 4300000 for Y.

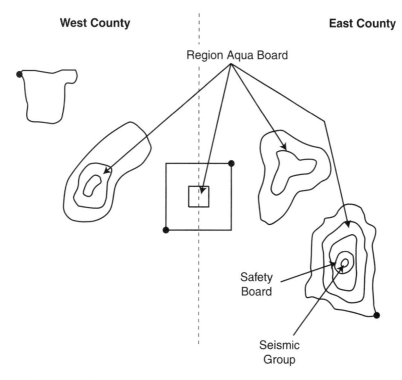

West County

East County

Region Aqua Board

Safety
Board

Seismic
Group

FIGURE 5-12   Five islands divided by region and agency

Press Enter. Press F6 to get the Absolute X,Y window back, and enter the southern coordinates: 515000 and 4160000. Press Enter. Right-click and select Finish Sketch. You should see two polygons where one was before. The lake will not be split, since it was not selected. Click Save Edits, then Stop Editing.

Merging Multipart Polygons

Now we have separate polygons for all the relevant areas. What we want are all of the polygons of a given jurisdiction as a multipolygon feature. We do this by merging the polygons that fall under each jurisdiction.

_____ **21.** _First select all the land polygons in the West County:_ Use the Select Feature tool. Hold down Shift and click each _land_ area in the western part of the island chain. (This area includes an island that is _in_ a lake _on_ an island.). You should have four polygons selected. Verify this with the attribute table. Click Editor, then Start Editing. Click Editor again, then Merge. Check the information in the Merge window, click the lowest polygon number as the one to keep, and click OK. Clear selections with Selection > Clear Selected Features.

_____ **22.** In the same way, select all the lakes and merge their polygons.

_____ **23.** Merge the polygons of the land areas in the East County except for the volcano and its cauldron. Check your work with the Identify tool, by flashing the polygons that belong to each jurisdiction. If you find an error, stop editing _without_ saving the edits, and start again.

____ **24.** The table should now show you five features, one for each jurisdiction. Click Save Edits, then Stop Editing. Dismiss the attribute table. Make sure ArcCatalog is not running. Save the map as North_Islands_map2 in Digitize&Transform.

____ **25.** Open the North_Islands_Polygons attribute table. Click Options > Add Field. Name the new field `Jurisdiction`. Make it a text field of length 20.

____ **26.** Click Start Editing. Select a record. In the Jurisdiction field of that record, place one of the following, as appropriate: East County, West County, Aqua Board, Safety Board, or Seismic Group.

____ **27.** Fill in the other four table values. Click Save Edits, then Stop Editing. Close the table.

____ **28.** Use the Identify tool to make sure that the jurisdictions of each polygon are correct. Clear all selections. Save the map in Digitize&Transform as North_Islands_map3. Dismiss ArcMap.

Exercise 5-5 (Warm-up)

Digitizing Geodatabase Polygons and Coping with Topology

Previously, in Exercise 5-4, you digitized lines and then converted them into polygons. The conversion process took care of the topology issues of overlaps and gaps. You can also digitize polygons directly, as you will see in this exercise, but then you have to cope with topology more directly. We look at some examples first, examining two different ways of dealing with the topological problems that occur. Then, in Exercise 5-7, you will digitize the polygon features of a newly discovered sixth island.

____ **1.** Start ArcCatalog. In

___IGIS-Arc_*YourInitialsHere*\Digitize&Transform.

highlight the Catalog entry of the PGDBFD UTM_Zone_2, in Islands.mdb. Choose File > New > Feature Class. Type Small_Squares for the name of the New Feature Class. Click Next, and on Next again. Click Geometry to be sure the Geometry Type is Polygon. Click Finish.

You are going to make a square 3 meters on a side as one polygon. Within it you will make a square 1 meter on a side.

____ **2.** Launch ArcMap. Add Small_Squares as data. Click `Start Editing`. The Task should be Create New Feature. The Target should be Small_Squares. Click the Sketch tool, and move the cross-haired blue dot onto the map. Right-click. Pick Absolute X,Y from the menu. Type 500000 for X and 6200000 for Y to make the beginning vertex. Press Enter.

____ **3.** Right-click and pick Delta X,Y, which will let you provide relative coordinates for the next vertex. Type 3 for X and 0 for Y.

____ **4.** At this point all you will see is a red dot. The problem is one of extent. Choose the `Zoom In` tool on the Tools toolbar. Repeatedly drag a small box around the red square until you see a line of reasonable length. Then click again on the Sketch tool.

____ **5.** Make the third vertex: This time, use Ctrl-D to bring up the Delta X,Y window, type 0 for X and 3 for Y. Press Enter.

In this digitizing process, the software understands that you are making polygons (because that is the type of feature you set up when you made the feature class), so the lines will automatically close to form a polygon. Thus, you will see a polygon outlined in the sketch.

If things are going wrong during the construction of a sketch, you can press Ctrl-Delete on the keyboard to delete the entire sketch. To delete a particular vertex, click the Edit tool (leftmost button on the Editor toolbar), place it over the vertex, right-click, and choose Delete Vertex.

6. Make the last vertex: −3 for X and 0 for Y. Right-click and select Finish Sketch, then Zoom To Layer. Save your edits. Open and resize the Small_Squares attribute table.

If an unwanted polygon is already completed, you can delete it by clicking the Edit tool, clicking the offending polygon, and pressing Delete (either on the keyboard or on the Standard toolbar).

7. Click the Sketch tool again. Make a square 1 meter on a side whose southwest corner is at 500001 and 6200001, which places it inside the square you just made. Don't forget to finish the sketch. Save your edits and stop editing. The image should look like Figure 5-13. Dismiss ArcMap.

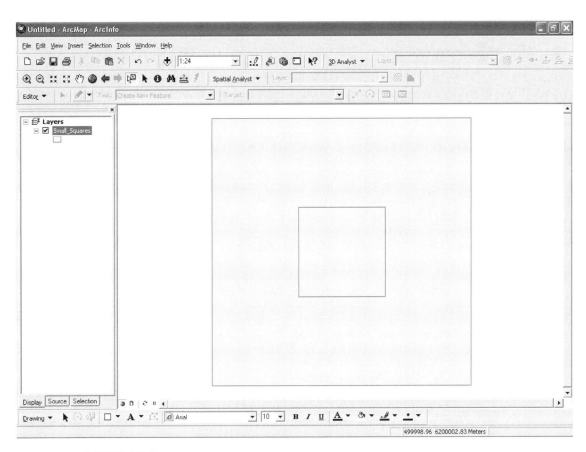

FIGURE 5-13

Making Copies of the Feature Class

___ **8.** Using ArcCatalog, in the Catalog Tree, highlight Small_Squares. Press Ctrl-C. Highlight UTM_Zone_2. Press Ctrl_V. Okay the Data Transfer window. Check that Small_Squares_1 has been added to the Catalog. In this way, also make Small_Squares_2.

___ **9.** Examine the geography of Small_Squares. What you are seeing is a large square with a small square *laid on top of it*. ESRI refers to these features as non-planar. By this it is meant that these features do not tessellate a plane. That is, here it is not the case that the larger square has a hole in it in which the smaller square fits. The larger square is a complete square; the smaller square is also a complete square. Verify this by looking at the table and examining areas and perimeters. Fill in the following, rounding to the nearest integer.

Large polygon perimeter: _____ Large polygon area: _____

Small polygon perimeter: _____ Small polygon area: _____

So you can see by the areas and perimeters that each of these squares is independent of the other, although they, in part, cover the same part of the Earth. We couldn't, for example, grow corn on one and wheat on the other. In general, in a given feature class we want disjoint polygons, so this is a bit of a challenge. Not having disjoint polygons is a problem that occurs whenever we digitize polygons inside other polygons.

___ **10.** The second thing to notice is that the numbers shown in the table are not integers.

Even though any fifth grader could properly calculate the area of a 3×3 square to be 9, the computer can't. The reasons are discussed in Chapter 6. Don't count on exact arithmetical answers from a computer, ever.

Using CLIP to Remove Overlaps from the Feature Class

There are at least two different ways to solve the problem of the non-planar features. The first one is to use the small square as a "cookie cutter" to clip away the duplicated area of the large square that is under it.

___ **11.** Start ArcMap. Add as data Small_Squares_1. Open its attribute table and reduce its size it so you can see the map. Make the color of the polygons Hollow or No Color so you can see the boundaries when you are editing. Sometimes the large square blocks out the small square.

___ **12.** Start editing. Click the Edit Tool. Click the map somewhere away from either square. Click the center of the small square, and notice one square or the other is selected in the table. By clicking where you did, you potentially could select either square. Now press the N key on the keyboard. This selects the Next feature, namely, the other square. Repeated pressings of the N key alternates the selections. If the N key doesn't seem to work well for you, use the attribute table and select the features with it. *End up with the small square selected.*

___ **13.** Drop the Editor menu down and click Clip.

This tool lets you create a new area, derived from the large square. What will remain from the large square is either (a) that area that intersects the small square or (b) from what is left of the large square after an area equivalent to the small square is taken out. To return to our cookie-cutter analogy, we can either keep the cookie or keep the dough around the cookie.

___ **14.** Click `Preserve The Area That Intersects` and click OK. Notice that the large square disappears and, as you can verify from the attribute table, you are left with two small squares. Not what you want. You are looking for two nonoverlapping polygons. Choose Edit > Undo Clip (or press Ctrl-Z) to return to the previous configuration. Again make sure the small square is selected. Clip again, but this time select `Discard The Area That Intersects`—that is, throw away the part of the big square that is under the selected small square. Now look at the attribute table. Record the values in the following.

Large polygon perimeter: _____   Large polygon area: _____

Small polygon perimeter: _____   Small polygon area: _____

Contrast these values with those you wrote previously and you will see that you have genuine areal polygons that don't overlap.

___ **15.** Click Save Edits, then Stop Editing. Dismiss ArcMap.

Using Topology to Remove Overlaps from the Feature Class

Here we explore a second way of disentangling these two polygons. We make use of the topology rules that accompany geodatabases.

___ **16.** Back in ArcCatalog, highlight UTM_Zone_2 in the Catalog Tree. Select `File > New > Topology`. Read the discussion in the `New Topology window`. Click Next. Name the new topology Squares_Topology. Click Next again. For the feature classes that will participate in the topology, select only Small_Squares_2. Click Next. Since only one feature class will be involved, there is no need to change the rank values. Click Next. Add the rule `Must Not Overlap`, which will give us polygons that tessellate the plane. The New Topology window should look like Figure 5-14. Click Next. Read the summary. Click Finish, then click Validate The New Topology.

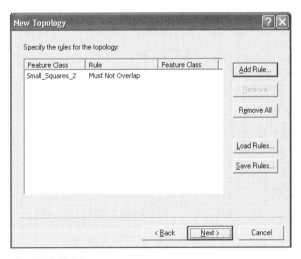

FIGURE 5-14

329

____ **17.** Launch ArcMap. Add as data Squares_Topology. When asked if you also want to add the feature classes that participate in the topology, select Yes.

____ **18.** Polygons with errors are shown with red borders. Areas of overlap are shown with pink. Turn the two entries in Layers off and on so you can get an idea of what is being shown. (If both are off and there is still an image in the map pane, press the Refresh View button at the lower left of the map window.) Turn both on.

____ **19.** Open the attribute table of Small_Squares_2. (The topology has no attribute table.) Reduce its size so you can see most of the rest of the screen easily. Experiment with selecting the two features. It should be clear that these are still non-planar polygons.

____ **20.** Use View > Toolbars > Topology to bring the topology toolbar onto the screen. It is inactive because you are not yet editing. Click Start Editing and notice that the toolbar becomes active. Find the Fix Topology Error Tool and press it. Click the pink area. Right-click and pick Show Rule Description to see what rule is being violated. Click OK.

____ **21.** Right-click again and notice that you have several choices. The ones of main interest are as follows:

❏ Subtract

❏ Merge

❏ Create Feature

To find out what these do, press F1 to get the ArcGIS Desktop Help, click Index, and type topology. Arrow down to fixes and press Display. For the topic, pick Topology Error Fixes and display that. Read, then click Expand All. Like the descriptions of topological errors that you saw in Chapter 4, there is a lot of material here. Under Polygon Error Fixes, read the potential fixes for the Topology rule: Must Not Overlap. Dismiss Help.

Use a second way to get help for the three possibilities. Right-click the pink square. Highlight the word Subtract by placing the cursor over it. Press Shift-F1. Read the information box that appears. Press Esc to close the box. Repeat for Merge. Repeat for Create Feature.

It turns out that you want Create Feature. What will happen is that the original small square will be discarded, as will the part of the large square that it overlapped. In their places. A new polygon will be generated that is congruent to the area of overlap (i.e., the equivalent of the small square).

____ **22.** Right-click again on the pink square and choose Create Feature. Note the change in the display. Press Ctrl-Z to undo the change you just made. Use Create Feature again and this time observe the change in the attribute table. Write the results in the following:

Large polygon perimeter: _____   Large polygon area: _____

Small polygon perimeter: _____   Small polygon area: _____

____ **23.** Click Save Edits and click Stop Editing. Dismiss the Topology toolbar. Dismiss ArcMap.

Learning all about editing in ArcMap is a course by itself. The Editor has many tools, particularly computer-aided design (CAD) tools, that are quite sophisticated. The ESRI manual on the subject is a tome of almost 500 pages. So the best I can do here is give you a few operations that you will surely need and the flavor of what editing in ArcMap is like. In the next few steps, you will simply splash around in the Editor. Actually learning to swim you will have to do on your own or in another course.

Exercise 5-6 (Project)

Learning Some Editor Fundamentals

Now that you have seen something of the Editor, let's practice using some of its capabilities, and hint at some others. As I indicated, one could have a whole course on just how to use the Editor. So what you learn here will necessarily be incomplete. You will be saving very little of your work and nothing that is significant. This exercise is simply to acquaint you with some of the editing facilities.

The Concept of the Edit Sketch

The Edit Sketch (or Sketch, as I will usually refer to it) is a companion to every line feature or polygon feature (or feature to be created) in a geodatabase feature class or shapefile.[17] When you want to create a feature with the Editor, you create a sketch. Once you save your edits, the sketch becomes a feature. If you are modifying an existing feature, when you make it a target of editing, it is the sketch of the feature that you are editing. Not until you stop editing the sketch do the modifications take effect on the feature.

Making Sketches with Snapping

_____ **1.** Start ArcCatalog. In

    ___IGIS-Arc_*YourInitialsHere*\Digitize&Transform.

highlight the Catalog entry of the PGDBFD named UTM_Zone_2. Select File > New > Feature Class. Type Edit_Play_Lines for the name of the new feature class. Click Next and on Next again. Click Geometry to set the geometry type to Line. Click Finish. Now make Edit_Play_Polys in the same way, except the Geometry Type should be Polygon.

_____ **2.** Start ArcMap. Add as data Edit_Play_Lines. Click Start Editing. Let's start by making a 100 meter long east-west line. The Task should be Create New Feature. The Target should be Edit_Play_Lines. Click the Sketch tool and move the crosshaired blue dot onto the map. Press F6 to bring up the Absolute X,Y window. Type 500000 for X and 6200000 for Y to make the first vertex. Press Enter.

_____ **3.** Right-click and pick Delta X,Y, which will let you give relative coordinates for the next vertex. Type 100 for X and 0 for Y. Press Enter.

---

[17]The ArcMap Editor may not be used to edit coverages. You have to use a command-line-driven package called ARCEDIT, which you find by choosing Start > Programs > ArcGIS > ArcInfo Workstation and typing ARCEDIT at the arc prompt.

___ **4.** At this point all you will see is a red dot. The problem again is one of extent. Choose the Zoom In tool. Repeatedly drag a small box around the red square until you see a line of reasonable length. Then click again on the Sketch tool.

___ **5.** Make a few random vertices by clicking. Click the undo arrow in the Standard toolbar. Click again. What is its effect? _____ Make some more random vertices, ending with a double-click to finish the sketch. Again click Undo. What is the effect? _____

___ **6.** Use a different method to again make an east-west line of about 100 meters: This time click (in about the center of the lower-left quadrant of the window) to make the first vertex, then observe the status bar to decide where to place the next vertex. The distance should, of course, be about 100. The direction will probably be approximately *0*. Click.

This is one of those instances in which two standards come into play. Both geographers (e.g., cartographers, navigators) and mathematicians commonly use 360 degrees to describe a complete revolution. There the similarity ends. Northern hemisphere geographers assign zero degrees to north and the number of degrees increases clockwise; mathematicians (using the polar coordinate system) assign zero degrees to the positive Cartesian x-direction (east) and the number of degrees increases counterclockwise. GIS relies on mathematical depictions of geography. So we have to deal with this inconsistency. ArcMap allows us to choose which system we want.

___ **7.** Choose Editor > Options > Units. Change the `Direction Type` of the `Angular Units` to `South Azimuth`. Click Apply, then OK, and look at the status bar as you move the cursor. Experiment with the other possibilities under Units, ending with North Azimuth, Decimal Degrees, and two decimal places.

___ **8.** Press F2 to finish the sketch.

___ **9.** Open the attribute table of Edit_Play_Lines. Notice that a polyline of about 100 meters is the only feature referenced by the table. Reduce the size of the table to get it mostly out of the way.

Next you will make a multipart polyline.

___ **10.** Make three more lines of about 100 meters parallel to the first (each 20 or so meters north of the one before it), but at the end of each, click just once, then right-click, and choose Finish Part. When all three lines are done, press F2 to finish the sketch. All three lines will turn cyan, indicating that they are all part of a single selected feature. Now observe the attribute table. You should see the additional feature as a single feature of length about 300.

Once you make a multipart feature (as you just did, and as you did earlier by merging features of the islands), you may disaggregate them if you wish.

___ **11.** Choose the Edit tool. Click the original horizontal line to highlight it. Then change the selected feature by clicking the three-line feature you just made so it is selected. Choose View > Toolbars > Advanced Editing. On that toolbar, using the status bar or ToolTips, find the tool that "explodes" a multipart feature. Click it and the three lines become separate features, as you will be able to see from the attribute table.

___ **12.** Pick two of the parallel lines (using Shift-click) and merge them back into a single feature. Save your edits.

_____ **13.** While you have the Advanced Editing toolbar up, run the cursor over the tools and read about each on the status bar. Write the ToolTips description, here:

_____

_____

_____

_____

_____

_____

_____ **14.** Get more information about each Advanced Editing icon by clicking the What's This? icon on the Standard toolbar. Making reference to this Help information, experiment with these tools as your interest (or instructor) dictates. When you are through, stop editing but don't save edits so you can continue below with the parallel lines. Dismiss the Advanced Editing toolbar by right-clicking one of its tools and clicking Advanced Editing to turn off its check mark.

_____ **15.** Choose Editor > Snapping. Set snapping so that lines in a developing sketch will snap to the Ends of features in Edit_Play_Lines. Leave the snapping pane visible. Choose Editor > Options > General and set the Snapping Tolerance to 5 Map Units (meters). Turn on Show Snap Tips. Click Apply, then OK.

_____ **16.** Start the Sketch tool. Starting in the southwest corner, move the cursor close to the west end of the line. You have used snapping before, so you are not surprised that while the crosshairs stay where the mouse put them, the blue dot jumps to the exact end of the line. Note the information provided by the Snap Tip when the blue dot locks onto the end of a line feature. Click. Now make a vertical line from there to the west end of the parallel line just above, and click. Continue to make vertical line segments to all four horizontal lines. Doubleclick the last vertex.

_____ **17.** In the snapping pane, click Edge. Now move the cursor around; when you get close to any part of a line, the blue dot jumps to it. Note three kinds of results indications: (a) graphically, (b) with the snap tips bubble, and (c) on the status bar.

_____ **18.** Select the Edit tool. Draw a box that contains or crosses all the lines you have made _except_ the original 100-meter line. Everything should turn cyan except that original line. Press the Delete button on the Standard toolbar or on the keyboard. The features will disappear, as will the associated rows in the attribute table.

The upper pane of the snapping window controls the action of sketch lines as they relate to the target features. The lower pane controls actions of edit sketch lines as they relate to other elements of the sketch itself, as differentiated from features already complete. You can guess what Edit Sketch Vertices and Edit Sketch Edges do. Perpendicular To Sketch is less obvious, until you try it.

_____ **19.** In the lower part of the snapping pane, turn on Perpendicular To Sketch. Make some line segments with the Sketch tool. Then swing the cursor around the red dot of the last vertex so that it makes approximately a right angle with that last line. The cursor will snap to locations on a line perpendicular to that last line, and the Snap Tip will announce: Perpendicular To Sketch. Click. Make a number of segments, each at right angles to the one before. (Also note that the blue dot snaps to a straight-ahead position—which is certainly not perpendicular, but handy nonetheless.) Right-click and select Delete Sketch.

___ **20.** Sketch a line with several segments but do not finish the sketch. Click the Edit tool. Put the cursor over a vertex. When the cursor changes to a square and four triangles, drag that vertex to a new location.

___ **21.** Put the cursor over another vertex. Right-click and choose Delete Vertex.

___ **22.** Complete the sketch with F2. Now use the Edit tool to select the original horizontal line by clicking it. Drag the entire line a few meters away from its current position. Now select and drag the feature you just completed. Click Save Edits, then click Stop Editing.

Now you at lease have some ideas of the editing capabilities of ArcGIS with regard to lines. You look at editing polygons next.

Experimenting with Editing and Polygons

___ **23.** Remove Edit_Play_Lines from the Table of Contents. Add as data the blank feature class Edit_Play_Polys. Click Start Editing. Using the Sketch tool and zooming, make a triangle with approximately 1000 meter sides and one vertex at (500000,6300000). Make the layer color No Color or Hollow. Open the attribute table. Click Save Edits.

Experimenting with the Editor's Union

Basically, a *union* of two polygons, say, A and B, is a polygon whose area includes both the area of A, the area of B, and any area that is common to both A and B.

___ **24.** Make a second polygon that partly overlaps the triangle.

___ **25.** Select both polygons. (Use either the attribute table with Ctrl-click or the Edit tool with Shift-click.) Using Union in the Editor menu, make a third polygon.

___ **26.** Prove to yourself that a new polygon has indeed been created by (a) looking at the attribute table, and (b) dragging it with the Edit tool. You could, of course, make this third polygon effectively replace the first two, by deleting them, but *don't*. Instead, select Stop Editing *without* saving edits.

___ **27.** Start editing. Make a second polygon that lies outside the triangle. Again perform the union operations. Observe the result. Stop editing without saving edits.

___ **28.** Start editing. Make a second polygon that lies inside the triangle. Again perform the union operations. Observe the result.

Experimenting with the Editor's Intersect

Basically, an *intersection* of two polygons, say, A and B, is a polygon whose area includes *only* the areas of A and B that are common to both A and B.

___ **29.** Repeat Steps 24 through 28, but use the Intersect command on the Editor menu.

Experimenting with the Editor's Buffer

____ **30.** Repeat Steps 24 through 28, but use the Buffer command on the Editing menu. When asked for a distance, use 100 meters.

Using Undo, Copy, and Cut

____ **31.** Start editing. Add a second polygon that overlaps the triangle. Union the two. Move the new polygon. Now place the cursor on the Undo tool. What does it say? _____. Press the Undo tool. Place the cursor again on the Undo tool. What does it say? _____. Place the cursor on the Redo tool. What does it say? _____. Press it. Place the cursor on the Redo tool. What does it say? _____. Press it. Stop editing *without* saving edits.

____ **32.** Start editing. Select the triangle. Press Ctrl-C. Press Ctrl-V. Verify that an identical polygon has been created. Select Stop Editing *without* saving edits.

____ **33.** Start editing. Under `Task`, `select Cut Polygon Features`. Select the triangle. Use the Sketch tool to make a polygon that starts outside the triangle, goes inside, and returns to the outside. Note what happens as this polygon is completed. Now use the Edit tool (selecting with the records in the attribute table) to move the resultant polygons.

Working with Editing Lines Again

____ **34.** Add back Edit_Play_Lines. Zoom to layer. Make some lines. Experiment with Move, Split, Divide, Buffer, and Copy_Parallel.

____ **35.** Close ArcMap.

In the previous two exercises, you learned how to digitize polygons directly into a personal geodatabase feature class, and also learned two different ways to cope with the topological problems that result.

In the next exercise you use these methods to add a sixth island to the UTM_Zone_2 feature class North_Islands_Polygons.

Exercise 5-7 (Follow-on)

Adding the Sixth Island

It turns out that the drawing you were working with Figure 5-10 in Exercise 5-4 was very old. A volcano has erupted since, making a new island. A digital orthophoto, converted into a grid named MSH, is available. (In the first step you will will copy one line feature class, and you will make a second, empty line feature class, both of which you will use later.)

____ **1.** Using ArcCatalog, make sure the 3D Analyst extension is turned on. *Begin* to make a geodatabase Feature Class named Trail in

___IGIS-Arc_*YourInitialsHere*\Digitize&Transform\

Islands.mdb\UTM_Zone_2.

When you come to the New Feature Class window that lets you define the Data Type and the Field Properties, make the geometry type Line and change the Contains Z Values field from No to Yes. Click Finish.

From [__] IGIS-Arc_AUX\D&T\ Islands.mdb\UTM_Zone_2

copy the PGCBFC named Trail_horizontal to

___IGIS-Arc_*YourInitialsHere*\Digitize&Transform\

Islands.mdb\UTM_Zone_2

____ 2. In ArcMap open the map North_Islands_map3. Add the geotif file

___IGIS-Arc\Image_Data\MSH.

Don't build pyramids. Zoom to the MSH image.

____ 3. As you observe the image, you can imagine that you see two triangular harbors along the south coast. Using the Georeferencing toolbar (don't forget to turn Auto Adjust off), move the apex of the southwest harbor to (Easting 630000, Northing 4100000) and the southeast harbor apex to (660000, 4100000) . When the move is complete, zoom to the image. Then zoom to full extent to see where this new island is with respect to the others.

Recall that you previously digitized the five islands as lines and then converted the lines to polygons. When you did this, the polygons that were created obeyed rules that resulted in no overlaps and no gaps. There are several ways to add the new island to North_Islands_Polygons, but let's use one in which you digitize *polygons* directly into the geodatabase feature class.

____ 4. Start editing. The target should be North_Islands_Polygons. You want to Create New Feature. Digitize first the outline of the island. When digitizing polygons with curved lines, use more points along the lines where the curvature is greatest. See Figure 5-15.

When you create polygons, you see a closed figure after you have put in the first three vertices. To complete the figure, double-click the last vertex, or single-click the last vertex and press F2 complete the sketch.

____ 5. If necessary, rearrange the entries in the Table of Contents so that the image is again on top. Digitize the cauldron of the volcano. There appear to be four lakes on the island; zoom in and digitize them. Digitize what appears to be new lava flow that covers about a third of the island, and in some cases, runs down to the ocean. Click Save Edits. Click Stop Editing. Turn off the image layer MSH. Make the North_Islands_Polygons layer hollow so you will be sure to see the boundaries of all the polygons. The map should roughly resemble Figure 5-15. Don't dismiss ArcMap, but save the map as North_Islands_map4 in your *YourInitialsHere* folder. Launch ArcCatalog

From this point on in the exercise the instructions will become very general. You should use the tools you learned in the previous exercises to accomplish the tasks.

____ 6. Using ArcCatalog, make two copies of North_Islands_Polygons.

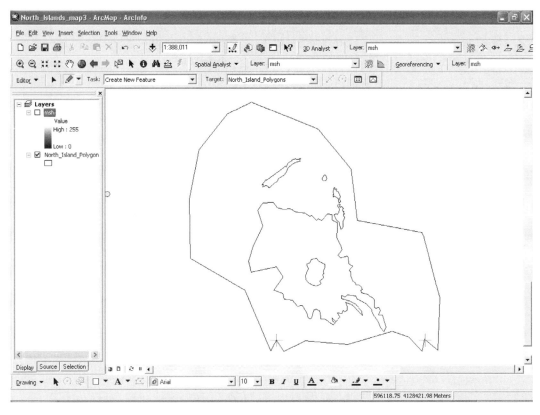

FIGURE 5-15   The sixth island

___ **7.** Use clip operations on one copy to make separate polygons of the lava flow, the lakes, and the cauldron. Since polygons are nested within polygons that are themselves nested this can be a little tricky. Things seem to go better if one does two things: (1) start with the most deeply nested (smallest) polygons first, and (2) open the attribute table and use it to select the polygons, rather than relying on selecting them graphically.

___ **8.** Merge the cauldron into the Seismic Group.

___ **9.** Merge the lava flow area with the Safety Board.

___ **10.** The lakes go to the Aqua Board.

___ **11.** The remaining land area goes to the East County.

___ **12.** Remove the first copy of North_Islands_Polygons from ArcMap. Add the second copy of North_Islands_Polygons to ArcMap.

___ **13.** Use topology on the second copy. Perform the same merging operations as immediately above.

___ **14.** Not that you ever make errors, but if you had merged polygons together that should not be, what toolbar and tool would you use to separate them again? _____, _____.

Creating a 3-D Feature

____ 15. Turn on the image layer MSH. Zoom in on the sixth island. You should see the image of the volcano. (If not, you will have to Georeference the image again.)

____ 16. In ArcMap add the Feature Class named Trail_horizontal from

___ IGIS-Arc_YourInitialsHere\ Digitize&Transform \Islands.mdb\UTM_Zone_2.

Make it a red line of width 3.

Trail_horizontal is a trail from the water's edge to the rim of the volcano, but with all segments in the horizontal, sea level plane. What we will do is make a three dimensional trail, named Trail, with the Editor, using the pattern of Trail_horizontal.

____ 17. In ArcMap add the empty Feature Class named Trail that you made earlier in this exercise, from

___IGIS-Arc_*YourInitialsHere*\Digitize&Transform\

Islands.mdb\UTM_Zone_2.

____ 18. Start editing. Make sure the target is Trail. Make a polyline with five segments that overlays the segments of Trail_horizontal. End each segment, *including the last*, with a *single click*.

____ 19. Find Sketch Properties[18] on the Editor toolbar. Click it. You will see the three dimensional coordinates of the six vertices you have made. All the z-coordinates are presently zero. Starting with vertex zero, click each z-coordinate to change the values to:

0, 400, 900, 1500, 3000, 4000

____ 20. Click Finish Sketch. Dismiss the Edit Sketch Properties window. Click Save Edits, then click Stop Editing. Change the Trail symbol to a black line of width 2.

____ 21. Start ArcScene by selecting Start Programs > ArcGIS > ArcScene. Add Trail and North_Islands_Polygons from

___IGIS-Arc_*YourInitialsHere*\Digitize&Transform\

Islands.mdb\UTM_Zone_2.

____ 22. Add Trail_horizontal from

[___] IGIS-Arc\Digitize&Transform\Islands.mdb\UTM_Zone_2.

____ 23. Explore, using the Navigate cursor (with the left mouse button) and the zoom control (use the right mouse button). Convince yourself that the polyline of Trail does have a three-dimensional component. Close ArcMap, ArcScene, and ArcCatalog.

---

[18]The status bar will say: "Shows a dialog for editing properties of the edit sketch geometry."

Exercise 5-8 (Project)

Obtaining Field Data and Joining Tables[19]

A Discussion of the Project

This is a cooperative class exercise with several objectives:

- ❏ To collect "field data" in the classroom by making measurements

- ❏ To understand some coordination issues associated with collecting data

- ❏ To gain an appreciation of how errors can occur in GIS data collection

- ❏ To enable each student to work in a small team of classmates

- ❏ To learn to enter nonspatial data into ArcGIS

- ❏ To learn to join tables associated with a spatial geodata set with nonspatial data

- ❏ To discover problems and errors that can occur with data collection

- ❏ To do something different (and fun) in lab.

The process will go something like this:

- ❏ Each student will be assigned to one of four teams (A, B, C, or D).

- ❏ Each student will be assigned a three digit identifying number, ID (e.g., 612).

- ❏ Each student will write the ID number on a sticky note, along with an "X" (see Figure 5-16) and will place that on the top of her or his computer terminal.

- ❏ Each team will perform a measurement or recording task.

- ❏ Each team will perform a second measurement or recording task.

- ❏ The data from the teams will be shared by all students

- ❏ Each student will enter data in ___IGIS-Arc_*YourInitialsHere*\Student_Measurements.

- ❏ The spatial data will go into a text file table and the nonspatial data into a database table.

- ❏ Each student will make a map showing the locations of the computer terminals.

- ❏ The tables of spatial and nonspatial data will be joined so that the map can provide all recorded information about each student.

---

[19]This is an exercise that requires about an hour of class time in which students in teams collect data in the classroom. Instructors should see the Wiley web site for advice on how to conduct the exercise. If you decide not to collect the data but just do the software part of the exercise some sample data is in [___] IGIS-Arc_AUX in an Excel spreadsheet named Exercise_5-8_Student_Computer_Data.xls

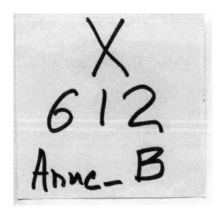

**FIGURE 5-16   Sticky note
identifying computer location**

Organization

❑   Students: Assign yourselves sequential numbers, beginning with number 601. Remember your number, which will serve both the computer ID number and your student ID number. Write the number on a sticky note, along with an "X." Stick it on top of your computer terminal.

❑   Students: Assign yourselves to teams (A, B, C, or D) with the same number of students in each team, plus or minus one.

❑   Students: Get together with your teammates in a vacant corner of the room and decide on a plan for accomplishing the tasks.

Environment and Measurements (Spatial Data)

❑   Use the back wall of the lab as southern latitude.

❑   Use a Cartesian coordinate system.

❑   Consider the intersection of the back wall and the "west" wall the Cartesian point (1000,2000).

❑   For eastings measure perpendicularly from the west wall to the center of the cross on the sticky note.

❑   For northings measure perpendicularly from the south wall to the center of the cross on the sticky note.

❑   Measure to the nearest 0.01 (one hundredth) of a foot.

Measurements (Nonspatial Data)

❑   Record the name of each student in the class in this form: FirstName_LastInitial (e.g., Anne_B).

❑   Measure the *armspread* of each student. (Armspread is the distance measured *across the person's back*, from finger tip to finger tip, with arms outstretched.)

❏ Measure the *height* of each student to the nearest 0.1 (one-tenth) foot.

❏ Assess the eye color of each student

Recording Data

Record the data on the black board or white board in two tables. The first table should have the headings

Computer Number, Northing, Easting.

The second table should contain

Student Name, Armspread, Height, Eye Color, Student Number.

Each student should make a paper copy, or computer text file copy that can be printed, of these data.

Also record the data in two separate computer text files. (Two people on the recording team could work on the boards, and two others type into the text files.) The text files will be mailed to the instructor for distribution to the class.

Team Assignments

All measurements will be taken twice and recorded twice. First,

❏ Team A will measure the x-coordinates (the eastings) of the set of computers (i.e., the distances from the west wall to the "X" on the sticky notes).

❏ Team B will measure the y-coordinates (the northings) of the set of computers. (i.e., the distances from the south wall to the "X" on the sticky notes).

❏ Team C will measure the armspreads and heights of the people in the class.

❏ Team D will record the data from the preceding teams on the board and in computer files.

When this process is complete, the boards will be erased and the computer files sent to the instructor. Then:

❏ Team A will measure the armspreads and heights of the people in the class.

❏ Team B will record the data from the preceding teams on the board and in computer files.

❏ Team C will measure the x-coordinates (the eastings) of the set of computers and will measure the width of the room.

❏ Team D will measure the y-coordinates (the northings) of the set of computers and will measure the length of the room.

These data will also be e-mailed to the instructor.

The instructor will put the four data files in a shared area on the class computer disk. Students will examine the pairs of data files for errors and inconsistencies.

Undertaking the Data Entry Process

_____ **1.** In ___IGIS-Arc_*YourInitialsHere*, make a new folder called Student_Measurements.

_____ **2.** Using Notepad, or other text processor, make a text file that contains the coordinate data. Call it Coordinates.txt. Each line should contain three values, using CSV (comma separated values) format:

❑ Each student's three-digit ID

❑ The easting value (x-coordinate) of the computer terminal

❑ The northing value (y-coordinate) of the computer terminal

For example:

```
612, 10.28, 20.05
606, 10.99, 21.22
```

and so on.

As the first line of the text file, use the following exactly as shown:

```
"Comp_ID","Easting","Northing"
```

_____ **3.** Save the file as Coordinates.txt in

___IGIS-Arc_*YourInitialsHere*\Student_Measurements.

_____ **4.** Start ArcMap. Add Coordinates.txt as data. The Table of Contents will show the path to the file. (Notice that the Source tab in the Table of Contents is active. If you click the Display tab, the Table of Contents will appear as nothing, since this is not—yet—a geodata set. Leave the Source tab active.)

_____ **5.** Right-click Coordinates.txt and choose Display XY Data to bring up a window of that name. In the X Field drop-down menu choose, of course, Easting. Make the right choice for the Y Field. When you OK the window, you should see point symbols for each computer. The Table of Contents will show a geodata set named

Coordinates.txt Events[20]

Open the attribute table of Coordinates.txt Events. You should see the text file you created, but here in attribute table format.

Making a Table That Will Contain the Student Data

_____ **6.** In ArcCatalog, highlight the folder

__:\IGIS-Arc_*YourInitialsHere*\Student_Measurements.

Select File > New > dBase Table. In the Table of Contents, change the name of the table to Student_Info.dbf.

---

[20]Event has a specific meaning in ArcGIS. It refers to a geographic location that is stored in a table.

——— **7.** Preview the table. You will see only two column heads: OID and Field1. In that window, click Options and pick Add Field. In the window, type St_Name in the Name field. For the Type, choose Text. For Length type 20. Click OK.

——— **8.** Add another field called Armspread. Make it a floating-point number that can contain two digits overall (precision value: 2) with one decimal place (scale value: 1). Add another field, Height, with the same characteristics. Add Eye_Color as a Text Field of length 8. Finally add St_ID as a Short Integer.

——— **9.** Right-click the column heading Field1. Delete the field.

Populating the Student_Info Table with Data

——— **10.** Close ArcCatalog. Going back to ArcMap, add Student_Info.dbf as data. Open the table and start editing. In the Start Editing window, click the choices until Student_Info shows up in the lower pane. Click OK. The table should now allow you to enter data. Start with the first student name, in the form Anne_B. Fill in the rest of the data for this student. Fill in data for the rest of the students.

——— **11.** Click Save Edits, then Stop Editing. Dismiss the table.

Joining the Two Tables to Make a Single Table

What you have now are two data sets. The first, the Events layer, has the geographic location of the computer terminals. Use the Identify Results cursor to look at one point. In addition to the Easting and Northing, you have a key value, the computer ID that also references the student using the computer. The second data set is not a spatial one but contains information about students. It also has a key field, St_ID. What you want to do now is to create a single table with both the geographic and the nonspatial information together. There are several ways to do this. We will pick the most straightforward one: the join.

——— **12.** Right-click Coordinates.txt Events. Choose Joins And Relates > Join to bring up a Join Data window. You want to Join attributes from a table. The key field in the layer is Comp_ID. The table to join to the layer is Student_Info.dbf. The key field in that table is St_ID. Click OK.

Seeing the Results of the Join

——— **13.** Using the Identify cursor, click a point in the map display. Notice that you get information not only about that point but the associated student information as well.

——— **14.** Open the attribute table of Coordinates.txt Events. Notice that all the information has been put together. The fields that came from the Student_Info table are identified as such.

If you now removed Coordinates.txt Events from ArcMap, or closed ArcMap (don't do either), the joined data would no longer be part of the table should you ask for it again. There are two ways to make the join permanent. The first is to save it a layer file. The second is to export it as a geodata set (geodatabase feature class or shapefile). Do both in the steps that follow.

——— **15.** Right-click Coordinates.txt Events and choose Save As Layer File. Navigate to the Student_Measurements folder. See Figure 5-17. In the Save Layer window, click Save.

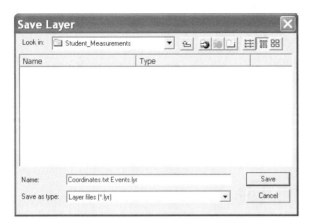

FIGURE 5-17

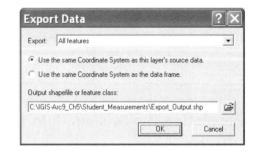

FIGURE 5-18

_____ **16.** Right-click Coordinates.txt Events and choose Data, then on Export Data. See Figure 5-18. Click OK. Do not add the exported data to the map.

_____ **17.** Remove all layers from the map. Ask for a new map. Add as data

IGIS-Arc_*YourInitialsHere*\Student_Measurements\Coordinates.txt Events.lyr

and check to see that the geographics and the table are present and contain all the information. Verify that each student's information is correctly hooked up with her or his computer location.

Add as data

\Student_Measurements\Export_Output

and check to see that the geographics and the table are present and contain all the information.

_____ **18.** Make a layout of your own design. The map should label each computer with the student's name. Print a map showing the Lab.

Exercise 5-9 (Review)

Checking, Updating, and Organizing Your Fast Facts File

The Fast Facts File that you are developing should contain references to items in the following checklist. The checklist represents the abilities to use the software you should have upon completing Chapter 5.

Important note: This checklist is on the CD-ROM that accompanies the book. It is available in Microsoft Word format. Rather than typing or writing by hand the text that follows, you can copy and paste it into your Fast Facts File from the CD-ROM file.

___ ArcGlobe data is located in

___ You can see Globe View, along with Geography and Table Views, in

___ Reference system information is locate in

___ Geographic coordinate systems information may be found in

___ A problem will occur if you attempt to combine date from UTM based on NAD 27 with data based on NAD 83 or WGS 84 because

___ Several difficulties in using state plane coordinates occur because

___ Four ways of getting (geo)graphic map data into digital form are

___ To do heads-up digitizing

___ To make a blank shapefile

___ To load an image file into ArcMap

___ The toolbar used in making lines in the shapefile is

___ The purpose of setting the snapping distance is

___ To use the Editor, you need to set a number of parameters:

___ Two ways to finish a sketch are

___ To convert a shapefile into a personal geodatabase feature class

___ Getting the extent of a personal geodatabase feature data set correct to begin with is important because

___ To move a personal geodatabase feature class to a new location

___ To convert a shapefile into a personal line coverage

___ Tics facilitate transformation by

___ To move a coverage to a new location

___ The toolbar to use when moving an image file is

___ Before you attempt to move an image, it is important to turn Auto Adjust off because

___ To make polygons of digitized lines

___ To make multipart polygons from single part polygons

___ To merge multipart polygons

___ Two ways of providing coordinates by typing in the editor are

___ In the Editor, to delete a sketch under construction

___ In the Editor, to delete a vertex

___ In the Editor, to delete an already formed feature

___ To copy a feature class

___ Two ways of removing polygon overlaps are

___ An "Edit Sketch" is

___ Finish Part makes

___ Finish Sketch makes

___ Snapping can be used to precisely locate coordinates in a variety of ways:

___ The upper pane of the snapping window

___ The lower pane of the snapping window

___ Snap tips help by

___ A vertex can be moved by

___ A feature can be moved by

___ A union of two polygons includes

___ An intersection of two polygons includes

___ A buffer around a polygon includes

___ The cut polygon feature tool allows

___ To make a LINE feature class, when the assumption is a polygon feature class:

___ To make a LINE feature class that can have a third coordinate, indicating vertical displacement:

___ To put in 3-D coordinates

___ To join two tables

___ To make the joined table permanent

PART II

Spatial Analysis and Synthesis with GIS

Analysis of GIS Data by Simple Examination

OVERVIEW

IN WHICH, by inspecting or simply viewing data in a variety of ways, you take small steps toward analysis.

Information

There are two simple ways of looking at a packet of information:

❑ A small amount of matter-energy with the potential to produce large effects

❑ A sequence of 1s and 0s—for example, 10100011

The poem *Paul Revere's Ride* (excerpted) by Henry Wadsworth Longfellow, illustrates both:

> LISTEN, my children, and you shall hear
> Of the midnight ride of Paul Revere,
> On the eighteenth of April, in Seventy-Five;
> Hardly a man is now alive
> Who remembers that famous day and year.
> He said to his friend, "If the British march
> By land or sea from the town to-night,
> Hang a lantern aloft in the belfry arch
> Of the North Church tower, as a signal light, —
> One, if by land, and two, if by sea;
> And I on the opposite shore will be,
> Ready to ride and spread the alarm
> Through every Middlesex village and farm,
> For the country-folk to be up and to arm."
>
> [. . .][1]
>
> Meanwhile, impatient to mount and ride,
> Booted and spurred, with a heavy stride
> On the opposite shore walked Paul Revere.

---

[1]Indicates omission.

Now he patted his horse's side,
Now gazed on the landscape far and near,
Then, impetuous, stamped the earth,
And turned and tightened his saddle-girth;
But mostly he watched with eager search
The belfry-tower of the Old North Church,
As it rose above the graves on the hill,
Lonely and spectral and somber and still.
And lo! as he looks, on the belfry's height
A glimmer, and then a gleam of light!
He springs to the saddle, the bridle he turns,
But lingers and gazes, till full on his sight
A second lamp in the belfry burns!

A hurry of hoofs in a village street,
A shape in the moonlight, a bulk in the dark,
And beneath, from the pebbles, in passing, a spark
Struck out by a steed flying fearless and fleet:
That was all! And yet, through the gloom and the light,
The fate of a nation was riding that night;
And the spark struck out by that steed, in his flight,
Kindled the land into flame with its heat.

[. . .]

You know the rest. In the books you have read,
How the British regulars fired and fled, —
How the farmers gave them ball for ball,
From behind each fence and farm-yard wall,
Chasing the red-coats down the lane,
Then crossing the fields to emerge again
Under the trees at the turn of the road,
And only pausing to fire and load.
So through the night rode Paul Revere;

And so through the night went his cry of alarm
To every Middlesex village and farm, —
A cry of defiance and not of fear,
A voice in the darkness, a knock at the door,
And a word that shall echo forevermore!
For, borne on the night-wind of the Past,
Through all our history, to the last,
In the hour of darkness and peril and need,
The people will waken and listen to hear
The hurrying hoofbeat of that steed,
And the midnight-message of Paul Revere.

Excerpted from Paul Revere's Ride *by*
Henry Wadsworth Longfellow.
Publicly appeared first in the Atlantic Monthly
January 1861

Certainly in this age—at the beginning of the twenty-first century, when what computers do seems indistinguishable from magic—it is difficult to remember or believe that a computer is, at its heart, a conceptually simple device. The way in which it is simple has a profound impact on its use for GIS. Further, the simplicity imposes constraints on what we can do.

The elementary nature of a computer is based the idea that just the existence or nonexistence of a single thing, when put together with other such things, can be a powerful mechanism for representing meaning—a code, if you will, consisting of 1s and 0s. In the case of *Paul Revere's Ride*, the things were lit lanterns—two of them hanging in the belfry arch. If a lamp burned we could call that a "1." If not, a "0." Therefore the code was as follows:

00:	No British
10 (or 01):	British by Land
11:	British by Sea

As for the other definition of information—a small amount of matter-energy with the potential to produce large effects—it's pretty clear, if the poem can be believed, that in some sense the United States was saved that night. The beginning of the battle has been described as "the shot heard 'round the world."

Computer Hardware—What a Computer Does

The information in this section is intended to demystify computers for you. As computer power increases—with abilities like speech and facial recognition and championship chess playing—the tendency is to regard a computer as basically incomprehensible. Since the fundamental logical structure behind the operation of a computer is quite simple, I believe such understanding should be part of a student's knowledge base. It's not that hard, and you may find it interesting. And the knowledge may help you to figure out why things go wrong sometime when you are using a computer to work on a GIS project.

All the operations of a computer are based on the concept that combinations of binary states can represent information. Binary states can be: on or off; A or B; tied or untied; yes or no; exists or does not exist. Customarily, we represent those two states by **BI**nary digi**TS** (BITS) With these a computer does three things:

❑ Strings of 0s and 1s are fed in (input) to the machine and placed electronically into a "store" or "memory."

❑ Bits from the store are manipulated according to a number of exact rules (operations of arithmetic making up a good-sized subset of those rules), and the results of those manipulations are placed back into the store.

❑ Strings of 0s and 1s are sent out from the store to output devices.

That's it! Everything else is simply elaboration on this basic theme.

Why binary states? Because, in the physical world, it is easier to identify the existence (1) or nonexistence (0) of something than it is to identify the degree to which something is present. For example, it is easy to tell by looking at a lightbulb whether the switch that feeds the lightbulb is on or off (i.e., whether the switch that controls it is set to 1 or 0). It is not so easy to tell, by looking at the brightness of a bulb, the position (say, 1, 2, 3, or 4) to which a lightbulb dimmer (rheostat) is set.

Input

When you press and hold the Shift key and then press the K key on a computer keyboard, a string of bits[2] is sent to the computer. When you move the mouse pointing device, sequences of bits are sent to the computer. When you speak into a microphone attached to your computer's sound card, a string of bits is generated by the sound card (also a computer, by the way) and winds up in the computer's store.

Computation

For a computer to do computation, it has to represent decimal (base ten) numbers in binary (base two). Here are some decimal numbers and their binary equivalents:

Base Ten	*Base Two*
0	00000
1	00001
2	00010
3	00011
4	00100
5	00101
6	00110
7	00111
8	01000
9	01001
10	01010
11	01011
12	01100
13	01101
14	01110
15	01111
16	10000

Given a number in base ten, you may interpret it in this way: Starting from the right and summing the values, the first position represents the number of 1s, the second position represents the number of 10s, the third position the number of 100s, the fourth the number of 1000s, and so on. So the number 342 means the sum: 2 times 1, plus 4 times 10, plus 3 times 100, which is 342.

The binary number system is essentially the same, but much simpler. When you have a number in base two, it may be interpreted in this way: Starting from the right, the first bit position represents the number

---

[2]Assuming that the computer uses the most commonly accepted convention of the relationship of characters, such as "K," to bits—namely the American Standard Code for Information Interchange (ASCII)—the string of bits sent from the keyboard to the computer would be 11010010. Such a string of 8 bits is called a byte. Computer memory and disk storage is commonly expressed in terms of bytes—always a whole lot of them. A kilobyte is 1024 bytes. A megabyte is 1024 kilobytes. A gigabyte is 1024 megabytes. A terabyte is 1024 gigabytes. A pentabyte is 1024 terabytes. Some hard drive manufacturers cheat and call a kilobyte 1000 bytes, a megabyte 1000 kilobytes, and so on, thus inflating the advertised hard drive capacity. Check the Internet for more information on this scuffle.

of 1s, the second bit position represents the number of 2s, the third position the number of 4s, the fourth the number of 8s, and so on and so on, doubling each time you move a position to the left. The binary number 01110 represents the decimal number fourteen, calculated as follows: the sum of zero times 1, plus one times 2, plus one times 4, plus one times 8, plus zero times 16, which is 14.

Addition

The table to give the results of addition of two binary numbers is quite simple, compared to the addition table for decimal numbers. It is as follows:

$0 + 0 = 0$
$0 + 1 = 1$
$1 + 0 = 1$
$1 + 1 = 0$ with a carryover to the next column to the left

For the computer to add two numbers together, say, the integers 3 and 9, the computer has them represented in its store as 00011 and 01001. So:

```
  00011
+ 01001
  01100
```

Most fundamentally, the central processing unit (CPU) of a computer can

1. Add, subtract, multiply, divide, and so on two sequences of bits

2. Compare one sequence of bits with another to determine if they are the same, or if one is numerically greater than the other. Based on the result of the comparison, the computer can begin executing one set of instructions, or another set of instructions.

The CPU has other capabilities, but these are the principal ones.

Output

When you see a color image on a computer monitor screen, it is composed of, approximately, a million little dots—called pixels (picture elements)—the color of each is controlled by, say, 32 bits. Thirty-two bits allows a large number of combinations, so that many shades of color can be presented. When you hear music coming from your computer's speakers, the sound is generated by strings of bits sending impulses to a speaker cone at varying frequencies—just as sound from a CD is generated by bits: A hole in the surface of the CD is a "1" bit, while no hole is a "0" bit. Whether or not a hole exists in the CD is determined by a laser beam in the CD reader.

(You can contrast this technology, called "digital," with another technology, called "analog," in which the elements are not limited to two states, 0 and 1, but, as discussed previously, may have many more. An example of an analog system is a vinyl music recording, where a wiggly grove in the disk creates an equivalent wiggle in the speaker cone.)

Why do you need to know this rather arcane stuff? ("Hey, I just want the car to go when I push down on the accelerator and stop when I press the brake. I don't care what makes it happen.") You should know because the "real world" is a mixture of continuous phenomena (water flowing in a river) and discrete phenomena (number of people living in a school district). A computer can only store discrete values. Therefore, results from analysis of GIS data that come from computer storage and manipulation may be in error—by amounts that may make an important difference.

Continuous and Discrete Phenomena

For this subject you have to think somewhat abstractly. Consider any system you like.[3] At any distinct moment in time, the elements of the system may be characterized by an exact condition, or "state." The state may be viewed as the values of a set of variables, which could (practically or theoretically) be measured. We can talk about those variables as being independent or dependent. For example, if the subject under consideration is elevations of the Earth's surface, the elevation of a point might be described by the dependent variable "height above sea level" based on the independent variables "position coordinates" (e.g., latitude and longitude) and "time." The terms independent and dependent are a little misleading, because cause and effect is suggested, where none exists. The idea is, rather, that we (independently) specify place and time, which corresponds to a particular elevation. The answer depends on the specifications, but the place and time do not *cause* the elevation.

(I should say here that the difference between discrete and continuous phenomena has been the source for much debate in philosophy, mathematics, and physics for centuries. Look, for instance, at Zeno's Paradox (check the Internet). At very small sizes, a mixture of the continuous and the discrete apply to many phenomena. Electromagnetic radiation (e.g., light) comes both in packets (discrete objects) and waves (continuous phenomena). In theory, every moving object has associated with it a wave, called the de Broglie wave. It applies primarily to subatomic particles (e.g., electrons), but one could calculate the frequency of a de Broglie wave for an SUV moving at 30 miles per hour. We won't. Instead, my intention here is to illustrate the difference between continuous and discrete in the practical, human-size world.)

So, a system (phenomenon) may be thought of as consisting of (or being in) a given state at a given moment in time. Following are some examples.

Discrete: A chess game has each player's pieces on particular squares of the board after a particular move. It does not matter, in terms of the game, where in a square a piece is. When the last piece was moved, the path it took or the length of time required for the move is of no consequence.

Continuous: A billiard ball on a table has, at a particular moment in time, a certain position. When it moves, its velocity, acceleration, jerk, direction, and so on, are vital to the outcome of its ending position, and the ending positions of other balls. The smallest difference in position, velocity, and spin can make a major difference.

Continuous phenomena are characterized by the following:

1. The existence of an infinite number of states over independent variables, for instance, time.

2. When there is a finite but extremely small (infinitesimal) difference between values of an independent variable, there is at most an infinitesimal difference in one or more dependent variables. (A tennis ball will change position very slightly in a fraction of a second.)

3. No matter how carefully measurements are undertaken, the state of the system can never be determined exactly.

---

[3]System: a collection of elements (things: objects, ideas, organs, equations, banks, planets, and so on) with connections of some sort between (proximity, gravity, pedagogy, electrical, chemical).

Discrete phenomena are characterized by the following:

❑ A finite number of states (There are only so many combinations of pieces in positions on a chess board.)

❑ The smallest possible difference in an independent variable may result in a significant finite difference between states. (If "move number" increases by one, the positions of pieces on the chessboard will be in a distinctly different state.)

❑ The exact state of the system can be determined.

To further illustrate the difference between continuous and discrete phenomena, let's look at graphs of each. Figure 6-1 might illustrate the amount of water in a stoppered sink as it is being filled from a faucet, using time as the independent variable. Note the connectedness of the line.

Figure 6-2 shows the number of students "in" a classroom as they enter in the minutes before the class starts. Since students come in packages of one, you see jumps in the graph.

With discrete phenomena, exactness is possible. With continuous phenomena, it is not. A basket may contain exactly eight eggs. A bathtub may not contain exactly eight liters of water.

Table 6-1 shows some examples of continuous and discrete phenomena. We might argue over some of the categorizations. Further, discrete phenomena may have continuous parts and vice versa. But you will probably get the idea from the list.

With our languages, we respect the difference between continuous and discrete. "How much" applies to continuous things. "How many" applies to things of a discrete nature. We wouldn't talk about an amount of votes—we would speak of a number of votes.

If we can use computers with enough precision to represent the world, does it matter that they are discrete machines and that that world is a mixture of continuous and discrete? (Can you tell if recorded music comes from a CD or an LP? Actually, some people can.) Is there an impact in using discrete machines to represent continuous phenomena? Usually only a little—but sometimes a lot. The idea of this section is to be sure that you understand that the potential exists for inaccuracies and errors when we use completely discrete machines to represent continuous phenomena. *Punch line: computers are completely discrete machines!*

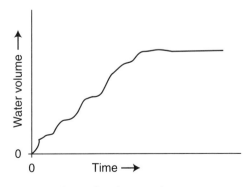

FIGURE 6-1   Continuous phenomena

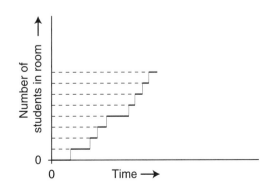

FIGURE 6-2   Discrete phenomena

TABLE 6-1

Some Continuous Phenomena:	Some Discrete Phenomena
Ramp	Steps
Water	Ice cubes
Time	Days
Real numbers	Integers, rational numbers
Musical representation—LP vinyl disk	Musical representation—compact disc
Musical performance	Musical score
Facial expressions	Wink
Sloping land	Terraces
Scrambled eggs	Soft-boiled eggs
Ping–Pong	Checkers
Meaning	Words
Analog (rotary) watch	Digital watch
Pond scum	People
Poison gas	Bullets
Belt-type people mover	Automobile
Light dimmer (rheostat)	Toggle switch
Steering (automobile)	Turn signal (automobile)
Brake (automobile)	Horn (automobile)
Pile of sulfur	Load of bricks
Handwriting	Typing
Slide rule (see Figure 6-3***)	Abacus
Analog computer	Digital computer

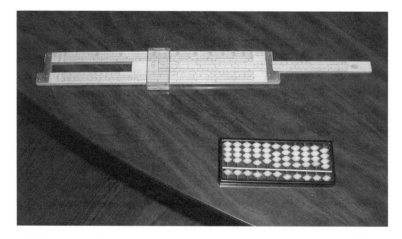

FIGURE 6-3   Early computers: abacus and slide rule

Some Implications of Discrete Representation for GIS

Because, with GIS, the world is represented in a discrete machine, everything a GIS tells you is an approximation—sometimes a bad approximation—of reality. For example, a GIS may represent a curving road by a series of straight-line segments, which are themselves represented by numbers. Sets of numerical coordinates (x and y pairs) placed along the centerline of the roadway define the beginnings and endings of the segments. The sum of the lengths of the segments will, unfortunately, underrepresent the true length of the curving road. The degree of error may be reduced by using shorter, and therefore more, segments, but the fundamental problem remains the same.

Further, the "real world" is virtually infinite in the level of detail that exists (look through a microscope if you doubt this), while a computer store is finite, and in many ways, quite small—compared, say, with what is in your own head.

How does a digital machine store information about a continuous environment? By digitizing—using this term in the most general way. We describe the world with numbers—integers (such as 7 and –383) and rational numbers (such as 1.618034 and 6.626 times 10 to the negative 34th power). We also use strings of letters and other symbols: A, a, B, #, %.

To even have a chance at being sufficiently accurate when we specify a location on the Earth's surface, we need to use a lot of digits. For example, the location of a particular fire hydrant near Vancouver, British Columbia, Canada, was reported by a GPS receiver to be 49.2773361 north latitude and 122.8793473 west longitude. Since each decimal digit requires some two and a half bits, the resultant binary number can be quite lengthy, especially when you must add more bits for the exponent needed for floating-point numbers. The capability to represent very precise numbers is sometimes called *double precision* or *extended precision*, and is frequently required in GIS. With regular "real" numbers, called *floating-point*, you can count on six significant digits. With double-precision numbers you can count on 16 significant digits. Another approach, in use with ESRI geodatabases, is to use long integer numbers for storage and to convert them to floating-point numbers when needed for computation and output.

Scientific Notation, Numerical Significance, Accuracy, and Precision

How Old Is the Dinosaur?

The curator of a natural history museum had a habit, from time to time, of walking around and listening to the guides give their lectures at the various exhibits. One day he arrived just in time to hear that the museum's Tyrannosaurus rex was sixty-five million and three years old. He went back to his office and told his secretary: "Gladys, tell George I want to see him as soon as he's done with his tour." When George appeared, the curator exasperatedly asked him, "What do you mean telling people that the Tyrannosaurus rex fossil is sixty-five million and three years old?" George looked abashed, but said confidentially, "Well, you hired me three years ago and you told me then that the skeleton was sixty-five million years old, so . . ."

A GIS will deal with very precise numbers—that is, numbers that contain many digits. For example, locating the longitude of a point on Earth's surface within a centimeter (not at all an unrealistic expectation nowadays) requires a number with ten significant digits—for example, –123.4567890. Because of the number of bits devoted to "ordinary" numbers by most computers, and because fractional decimal numbers may not be represented exactly by fractional binary numbers, GIS frequently use "double-precision" numbers to represent positions.

Also, a GIS may deal with very big and very small numbers. Very large and very small numbers are stored in the computer in a fashion similar to scientific notation: The number is represented as a mantissa (which contains the significant digits of the number) and an exponent (which tells how many digits and in which direction to move the decimal point). For example, the number in scientific notation:

2.0×10^{-7} (which is two times ten raised to the negative seven power) represents the number 0.0000002

To use this mantissa-exponent notation to represent 6,281,823,925, you would write the following:[4]

$6.281823925 \times 10^{9}$

Precision vs. Accuracy

The terms "accuracy" and "precision" do not refer to the same idea. Both are important to GIS. A classic example of the difference follows:

Weather forecaster A indicates that it will be between 40 and 50 degrees tomorrow at 4 P.M. The actual reading turns out to be 43. Thus, the forecast was *accurate*, but not very *precise*. Forecaster A provided a true statement but without much detail. Forecaster B states that it will be 52.47 degrees at 4 P.M. tomorrow. The temperature turns out to be 43 degrees. Forecaster B was very *precise*, but not *accurate*.

The terms accuracy and precision can also be applied to a *set* of readings or measurements. Again, a classic example:

Darts are thrown at targets. The target of person "R" looks like Figure 6-4,

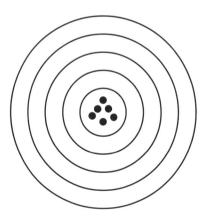

FIGURE 6-4   Good accuracy and good precision

---

[4]The estimated world's population as of the first day of spring 2003 at 20:08:53 Greenwich Mean Time (GMT).

Her dart throwing was both accurate and precise. The darts are where they are intended to be (accurate), and very close together (precise).

Person S's target looks like Figure 6-5.

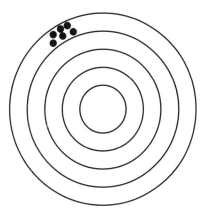

FIGURE 6-5   Poor accuracy but good precision

The darts are tightly clustered (they hit a spot almost precisely) but lack accuracy, as they missed the intended spot. (Perhaps some sort of systematic error is present—such as a breeze blowing up and to the left.)

Person T's target is as shown in Figure 6-6.

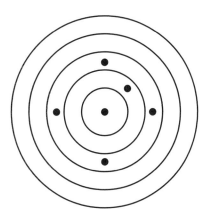

FIGURE 6-6   Good accuracy but poor precision

Figure 6-6 indicates a lack of precision, but when the dart positions are averaged out, they are quite accurate—they come quite close to the intended spot. (Note that this doesn't count for much in the game of darts, but would be useful in finding an intended spot if one had a number of readings, say, latitude and longitude, from a process or piece of equipment, such as a GPS receiver, known to be accurate.)

Well, what can we say about the target of Person U?

The darts are not clustered (not precise). They do not average out to the intended spot (inaccurate). See Figure 6-7.

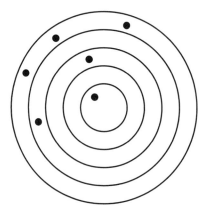

FIGURE 6-7   Poor accuracy and poor precision

To summarize: *Precision* refers to the degree of specified detail. A number with many significant digits is a precise number. (Unfortunately, the "significant digits" may not truly be significant.) *Accuracy* refers to the correctness of the specified number. What counts is its truth. It may be precise or imprecise.

For information to be useful it must have substantial degrees of both precision and accuracy. "How much" of each? Obviously the issue is complex.

Basic Statistics

Statistics may be defined in very erudite ways, but fundamentally, it is the art and science of making a big bunch of numbers into a little bunch of numbers so you can understand stuff about the big bunch of numbers. A GIS can usually do statistics without even breathing hard. Here are some basic statistical measures, most of which you probably know:

Average

The *mean* or *average*: add up all the numbers in a set and divide by the number of numbers. Expressed as a formula the average, called x-bar here, because of the x with the bar over it on the left-hand side, looks as shown in Figure 6-8.

$$\overline{x} = \frac{\sum\limits_{i=1}^{n} x_i}{n}$$

FIGURE 6-8 Formula for average of a set of numbers

$$\sum\limits_{i=1}^{n}$$

FIGURE 6-9 Sigma: symbol that means to sum a set of numbers

This formula scares some people. It shouldn't. It simply says what was said in the preceding text. More specifically: Assign integers to each number, like 1 to the first number, 2 to the second, and so on until you reach the last number, which we refer to as n. (If there were 12 numbers, n would be 12.) The numbers then are $x_1, x_2, x_3, \ldots x_{12}$. So when we say x_i and $i = 7$, we are referring to the seventh number, named x_7. The symbol shown in Figure 6-9

means "take the sum of" whatever follows it. So in the case of the symbol in Figure 6-9, what follows it is x_i. What values of i do you use? Those that run from 1 through n, namely, 1, 2, 3, . . . 12. Finally, the formula in Figure 6-8 says divide the sum by the number of numbers, which is n, or, in this case, 12.

Median

The *median* number is simply the middle number (or the average of the two middle numbers) in a set of numbers that has been sorted from smallest to largest (or largest to smallest—it makes no difference). The median is not sensitive to very large or very small values, as is the mean.

Mode

The *mode* of a set of numbers is the most frequently occurring number. A set of numbers could have several modes or none at all.

Range

The *range* of a set of numbers is the largest minus the smallest. This gives you an idea of the spread of the numbers. A better idea of the spread is perhaps the standard deviation, shown next.

Standard Deviation

The formula for *standard deviation* is shown in Figure 6-10.

$$\sqrt{\frac{\sum (x - \overline{x})^2}{n - 1}}$$

FIGURE 6-10 Formula for standard deviation

The formula says this: (1) Find the average (that's x-bar, from Figure 6-8), (2) make a new set of numbers by subtracting that average from each of the numbers in the original set, (3) make a third set of numbers by squaring each of those in the second set, (4) add up this third set, (5) divide the sum by $n - 1$, and (6) take the square root.

If the data are normally distributed—as, for example, the heights of adult men in a town *might* be—then the standard deviation has a graphical meaning as well as a numerical one. Suppose Figure 6-11 is the curve for the heights, with each height rounded off to the nearest inch.

The x-axis direction represents the possible heights and the y-axis direction represents the number of individuals who have each given height. The curve is "bell shaped about the mean"; the mean is 65 inches. Graphically, the standard deviation is defined by the x values on the curve at its points of inflection— where the curve goes from being cupped down to being cupped up. The standard deviation is 4 inches. This means that approximately 68 percent of the individuals are between 61 inches tall (65 – –4) and 69 inches tall (65 + 4). The idea behind the standard deviation is to give you an idea of how spread out your data values are. The standard deviation is meaningful, but only if the data are close to a normal distribution are you able to make statements like: 47 percent (34 percent + 13 percent) of the individuals are between 65 inches and 73 inches tall (65 + 4 + 4).

Putting Values into Classes

Another way of reducing a large group of numbers to a more manageable size is to put similar values into separate classes or categories. For example, we might have the data on real estate parcel sizes. We could put each parcel into one of several categories: (1) greater than 10 acres, (2) 10 acres to 5 acres, (3) 5 acres to 1 acre, and less than 1 acre. In the Step-by-Step section of this chapter you will do an exercise on classification of values.

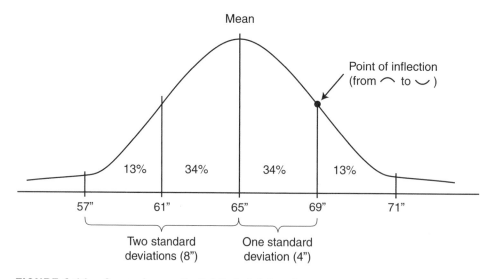

FIGURE 6-11   Curve of normally distributed data values

Measurement Scales

When you see a number, its context determines how you interpret that number and how you may use it in computation and comparison. A number that represents some description or measurement of something in the real world will fall into one of four categories:

❑ *Nominal*—A nominal data set is just a set of names, except the names take the form of numbers. If in an experiment you randomly assign numbers to test tubes (say, to prevent student experimenters from knowing what is in them), you are forming a set of nominal data. There is no numeric relationship between the numbers. Each is merely a name. The only operators you may use on the numbers are equal (=) and not-equal (π or =, or sometimes written as less than or greater than, or < >). For example, suppose the soil type found in area A is stored as x and the soil type found in area B is stored as y, you can ask: does $x = y$ (assuming x and y are integers or text strings)?

❑ *Ordinal*—The numbers in an ordinal data set indicate an order among the entities they represent. Perhaps 1, 2, and 3 indicate first-born, second-born, and third-born children. You know that, in a given family, a child numbered 1 was born before a child numbered 3. But you don't know how many months or years before. In addition to using the equal and not-equal operators on ordinal data, you may use less than (<), less than or equal ($\leq$), greater than or equal ($\geq$), and greater than (>). For example, house numbers on a given side of a street in the United States are generally in order. Suppose house numbers increased along the right side of a north-south road as you proceeded north. You could determine which of two houses was the northmost house by asking: is house number of X greater than house number of Y?

❑ *Interval*—An interval data set consists of numbers that come from a measurement scale where a unit amount is established and values along the scale are linear multiples of that amount. Examples are the common, nonscientific temperature scales: Celsius and Fahrenheit. With interval data you may use all the operators described previously and also the arithmetic operators addition (plus, +) and subtraction (minus, –) as well. For example, if it were 20°F in the morning and it became 60°F in the afternoon, you may say that it had become 40°F warmer (60 – 20 = 40).

You have to be very careful when using multiplication and division operators on interval data. For example, you may not say, given the preceding numbers, that it became three times as warm in the afternoon. The reason is that a set of interval data does not have a meaningful zero point from which to measure.[5] For example, if the morning temperature were negative 20°F and the afternoon positive 20°F, the difference would still be 40°F, but you can see the difficulty in saying that one is a multiplicative factor warmer than the other. (Three times –20 is –60.) The use of multiplication (*) and division (/) with interval data is not completely ruled out, however. As long as the answers are understood to be in the context of the arbitrary zero, you can make limited use of * and /. For example, if the morning temperature were 20°F and the afternoon 60°F, you could say that the mean temperature was 40°F ((20 + 60) / 2).

❑ *Ratio*—A set of ratio-scale data is like an interval-scale data set, with the additional proviso of a non-arbitrary zero point. The simplest example would be measured distances. A ruler has an absolute zero and a unit value of 1 inch. I can say both that, if desk A is 30 inches long and desk B is 45 inches long, desk B is 15 inches longer than table A, and also that desk B is 1.5 times as wide as table A.

---

[5]The zero in the Fahrenheit scale was originally determined by the temperature of a mixture of ice and salt. Check the Internet for the bizarre determination of 100 in that scale. The zero in a Celsius (centigrade—changed in 1947 to honor the inventor of the Celsius thermometer) scale is arbitrarily defined to be the freezing point of water and 100 is its boiling point.

With ratio data you may use any comparison or arithmetic operators (=, < >, <, ≤, ≥, >, +, −, *, /, and exponential (^ or **)) plus a whole batch of functions, such as square root. Of course, you must be careful to know the limits and characteristics of your data. You may not, for example, take the logarithm of a negative number. And when taking the cosine of 450° you must understand what that means.

Analysis of GIS Data by Simple Examination

__ **Open up your Fast Facts text or document file.**

Reviewing and Learning More of ArcMap

The goal of this assignment is to review ArcMap and to become familiar with more of its features and complexities. This will begin to lead us toward using GIS for analysis of spatial data. These steps involve basic statistics, thematic mapping, changing symbology for emphasis, and classification. You can probably breeze through the steps very quickly, but you shouldn't. Each step is one that you will have to know how to do in the future, without much explanation. After doing each step, you should ask yourself: Will I know how to do this in a month, if I don't use it until then and simply rely on my memory? If not, write it in your Fast Facts File. Later you can reorganize the information you record during these exercises.

To make sure that everyone is working from the same geodatabase, you will copy a pristine version from the folder IGIS-Arc_AUX. Then you'll load the personal geodatabase featureclass Landcover onto a new empty map:

__ **1.** If necessary, close ArcCatalog and ArcMap. Use the operating system of the computer to delete the folder

___IGIS-Arc_*YourInitialsHere*\Wildcat_Boat_Data.

Launch ArcCatalog. Find the folder

[___] IGIS-Arc_**AUX**\Wildcat_Boat_Data

noting carefully that it is the AUX folder you want, not the one you have been working with.

Copy the folder Wildcat_Boat_Data from [___] IGIS-Arc_**AUX**

to ___IGIS-Arc_*YourInitialsHere*.

—— **2.** Launch ArcMap and make the window occupy the full screen. Click the Add Data icon. Pull down a list from the Look in field. Push the slider bar to the top so that you see Catalog Select it. In the window, navigate to

___IGIS-Arc_*YourInitialsHere*\Wildcat_Boat_Data\ Wildcat_Boat.mdb\Area_Features\Landcover

and press Add. A map appears, consisting of a homogeneous color background with lines dividing polygons.

—— **3.** In Data Frame Properties > General, make sure Display Units are set to Meters. Map Units are already set to meters because of the coordinate system of the feature class.

Examining the Toolbars

—— **4.** If you right-click anywhere in the areas containing toolbars (or use View > Toolbars), you get a list of all those available, with an indication of which ones are displayed. To get an idea of the extensive capability of ArcMap, and for future reference, in the blanks that follow, write the names of the available toolbars for your system. Give a moment of thought to each one, regarding what sort of capability it might provide.

_____, _____, _____

_____, _____, _____

_____, _____, _____

_____, _____, _____

_____, _____, _____

_____, _____, _____

_____, _____, _____

_____, _____, _____

_____, _____, _____

_____, _____, _____

_____, _____, _____

_____, _____, _____

_____, _____, _____

_____, _____, _____

At the bottom of the list notice that there is a toolbar named Customize. ArcMap is software that you can "adjust" to your own specifications. You can customize toolbars and commands. Even when you find out how (which you won't from this book), do this with care. The more you customize an interface, the more difficulty you will have when confronted with the standard interface, say, on another computer. Further, if you share your computer with another person, he or she may have trouble operating the interface.

____ **5.** Turn off all toolbars except Main menu, Standard, Draw, and Tools.

____ **6.** Click the Identify button. Point to a polygon with the mouse cursor and click. Examine the Fields and Values that appear in the Identify Results window. LC_CODE[1] is the number of the land cover type (see table below); COST_HA is the cost in US dollars per hectare.[2] Dismiss the Identify Results window.

LC-CODE	LC-TYPE
100	Urban
200	Agriculture
300	Brushland
400	Forest
500	Water
600	Wetlands
700	Barren

____ **7.** Open the Landcover attribute table. How many records are in it? _____ So, how many land cover polygons are there? _____ Sort the table by polygon area by right-clicking the heading Shape_Area and choosing a sort command. What is the area of the largest polygon that is *not* land cover code 500 (which is water)? _____ (The numbers are square meters.) What is the land cover code of the smallest polygon? _____ What sort of land cover is it? _____ If any records or polygons have been selected, press Clear Selection.

____ **8.** *Calculate statistics on areas:* Again right-click the heading Shape_Area and click Statistics to get a Statistics Of Landcover window. What is the average of the areas of the polygons (to the nearest tenth of a square meter)? _____ While you have the statistics window open, choose Shape_Length in its Field text box. Shape_Length gives the perimeters of the polygons. If you added the perimeters of all the polygons together what would be the sum? _____ What is the approximate cost per hectare of the most expensive land? _____ Dismiss the statistics window.

____ **9.** Sort the records back into their original order based on the Object Identifier (OBJECTID). Clear any selections.

Pointing at Records

____ **10.** Experiment with various ways of *pointing at* records: Click any cell of any record. Note that, in the gray box to the left of the record, a triangular marker appears. The Record indicator at the

---

[1]The land cover code explanation may be found in Exercise 1-1.
[2]A hectare is an area equivalent to a square 100 meters on a side—about two and a half acres.

bottom indicates the position of that record in the table. By pressing the buttons on either side of the Record indicator, you can control which record is pointed to. Only one record may be pointed to at a time. Pointing to a record has nothing to do with selecting a record or records.

Two Windows Are Available for Selecting

As with many operations with this software, there are multiple ways of accomplishing a task. No exception is using the Selecting by Attributes option. It's worth taking a look at the two windows you can use for this. You get to one window through the Options button on an attribute table. You get to the other one through the Selection menu from the Main menu. As a side trip here, look at both.

_____ **11.** On the attribute table, click Options > Select By Attributes. Move the window down a bit and to the right. From the Main menu, choose Selection > Select By Attributes. Move this window off to the left. Compare the two windows. See Figure 6-12. If you use the window from Options, then the Layer is obviously set for you, since you picked Options off an attribute table of a layer. In the other window you must pick the layer. Depending on the

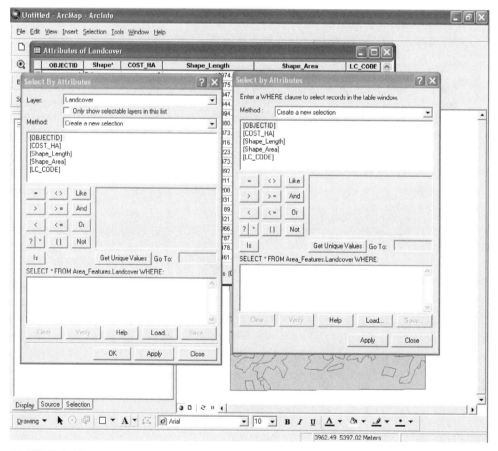

FIGURE 6-12

version of ArcGIS that you are using, the window from Selection on the Main menu may give you access to a Query Wizard that will step you through the selection process. A wizard is composed of a set of steps that asks you questions about the process you are doing and tries to make sure you fill in all the necessary blanks. Close both selection windows.

Selecting Records (and, Thereby, Features)

A record is selected if it is highlighted in blue. Any number of records may be selected—from none to all or any combination in between.

____ **12.** Experiment with various ways of *selecting* records: Click the gray box at the left of any record. The entire record is selected. Click another box of another record and that record is selected instead. If you hold down the Ctrl key and click the box of an unselected record, it is added to those already selected. If you hold down the Ctrl key and click the box of a selected record, it is deselected. If you hold down the Ctrl key and drag the cursor over consecutive boxes, the selected records are deselected and the unselected records are selected. The number of selected records is shown at the bottom of the table. Also at the bottom of the table, you can choose that the window displays all the records or only those selected. Choose records with OBJECTIDs 2, 3, 5, 7, 11, 13, 17, 19, and 23. Show only those records. Then show all records.

____ **13.** *Use simple logic to select records:* Press the Options button on the table window. (If the table window isn't wide enough, the button may be hidden.) Press Switch Selection. Note that all records except 2, 3, 5, 7, 11, 13, 17, 19, and 23 are selected. Switch again. Choose Clear Selection, then Switch Selection, then Clear Selection, then Select All, and finally, Clear Selection.

____ **14.** *Select records by attributes:* Click Options > Select By Attributes to bring up a window of the same name. Type the following expression (observing capitalization) in the large blank in this window:

[Shape_Length] <= 201

and click Apply. This selects all those records that have a Shape_Length value less than or equal to 201. Dismiss the Select By Attributes window. Examine the table. Note that only those records of polygons whose perimeters are less than or equal to 201 meters are selected. Show only those records. How many are there? _____ Run Statistics on the Shape_Length field. What is the average? _____. Now select Show All Records and run Statistics again. What is the average? _____. Choose Clear Selection. Run Statistics again. What is the average? _____. Make a note to yourself in your Fast Facts File that if any records are selected, no matter what records are shown, statistics are calculated *only on the selected records*. Forgetting this fact is a good way to get wrong answers. Dismiss the statistics window.

____ **15.** Again click Options > Select By Attributes. Erase the expression (there is a Clear button). Experiment in building the partial expression

[Shape_Area] <=

by clicking or double-clicking various buttons and values in the window. This *query building* is a quick way to create an expression while avoiding the possibilities of mistyping. (Even if you use the query builder, you still can type parts of an expression, and at times you may have to.)

Press Get Unique Values to bring up all the possible Shape_Area values in the Unique Values pane. Scroll down to the value that begins 7710.087 and double-click. This will complete the expression:

[Shape_Area] <= 7710.08730661013

Click Verify to determine that the expression is valid. Click OK, then on Apply. Dismiss the Select by Attributes window. Note what happens when you show all records and sort them in descending order by Shape_Area. Determine that the query worked.

_____ **16.** Make the table narrower and move it over so you can see the map. You will notice that many of the small polygons on the map have a border around them. They are the polygons that correspond to the selected records—those with area less than or equal to 7710.08730661013. The arcs that define a selected polygon are shown in cyan (cyan, by default—you can change the color if you wish).

_____ **17.** Using Selection on the Main menu, clear all selected features. Sort according to OBJECTID. Select the record with FID number 15. Note that on the map you see two cyan rings. Why? Use Identify if you need help.

_____

_____ **18.** If the Query Wizard exists in the Selection > Select By Attributes window (it does in version 9.0 and earlier), run through the previous steps (from 15 to 17) again, with three changes: (1) close the attribute table, (2) use the Query Wizard (get to it with Selection > Select By Attributes) to construct the expression, and (3) observe the map as you do so. You may have to slide the window around to see the map.

Looking at the Other Capabilities of the Options Menu

_____ **19.** Open the attribute table again if you closed it. Besides the four ways of selecting records that you just explored, the Options button leads to additional capabilities. The entries are:

_____, _____, _____

_____, _____, _____

_____, _____, _____

_____, _____, _____

_____, _____, _____

Selecting Features (and, Thereby, Records)

Selecting works the other way too. If you select features on the map the corresponding records are selected. A myriad of ways of selecting geographic features are supplied by the software.

Quick Selection of Features

___ **20.** Set the attribute table so it shows all records. Clear all selections. Set the table so it shows only `Selected Records`. Find the `Select Features` icon on the `Tools` toolbar. Note what the status bar has to say about the operation of the tool. Press the button. Click a polygon and notice that its record is selected. Click another. To select several polygons, hold down the Shift key (rather than Ctrl as when selecting multiple records in a table) and click the intended polygons.

___ **21.** Drag a box inside a polygon. Notice that that polygon is selected. Draw a box that is encompasses parts of several polygons. Note that they are all selected.

___ **22.** Under `Selection` in the Main menu, pick `Zoom To Selected Features`. So here you have a technique of zooming in on exactly what you selected (whether selected by attribute or location). Zoom back to `Full Extent`.

___ **23.** To clear all selections, choose `Selection` > `Clear Selected Features` (or click somewhere off the map with the `Select Features` tool). Close the attribute table.

Selecting by Location

In what follows you will examine some of the interesting and useful ways ArcMap has of selecting features by their locations.

___ **24.** From___IGIS-Arc_YourInitials Here\Wildcat_Boat_Data\ Wildcat_Boat.mdb\Area_Features

add the feature class Soils to the map. Make the symbol color of soils `Light Apple` green. Make sure Soils appears above the Landcover layer.

From ___IGIS-Arc_*YourInitials Here*\Wildcat_Boat_Data\ Wildcat_Boat.mdb\Line_Features

add the feature class Roads. Using the `Symbol Selector` window, represent the roads as Highway (red, 3.40). Make sure it appears at the top of the table of contents.

___ **25.** Choose `Selection` > `Select By Location`. Read the text at the top of the box. You may select feature by completing the sentence `I want to`. List the contents of the first drop-down text box to fix in your mind the possibilities.

_____

_____

_____

_____

Pick the option that says "`select features from.`"

In the text box headed `the following layer(s)` put a check by Roads. In the "that" text box pick `are completely within`. In the `features in this layer` text box, pick Soils. Now read the whole sentence:

"I want to select features from the following layer(s): Roads that are completely within the features of this layer: Soils."

Read the Preview text. Since Roads consists of lines, look at the Lines graphic that accompanies the description. Understand the text and graphics.

___ **26.** Click `Apply`, then `Close`. From the map you can see that the segments of road that are completely contained within any single polygon are highlighted in cyan. Those segments that cross polygon boundaries are not highlighted.

___ **27.** Open the Roads attribute table and note, using `Identify` on a few of the features, that the corresponding records are selected. Close the attribute table.

___ **28.** If necessary, drag the layer Landcover to the bottom of the Table of Contents. Clear any selection using `Selection > Clear Selected Features`. Turn off Roads, but leave both Landcover and Soils on. What layer appears on the map? _____

___ **29.** Again choose `Select By Location`. This time `select features from` Landcover that `are contained by` the features in Soils. Click Apply. Click Close. Dismiss the `Select By Location` window. Notice that although the Soils layer obscures the Landcover layer, the selected Landcover polygons are shown in cyan. Open the attribute table of Landcover. How many polygons are selected? _____

___ **30.** Open the `Select By Location` window one more time and look at the possibilities, from the drop-down menu, that follow the word "that" in the window. List them in the blanks that follow. Think what each means as you write it down.

_____ , _____ , _____

_____ , _____ , _____

_____ , _____ , _____

_____ , _____

___ **31.** Click each of these methods of selection to get a graphic representation (at the bottom of the window, under Preview) and read the text to get a general idea of what the method of selection is. Notice that you can use this window to select points, as well as lines and polygons. Notice also that some don't "make sense" (e.g., points cannot contain lines) and are thus shown with a "do not" symbol.

___ **32.** Do Step 31 again, but change the "`following layer(s)`" field to Roads. Again, you are not expected to know what all of these possibilities are, but to realize that just about any combination of geometric possibilities for selection is available to you. When finished, close the `Select By Location` window. Clear all selected features. Remove any attribute table.

Reviewing and Understanding Actions on the Table of Contents

____ **33.** *Look at getting information from entries in the table of contents (T/C):* Drag Soils to the bottom of the Table of Contents. Click the box containing the check mark next to Landcover. Note the result. Click it again so the layer is on. Click the box with the little minus sign in front of Landcover and note that the symbol goes away. Click again. Click the `Source` tab at the bottom of the panel. Note that you are given the path to each data set. Experiment by clicking the boxes that contain either plus (+) or minus (–) in the T/C. Understand these actions. Leave all entries fully expanded.

____ **34.** Press the `Selection` tab at the bottom of the Table of Contents. The check boxes here let you set the layers that are selectable. That is, if a layer's box isn't checked, the `Select Features` tool cannot select features from that layer.[3] Prove this to yourself by displaying the Landcover layer (Source tab), but making it unselectable (Selection tab). Now click the map with the Select features tool. Features will be selected, but they will be from the soil layer underneath. Turn off the display of the soils layer. Now notice that Select Features has no effect. Lesson: For `Select Features` to work, a layer must be both displayed and selectable. Make all layers selectable. Clear any selections.

Layers and the Data Frames

____ **35.** *Understand layer properties verses data frame properties:* In the layers pane, click the `Display` tab. In the table of contents make all the +/– boxes say minus, so that nothing is hidden. Right-click the text Landcover. Click `Properties`. A `Layer Properties` window, with a number of tabs, appears. The `Layer Properties` window tells you (or lets you specify) the properties of the *particular* layer that you right-clicked on, such as how it is symbolized, what the fields of the table are, what its extent is, how selected features are to appear, aspects of feature labeling, the coordinates and characteristics of its data source, and several other parameters.

Write here the names of the tabs:

_____, _____, _____

_____, _____, _____

_____, _____, _____

Dismiss the `Layer Properties` window for Landcover.

____ **36.** Now right-click in the right pane of the overall window—where the map is shown. Click `Properties`. The `Data Frame Properties` window tells you (or lets you specify) the properties of the display of the map, such as the scale, the map units, the coordinate system, whether a reference grid appears and so on.

---

[3]Features from unselectable layers can still be selected by selecting their records from the attribute table

Write here the names of the tabs:

_____, _____, _____

_____, _____, _____

_____, _____, _____

_____.

Changing Layer Properties

As noted, Landcover is displayed using a homogeneous color background with lines dividing polygons. The polygons represent different land uses. We can make the map display these different uses in a variety of ways.

____ 37. Bring up the `Landcover Properties` window again—this time by double clicking the Landcover layer name. Choose the `Symbology` tab. You have your choice of five possibilities of what to show. List them:

_____, _____, _____

_____, _____

Note that features are displayed with a `Single symbol`. (If you don't see this, click the word `Features`.)

____ 38. Instead, let's display the layer using `Categories > Unique Values`. For the `Value Field`, pick LC_CODE, which, as you may recall, is an integer (100, 200, and so on). Click `Add All Values` to bring up the list of values and a random list of colors with which those values will be symbolized. Click `Apply` and OK the `Layer Properties` window. A garish-looking map appears. Observe the map. Rather than being displayed with a single color, different polygons are being shown with different colors, based on the value in the LC-CODE field.

____ 39. *Label the polygons with text:* Bring up the `Landcover Properties` window again. Click the `Labels` tab. Click the box that says `Label features in this layer`. Make the Label Field read LC_CODE. Click Apply. Click OK. The land cover code appears in each polygon. Right click on the Landcover name in the Table of Contents. Remove the check beside `Label Features`. The labels go away. Restore them by turning `Label Features` back on.

____ 40. Suppose we want the cost per hectare shown instead. Bring up the `Layer Properties` window and click the `Symbology` tab. In the `Value Field`, drop-down menu, pick COST_HA. Add all values. Click Apply, then on OK. The map colors have changed, *but the labels remain the same*. So go into properties again and select the `Labels` tab. In the `Text String Label Field` box select COST_HA. Click Apply then on OK. The cost figures appear.

____ 41. The numbers are a bit too big for the map. We don't need all those zeros. In the `Layer Properties` window, `Labels` tab, click `Expression`. In the `Label Expression` window, add / 1000 to

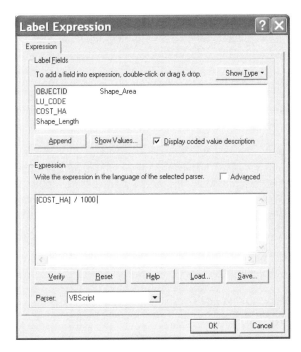

FIGURE 6-13

[COST_HA] (do type the blanks). The window should look like Figure 6-13. When executed this expression will cause each label value on the map to be divided by one thousand.

Click OK, Apply, and OK. Now only two-digit numbers should appear in the polygons.

____ **42.** ***Change the color symbology for a subset of features:*** the Land cover code 500 is for water. Water in large quantities is usually blue. Change the field on which to base the Symbology of the Landcover layer back to LC-CODE. Display the map with random colors as before. In the Table of Contents, right-click the color rectangle next to 500. Pick a blue from the color pallet that shows up. Note the effect on the map.

____ **43.** Experiment with other layer properties and data frame properties. These windows allow you a great deal of control over what is displayed. You should become familiar with them. A large part of successfully operating ArcMap is knowing where to find and use the various controls. (Note that to change the label field back to LC-CODE you have to put [LC-CODE] by itself in the expression text box.)

____ **44.** Right-click the Landcover layer and remove it from the map. Remove Roads as well. The map now appears consisting of a homogeneous color background with lines dividing polygons of Soils.

As before, you could identify individual polygons with the Identify tool. You could place text in each polygon with the labeling facility. You could open the attribute table and, by selecting polygons, see the corresponding rows in the table, and vice versa. All of these techniques are useful, but none give you a very good picture of what features are where, and it doesn't let you compare various features. In the steps that follow, you change this by presenting the polygons as a thematic map, with different colors for each type of soil. First you label the polygons with SOIL-CODE.

____ **45.** Bring up the layer's properties window. (What are two ways to do this? _____, _____.) Click the Labels tab. Check the box that says Label Features In This Layer. Use SOIL-CODE in the Label field. Click Apply and on OK. Check out the resulting display.

____ **46.** Bring up the layer's properties window again. Click the Symbology tab. In the Show box you will probably see Single Symbol highlighted under Features. Click Categories and make sure Unique Values is selected. Make the Value Field SOIL-CODE. Click Add All Values. Click Apply, then OK. Notice that all polygons labeled Sg are the same color. Change the symbol used to depict the water to Atlantic Blue or Lake Blue. If there are other polygons with a color close to the blue you picked, change them to some other color.

____ **47.** Now you can clearly get an impression of the locations and amounts of the various soil types. What (commonly named) color is associated with Ko? _____

____ **48.** Change the labeling so that soil suitability (SUIT) is shown. By inspecting the map and the legend, and using the Identify tool if necessary, determine the codes of the soils that have suitability 1. _____.

Suitability 2? _____.

Suitability 3? _____.

What is the suitability code for water? _____.

____ **49.** Now switch the labeling and the color representation, that is, label the polygons again with the soil type and show the suitability with different colors. Check your answers from the last step.

____ **50.** There are practical and visual limits to the numbers of color categories you can add to a map. So ArcGIS has the capability to display only the categories you choose. To see how this works, under Symbology in the Layer Properties window, click Remove All. Then click Add Values. Now you have the option of adding only some of the values. Since soils suitable for the Wildcat Boat facility are only 2 and 3, add only those values (use Ctrl-click). Click Apply, then OK. Note that the rest of the map is colored with the <all other values> symbol.

____ **51.** Under the Symbology tab in the Layer Properties window, click Remove All. Now click Unique Values, Many Fields. Note that there are now three possible value fields instead of just one. Put SUIT in the first of these text boxes and SOIL-CODE in the second. Now click Add All Values. Click Apply, then OK. Different colors for all the unique combinations of suitability and soil codes that exist in the table will be shown in the Table of Contents and on the map.

____ **52.** You can also fix it so that the labels reflect both suitability and soil code. Under the Labels tab in the Layers Property window, click the Expression button. Arrange it so that the Expression text box says, exactly:

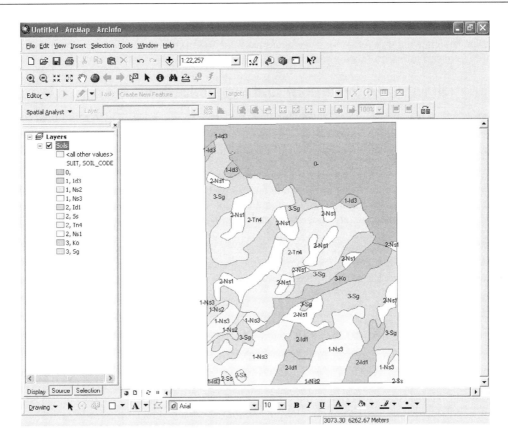

FIGURE 6-14

```
[SUIT] & "-" & [SOIL_CODE]
```

by typing and/or clicking. This expression says concatenate three values, the middle one of which is a hyphen. The double quote marks let you place any text string in the expression. Click OK, then Apply, then OK. The result should look something like Figure 6-14***. Close ArcMap.

Exercise 6-2 (Project)

Categorization and Symbolization

Thinking About Maps Again

Now that the connection between the table and the map is firmly re-established in your mind, we turn back to the map. You could broadly classify maps into two categories: general-reference and thematic. Many maps, like road maps or topographical maps are used for general information, navigation, topology, and to show locations of features. Thematic maps (sometimes called statistical maps), on the other hand, show polygons that divide up the landscape into distinct areas; the areas are coded—with color or other symbolization—to indicate the value of a characteristic or attribute about each individual area. So, in a sense, attribute information (in GIS you find this information mostly in tables) is reflected directly, visually, on the map. (This is somewhat true for general-reference maps as well, but in thematic maps this information is the primary reason for the existence of the map.)

Classification (or Categorization) and Symbolization

Basic statistics is one way of understanding bunches of numbers. Another way is to place things in categories, based on some pertinent number associated with the thing. For example, we could categorize (i.e., place into separate groups) the students in a class based on their heights. Put everybody less than 4 feet tall in group A. Those 4 feet or more but less than 4 feet, 3 inches belong in category B. Those who are 4 feet, 3 inches or more but less than 4 feet, 6 inches go in category C, and so on. So here's the general concept: We have k objects and we place each one of them in one, and one only, of n categories, where n is less than (or, in a trivial case, equal to) k. Almost without saying, the first category consists of a set of smallest numbers, the next category consists of the set of next smallest of numbers, and so on. That is, we could sort all the numbers from low to high, say, in a text string from left to right. Then each number in a given category would be adjacent to one or two other numbers in the category, and one could assign obvious breaks between categories. For example, if we wanted three categories, and the numbers were

1 1 2 3 3 4 5 7 8 10 10 11 12 15 19 19 22 23 25.

we might partition them as follows:

1 1 2 3 3 4 5	7 8 10 10 11 12 15	19 19 22 23 25
Category 1	Category 2	Category 3

Here the simplicity ends. The goal is to arrange things so that humans can best understand the nature of whatever is being studied. To do this, three fundamental questions must be addressed: (1) how many categories are appropriate for a given set of data, (2) what approach should be used to partition the objects into categories, and (3) how might the results be effectively presented or displayed.

Looking first at the issues raised by question (2), ArcMap allows you to use several techniques to plunk objects into categories, or classes, based on the numbers associated with the objects:

❑ Manual

❑ Equal interval

❑ Defined interval

❑ Quantile

❑ Standard deviation

❑ Natural breaks

We look at classification not only as putting numbers in categories but by classifying areas to which they relate. For example, in what follows I've made a cartoon grid of equally sized areas which has different populations of females, males, and wombats in each area.

—— **1.** Start ArcMap. From ___IGIS-Arc_*YourInitialsHere*\Trivial_GIS_Datasets add as data a shapefile named Classify_This.shp. Ignore a warning about projection. Open the attribute table. Sort the field FEMALES (which indicates the number of females in each rectangle) so it is in ascending order. How many records are there? _____ How many attribute values are in the vicinity of each of the following numbers? (*Hint:* Select records by dragging; read the bottom of the window.)

1,000   _____
5,000   _____
10,000   _____
15,000   _____
25,000   _____

Close the table. Clear any selections.

____ **2.** Label each polygon with the appropriate attribute value in the field FEMALES. Examine the map. Try to get an idea of where the clusters, and highest and lowest, female population values are. You may find this difficult.

____ **3.** Bring up the Layer Properties of Classify_This. Click Symbology > Quantities > Graduated Colors. Choose the Fields Value FEMALES from the drop-down menu. Find a color ramp that goes from light green to dark green, left to right.[4]

____ **4.** Press the Classify button. In the Classification window set the Classification method to Equal Interval. Choose seven classes. Observe the Classification Statistics. What's the minimum? _____ Maximum? _____

____ **5.** Look at the histogram. You see gray columns that represent the number of values that are in the vicinity of the abscissa (x-coordinate) value. Change the number of columns to 20. By looking at the vertical axis, indicate how many values (around 1003) are represented by the first column on the left. _____ How many by rightmost column? _____ (Notice the tick marks and reference numbers at the bottom of the histogram. They divide the range of the data set— that's 25034 minus 1003—up into intervals of fourths.)

____ **6.** Examine the Break Values box. These numbers—4436, 7869, . . . , 25034—represent the top end of each category. They divide the range of the data set (that's 25034 minus 1003) up into seven equal subranges. If you click one of these numbers, the break line between its class and the one above it turns red. At the bottom of the window, you can read the number of numbers (elements, as they are called) in that class. How many numbers are in the class whose top value is 7869? _____

By clicking the % button, determine the percentage of elements that fall in the first four categories. _____. Press % again to return to the numerical break values.

____ **7.** Make sure the Classification Method is still Equal Interval. Change the number of classes to 5.

User Selection of Classes

____ **8.** Notice each vertical blue line. The value at its top end indicates the break value as well. So the first category consists of 23 values, 19 of which are at the lower end and 4 of which are toward the upper end of the category. The range of this category is from 1003 to 5809.

---

[4]If you are more comfortable with a name for the color ramp right click on the ramp, uncheck Graphic View, and pick "Green Bright."

___ **9.** Use the mouse cursor to drag one of the lines. Notice that the Classification Method has changed to Manual—the method selected when the software realizes that you are going to be in charge of where the class breaks are. By moving the lines around, you can choose the Break Values (look in the pane to the right). Also observe, as you move a break line, that the number of values in the class *to the left of the break line* is dynamically displayed at the bottom of the window. Set the break values—the value at the high end of each category—between classes to *approximately* 2000, 6000, 11000, 16000, 26000. By doing this, you see that you are putting each natural group of values in a class by itself. Click OK. In the Layer Properties window, click Apply, then OK.

___ **10.** Examine the Table of Contents. The five categories into which you have put the populations of FEMALES are represented by increasingly darker colors of green. Note that the range of each class is shown next to its symbol. Look at the map. Notice how much easier it is to see the clusters of similar values.

___ **11.** In Layer Properties, click Classify again. By highlighting and typing in the Break Values pane, make the break values *exactly* 2000, 6000, 11000, 16000, and 26000. Click OK, click Apply, then OK again. Observe now the Table of Contents and the map. You see that you can easily make divisions between sets of values more comprehensible. The Table of Contents shows the low and high values in each category.

___ **12.** Classify the values again. Click the box Snap Breaks To Data Values. Each break line moves to the closest data value. Click OK, then Apply. Now the Table of Contents indicates the upper bound of each class. In the Layer Properties window, turn on Show class ranges using Feature Values. Click OK. Now the Table of Contents shows the lower and upper bound of each class, based on actual values in the class. See Figure 6-15***.

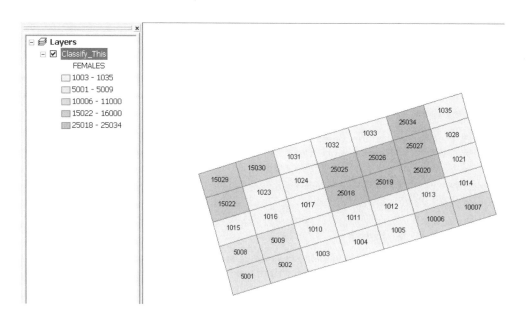

FIGURE 6-15

A More Careful Look at Equal Intervals

____ **13.** Classify the values again. Choose the `Equal Interval` method again. Notice that the break lines move back to the positions they had before. These positions are calculated so that each class occupies the same distance along the horizontal axis. What is that distance? _____. The calculation takes the form of "greatest value in the data set minus the least value, divided by the number of classes." Click OK, Apply, OK. Notice that the resulting map has five categories in the Table of Contents, but only four colors show up in the map, because one class is empty. What is the range of the class that has no values in it? _____ – _____ Notice that the values around 1000 and those around 5000 are lumped together. Look at the `Classification` window again to notice that one class contains two clusters of values that beg to be separated while another class is empty. This is a difficulty with equal intervals.

Defined Interval

____ **14.** Choose the method `Defined Interval`. This is an equal interval method, but instead of the interval size being chosen by the formula (High – Low)/N, it is chosen by the user and adjusted so the numbers look nice. Type 4000 in the `Interval Size` box and click the lower right corner of the histogram. Note that defining the interval determines the number of classes. How many? _____ Click OK. Click off `Show class ranges using feature values`. Click Apply, then on OK. Note the results on the map and in the Table of Contents.

Quantiles

The problem of empty classes may be solved by using another approach: quantiles. Quantiles places approximately the same number of values in each category.

____ **15.** Classify and select the `Quantiles` method. Change the number of classes back to 5 (this gives you quintiles; other popular classifications are quartiles, deciles, and percentiles). Click OK, Apply, and OK. Notice the map and the Table of Contents. Again the pattern looks somewhat similar to previous ones, but here the values around 1000 are broken into two groups. And two of the values around 5000 are lumped in with some 1000s, while the other two are in with the 10000s and the 15000s. Again, the divisions seem inappropriate.

Standard Deviation

The standard deviation method is not really appropriate for this set of data, because the values are not distributed "normally"—that is, in a bell-shaped curve. But just because a method is inappropriate doesn't mean we can't have the software apply it. It just means that we get bad results. (Garbage in, garbage out, as the saying goes.)

____ **16.** Select the `Standard Deviation` method and make the `Interval Size` one half of a standard deviation. Find and click boxes to show both the `Standard Deviation` and the `Mean` on the histogram. Click OK. Notice that the color ramp changes so that the possibilities reflect a change at the mean, rather than gradual change from one end to the other. Click Apply, then OK. Observe. You might find the display in the Table of Contents confusing. That's because the "–" sign is used both as a minus and as a separator.

Natural Breaks[5]

To understand "natural breaks" (also known as the Jenks method), you need to comprehend the idea of "variance from a mean." If you have a cluster of values, they have a mean, or average, value. If you take each value and subtract the mean from it, you get another set of numbers. The variance used by the Jenks approach is calculated by squaring each of these new numbers, taking their sum, and dividing by the number of numbers. You can see that the variance would be larger the further the original values were from the mean. So the variance is an indication of the dispersal from the mean of the set of values. If we had, say, five clusters of numbers, each with its own variance, we could add those five variances together to get a total variance. The idea behind the Jenks approach is to minimize the total variance by moving values around from cluster to cluster. At the end of the process, it should be the case that no value can be moved from one class to another without raising the total variance. This method works so well that it is the default that ArcMap uses. We illustrate it by using a different data set.

_____ **17.** In `Layer Properties > Labels`, change `Label Field` to WOMBATS (click Apply) and, under the `Symbology` tab, change the `Show` box to `Categories–Unique Values`. Select `Add All Values` Click Apply, then click OK. Look at the quilt. Open the attribute table and sort the WOMBATS field into ascending order. Note that there are seven groups of values (100s, 200s, and so on). Each group has five values, except for the first group, which has six, and the second group, which has four.

_____ **18.** Go to the `Layer Properties` window and pick `Quantities > Graduated Colors`. Now classify areas according to the number of WOMBATS contained in each, using seven categories and the equal interval method. Looking at the `Classification` window, you can note that in two categories two groups are placed together. Also note that there are two empty classes. Display the labeled map.

_____ **19.** Classify according to septiles (quantiles, seven classes), showing class ranges using feature values. Show the labeled map. Look at the ranges shown in the Table of Contents. What is the problem with this method? _____

_____

_____ **20.** Classify according to `Jenks`, using seven classes. Problem solved? _____

Normalization

Sometimes we are interested in seeing values that have been divided by other values. The first set of values is said to be "normalized" by the second set of values. For example, if we were interested in population density of counties, we would be interested in the value of population _divided by the area_ in which it resides. Or, as in the next step, we might want to display the ratio of females to males, so we would apply normalization.

_____ **21.** Normalize the number of females by the number of males, using `Layer Properties` and the Fields area under `Symbology`. Classify into seven classes using `Jenks`. Label the areas using the expression `[FEMALES] / [MALES]`. Look at the map. Notice that the labeling makes an unreadable mess of things, with ratios carried out to 14 decimal places.

---

[5]A method first suggested by Professor George Jenks, Department of Geography, University of Kansas. If you use the Natural Breaks method in ArcGIS check your answers carefully. I have noted differences between the results produced by Natural Breaks and other Jenks programs.

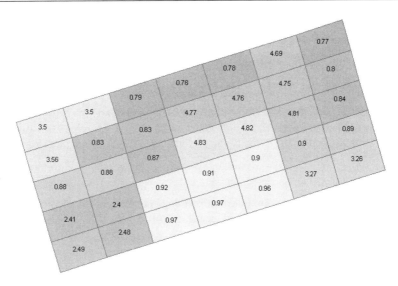

FIGURE 6-16

_____ **22.** Go back to the Labels tab and use the expression

Round ([FEMALES] / [MALES], 2)

which has the effect of rounding the ratios off to two decimal places, and should give you the result shown in Figure 6-16***.

Using Charts and Graphs

_____ **23.** *Examine another way of comparing values in an area:* In Layer Properties > Symbology > Charts, click Pie. Highlight FEMALES and move it to the right pane with the > button. Do the same with MALES. Make the symbol colors distinct: very light for females, very dark for males. Click Apply, then on OK. You see a disk for each area that indicates the ratio.

_____ **24.** Change the chart type to Bar/Column. Notice that you get a sense of the relative numbers of females to males, and also a sense of the total population of the area.

_____ **25.** Change the chart type to Stacked and observe.

Making a Layout

_____ **26.** Switch to Layout View. Make a layout with the pie charts, a title, the legend, your name, and the top portion of the attribute table. You can reposition the charts with the pan tool on the Tools toolbar. The Insert menu works to get you the legend, the title, and text for your name. To get the table onto the layout, on the Attributes Of Classify_This window, select Options > Add Table to Layout. Print the layout. Close ArcMap.

Exercise 6-3 (Short Project)

Comparing Data Sets: Medically Underserved Areas (MUAs) and Health Professional Shortage Areas (HPSAs)

The U.S. Government's Health Resources and Services Administration (HRSA) has two (of several) databases that relate to areas that do not have sufficient medical resources to meet the needs of the people in those areas. The databases relate to Medically Underserved Areas (MUAs) and Health Professional Shortage Areas (HPSAs). Information about them, and a lot of other interesting data, may be found at www.hrsa.gov.

Since each database deals with areas that have something in common, it would be interesting to contrast them.

_____ **1.** Start Catalog. Select: Edit ~ Search. You will search the Catalog. Look >in ___IGIS-Arc. You are looking for a Personal Geodatabase named HRSA_Geography. When the fields are all properly filled in press Find Now. Once the magnifying glass stops moving dismiss the search window. The shortcut to HRSA_Geography should show up in Search Results ~ My Search in the Catalog Tree. Right click on the database name and Go To Target. In what folder is the geodatabase? _____.

In the geodatabase you will find four freestanding pgdb feature classes. Name them:

_____

_____

_____

_____

_____ **2.** Click through each of the four, looking at the geography. Form an impression of the amount of area in each, compared to the others. Then look at each attribute table, writing the number of records for each data set next to the preceding blanks. Does the number of records seem to correlate with the areas involved? _____. What is the coordinate system used? _____.

_____ **3.** Copy the folder [____] IGIS-Arc\Health_Areas_Data to

___IGIS-Arc_YourInitialsHere. Make a folder connection to it. Work with this folder for the remainder of the exercise.

Suppose you are interested in the total of the medically underserved areas in the U.S. You might imagine you could run statistics on the Shape_Area column.

_____ **4.** Display the MUAs attribute table. What is the smallest (according to the values in Shape_Area) area? _____. The largest? _____. What are the units of these areas? _____. Not much help here.

A problem is that the "areas" are in decimal degree angular units—which are useless. To determine actual areas, you have to convert the MUAs feature class to a projection that preserves area. Recall that no projection preserves all of the four measures of major interest: area, shape, distance, and direction. One projection that works well for the United States for area is the Albers Equal-Area Projection.

_____ **5.** In ArcToolbox invoke the `Project tool` by selecting `Data Management Tools > Projections and Transformations > Feature`. Make the Input US_20021105_MUAs. Make the output name MUAs_Albers. For the `Output Coordinate system`, navigate to `North America Albers Equal Area Conic.prj`. (*Hint:* Select `Projected Coordinate Systems > Continental > North America`.). Run the tool.

_____ **6.** Start ArcMap. Add MUAs_Albers. Examine the data frame properties. What are the map units? _____. Make the display units `Miles`. What is the central meridian on which the projection is based? _____. What are the two standard parallels for the projection? _____, _____.

_____ **7.** Just to reinforce the idea that ArcMap does on-the-fly projections, add the US_20021105_MUAs feature class. Notice that it lies on top of the Albers projection. Flip back to ArcCatalog to see it in its decimal-degree form. Close ArcCatalog. Flip back to ArcMap. Remove the US_20021105_MUAs feature class.

Examine the attribute table of MUAs_Albers. As you determined previously, the units are meters. So the calculated areas are in square meters—not very useful numbers because they are so large. Suppose we prefer square miles. A square mile contains about 2,589,988 square meters.

_____ **8.** In the attribute table, under `Options`, click `Add Field`. Call it Area_SQMI. Make it a floating-point number. Click OK. Select the column and `Calculate Values`. Ignore the warning—if you make a mistake, you can just do the calculation over. What expression should you use in the `Field Calculator`? _____ Apply it. Run `Statistics`. Is the largest area about 13,678 square miles? (If not, look at your expression and correct it.)

_____ **9.** How many square miles are there of Medically Underserved Areas? _____ If the area of the coterminous ("lower 48") United States is about 3 million square miles, what percentage of the country fits the "medically underserved" designation, according to these data? _____

Geographically Comparing Two Data sets

_____ **10.** Create HPSAPrimaryCare_Albers from US_200301_HPSAPrimaryCare. Add it to ArcMap.

Since the two data sets (MUAs_Albers and HPSA_PrimaryCare_Albers) tend to get at the same concept—places where there isn't sufficient medical care—it might be interesting to see how they compare.

_____ **11.** Start visually by flipping between the two data representations, then turning them both on.

One measure of how the areas compare is whether or not there is overlap between them—that is, do they intersect?

_____ **12.** Under `Selection`, clear any selected features. Use `Selection > Select By Location`. Select features from MUAs_Albers that intersect HPSA_PrimaryCare_Albers. See Figure 6-17.

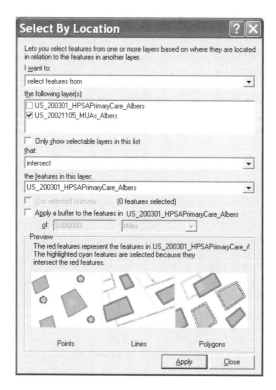

FIGURE 6-17

_____ **13.** Examine the resulting MUAs_Albers map. Verify that there seems to be a sizable amount of overlay. From the attribute table of MUAs_Albers, determine how many records are selected.

_____

_____ **14.** Determine how many MUAs_Albers polygons relate to the HPSA_PrimaryCare_Albers polygons in the following ways—being careful to clear selections before each test.

Intersect _____

Completely contain _____

Are completely within _____

Are identical to _____

Contain _____

Are contained by _____

_____ **15.** Reverse the questions posed previously. That is, determine how many polygons of HPSA_PrimaryCare_Albers relate to the MUAs_Albers polygons:

Intersect _____

Completely contain _____

Are completely within _____

Are identical to _____

Contain _____

Are contained by _____

That these numbers agree generally is comforting. That some don't agree exactly should give you concern. If the number of A polygons that completely contain B polygons is N, shouldn't the number of B polygons that are completely within A polygons also be N?

More dramatically, notice that the number of A polygons that are identical to B polygons is different from the number of B polygons that are identical to A polygons. Recall that previously I alleged that asking the question "are two floating-point values calculated by a computer exactly equal?" was folly. This is a good example of that principle. "Identical polygons" is a kind of "equal," and you notice it doesn't work.

For another shocking illustration of the effects extensive computation can have, do the following:

____ **16.** In ArcMap start a new map. Add as data the original, unprojected data sets:

US_20021105_MUAs and US_200301_HPSAPrimaryCare.

Rerun the tests that compare the two and fill out the tables—being careful to clear selections before each test.

How many MUAs_Albers polygons are selected?

Intersect _____

Completely contain _____

Are completely within _____

Are identical to _____

Contain _____

Are contained by _____

How many HPSA_PrimaryCare_Albers polygons are selected?

Intersect _____

Completely contain _____

Are completely within _____

Are identical to _____

Contain _____

Are contained by _____

Combining Demographic and Geographic Data

Suppose we have decided to open a day care center on Galbraith School Road in Knox County, Tennessee. We have a banker who tells us that a loan can be obtained if we can show that 25,000 people reside within 3 miles of the road on which we plan to build the center. We make a plan as to how to proceed: We will combine data obtained over the Internet, first from ESRI and second from the U.S. Bureau of the Census. The census data is in the form of a spreadsheet that contains information about the population in census blocks in our area of interest. It is strictly tabular data, with no geographic component.

The data we will get from ESRI consists of a modified TIGER street file and a modified TIGER census block file. (Actually this information came originally from the Bureau of the Census but has been converted to shapefile format for geographic display and analysis.) The project will involve combining these data sets so that we can determine how many people reside within 3 miles of Galbraith School Road.

This project has several stages. The initial ones (which are optional) require access to the Internet, getting data from the ESRI Web site, figuring out how to get data from the www.census.gov Web site, saving a table in Microsoft Excel spreadsheet format, modifying the spreadsheet, and saving the spreadsheet in dBASE format. There are a lot of steps, but the whole process is quite instructive, given that a lot of spatial data is available from ESRI and the Bureau of the Census.[6] Here are the steps:

❏   Get data from ESRI (optional for the exercise).

❏   Get data from the Bureau of the Census (optional for the exercise).

❏   Put the Census data in a Microsoft Excel spreadsheet (optional for the exercise).

❏   Modify Census data (optional for the exercise).

❏   Combine Census data and ESRI Data.

❏   Display the data.

❏   Analyze the data.

A warning: This is an exercise that requires a lot of pointing, clicking, and *waiting*. Its annoyance component is high and there are lots of places to go off the rails. As such, it is emblematic of many of the activities you will undertake when doing real GIS projects.

——   **1.** In ArcCatalog make a new folder named Day_Care_Data in

   ___IGIS-*Arc_YourInitialsHere*. Make a folder connection to it.

Getting TIGER-Based Street and Block Shapefiles from ESRI

The TIGER/Line feature files, as they come from the U.S. Bureau of the Census are what are called "flat files." That is, they have no hierarchy. Each file consists of a large number of records, each consisting entirely of text or numbers. Included in each record is a pair of coordinates, latitude and longitude, in

---

[6]If you don't have the access or tools to do these steps (or lack the inclination) you can skip them and still do the analysis part in ArcGIS by using data files in the [___]IGIS-Arc_AUX\DCD folder.

decimal-degree format. ESRI has taken these records and used the position information to create shape-files that users can download for free. As you might guess the files are large, so they are zipped up. The steps that follow indicate how you can obtain them. If you are unable to download and unzip the files you can use a workaround that will be prescribed shortly.

2. Point your Internet browser at[7] http://www.esri.com/data/download/census2000_tigerline/index.html

Under Free Download click Preview and Download. (You may have to register.) You should see a window with a graphic of the U.S. (Alaska and Hawaii will be somewhat out of place.) Select Tennessee (TN). Under Select by County pick Knox. Submit selection. Put checks by Census Blocks 2000 (1.2 MB) and Line Features — Roads (2.0 MB). At the bottom of the page click Proceed to Download.

After a brief wait you will be told that your data file is ready. Click Download File.

Here is a procedure that worked in Internet Explorer:

Click Save. In the Save As window navigate to

Day_Care_Data in ___IGIS-*Arc_YourInitialsHere*.

Keep the file name the same. Press Save. When the download is complete, close all browser functions.

Here is a procedure that worked in Firefox:

Pick Save to Disk. OK. In the Enter name of file to save to window leave the file name and type unchanged but browse to

___IGIS-Arc_YourInitialsHere\Day_Care_Data for the Save in field. Click Save.

When the download is complete, press Close. Then close all browser functions.

3. Examine the Day_Care_Data folder. You should find a ZIP file there with a name something like at_tigeresri1234567890.zip. Unzip the file, using either a commercially available unzipping program (like WinZip or PKZIP) or the unzipping facility in Microsoft Windows (right click the zip folder, Extract All). Unfortunately the structure of downloads from ESRI is somewhat like Russian nested dolls: You now have two additional zip files to contend with. Unzip the folder blk0047093 (the census block file) and you should get the three basic components of a shape file:

tgr47093blk00.dbf
tgr47093blk00.shp
tgr47093blk00.shx

(47093? Tennessee is state number 47, and Knox County is county 93 in the state, according to the government's FIPS classification).

---

[7]Internet sites change frequently so this procedure may not be quite right. However the fact that ESRI will provide TIGER-based data files for free is unlikely to change. So you may have to be diligent in finding them if this procedure doesn't work, but they will be there.

Also unzip the roads folder IkA47093.zip to produce tgr47093lkA.dbf, tgr47093lkA.shp, and tgr47093lkA.shx.

____ **4.** Back in ArcCatalog, refresh the Catalog Tree. Find the two shapefiles somewhere in the tree of Day_Care_Data. If necessary, move them directly under Day_Care_Data. Change the name of tgr47093blk00.shp to Blocks.shp. Change the name of tgr47093lkA.shp to Roads.shp. Look at both the tables and geography. Dismiss ArcCatalog.

Obtain Data from the U.S. Bureau of the Census

You are pretty much on your own in getting the census table because the site changes occasionally. But here is a procedure that worked at one time. If it doesn't work for you regard the following as hints. Or use the data in IGIS-Arc_AUX\DCD.

____ **5.** On the Internet browse to www.census.gov.[8] Click `American Fact Finder`. Digression: Click on `More Population Clocks`. What time is it? _____ What is the U.S. Population? _____. Determine what the world's population is. _____. Go Back to the `American Fact Finder` page of www.census.gov.

____ **6.** Click on `Data Sets`. Find and click on `Enter a table number`, type P1 in the box, and `Go`. Under `Choose a selection method` click `geo within geo`. Under `Show me all` click `Blocks`. In the `Within` box pick `County Subdivision`. For a state pick `Tennessee`. Pick `Knox County`. Pick `Knoxville CCD`. Wait. Pick `All Blocks` and `Add` it. Wait some more.

Next to Current geography selections click `Show Result`. Wait. In the `Print/Download` menu click `Download`. Under `Database Compatable` choose `Microsoft Excel (.xls)`. OK. Wait for a `File Download` window. Save the file named `output.zip` to Day_Care_Data. Wait. (I'll stop saying "wait" but you still have to do it.) When the download is complete open the folder. Right click on `output.zip` and `Extract All`, or use some other unzipping program. Finally you should have two text files and two .xls files as a result.

____ **7.** Before leaving the Census site look at the population clocks again. (You might have to refresh the page: F5.) What time is it? _____ What is the U.S. Population? _____. What is the world population. _____. How many people per hour are added to the U.S. population? _____. How many people per hour are added to the world population. _____. In 1950 the world population was about 2 billion. Now it is more than 6 and a half billion. We probably need to do something about the dramatically increasing population. Maybe using our GIS skills. Leave the Census site.

Converting the Census Data Spreadsheet to dBASEIV Format

The dBaseIV format is the one that goes with the TIGER shapefiles we downloaded earlier. We will be joining the TIGER data with the census population data.

____ **8.** With the computer's operating system find and pounce on the `dt_dec_2000_sf1_u_data1.xls` file. The spreadsheet should open. How many thousands of

---

[8]At the time this was written this was the approach to getting this sort of census data. You may have to make modifications but the data is almost certainly available somewhere on the census site.

rows are there in the spreadsheet? _____. That's about how many census blocks there are in the Knoxville CCD. (A CCD is a Census Collection District. In this case it is a subset of the Knoxville metropolitan area.) What are the five headings of the spreadsheet columns (row 1)? _____, _____, _____, _____, _____.

You are interested in the GEO_ID and the P001001 columns. The Geography Identifier is (a part of) the Block Number. The population column cells indicate the number of people living in the block.

____ **9.** Click the A1 cell, to make sure the spreadsheet is selected. Choose: `File > Save As`. Save it in `Day_Care_Data` with the name `Census1.xls`. Now Save As again, this time with the name Census2. (Whatever happens, whatever mistakes you make in the future with this data set, you certainly don't want to go through the download procedure again.)

When we convert the spreadsheet to a dBASE4 table, the top row becomes the headings and the remainder of the table go into the cells of the attribute table. Therefore row number 2 is in the way. So we will delete it.

____ **10.** Click on the 2 of row 2. Right click. Pick `Delete`.

The next step is to get a column we can use to join this information with the ESRI blocks shapefile table.

____ **11.** Working with `Census2.xls`, put your cursor on the divider that separates the A column head from the B column head and drag it to the right. (Clever of them to hide all those other digits in GEO_ID!) What we want are just the last four digits (L4D) of GEO_ID. So we will make column F our working `GEO_ID` and put the desired digits there. Click in cell F1. Type the name GEO_ID_L4D and press Enter.

____ **12.** Click in F2 and type:

```
= RIGHT(A2,4)
```

which invokes a spreadsheet function that grabs off the last four characters of the A2 cell and puts them in cell F2. Hit Enter. The number 1000 should appear in F2.

____ **13.** Click on cell F2. Press `Ctrl-C` to copy the value to the clipboard. Cell F2 should be activated. Click on cell F3. With the vertical slider bar slide the rows so you see the last one (6022). Hold down `Shift` and click cell F6022. Press Ctrl-V to paste the formula in that is in F2 into all the selected cells of column F. Check the results. Each cell F value in a given row should equal the last four characters of the GEO_ID in the A column of that row. Press Esc to clear the activation of F2.

____ **14.** *Click in the* A1 *cell.* Save the file with `File > Save As`, this time using the name Census3.xls.

Our spreadsheet is now ready to be converted to an attribute table.

____ **15.** Click `File > Save As`. Set things up so the file goes directly into Day_Care_Data. Type Census4 in the `File name` box. From the `Save as type` dropdown menu pick

DBF 4 (dbase IV)(*.dbf). The `Save As` window should look something like Figure 6-18.

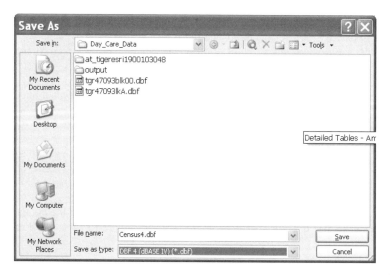

FIGURE 6-18

Ignore the Microsoft Excel warning about losing features. Click Yes. Dismiss Microsoft Excel—no need to save changes, since you just did.

Now you have everything you need to determine how many people live within 3 miles of the road of the proposed day care site. But since Web sites change and data sets get updated things may not have gone smoothly up to this point. Again, that's the real practice of GIS. Frustration and workarounds are part of the job. But to assure that the rest of the project goes smoothly, and that we all get the same answers, I'm going to ask you to delete some of the work you have done and substitute some data files that I know will work.

_____ **16.** Start ArcCatalog. From

_____IGIS-Arc_*YourInitialsHere*\Day_Care_Data

delete Roads.shp, Blocks.shp, and Census4.dbf if they exist there.

From [_____] IGIS-Arc_AUX\DCD copy Roads.shp, Blocks.shp, and Census4.dbf to

_____IGIS-Arc_*YourInitialsHere*\Day_Care_Data.

_____ **17.** Launch ArcMap. From

_____IGIS-Arc_*YourInitialsHere*\Day_Care_Data

add Roads.shp, Blocks.shp, and Census4.dbf.

If you slide the cursor around the map you notice that the coordinates are in degrees, minutes, and seconds. The display would look more like the real world if we used a Cartesian projection.

_____ **18.** Right click the data frame and pick Properties. Change the Coordinate system to Predefined > Projected Coordinate Systems > State Plane > NAD 1983 (Feet) >

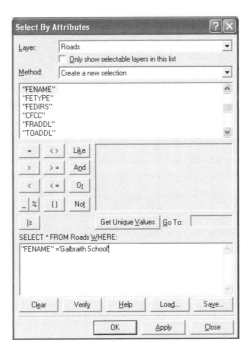

FIGURE 6-19

Tennessee FIPS 4100.
Ignore any Warnings. Notice now the coordinates are in Feet. Be aware that, as you recall, you have only changed the display, not the underlying data.

____ **19.** Using `Selection > Select By Attributes` look for an FENAME of 'Galbraith School' in Layer Roads. (Be sure you use Galbraith School as the FENAME. Don't include the word Road. Don't leave off the word School (there is a Galbraith road—it's not the one you want.) It is probably best not to ask for Unique Values since there are a ton of them. Waiting for them to be enumerated and then finding the right one is more work than typing. But be careful of the quotes. See Figure 6-19.

____ **20.** Click Apply, then Close. You should see a little patch of selected road on the map. Choose `Selection > Zoom To Selected Features`, and then zoom out some. Have ArcMap label the roads with FENAME. See Figure 6-20***.

Assessing What We Have and What We Need to Solve the Problem

What we are going to do is find the Blocks polygons that lie within three miles of Galbraith School Road. But since our objective is to find population we have to first join the census table to the Blocks table, so that the number of people in each census block will be available to us. To join these two tables we need to find a key field in each table whose values serve as block identifiers. In the census table that key field is GEO_ID_L4D. In the Blocks table the key field is BLOCK2000.

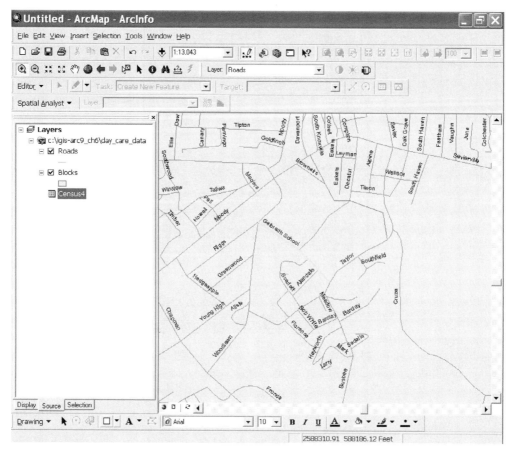

FIGURE 6-20

_____ **21.** Look at the table of Blocks. In the column BLOCK2000 we have the four-digit block number. In the Layer Properties window check out Fields. Note that BLOCK2000 is a String field of Length 4.

_____ **22.** Look at the table Census4. In the GEO_ID_L4D column you see four-digit block numbers. In the Layer Properties window check out Fields. Note that GEO_ID_L4D is a String field of Length _____. (We will hope that it will match up with BLOCK2000 string field of length 4 when we join the tables.) Also we have P001001 which is the population of each block. We need to join the Blocks table and the Census4 table so we can combine the geographic information of the shapefile with the tabular information in the census file. Dismiss both tables.

_____ **23.** Right click on Blocks and choose Joins and Relates > Join. Step 1 should be BLOCK2000. Step 2 is Census4. What's Step 3? _____. OK. Read the Create Index box if it shows up.

Since we are dealing with large tables let's take ArcMap's word for it that indexing the join field will improve performance. Click Yes.

____ 24. Right click Blocks, open, and examine the table. The important thing to note is that we now have P001001 as part of the shapefile table. Dismiss the table.

____ 25. Launch the Identify tool, set it to Blocks, and check out the two blocks northeast of Galbraith School Road. What are their populations? _____, _____. Dismiss Identify Results.

____ 26. Start the Select Features tool and, using Shift-Click, select the same two blocks. In the Attributes of Blocks window Show Selected Records. Verify the populations.

____ 27. *Clear all selections.* (Otherwise when we export the table, as we will below, only the selected records will be exported!!) Dismiss the table.

You might recall that, even though you have the shapefile together with the table, the connection between the original shapefile table and the census data is pretty tenuous. All you would have to do is leave ArcMap and, poof—no join. So set this shapefile and its expanded table in concrete.

____ 28. Right click Blocks and go Data > Export Data. From the Export Data window browse to ___IGIS-Arc_*YourInitialsHere*\Day_Care_Data. In the Saving Data window change the Name Export_Output to Blocks&Pop.shp. Make sure you Save as type Shapefile. Click Save. Click OK. Add the exported data to the map as a layer. Remove Blocks. Remove Census4. Check out the new table of Blocks&Pop. How many census blocks are there in the Knoxville CCD? _____

Converting the Relevant Files to Cartesian Coordinates

You are going to be working with distances, and that suggests that you should probably convert our files, which are stored in decimal-degree format, to a coordinate system where you can be surer that distance calculations are done simply and properly. You are displaying in the State Plane system, but for the sake of neat and tidy convert the basic data and remap it.

____ 29. Ask for a new map in ArcMap, using the New Map icon. Don't save the previous map.

____ 30. In ArcToolbox invoke the Project tool in Data Management Tools > Projections and Transformations > Feature. Make the Input Blocks&Pop.shp. Call the Output Blocks&Pop_TN_SP.shp, because we plan to put the data in the Tennessee State Plane Coordinate System. For the Output Coordinate System navigate to the

NAD 1983 StatePlane Tennessee FIPS 4100 (Feet).prj

coordinate system. Add. Apply. OK the Project window. After the tool executes the new layer is automatically added to the map.

____ 31. Use this same projection to Project the Roads shapefile to Roads_TN_SP.shp and notice that this layer is added to the map. Hide ArcToolbox.

Finally

With all this preparation you may have forgotten the goal of this exercise: to determine the population that resides within 3 miles of Galbraith School Road. Now that everything is in place we can use Select by Attribute to select 'Galbraith School' road. Then we can use Select by Location to find the census blocks that are within 3 miles.

32. From the `Selection` menu pick `Select By Attributes`. As you did before, select Galbraith School. Use the measure tool to determine how long the road is. _____.

33. From the `Selection` menu pick `Select by Location`. We want to select features from `Blocks&Pop` that are within a distance of the *selected* features of `Roads_TN_SP`, applying a buffer of 3.0 Miles to the selected roads features. Apply. Close the `Select By Location` window and zoom to selected features in the `Blocks&Pop` layer.

34. Open the `Blocks&Pop` table. How many records were selected? _____. Run statistics on the P001001 column. Double-check that the number of records you are calculating from (Count) is the number selected. What is the population in the selected census blocks? _____ Minimize the table.

Since the population is well over the 25,000 that the banker required we are elated. So we invite her over to see the results. After being suitably impressed with the operation and the software, she goes, "Gee, knowing that area as I do, the spread of census blocks looks like more than 3 miles." With slight trepidation, immediately confirmed, we take a measurement across the area.

35. What is the greatest distance across the selected blocks? _____ feet. How many miles is that? _____

The problem here appears to be that any part of any census block that was within the 3-mile radius causes that entire block to be selected—thus probably overestimating the population that lies within the buffer.

As with most operations with this software, there is more than one way to achieve a result. We can make a graphic circle with the Drawing toolbar and then use Select by Graphics.

36. Restore the `Blocks&Pop` table. With `Options` clear the selections. Dismiss the table. Make sure the four segments of Galbraith School Road are selected. Select them again if necessary. Change the `Display Units` of the data frame to `Miles`. Zoom to the approximate area of about 10 miles across in the area around the road.

37. Make sure the `Drawing` toolbar is available. On the `Drawing` toolbar, toward the left, you can find a drawing icon with a triangle next to it indicating a dropdown menu. Choose the circle. Place the cursor crosshairs in about the middle of the selected line of Galbraith School Road. Drag a circle of *Radius* 3.00 miles.

38. Under `Selection > Options > Interactive Selection` pick `Select features completely within the box or graphic(s)`. Click OK.

This will satisfy the banker's objection that areas of census blocks that don't lie within the three-mile radius are being selected.

39. Under `Selection > Set Selectable Layers` pick only `Blocks&Pop_TN_SP` and close that window.

40. With the `Select Elements` tool, click on the circle to select it. Handles and a boundary should appear. Under the `Selection` menu pick `Select By Graphics` and wait. The boundaries of the selected blocks should appear shortly. Zoom in on an area in the northwest where the circle meets the rest of the map. Assure yourself that no census block has any part outside the circle.

41. Now open the `Blocks&Pop_TN_SP` attribute table. How many records are selected? _____. Run statistics on the selected records. What is the population within the blocks that fall completely inside the 3-mile circle. _____.

Whoops. Now we don't have the 25,000 people we need. After a brief period of despair we re-examine the requirement. It says "a loan can be obtained if we can show that 25,000 people reside within 3 miles of the road on which we plan to build the center." Since we centered our circle around the midpoint of the road, maybe we can get a few more blocks by looking at two circles, each centered at opposite ends of the road. (If this doesn't work we can argue that while our first method overestimated the population, the current one probably underestimates it.)

42. Click the circle with the `Select Elements` tool. Press the `Delete` key to make it go away. Under `Selection` click `Clear Selected Features`. Select the Galbraith School Road segments again and zoom to leave about five miles all the way around them. Click the circle icon and again draw a 3.00 mile radius circle, but center it on the southeast end of the road. Then click the circle icon again and drag a 3.00 mile circle centered on the northwest end. Pick the `Select Elements` tool. Hold down `Ctrl` and click within each circle so that they are both selected.

43. Set Selectable Layers to be only `Blocks&Pop_TN_SP`. `Select By Graphics`.

44. Now open the `Blocks&Pop_TN_SP` attribute table. How many records are selected? _____. Run statistics on the selected records. What is the population within the blocks that fall completely inside the pair of 3-mile circles? _____. Call the banker back!

45. Close ArcMap, saving the map as `Day_Care_YES.mxd` in Day_Care_Data.

Exercise 6-5

Determining Proximity of Points to Points and Lines

For this exercise we return to our fictional islands of Chapter 5. You may recall that you digitized them in

___IGIS-Arc_*YourInitialsHere*\Digitize&Transform\Islands.mdb\UTM_Zone_2

The data set became North_Islands_Polys. You stored a map named North_Islands_map3. Houses are being built on the square island. A power generation plant is being constructed on the northwest corner of the lake and a power line run from the plant around the island. Also, wells are being dug to supply water to the houses. These feature classes are called Houses, Power_Line, and Wells. In order to bring the appropriate lengths of electrical wire and water pipe to the island, we want to know the distances from the houses to the power lines and the distance from the houses to the wells. This information will be stored in the attribute table of the Houses feature class.

_____ **1.** Start ArcCatalog. From

[__] IGIS-Arc_AUX\D&T\Islands.mdb\UTM_Zone_2 copy the PGDBFC named Power_Lines to

___IGIS-Arc_*YourInitialsHere*\Digitize&Transform\Islands.mdb\UTM_Zone_2.

Now also copy the feature classes Houses and Wells. Close ArcCatalog.

_____ **2.** Navigate to the map North_Islands_map3.mxd in

___IGIS-Arc_*YourInitialsHere*\Digitize&Transform. Double click the name to start ArcMap. Zoom in on the square island.

_____ **3.** Because of your previous copying actions Houses, Power_lines, and Wells are feature classes in

___IGIS-Arc_*YourInitialsHere*\Digitize&Transform\Islands.mdb\UTM_Zone_2.

Add them as data. Show the wells as green circles of size five. Accept ArcMap's choice of symbol for the power line. Change the color of the houses to black. Looking at the attribute tables: How many houses are there? _____. How long is the power line? _____. How many wells are there? _____.

_____ **4.** Examine the attribute tables of houses and wells. Note that both are dull, as befits basic point feature class tables.

_____ **5.** Measure the diagonal distance across the island. Obviously, the distance from any feature to any other feature cannot be greater than this number. Write it here. _____.

_____ **6.** Open ArcToolbox. Use the Index button to Locate the Near (analysis) tool. What is the path to it?

ArcToolbox > _____.

_____ **7.** Right-click the tool and click Open. Show Help for the Near tool, if it is not already there. Click on Help within the Help panel. Expand and read the Usage Tips. Dismiss the ArcGIS Desktop Help window.

_____ **8.** For Input Features, click the drop-down menu arrow (since the layers are already represented in ArcMap, they will appear) and choose Houses. For Near Features, pick Wells. For the search radius, type the number you found previously for the diagonal distance across the island. Check to see that the units are meters. Click the boxes that will specify the location of the well and the angle from the house to the well. Run the tool. Close it when the Close button appears. Dismiss ArcToolbox to get more room on the screen. Make sure ArcCatalog is closed.

_____ **9.** Again examine the attribute table of Houses. What are the field names?

_____, _____, _____, _____, _____, _____.

___ 10. Run statistics on NEAR_DIST. What is the total number of meters of pipe that would be required to connect each house to its nearest well? _____.

___ 11. Label each well with its OBJECTID. Label each house with its near neature ID (NEAR_FID) attribute value. "Eyeball" the results to verify that correct houses were assigned to wells. Label each house with its NEAR_DIST. Use the measure tool on a couple of house-well pairs to verify that the Near command got it right in the attribute table.

___ 12. Label the houses with NEAR_ANGLE. Check to see that things look right. The angle given is the angle from the house to the well, using the form given by the help file that you read. Realize that the angle given is a modification of the "math" version (0 degrees is the x-axis, the angle increases counterclockwise), rather than the compass version (0 degrees is the y-axis, the angle increases clockwise). Just one of the wrinkles one encounters in knowing about GIS—the kind of thing that gets you paid the big bucks when you graduate—unless you go into teaching, that is.

___ 13. Extra credit problem: Could you reduce the cost of putting in water lines by using mains (larger pipes to some intermediate junction) and then smaller pipes from there? What is the shortest length of pipe you can find that would do the job? _____. Of course, this makes the problem more complicated, since you have to figure in the greater cost of the larger pipe, the additional connections, and so on.

___ 14. Run the Near tool again, this time determining connections between the houses and the power line. (Re-running the tool will overwrite the previously stored attribute values NEAR_DIST, NEAR_X, and NEAR_Y, and so on; there is no warning to this effect.) What length of electric cable would be required? _____.

Exercise 6-6 (Review)

Checking, Updating, and Organizing Your Fast Facts File

The Fast Facts File that you are developing should contain references to items in the following checklist. The checklist represents the abilities to use the software you should have upon completing Chapter 6.

Important note: This checklist is on the CD-ROM that accompanies the book. It is available in Microsoft Word format. Rather than typing or writing by hand the text that follows, you can copy and paste it into your Fast Facts File from the CD-ROM file.

___ Double-clicking a toolbar's left end

___ Double-clicking a toolbar's title

___ To get a list of all toolbars

___ Shape_Length is

___ Shape_Area is

___ Pointing at records

___ Two windows for selecting records are available at

___ When a record is selected

___ When a feature is selected

___ Once some records are selected, other records, which are not selected, may be added to the selected set by

___ In selecting records, logical and arithmetic expressions

___ In the Options window, the Statistics button produces

___ When some records are selected, Statistics are calculated only on the

___ It is not usually necessary to type all of an expression because

___ When a user is selecting records, all the values of a variable may be seen by

___ When a polygon feature is selected, it is shown on the map by

___ When a polygon feature has islands within it, it is displayed

___ The Options button on a table also provides these capabilities

___ The difference between the Select Elements button and the Select Features button is

___ Selection of features based on two different layers is accomplished by

___ A graphic representation of Select by Location is available

___ Plus and minus signs in the table of contents

___ A data frame is

___ A layer file

___ To get layer properties

___ To get data frame properties

___ Under the Symbology tab, the user the following options exist

___ To change the field used for labeling

___ To symbolize features based on values in an attribute field

___ To label features with values from more than one attribute

___ ArcGIS has a number of ways to put data into classes:

___ The windows involved in the classification process are

___ A histogram shows

___ When selecting a ramp a user may see either

___ The Break Values box shows

___ On-the-fly projecting means

___ Select by Location, used with intersect, produces

___ TIGER/Line street files

___ To convert a spreadsheet to database format

___ To make a join permanent

___ TIGER/Line Files usually have ____ coordinates

___ Graphics may be used to select features by

___ The Jenks method

___ Normalization means

___ To round off numbers

___ Three sorts of graphs that ArcMap provides are

___ To add a column to a table

___ To calculate values in a column

___ The Internet site of the Bureau of the Census is

___ To determine distances between various features, use

Creating Spatial Data Sets Based on Proximity, Overlay, and Attributes

OVERVIEW

IN WHICH you learn three of the fundamental tools of GIS analysis and use them to solve problems. Also you are introduced to the ESRI model building software

Generating Features Based on Proximity: Buffering

Proximity is a word that implies nearness in terms of physical distance. It is not quite the inverse of distance, but greater proximity implies smaller distance. Proximity is a concept that strongly impacts our lives and activities. We, along with most of the animals of higher intelligence, innately understand the idea. We want to be close to pleasant and useful things, or to those things we must access on a regular basis, like friends, places of employment, shopping. We want to be far from those things that are unpleasant or noxious, like smelly dumps or plants, irritating people, or dangerous environmental conditions. Much of the law regulating land use is written with the concept of proximity in the background—it sets limits of acceptable proximity, usually as a threshold distance. An example would be a law that says a liquor store may not exist within 250 feet of school grounds.

The analytical tools of GIS that implement the concepts of proximity create what are called buffers. To create a buffer, a set of features, say X, is specified. A distance, say Y, is specified. The software then generates a buffer: a set of areal features that have the property that every point on or within their borders lies within the specified distance of the original features. Formally, the buffer consists of the locus of points whose distance from each of the features in X is less than or equal to the specified distance Y.

Because we have such an intuitive idea of proximity, understanding what buffers are is not nearly as fearsome as the definition suggests. You encountered the buffer concept in the first exercise of this book, when you solved the Wildcat Boat problem by manual means. Recall that the testing facility had to be within 300 meters of a sewer line and could not be less than 20 meters away from streams. That is, the facility had to be within a buffer of the sewers and outside a buffer of the streams.

Buffering Points

Figure 7-2 shows what a buffer around the set of point features in Figure 7-1 would look like.

You don't see the point features in Figure 7-2 because the *point features are not part of the buffer.*

FIGURE 7-1   Points to be buffered   **FIGURE 7-2   Buffers around points**

In addition to the (geo)graphic portrayal of the buffered areas, an attribute table is created. Since the production of areas inside the buffer can also create polygons that are outside the buffer, the attribute table should contain a field that indicates the status of each polygon: inside or outside. If the buffering is done with coverages, the default item is called INSIDE. It will have a value of 100 for polygons inside the buffered area and 1 for those outside. The tool you will use for buffering with geodatabase feature classes handles the issue of what's inside and what's outside in a somewhat different way: A field is generated that contains the distance used in creating the particular polygon. If that distance isn't greater than zero, the area is not within the buffer.

Buffer Lines and Polygons

Line features may be buffered as well. The area of the polygons generated is that covered by the locus of points that lie within the threshold distance of any point on the line. Examine Figure 7-3, and then Figure 7-4.

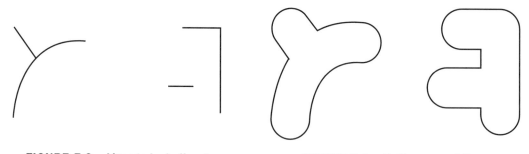

FIGURE 7-3   Lines to be buffered   **FIGURE 7-4   Buffers around lines**

You could now see that buffering polygons is merely an extension of the preceding ideas. Again, the polygons that are buffered, although they are areal features, are not present in any way in the buffer. Examine Figure 7-5, and then Figure 7-6.

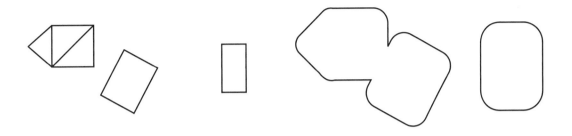

FIGURE 7-5   **Polygons to be buffered**   **FIGURE 7-6**   **Buffers around polygons**

There are many variations on the buffering theme. You can buffer only one side of a line. You can ask that line buffers have flat ends, in which case areas past the end of the line are not included in the buffer. Very importantly, you can buffer features by different thresholds. The thresholds are taken from information in the attribute table of the feature being buffered. For example, suppose you wanted to establish "no smoking" zones around wells. Perhaps your features consist of water wells, oil wells, and natural-gas wells. You could set things up so that water wells were buffered by 30 meters, oil wells by 100 meters, and gas wells by 400 meters.

Generating Features by Overlaying

Pick any point of Earth's land area. A large number of attributes might describe that point. It may be within the area owned by a person. The point will be in some country or other. It might be part of a floodplain. It will be associated with a certain soil type. The representation of the point in a GIS might indicate that it is part of a roadway or stream. Its GIS representation show it to be an oil well.

In doing GIS analysis, we are frequently interested in combinations of attributes that relate to points, lines, or areas. For example, we might be interested in those areas that have a certain land use zoning and that are also for sale. Suppose we have one GIS layer whose polygons show zoning and another, different layer that shows properties for sale. By a process called overlaying, we can create a third layer from which we can identify those properties with the zoning we want that are also for sale.

The term "overlay" comes from the physical process of laying one map of transparent material on top of another—say, on a light table—and examining the effect of the combination. To continue the preceding example, suppose you wanted to buy some property to start a business. The zoning had to be "B1" and, of course, the property had to be for sale. You might take a clear Mylar map of the area and use a marker to blacken all of the zones that were not B1. On a second such map you might blacken all of the properties that were not for sale. If you then laid both maps on a light table (assuming that they covered the same area, were at the same scale, were based on the same geographic datum, had the same projection, and so on), light would shine through the areas that were suitable for your intended business. With GIS software we can simulate this activity, with much greater efficiency and garnering much more information in the process. You encountered the essence of overlay, working the Wildcat Boat problem, when you found combinations of suitable soils and land use.

Let's look first at the idea of overlaying a polygon feature class with another polygon feature class. We get two types of results: one (geo)graphic and the other tabular.

Figure 7-7 and Table 7-1 depict a feature class named A, consisting of four polygons. In feature class A there is an attribute named Zone which has values p, q, r, and s, as shown on the diagram and in the table. (In Table 7-1, values and columns not germane to the discussion have been omitted. These include feature identifier numbers, which would be unique. Also, no values have been given for Shape_Length and Shape_Area, since these are not important to our discussion.)

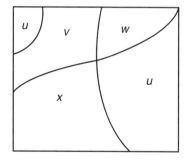

FIGURE 7-7   Feature class "A" with ZONE attribute

TABLE 7-1

Shape_area	Shape_length	Zone
		p
		q
		r
		s

Figure 7-8 and Table 7-2 depict feature class B, of five polygons. Feature class B has an attribute named "Jurisdiction." Values of Jurisdiction are u, v, w, and x. Note that two different polygons are characterized by u.

FIGURE 7-8   Feature class "B" with JURISDICTION attribute

TABLE 7-2

Shape_area	Shape_length	Jurisdiction
		u
		v
		w
		x
		u

406

Suppose feature classes A and B are overlaid to produce polygon feature class C. Then C would consist of ten polygons, as shown in Figure 7-9. Each of those polygons would be homogeneous in pairs of attribute values, as illustrated. The attribute table, Table 7-3, of C would, of course, consist of ten records. It would contain the attributes, Zone and Jurisdiction, *from each of the constituent feature classes.*

This is the fundamental essence of polygon overlay. You generally get more polygons and more attributes than in either of the input feature classes. You can identify each polygon by the attributes in each of the input feature classes.

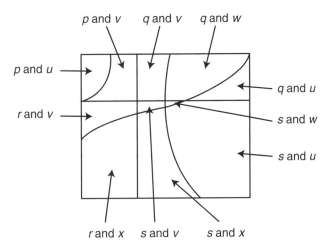

FIGURE 7-9   Overlay of "A" and "B"

TABLE 7-3

Shape_area	Shape_length	Zone	Jurisdiction
		p	u
		p	v
		q	v
		q	w
		q	u
		r	v
		r	x
		s	u
		s	v
		s	x

Overlaying with Line and Point Feature Classes

When you do an overlay, one of the two feature classes or coverages must be of the polygon type. The other may be point, line, or, as you saw above, polygon. What does it mean to overlay a polygon feature class with a point feature class? Basically, the change is only the tabular output (although some points may be omitted if they lie outside the scope of all polygons, depending on which command you use).

Overlaying Point Features

Let's suppose that some point features are parking meters, each of which has a number. See Figure 7-10. In the attribute table, the field Meter_number records this number. See Table 7-4. Also assume that the polygon feature class A examined previously depicted zones of the city, and it was desired to know which parking meters fell into which zones, so that the income from each meter would go to the correct budget.

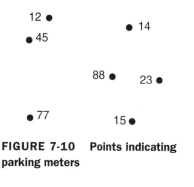

FIGURE 7-10   Points indicating parking meters

TABLE 7-4

Meter_number
12
45
77
88
23
15
14

Now if the point feature class is overlaid on the polygon feature class A (see Figure 7-7), the graphical result of the new point feature class will be the same as Figure 7-11. However, the attribute table of the new point feature class will look like Table 7-5. Figure 7-11 indicates why, with the dashed lines of Figure 7-7.

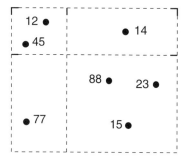

FIGURE 7-11   Points overlaid with feature class "A"

TABLE 7-5

Meter_number	Zone
12	p
45	p
77	r
88	s
23	s
15	s
14	q

Overlaying Line Features

A similar effect occurs if lines are overlaid in polygons. The records associated with the lines in the output table acquire the attributes of the polygons in which the lines lie. The output line feature class usually will consist of more lines than the input because when a line from the input crosses a polygon boundary it is cut, forming a line on either side.

Suppose that we have three roads in the area of feature class A. See Figures 7-12 and Table 7-6. For purposes of maintenance we would like to know what zone each stretch of road falls into.

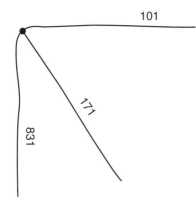

TABLE 7-6

Shape_length	Highway_number	Lanes
	831	4
	101	4
	171	2

FIGURE 7-12   Lines indicating roads

The overlay would produce a line feature class with seven lines as illustrated in Figures 7-13 and Table 7-7. The Shape_lengths would be the lengths of the newly created lines.

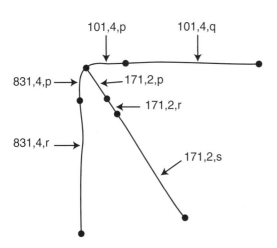

TABLE 7-7

Shape_length	Highway_number	Lanes	Zone
	831	4	r
	831	4	p
	171	2	p
	171	2	r
	171	2	s
	101	4	p
	101	4	q

FIGURE 7-13   Lines overlaid with feature class "A"

Spatial Joins in General

The term "spatial join" has been coined to describe overlay and similar processes. Basically, spatial joins share the idea of taking two or more spatial data sets, each with associated attribute tables, and creating another data set. Both the (geo)graphics and the attribute tables in the new set differ from the others. You find the tools to perform these actions in two places in ArcToolbox: Analysis Tools (for work with feature classes in general) and Coverage Tools > Analysis.

Union combines feature classes in such a way that the extent of the derived data set includes the extents of all the constituent data sets. The idea is that, in the union of P and Q to make R, say, the territory covered by R is the territory covered by P and also by Q including any overlapping territory. In logic, the union of two sets (A and B) is the set of all objects in A or B or both. If A is all red squirrels and B is all squirrels with fuzzy tails, then A union B is the set of all squirrels that are either red or have fuzzy tails or both. The union is equivalent to the logical *inclusive* OR. (The logical *exclusive* OR—XOR—would consist of those squirrels that were red or had fuzzy tails, but not those that were red and also had fuzzy tails.)

Intersect combines feature classes in such a way that the extent of the derived data set includes only the overlapping area of the constituent data sets. The idea here is that, in the intersection of S and T to make U, say, the territory covered by U is only the territory covered by P that is also covered by Q. In logic, the intersection of two sets (A and B) is the set of all objects that are in both A and B. If A is all red squirrels and B is all squirrels with fuzzy tails, then A intersect B is the set of all squirrels that are red and have fuzzy tails as well. The intersection is equivalent to the logical AND.

A number of other spatial join tools are available in the tool chests:

Identity

Erase

Update

Clip

Split

Symmetrical Difference (the union with the intersection taken out)

You will work with some of these in the Step-by-Step section of this chapter. Now we turn to what you can do once you have applied these buffers and spatial joins.

Deriving Geodata Sets by Selecting Attributes: Extraction

You have perhaps buffered some points and lines, and have the data sets to prove it. You have overlaid these buffers on some other polygon data sets. What you now have is a feature class that has dozens, hundreds, maybe even thousands of polygons. That feature class also has an attribute table with a list of fields as long as your arm, since it consists of almost all the attributes of the constituents tables. Now it is time to make a geodata set that consists only of those polygons that meet your requirements. You do this by writing a query that will be processed by the Select tool, within the Extract tool chest. Selection By Attributes is nothing new to you. The difference here is that you will make an entirely new geodatabase feature class, coverage, or shapefile based entirely on the selected features and records.

Let's pretend we are trying to find the appropriate polygons for the Wildcat Boat problem. Without being rigorous, consider a query like the one that follows, made to the composite table of land use, soil suitability, sewer buffers, and stream buffers.

Use Extract to make a feature class of the polygons where:

```
LANDUSE = "Brushland"

AND

SOIL_SUITABILITY = "Fair" OR SOIL_SUITABILITY = "Good"

AND

DISTANCE_TO_SEWERS <= 300

AND

DISTANCE_FROM_STREAMS > 20
```

In summary, The EXTRACT Wizard produces a data set B from a data set A by extracting features based on their *attribute* values. EXTRACT works very much like building a query to select records in a table (e.g., `AREA > = 500 AND COLOR = 'Green'`).

But instead of simply highlighting records in a table (and the corresponding features on the map), EXTRACT creates an entirely new data set (B) consisting of selected records.

The way to use EXTRACT is to construct, by repeated overlays (using ArcToolbox), a data set (A) that contains all the features that the overlay process produces, and then to extract just those features that meet your criteria, making data set B.

The concepts in this Overview pretty much cover the sort of geographic data analysis that is created by deriving data sets from other data sets. There are, of course, a lot of data management commands that go along with this form of analysis. And there are several other types of GIS analysis—geostatistical for one, network for another, covered in Chapter 9.

In the following Step-by-Step section, you will apply the buffer, overlay, and extraction principles and tools.

Finally, in the Step-by-Step section of this chapter I introduce the Model Builder, which is one way to link and automate the processes in GIS analysis. By building a model:

❑ You can document the flow through the processes of a project.

❑ You can modify those processes easily, change processes, change the inputs to the processes, and so on, in a friendly graphical environment.

❑ You can have the computer repeat some or all of the processes with trivial effort on your part.

Creating Spatial Data Sets Based on Proximity, Overlay, and Attributes

————— **Open up your Fast Facts text or document file.**

Exercise 7-1 (Warm-up)

Making a Trivial Buffer Around a Trivial Coverage

————— **1.** In ArcCatalog make or confirm a folder connection to

___IGIS-Arc_*YourInitialsHere*\Trivial_GIS_Datasets.

Use ArcCatalog to make a folder named Buffer&Overlay in

___IGIS-Arc_*YourInitialsHere*.

Also make a folder connection to the Buffer&Overlay folder.

————— **2.** Launch ArcMap. In the folder

___IGIS-Arc_*YourInitialsHere*\Trivial_GIS_Datasets

you will find a line coverage named TICTACTOE. Display the arc component and the node component of this coverage in ArcMap. Set the map units and display units to Meters. What is the width of the coverage? _____ meters.

————— **3.** In *ArcToolbox* find the Buffer tool in Coverage Tools > Analysis > Proximity. Use that Buffer tool to buffer the lines (arcs) of the TICTACTOE coverage, making a coverage named TICTACTOEBUF in

___IGIS-Arc_*YourInitialsHere*\Buffer&Overlay.

The Feature Type should be LINE. Make the Buffer Distance 5 meters. The buffer should be FULL and have flat ends. Examine the message window that appears, then dismiss it.

4. Turn off TICTACTOE node and TICTACTOE arc. Add the *polygon* component of TICTACTOEBUF to your ArcMap display. Open its polygon attribute table. How many polygons are there in TICTACTOEBUF? _____. Select records to determine which polygon is which. What is the area of the larger one? _____. The smaller? _____.

5. What is the value of INSIDE for the larger polygon? _____. The smaller? _____.

6. Use the Identify tool to explore each polygon, noting what flashes when you click it. What is the TICTACTOEBUF-ID number of the polygon whose area is within the buffer distance, as signified by the INSIDE value of 100? _____

Some points are really important here:

❑ The arcs and nodes of TICTACTOE are not part of the coverage TICTACTOEBUF, which is strictly a polygon coverage.

❑ The buffering process created a polygon with INSIDE value 100, indicating that the area of that polygon was within 5 meters of the arcs that were buffered. This value of 100 is the way, *with coverages only*, that you distinguish areas of the spatial field that are within the buffer distance.

❑ The buffering process also created an "island polygon," with INSIDE value 1, which indicated that the area of that polygon was *not* within 5 meters of the arcs that were buffered.

Examining the TICTACTOEBUF Polygon Attribute Table (PAT) with ArcInfo Workstation

7. ***Start ArcInfo Workstation:*** On the Windows taskbar, click Start > Programs > ArcGIS > ArcInfo Workstation > Arc. You will see a black window with an Arc: prompt. You need to identify the folder containing the coverage TICTACTOEBUF with a WORKSPACE command. Type the following:

WORKSPACE ___IGIS-Arc_*YourInitialsHere*\Buffer&Overlay

Display the names of the coverages that are in the workspace by typing LISTCOVERAGES. (After all this typing, you see again why point-and-click was developed?) You should see the coverage TICTACTOEBUF.

8. Display the TICTACTOEBUF polygon attribute table (PAT) by typing

LIST TICTACTOEBUF.PAT

Notice that there are three polygons listed here, where the ArcMap table showed only two. The polygon with the TICTACTOE# of 1 (this is the polygon that was omitted from the ArcMap table) is the outside polygon, which we discussed earlier. Of course, its INSIDE value is 1, since it is outside the buffer area. The polygon representing the area within the buffer is numbered 2, and has INSIDE value of 100. The island polygon that was created is numbered 3, and, as discussed previously, its INSIDE value is 1.

9. Quit Workstation ArcInfo by typing QUIT.

10. Dismiss ArcMap without saving changes.

Exercise 7-2 (Project)

Exploring PGDBFC Buffers with the Wildcat Boat Data

The concept of buffering is pervasive in GIS. Over the years, ESRI has developed several tools to do the operation. You just used the coverage Buffer tool. For geodatabases we will explore two others. The one we will use first is called the Buffer Wizard. It is not automatically available in ArcMap or ArcToolbox, so you will have to add it. In addition to getting the wizard, you will also learn something of the software's ability to be customized. The Buffer Wizard will be added to the Tools menu.

_____ **1.** Start ArcMap. Choose `Tools > Customize > Commands`. Click `Tools` in the `Categories` list. Click `Buffer Wizard` in the `Commands` list. Drag it to the `Tools` menu on the Main menu, but don't let go of it yet. The `Tools` menu opens. Drag it down to just barely below `ArcCatalog` and release the mouse button. The wizard is now one of the tools you can use. Close the `Customize` dialog box.

_____ **2.** In ArcCatalog, make or confirm a folder connection with

___IGIS-Arc_*YourInitialsHere*\Wildcat_Boat_Data

_____ **3.** In ArcMap, display the Personal Geodatabase Feature Classes that depict Sewers and Roads down in the tree of:

___IGIS-Arc_*YourInitialsHere*\Wildcat_Boat_Data

You may bring both feature classes in with a single step by selecting both (click with Ctrl) and pressing Add. Also add the Soils geodatabase feature class. Zoom to extent of all layers.

_____ **4.** A review: Rearrange the order of the entries in the Table of Contents so that both the layer Roads and the layer Sewers are at the bottom. (Drag each entry by its name.) What does this say about the order in which layers are drawn? Note the effect of drawing a polygon layer after line layers. Experiment with turning layers off and on. Click the Source tab at the bottom of the Table of Contents. Note that you cannot now rearrange the order. Click the Display tab again. Then rearrange the layers in a reasonable order, noting that, in some places, the sewers follow the roads.

_____ **5.** Open the attribute table for sewers. How many segments of sewer pipe are there? _____. What are their diameters? _____, _____. Close the table.

_____ **6.** Display the Roads in black, width 2. Display the Sewers in green, width 5. Arrange things so that where the roads and sewers are coincident you can see both.

_____ **7.** It may be that when you slide the cursor around the map the coordinates are given in unknown units. If so, go into the data frame properties window and, using the `General` tab, make the `Map Units` and the `Display Units` "meters." (ArcMap should have picked up the coordinate system, and the units, from the metadata, but sometimes this doesn't happen.)

Using ArcMap to Make Buffer Zones Around the Roads

_____ **8.** Use the `Buffer Wizard` in the `Tools` menu of ArcMap to make a buffer around the roads at a specified distance. In the next panel of the wizard, be sure the buffer distance units are meters and make the buffer distance itself 44. Click Next. Dissolve the "barriers" between the buffers. Accept the default name and location for the new layer. Click Finish.

_____ **9.** In the Table of Contents, move the name Buffer_of_Roads below Roads. Turn the Sewers layer off.

_____ **10.** Display the attribute table for the buffer layer. Because you dissolved the barriers between the buffers, there are only three records. Select records to identify the three buffers. Note that an attribute called `BufferDist` was created; for all the polygons, its value is 44, which will be useful when you want to identify the buffers. Notice that, unlike the buffer command that operates on coverages, the `Buffer Wizard`, operating on geodatabase feature classes, does not make explicit, separate polygons for the islands that are created. Clear selected features and dismiss the table.

_____ **11.** Zoom in on a "straight" portion of the road buffer. Use the `Measure` tool to determine the entire width of the buffer to two decimal places. What value do you get? _____. What should it be if you could measure exactly? _____. Zoom to full extent of the Buffer of Roads shapefile.

_____ **12.** Use the `Buffer Wizard` in the `Tools` menu of ArcMap to make another buffer around the roads—this time with a distance of 111 meters. Again dissolve the "barriers" between the buffers. Accept the default name and path. What is the name of this feature class? _____.

_____ **13.** Display the attribute table for this second buffer layer. What is the name of the attribute that identifies this buffer? _____. This feature class contains only one polygon instead of three. Why? _____. _Hint:_ Flip the 111-meter buffer off and on. Dismiss the table.

_____ **14.** Remove the Soils layer. Drag the other polygon layers below the line layers, with the 111-meter buffer at the bottom. Display the 44-meter buffer with a light yellow. Display the 111-meter buffer with a dark yellow. Using the Identify feature tool (with the `Layer` set to `Top-most layer`), click the various portions of the buffers, paying attention to what flashes.

_____ **15.** Dismiss ArcMap. Dismiss ArcCatalog if it is running.

Variable-Width Buffers

As you know, you frequently have features in a layer that have different attributes that distinguish features from each other. For example, in Roads there are some features that are associated with RD_CODE "1" and others with RD_CODE "2." Suppose we wanted to buffer the "1" roads with a distance "x" and the "2" roads with a distance "y." We can do the following:

_____ **16.** Start ArcMap. Add Roads from the Wildcat_Boat geodatabase in

___IGIS-Arc_*YourInitialsHere*. Represent the features by a red line of width 2.

17. Open the attribute table. Under `Options` add the field named `Buffer_by`. Make it a `Short Integer` field.

18. Start editing with the `Editor` toolbar. Note that a pencil shows up in the lower right-hand corner of the table. Suppose we want to buffer the `RD-CODE` = 1 roads by 20 meters and the `RD-CODE` = 2 roads by 40 meters. Notice that the first records have a 1 in `RD-CODE`, so we would want to put 20 in the `Buffer_by` attribute of those records. Click the `Buffer_by` field cell in the first record and type 20. Now do the same with the second record. And the third. Looks like a long haul. We'll use another way instead.

19. On the Main menu, choose `Selection` > `Select By Attributes`. Construct the expression `[RD-CODE]` = 1. See Figure 7-14. Click Apply, then OK to close the `Select By Attributes` window. Now you can see that all the records with 1 in the `RD-CODE` are selected.

20. Right-click the column name `Buffer-by` and pick `Calculate Values`. Type 20 in the `Buffer_by=` text box that appears and click OK. Now every `RD-CODE` = 1 road should have a 20 in the `Buffer-by` field. Use `Options` in the attribute table window to switch selection, which will select every `RD-CODE` = 2 road. Use `Calculate Values` to put a 40 in the `Buffer_by` cells of those records. Save edits, then stop editing.

Now that we have the proper distances in the attribute table we can proceed to make the actual buffers.

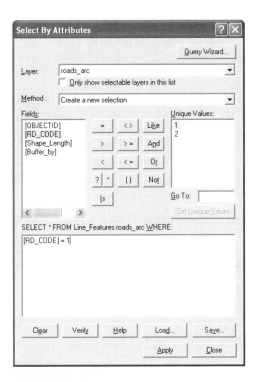

FIGURE 7-14

_____ **21.** Start the Buffer Wizard. The Input, of course, is Roads. Make sure Use Only the Selected Features is unchecked. Click Next. Choose Based On A Distance From An Attribute. Choose Buffer_by from the drop-down menu. Click Next. Pick Yes for dissolving barriers. Actually browse (using the little yellow file folder) to save the buffer in Area_Features. It's appropriate to save it there because, as you remember, buffers are always polygons regardless of the type of feature being buffered. For a name use Roads_variable_buffer. Click Save, then Finish.

_____ **22.** The PGDBFC Roads_variable_buffer will be added to the map. Pull its name below Roads. In the open attribute table of Roads, select the RD-CODE = 1 roads (hold down Ctrl and drag the cursor down the boxes at the left of the records) so that the line features show up selected in the map. Measure some RD-CODE = 1 roads buffers. _____ Measure some RD-CODE = 2 roads buffers. _____.

_____ **23.** Close ArcMap.

Exercise 7-3 (Project)

Overlaying One Coverage with Another Using UNION

_____ **1.** Start ArcCatalog. Make or confirm a Folder Connection to the

___IGIS-Arc_*YourInitialsHere*\Trivial_GIS_Datasets

folder. Examine the polygon component of the coverage TWO_STALKS. Look at both its graphic representation and its table. What are the values of the POSITION attribute? _____, _____. Using the Identify tool, determine the location, to the nearest integer, of the southwest corner of the left polygon: X = _____. Y = _____. Using the Spatial tab of the Metadata (Stylesheet: FGDC ESRI), determine the location of the northeast corner of the coverage. X = _____. Y = _____. Click on the name TWO_STALKS in the catalog tree. Right click to bring down a menu and select Properties. Using the Tics and Extents tab of the Coverage Properties, determine the location of the most northeast tic: X = _____. Y = _____.

_____ **2.** Again using the Metadata, write down the names of the attributes of the polygon component of TWO_STALKS:

_____, _____, _____, _____,

_____, _____, _____

_____ **3.** Examine the polygon component of the coverage THREE_BARS. Look at its Geography. Now see its table. What are the PLACEMENT item values of the three polygons?

_____, _____, _____. From the THREE_BARS polygon Coverage Feature Class Properties window, determine the type and width of the PLACEMENT attribute (item): _____, _____.

____ **4.** Examine the attribute table of the *polygon* component of THREE_BARS. What are the values of the attribute THREE_BARS-ID? _____, _____, _____. Using the Metadata, determine, for the polygon attribute table, the data type and width of the THREE_BARS-ID attribute: _____, _____.

____ **5.** Write down the names of the attributes of the polygon component of THREE_BARS:

_____, _____, _____, _____,

_____, _____, _____

____ **6.** Launch ArcMap. Add the *polygon components* of both THREE_BARS and TWO_STALKS from

___IGIS-Arc_*YourInitialsHere*\Trivial_GIS_Datasets.

Zoom to Full Extent. Notice the positions of the features relative to each other. Slide the cursor pointer around the map to verify the locations of the images.

____ **7.** In ArcToolbox use the Index to find Union (arc)—not Union (analysis). Press Locate. What is the path to it?

ArcToolbox > _____ Start the tool by double-clicking the name. Read the Help panel—clicking in each area where you can provide input, to see a discussion of that input. Particularly look at Fuzzy Tolerance. For the Input Coverage browse to

___IGIS-Arc_*YourInitialsHere*\Trivial_GIS_Datasets\TWO_STALKS

and add it. Specify the Union Coverage as

___IGIS-Arc_*YourInitialsHere*\Trivial_GIS_Datasets\THREE_BARS.

For the Output Coverage, browse to

___IGIS-Arc_*YourInitialsHere*\Buffer&Overlay

and specify STALKS_BARS. Click Save to insert the output coverage name. Click OK in the Union window and watch the messages go by. Close the Union window.

____ **8.** Hide ArcToolbox. In ArcMap turn off the THREE_BARS polygon and the TWO_STALKS polygon coverages. Add to the map the *polygon component* of

___IGIS-Arc_*YourInitialsHere*\Buffer&Overlay\STALKS_BARS. Make the layer color light green.

____ **9.** In the Layer Properties window for STALKS_BARS, click Labels. Click Label Features In This Layer. Click Expression in the Text String pane. Place the following expression in the box:

[POSITION] & vbNewLine & [PLACEMENT]

Click OK. Change the Font Size to 6. Click Apply, then OK.

This last step has the effect of labeling each polygon with the POSITION attribute and then, *starting on a new line*, the PLACEMENT attribute, using a font size of 6. Make sure ArcMap occupies the full screen. Even so, you will have to zoom in on the parts of the window to read the labels well.

_____ **10.** List *all the combinations* of POSITION and PLACEMENT. Also list the *number of polygons of each combination*. (Two are done for you as examples.) Use the STALKS_BARS attribute table and sorting to help you.

❏   POSITION is LEFT, PLACEMENT is blank. There are three such polygons.

❏   POSITION is LEFT, PLACEMENT is TOP. There is one such polygon.

❏   _____

❏   _____

❏   _____

❏   _____

❏   _____

❏   _____

❏   _____

❏   _____

❏   _____

❏   _____

❏   _____

How many *polygons* are there in total, according to what you wrote above? _____. How many polygons do you get by counting polygons that make up *the image*? _____.

I realize that was not a fun exercise, but the main point to be made is essential. Overlay usually creates a large number of polygons. Further all the attributes of the input polygons are carried over to the output. This produces lots of polygons and attributes. You should understand where each of the resulting polygons came from and how it is characterized by its attributes.

_____ **11.** Use Select By Attributes to highlight the two polygons specified as POSITION equals LEFT and PLACEMENT equals BOTTOM_OUTSIDE. Zoom in on the selected polygons to check their labels.

_____ **12.** Use Select By Attributes to *additionally* highlight the polygons specified as (blank and blank)[1] or (RIGHT and TOP). *Hint:* For Method, use Add to Current Selection. See Figure 7-15. Once the selection is done, close the window and examine the result on the map. How many polygons did this operation add? _____.

_____

[1]Blank is indicated under Unique Values by two single quote marks, which, unfortunately when placed next to each other look like one double-quote mark: ''.

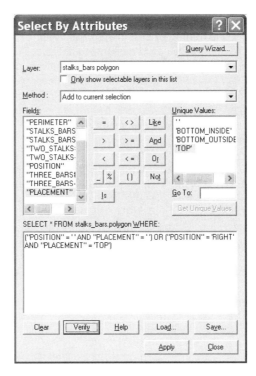

FIGURE 7-15

_____ **13.** Open the STALKS_BARS polygon attribute table. Write down all the attributes of STALKS_BARS. Make sure the fields are wide enough so you see all the characters of the attribute names.

_____, _____, _____, _____,

_____, _____, _____, _____,

_____, _____, _____, _____,

_____ **14.** Select the two columns entitled POSITION and PLACEMENT. Right-click and sort them into ascending order. From the table, now you should be able to easily verify the correctness of the blanks you filled out in Step 10. When you are done, clear all selections.

You should come away from this exercise understanding of the results of combining two polygon coverages with UNION and what makes up the attribute table. As I mentioned earlier, you usually get lots more polygons and lots more records in the result of a union than exist in in the constituent coverages.

Making a New Coverage from a Subset of Polygons: Extract

As you saw, the union of the coverages produced lots of output polygons from a few input polygons. As you also saw, those polygons could be identified by their attributes. Frequently what you want to do after an overlay is to select a subset of the output polygons and make a new coverage. For example, to solve the Wildcat Boat problem, you would want to extract the polygons that met the requirements of soil suitability, land cover, and so on. To do this, you build a query in what is called the Structured Query Language (SQL). There is a lot of similarity between using SQL and what you did above in selecting polygons. The difference is that *the result of the selection process is a new coverage* that contains the selected features.

Suppose now that we want to make a *new coverage* named TOP_AND_LEFT consisting of those polygons on the top plus those polygons at the left.

___ **15.** Choose `ArcToolbox > Coverage Tools > Analysis > Extract > Select`. Start the `Select` tool. Read the help panel. Then read the several different help discussions, obtained by clicking various places on the Select window. The selection method you will use is `Subset`.

When you begin to "build a query," it is assumed that all features (and therefore records) are automatically selected. Note that this is different from the previous assumption, made when you display tables, that none are selected. The queries you build selects a subset of records. That is, building a set of queries ultimately (usually) reduces the number of records selected, from all to some fewer.

___ **16.** Browse to the `Input Coverage` STALKS_BARS in

___IGIS-Arc_*YourInitialsHere*\Buffer&Overlay.

Call the `Output Coverage` TOP_AND_LEFT; it will go into the same folder.

The coverage to be produced will contain only the polygons labeled TOP and those labeled LEFT.

___ **17.** Press the `INFO` button to develop the INFO Expression[2] in a `Query Builder` window. Build the query:[3]

POSITION = 'LEFT' OR PLACEMENT = 'TOP'

Once the expression is built in the `Subset` of the `Current expression`, press the down arrow at the end of that text box to move the expression into the bottom panel. See Figure 7-16. Press OK to return to the `Select` window where your expression will be in place. Press OK. When the messages stop, close the `Select` window.

___ **18.** Add the polygon component of the coverage TOP_AND_LEFT. Make the layer color light blue. Verify that those polygons that were LEFT and those polygons that were TOP are both represented. Right-click the STALKS_BARS layer and click off `Label Features`. In the `Layer Properties` window of TOP_AND_LEFT, check the box to label features and make the `Label Field` read POSITION. Click Apply and OK. Observe. Change the `Label Field` to PLACEMENT. Observe. Turn off TOP_AND_LEFT.

---

[2]INFO is the name of the database management system used with coverages.
[3]In the Query Builder you single-click on an attribute name, rather than double-click as in the Select By Attributes. Just an anomaly in the software.

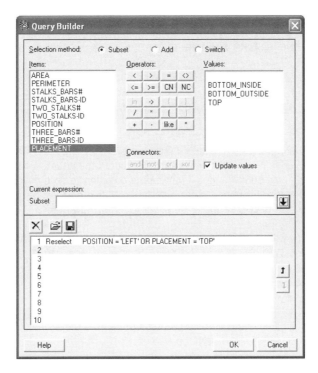

FIGURE 7-16

_____ **19.** _Make another coverage by extracting from **STALKS_BARS**_: For output accept whatever name that ArcGIS suggests. This time you want the polygon(s) that are designated TOP and are _also_ designated LEFT. Use the Query Builder, putting in the expression POSITION = 'LEFT' **AND** PLACEMENT = 'TOP'. Display the results, showing the layer in red.

_____ **20.** Notice that you have only one polygon. Label it, first with POSITION, then again with PLACEMENT.

More Complex Queries—ANDs and ORs

If you are confused about and's and or's—like why you get both the left polygons **and** the top polygons when you use the expression POSITION = 'LEFT' **OR** PLACEMENT = 'TOP'—I say, if you are confused, that's good! It will make you think carefully about the Boolean expressions you put in the Query Builder. It is amazingly easy to make mistakes.

The Query Builder allows you to use multiple expressions. You can either create additional subsets of subsets of records (and hence features) or you can add records. If you use multiple lines in the expression box, you are basically writing one long query with ANDs between the subexpressions. That is, any selections found must satisfy _all_ the criteria, not just some of the criteria. For further information check out the information in the Help, entering SQL at the Index tab, and then choosing Building Expressions.

Other Spatial Joins: INTERSECT and IDENTITY

____ **21.** Again overlay TWO_STALKS with THREE_BARS, but use the Intersect tool rather than Union.

Name the Output Coverage INTSCT_S_B and put it into

___IGIS-Arc_*YourInitialsHere*\Buffer&Overlay.

____ **22.** To a new map add the *arc* components of TWO_STALKS and THREE_BARS from
___IGIS-Arc_*YourInitialsHere*\Trivial_GIS_Datasets. Add the *polygon* component of INTSCT_S_B
from Buffer&Overlay. Verify that Intersect computed the geometric intersection of the coverages, producing new polygons only in the areas common to both coverages. Use Identify to
look at each polygon. Examine its attribute table. Notice that the attributes are the same as for
STALKS_BARS, but of course there are fewer records. How many? _____.

____ **23.** Perform Overlay with TWO_STALKS and THREE_BARS, but use Identity rather than
Intersect. Use TWO_STALKS as the Input Coverage and THREE_BARS as the Identity
Coverage. Call the Output Coverage IDNTY_S_B and put into Buffer&Overlay. How many poly-
gons are in the Output coverage? _____ Note that only polygons in the areas of the
TWO_STALKS polygons are present in IDNTY _S_B.

When using polygon-on-polygon overlay with the Identity option, it is the area covered by the *input coverage*
that determines the area covered by the *output coverage*.

____ **24.** Perform Overlay with Identity again, but this time use THREE_BARS as the input coverage and
TWO_STALKS as the identity coverage. Call the output coverage IDNTY_B_S and put into
Buffer&Overlay. How many polygons are in the output coverage? _____ Note that only poly-
gons in the areas of the THREE_BARS polygons are present in IDNTY_B_S.

Exercise 7-4 (Project)

Using Overlay with Trivial Point and Line Coverages

As discussed in the Overview section of this chapter, intersect and identity operations may be used with
point coverages and line coverages. When you use Overlay with a point coverage, the point coverage is
always the input coverage and the output coverage is always a point coverage. The overlay coverage is
always a polygon coverage. With Identity, all of the points in the input coverage are present in the output;
with Intersect, only those points that are in the areas covered by the overlay coverage polygons are
present.

The main thing that happens when you use Overlay with a point coverage is that each point picks up the
attributes of the polygon into which it falls. For example, if you had points representing parking meters and
a polygon coverage representing enforcement areas, an overlay of the two could tell you into which enforcement area each particular parking meter fell.

____ **1.** Start a fresh map in ArcMap and add the *point* component of POINTS_B from

___IGIS-Arc_*YourInitialsHere*\Trivial_GIS_Datasets

and the *polygon* component of

STALKS_BARS from

___IGIS-Arc_*YourInitialsHere*\Buffer&Overlay. Zoom to full extent. Examine the geography and table of each layer. Close the tables.

_____ **2.** Use ArcToolbox > Coverage Tools > Analysis > Overlay > Intersect to perform an overlay of POINTS_B and STALKS_BARS, using the POINT feature class. Name the Output Coverage POINT_IN_POLY and put it in

___IGIS-Arc_*YourInitialsHere*\Buffer&Overlay.

_____ **3.** Add the point component of POINT_IN_POLY to your map. Represent POINTS_B with a yellow dot of size 7; represent POINT_IN_POLY with a red dot of size 3. Arrange it so POINT_IN_POLY is at the top of the Table of Contents. How many points are there in POINTS_B? _____ How many points are there in POINT_IN_POLY? _____ Describe the differences in *geography* the operation produced, comparing the geography of POINTS_B and the geography of POINT_IN_POLY.

_____

_____ **4.** Describe the differences in *tables* that the operation produced, comparing the table of POINTS_B with the table of POINT_IN_POLY.

_____

_____ **5.** Again use ArcToolbox to perform an overlay of POINTS_B and STALKS_BARS, but this time use the *Identity* tool. Name the output coverage whatever you choose (write its name here _____) and put it in

___IGIS-Arc_*YourInitialsHere*\Buffer&Overlay.

_____ **6.** Remove the POINT_IN_POLY coverage from the map. Add the new coverage. Represent POINTS_B with a yellow dot of size 7; represent the new coverage with a red dot of size 3. Arrange it so the new coverage is at the top of the Table of Contents. How many points are there in POINTS_B? _____ How many points are there in the new coverage? _____ Describe the differences in geography the operation produced, comparing the geography of POINTS_B and the geography of the output coverage.

_____

_____ **7.** Carefully examine the new coverage table. How many items are present in the POINTS_B table? _____ How many in the table of the new coverage? _____

_____ **8.** Describe the differences in tables that the operation produced, comparing the table of POINTS_B and the new table.

_____

Intersect and Identity Operations Used with Line Coverages

As with point coverages, Intersect and Identity operations may be used with line coverages. When you use Overlay with a line coverage, the line coverage is always the input coverage and the output coverage is always a line coverage. The second coverage (the overlay coverage) is always a polygon coverage. With Identity, all of the lines in the input coverage are present in the output; with Intersect, only those lines that are in the areas covered by the overlay coverage polygons are present.

Two things happen when you use Overlay with a line coverage. First, the lines are divided where the polygon boundaries cut across them. Second, each line takes on the attributes of the polygon into which it falls. For example, if you had lines representing roads and a polygon coverage representing highway districts, an overlay of the two could tell you into which district each particular highway segment fell.

_____ **9.** Start a fresh map in ArcMap, add the *arc* component of LINE_B from

   ___IGIS-Arc_*YourInitialsHere*\Trivial_GIS_Datasets

   and the *polygon* component of

   STALKS_BARS from

   ___IGIS-Arc_*YourInitialsHere*\Buffer&Overlay. Zoom to full extent. Examine the geography and table of each layer. Close the tables.

_____ **10.** Use ArcToolbox to perform an overlay of LINE_B and STALKS_BARS, using the `Intersect` tool. Name the `Output Coverage` LINE_IN_POLY and put it in

   ___IGIS-Arc_*YourInitialsHere*\Buffer&Overlay.

_____ **11.** Add LINE_IN_POLY to your map. Represent LINE_B with a yellow line of size 6; represent LINE_IN_POLY with a red line of size 1. Arrange it so LINE_IN_POLY is at the top of the Table of Contents.

   Use the `Identify` tool to explore LINE_B and LINE_IN_POLY, which, as you see, overlie each other in part. How many arcs are there in LINE_B? _____ How many arcs are there in LINE_IN_POLY _____ Describe the differences in geography the operation produced, comparing the geography of LINE_B and the geography of LINE_IN_POLY. *Hints:* (1) Use the Identify tool, (2) zoom in, and (3) build and display the node component of LINE_IN_POLY.

   _____

   Describe the differences in tables the operation produced, comparing the table of LINE_B with the table of LINE_IN_POLY.

   _____

_____ **12.** Remove LINE_IN_POLY from the map. Use ArcToolbox to perform an overlay of LINE_B and STALKS_BARS, but this time use the `Identity` option. Name the `Output Coverage` whatever you choose (write its name here _____) and put it in

___IGIS-Arc_*YourInitialsHere*\Buffer&Overlay.

Add the new coverage to your map. Continue to represent LINE_B with a yellow line of size 6; represent the new coverage with a red line of size 2. Arrange it so the new coverage is at the top of the Table of Contents. How many arcs are there in the new coverage? _____ Describe the differences in geography that the operation produced, comparing the geography of LINE_B and the geography of the new output coverage.

_____

Describe the differences in tables the operation produced, comparing the table of LINE_B and the new table.

_____

In summary, with point and line coverages, the input coverage is always a point or line coverage; the overlay coverage is always a polygon coverage. The output coverage is always the same type (point or line) as the input coverage. With the Identity tool all of the geography of the input coverage is represented in the output coverage. With the Intersect tool only the geography of the input coverage that lies within the overlay coverage boundaries is represented in the output coverage. In all cases, the elements of the output coverage pick up the attributes and appropriate values of the polygon coverage.

Exercise 7-5 (Project)

Using Buffer and Overlay with Geodatabases

Many of the principles you just learned regarding buffering and overlaying with coverages extend to geodatabase feature classes as well. The tools are a bit different, but the concepts remain the same. In this exercise you solve a problem in which all the feature classes come from a feature data set in a personal geodatabase.

The Getrich Saga

You have found some potentially valuable information! In a western county, called Getrich_county, a miner long ago buried gold now worth *at least* 4 million dollars but was never able to come back to dig it up. He buried it within 1900 meters of the *edge* of one of the wagon trails that run throughout the county, but at least 300 meters away from the *edge* of the trail. He put it into sandy soil, but more than 2500 meters away from any of the old oil wells. It will cost three cents ($0.03) per square meter to search for the treasure—using metal detectors and Global Positioning System (GPS) receivers.

Unfortunately, you are penniless. But some friends have money to invest in the search if the cost is low enough and the maximum cost is well known in advance. There is not nearly enough money to search the entire county (besides, it would cost more than the gold is worth), but you have access to an ESRI ArcGIS software system and you have some data available. Your task is to display a map showing the areas where the gold might be buried, and to compute the total cost of searching those areas. You will do this by making a *personal geodatabase feature class (PGDBFC)* named Look_Here that contains only the polygons representing the areas to be searched.

The data sets for the problem are all contained in a personal geodatabase named PGDB_gold.mdb, located in a folder named [___] IGIS-Arc\Gold_Data. You have four PGDBFCs to work with:

Getrich_County—A very simple POLYGON PGDBFC of the county. The treasure is within the outline of the county.

Oilw—A POINT PGDBFC of all the sites of the abandoned oil derricks and wells.

Wagt—A LINE PGDBFC of the centerlines of the wagon trails. (Each trail is assumed to be 10 meters wide. Recall that the treasure is buried a certain distance from the *edges* of the trails.)

Soils—A POLYGON PGDBFC of the soils in the area. There are four types of soil: clay, dirt, sand, and rock. The feature class contains a key that will let you join a separate table that indicates the soil types located in a column with the heading of Characteristic.

_____ **1.** Start ArcCatalog. Copy the Gold_Data folder from [___] IGIS-Arc to your

_____IGIS-Arc_*YourInitialsHere* folder.

_____ **2.** Make a folder connection to the

_____IGIS-Arc_*YourInitialsHere*\Gold_Data folder.

The personal geodatabase PGDB_gold.mdb contains a personal geodatabase feature data set named PGDBFD_gold, which contains the personal geodatabase feature classes Getrich_county, Soils, Oilw, and Wagt. PGDBFD_gold also contains a table named soil_type.

_____ **3.** *Carefully* look at the *geographics* and *attribute tables* of each PGDBFC, so you have a good idea of what you have to work with. Identify the three named wagon trails. What are they? _____, _____, _____ How many oil wells are there? _____ How many different soils polygons? _____

_____ **4.** So we know precisely where the oil wells are, add the x- and y-coordinates to the attribute table of Oilw. (If you don't remember how to perform the ADD XY operation, check your Fast Facts File or the Help files. Remember you are using a geodatabase feature class, not a coverage.)

_____ **5.** Start ArcMap. Set the map units and the display units to meters.

The first step in solving this problem will be to make buffers. You will use the Multiple Ring Buffer tool partly because the wagon trail specifications require two buffers and partly because it puts a field in the table so you can isolate the buffered areas.

The first buffer will be around the oil wells. Even though we need only one buffer, we will use the Multiple Ring Buffer tool.

_____ **6.** Start ArcToolbox. In Analysis Tools > Proximity, launch Multiple Ring Buffer. Check the tool out with the help panel.

_____ **7.** For Input Features, browse to Oilw in

___\:IGIS-Arc_*YourInitialsHere*\Gold_Data\PGDB_gold.mdb\PGDBFD_gold

and add it. Put the Output Feature class in the same place, calling it Oilw_buf. For Distances put in the number of meters that is appropriate (check the problem statement) and press the + sign. Scroll down and make the Buffer Unit Meters. For Field Name type OilwNoGold. Press OK. When the tool stops close the window. This tool automatically adds the result to the map, so the polygons representing the oil well buffers should appear as a layer in ArcMap. (If not, add the layer.) Examine it. Also add Oilw to the map.

_____ **8.** Open the attribute table of Oilw_buf. Notice that the OilwNoGold field has the value 2500 in it. You will use this when you extract features from the overlay. Since the gold is not within this buffer, you will want to find those polygons where OilwNoGold is *not* 2500. Also notice the table has only one record. Use the Identity tool to click one circle and notice that they all light up. The Multiple Ring Buffer tool makes a multipart polygon. That is, all the separate circles are together considered as just one polygon. This one polygon might work okay for our analysis, but let's separate them just for the sake of neat and tidy. You did this before with the Advanced Editing toolbar of the Editor, but let's look for a different way. Close the attribute table and Identify window. Type Explode into the Index of ArcToolbox. Nothing useful comes up. Type it into the Search text box of ArcToolbox and press Search. Pay dirt (so to speak). You find Multipart To Singlepart. When you highlight it, and press Locate, you find that the path to it is:

ArcToolbox > _____

_____ **9.** Start the tool. Browse for Input Features: Oilw_buf—or, better yet, since it is on the map, click the drop-down menu arrow and find it there. Browse to name the Output Feature Class Oilw_bufs (note plural: bufs) and put it in the same location as Oilw_buf. Click OK. When it appears in the Table of Contents, drag it below Oilw. Now check out a few wells with Identify. Also look at the attribute table. Note that there is a buffer for each oil well. Remove Oilw_buf (note singular: buf) from the map. Add the feature class Wagt to the map.

_____ **10.** ***Use Multiple Ring Buffer to make the two buffers along the wagon trails:*** What would the appropriate distances be, considering the original problem statement concerning the width of the wagon trails? _____, _____.[4] Start the tool. Call the Output Feature class Wagt_bufs. Put in the appropriate distances, smaller first, clicking the + after each. Units are again meters. For the Field Name use Wagt_areas. Click OK.

_____ **11.** Drag Wagt_bufs to the bottom of the table of contents and zoom to the layer. Carefully examine Wagt_bufs with both the attribute table and the Identify tool, zooming up as necessary. Using the Layer Properties of Wagt_bufs, display the 305-meter buffer with light green and the 1905 meter buffer with some color close to gold. You can see that when the Multiple Ring Buffer makes multiple buffers they are independent of each other. Look at Figure 7-17. The amount you put in for the outside distance becomes the field value for that buffer. So what is the value of Wagt_areas that you will want in the attribute table of the final PGDBFC that you are creating? _____.

---

[4]If you didn't write 305 and 1905, check the problem statement again.

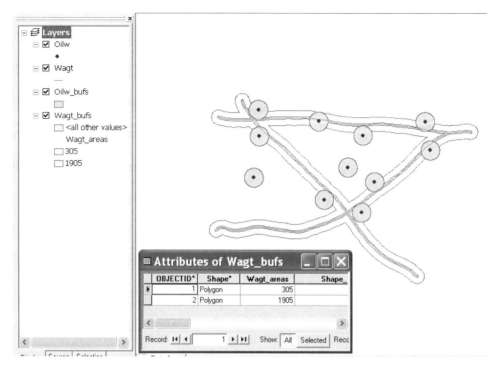

FIGURE 7-17

Deriving Information by Combining Tables

____ **12.** Add Soils to the map. Drag it to the bottom of the Table of Contents. Look at the Soils attribute table. Write the names of the fields in Attributes of soils.

_____, _____, _____, _____, _____

Note that you have no way of knowing which of the 20 polygon areas consist of sand. However, each polygon does have a value in a SOILS_ID field. This field will allow you to key into a second table that matches up the values of SOIL_ID with a similar field in another table.

____ **13.** In the PGDB_GOLD.mdb database, you will find a table named SOIL_TYPE. Add that to the Table of Contents and open it. What fields do you find?

_____, _____, _____

Close all attribute tables.

Here are some salient facts:

❑ The values in the field SOILS_ID (in SOIL_TYPE) correspond to the values in SOILS_ID (in the Soils feature class). Each is a key field in its table. Equal values identify the same polygons.

❑ When a table B is joined to a table A, the records of table B are concatenated with the records of table A so that the values in the key fields are the same.

❑ Joining the SOIL_TYPE table to the Soils feature class table, using the keys, will result in associating the correct polygon with its CHARACTERISTIC ("sand," "rock," and so on).

❑ The values in CHARACTERISTIC are the soil types you will need to identify the sandy soils.

Putting the table together with the feature class is a bit convoluted, so you may want to be especially detailed in your Fast Facts File.

_____ **14.** With the Display tab active, right-click the Soils entry. Under Joins And Relates, pick Join. In the Join Data window: you want to select Join attributes from a table.

_____ **15.** In Step One choose SOILS_ID from the drop-down menu. In Step Two browse to the table SOIL_TYPE in PGDB_Gold.mdb or pick it off the drop-down menu. In Step Three pick SOILS_ID from the drop-down menu. Click OK. Decline any offer to index the join table. (This is a time-saver for really big tables. It basically puts the keys of the join table in order so they can be located quickly during the joining operation. Our tables are so small that it doesn't matter.)

_____ **16.** Open the attribute table for the Soils feature class. List the field names.

_____, _____, _____, _____,

_____, _____, _____, _____

Note that the fields that came from the join table SOIL_TYPE are preceded by soil_type.—for example, soil_type.CHARACTERISTIC. We say that the name is "qualified" by the name of the table from which it came.

_____ **17.** Sort the soiltype.CHARACTERISTIC field. How many "sand" polygons are there? _____ What are the values of the key fields of the "sand" polygons? _____, _____, _____, _____, _____.

_____ **18.** Zoom to the Soils layer. Turn off all other layers. Dismiss the attribute table. Label the polygons in Soils with SOILS_ID. Observe. Now label the polygons with soil_type.CHARACTERISTIC. Observe.

You are going to use the information in the new Soils attribute table in an overlay command. Because the table is somewhat "ethereal" at the moment—it exists only in the fast memory of the computer—you need to put it onto disk, where it won't go away if you leave ArcMap.

_____ **19.** Right-click Soils in the Table of Contents. Slide the cursor to Data, then click Export Data. In the Export Data window, select Export All Features. Browse to

IGIS-Arc_*YourInitialsHere*\Gold_Data\PGDB_Gold.mdb\pgdbfd_gold

(being sure to select Personal Geodatabase feature classes from the Save as Type dropdown menu) and use the name Soils_with_Characteristics instead of Export_Output. Click Save, then OK. When asked, add the exported data to the map. Remove the layer based on the feature class Soils. Soils_with_Characteristics is now the feature class you will operate with.

_____ **20.** Open the attribute table of Soils_with_Characteristics. Notice that you have (almost) the same fields, except the qualifications are gone.

Overlaying the Feature Classes

_____ **21.** Start the Union tool in ArcToolbox by selecting Analysis Tools > Overlay. For input, use Soils_with_Characteristics, Oilw_bufs, and Wagt_bufs. Call the new PGDBFC Union_So_Ow_Wt_MPP[5] and put it in the feature data set pgdbfd_gold. After it is added to the map, examine its attribute table. How many polygons are represented? _____. What are the names of the fields you will use to extract the polygons where the gold might be? _____ _____ _____.

_____ **22.** Explode the multipart polygons that Union makes, using the Multipart To Singlepart tool. Call the resulting feature class Union_So_Ow_Wt.

_____ **23.** Since the gold lies within the county boundaries, you need to limit yourself to the part of Union_So_Ow_Wt that lies within Getrich County. Use ArcToolbox > Analysis Tools > Extract > Clip to produce the PGDBFC GR_Cnty_only. Display that PGDBFC by itself in ArcMap.

As the final analysis step, you will use ArcToolbox > Analysis Tools > Extract > Select.

_____ **24.** Start the Select tool. For input, use GR_Cnty_only. Name the output Look_Here. Before you run the tool, write in the space that follows the expression you will enter in the SQL section in order to extract the polygons that represent the areas that must be searched.

_____

_____ **25.** Enter the expression, making good use of the Get Unique Values button. Click OK. The expression should look like the one in Figure 7-18.

FIGURE 7-18

_____ **26.** Make the polygons of Look_Here bright yellow. Make the polygons of GR_Cnty_only a light pink. The map should look like Figure 7-19***. Verify, from both the Identify tool and the attribute table, that they satisfy the original problem requirements. How many areas are there? _____.

---

[5]Union makes multipart polygons, which we will convert to single-part polygons—hence the MPP designation.

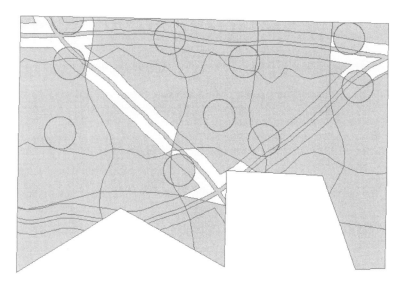

FIGURE 7-19

___ 27. From the attribute table create Statistics on the areas of Look_Here. In the worst case, where you have to look at all possible areas, how much would have to be searched? _____ How much would it cost to search it? _____ Would you invest in the search? _____ Why or why not? (Think about probabilities.)

_____

_____

___ 28. On the map turn on the following layers, adding PGDBFCs as necessary: Soils, Wagt, Wagt_bufs, Oilw, Oilw_bufs, Getrich_County, and Look_Here. Turn all other layers off. For point and line layers, pick distinctive colors. For all polygon layers except Look_Here, make the polygon color No Color or Hollow. For Look_Here use Mars Red. Make a Layout of the map. (You will need the Drawing toolbar.) Enhance the Layout to make a map that satisfies you. Print the map.

Exercise 7-6 (Project)

Building a Model of the Getrich Project Solution

Suppose you have worked the Getrich County problem and given the answer to the client. They come back with new information: Actually, instead of 1900 meters for the extent of the buffer for the wagon trails, the number should be 2000. So could you please redo the process. Well, okay, you agree. But are there likely to be more changes? The client doesn't think so but some better maps may soon be available, so maybe. At this point you begin to wish for a more automated way to run this problem. And there is: the ESRI MODEL.

433

A model will let you make a diagram of the steps to solve the problem and then execute (run) those steps at, literally, the click of a single button. In what follows you will make a model of the Getrich problem.

1. Start ArcCatalog. Hide ArcToolbox if it is open. Highlight the folder

___ IGIS-Arc_*YourInitialsHere*\Gold_Data. Go: `File > New > Toolbox`.

Highlight the newly created Toolbox and choose `File > New > Model`. A Model window opens.[6] Close the `Model` window. Expand the Toolbox icon. In the Catalog Tree, change the name `Model` to `Gold_Model`. Right-click the new name. In the menu that appears, click `Edit`. A Gold_Model window appears.

At this point, you will begin building the model in the window. What goes into the model primarily are data sets (called parameters) and tools. The combination of a tool, with its input and output data sets, is called an *operation*. The model will consist of operations in series and parallel configurations—where the output of one operation may serve as the input of the next operation.

You may recall that you began the gold finding project by buffering the oil wells with the Multiple Ring Buffer tool.

2. Open ArcToolbox. (Along the way in this project, you may have to move and resize some windows to get to everything you need.) Find the `Multiple Ring Buffer` tool and highlight it. *Drag it into the Model window.* Notice that boxes representing the tool and its output, with a directed link between them, come as a package, and the handles show that the package is selected. By clicking at a point away from the package, you can unselect it. By clicking one box or the other, you can select it individually. By dragging a rectangle around them, you can select both boxes together. You can move whatever is selected by dragging it. The connection arrow (link) stays with the pair.

3. ***Provide the Multiple Ring Buffer tool with real input and output:*** Right-click the box containing the tool and select `Open`. You will see a familiar window. For `Input Features` browse to

___Arc-IGIS_*YourInitialsHere*\Gold_Data\pgdb_gold.mdb\pdgbfc_gold\Oilw.

In the same location on disk, specify Oilw_buf_m for the `Output Feature` class. The `Distance` again is 2500. Use meters for the `Buffer Unit` and OilwNG for the Field Name. The Apply button has no effect when building models, so just press OK.

There is a dramatic change in the model diagram. An input bubble has been added. And the elements of the model are colored in: blue for initial data, yellow for the tool, and green for output, which will also server as data (called derived data) for the next operation. This is, by itself, a complete model.

4. From the Gold_Model Menu bar, click `Model` and select `Run`. The tool turns red while it is running. Another `Gold_Model` window appears, giving details of the process.

5. Close both `Gold_Model` windows, saving changes. In your folder, a new data set should have appeared: Oilw_buf_m. Check out its geography and table.

---

[6]If you need to bring up a model window later, you can expose the Model under the Toolbox, right-click it, and select Edit from the drop-down menu.

6. Bring up Gold_Model again by right-clicking the name and choosing Edit. Recall that the Wagon Trails required two buffers. Again drag the Multiple Ring Buffer tool into the Gold_Model window. To explore a different way of adding data, drag the PGDBFC Wagt into the Gold_Model window. On the Gold_Model window toolbar, click the LINK icon (a line connecting two little boxes). With the wandlike cursor, click Wagt and then on Multiple Ring Buffer(2) to forge a link between them. Open the tool and notice that Wagt has already been filled in for you. For an output feature class, put in Wagt_bufs_m. The distances to put in are 305 and the revised distance of 2005. Put in Meters and Wagt_areas. Click OK.

7. Run the revised model. Dismiss the window with the details. Minimize the Gold_Model window. Check out Wagt_bufs_m to be sure that the proper action was taken. Bring the Gold_Model window back. Note that the tools that have been run, and the PGDBFC bubbles, which have had data created for them, now have shadows. Click the Model menu and choose Delete Intermediate Data. The shadows disappear. Check the Catalog Tree and note that both the PGDBFC buffers you created have disappeared.

8. At this point, the diagram of the model may be getting a bit unwieldy. There are tools, accessible by buttons below the menu bar, that let you rearrange and resize the elements of the model. See Figure 7-20. Make the model window bigger. Slide all its elements around with the Pan tool. Experiment with the Full Extent tool, the Zoom tools and the Auto Arrange tool. Check out the Continuous Zoom tool: Hold down the left mouse button and move the mouse toward and away from you. For more information consult the Help file.

9. By clicking the rightmost icon on the model toolbar, rerun the model. When it finishes, you can see the buffers are there again. Is this not painless flexibility?

The model-making procedure involves mainly bringing in tools and then defining their input and output. Here is a process for doing that that works well: Drag the tool in. If it has inputs that will have been previously computed in the model, make links from those inputs. Then open the tool and specify the outputs.

10. *Make a union of the two buffer feature classes with the soils feature class:* Drag in the Union tool from ArcToolbox > Analysis Tools > Overlay. Drag in the PGDBFC Soils_with_Characteristics from the Catalog Tree. Use the Link icon to make connections from the PGDBFCs to Union. See Figure 7-21.

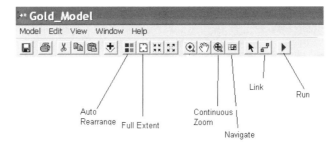

FIGURE 7-20

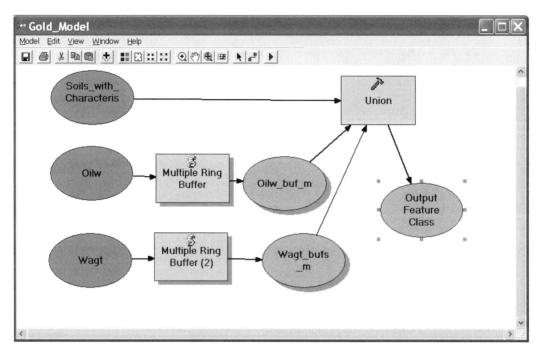

FIGURE 7-21

___ **11.** Open the Union tool in the Gold_Model window. For the Output Feature Class, specify U_WT_OW_S_m. OK. Delete Intermediate Data and run the model. Check out U_WT_OW_S_m.

___ **12.** Using the Pan tool, make room on the right for a new tool. In order to clip the union with the county, add the Clip tool from ArcToolbox > Analysis Tools > Extract. Add a connection from U_WT_OW_S_m. Drag in the Getrich_county PGDBFC. Make a connection there as well.

___ **13.** Open the Clip box. Check that the Input and Clip Features are correct. For the Output Feature Class, specify Just_Getrich_m. Without deleting intermediate data, run the model. Notice from the dialog box that only the Clip tool is run, since the data needed for it is already been derived. Close the dialog box. Save the model: Model > Save. Check out Just_Getrich_m from the Catalog Tree.

___ **14.** *Insert the final tool into the model:* Use, Select from ArcToolbox, Analysis Tools > Extract. Make a link from Just_Getrich_m. Call the Output Feature Class Look_Here_m. For the SQL Expression, use:

[Characteristic] = 'sand' AND [Wagt_areas] = 2005 AND [OilwNG] <> 2500

___ **15.** Run the model. Check out Look_Here_m. Congratulate yourself. Rearrange the model elements so they look like Figure 7-22[***]. Save the model. Call the client.

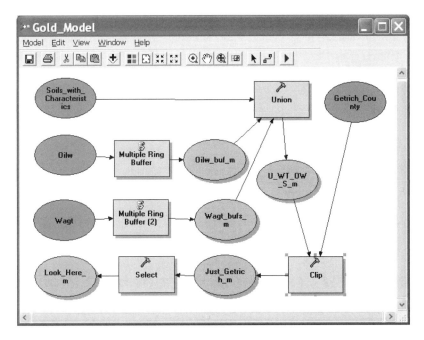

FIGURE 7-22

The client arrives, looking sheepish. It turns out that, of course, when the miner buried the gold, the units of measurement were yards, not meters. One yard is 0.9144 meters. So the distance from the oil wells is not 2500 but 2197 meters. And the buffers for the wagon trails are at 279 and 1834. Knowing that you have an easily modifiable model available you simply smile.

___ **16.** Open the `Multiple Ring Buffer` box for the oil wells. Delete the 2500, with the little cross under the plus icon, and put in 2197. Click OK. Fix up the distances for the tool dealing with Wagt so they read 279 and 1834.

___ **17.** Open the Select box and modify its SQL expression, with the new distances. `Delete Intermediate Data`. Run the model. Print the Gold_Model from the Model menu. In ArcMap, print the final map for the client.

There is a lot more modeling capability than you have seen here. And more complexity. To learn more, consult the ESRI manual *Geoprocessing in ArcGIS*.

Another way to automate geoprocessing activities is with scripts. These are programming instructions to languages that can be used to direct GIS processes. Scripts are beyond the scope of this book, but you should at least know of their existence. You can use a model to make scripts in three of the languages used by ESRI: Python, Jscript, and VBScript.

___ **18.** ***Make scripts of Gold_Model:*** Choose `Model > Export > To Script > VBScript`. In the Save As window, navigate to

___ IGIS-Arc_*YourInitialsHere*\Gold_Data

and name the file VB_Gold_Script. Save. Using a text or word processor, open VB_Gold_Script.vbs. As you scan over the text, you can probably identify the processes that were used to solve the Getrich problem. Comments in the file begin with a single quote mark. Other lines are instructions to the VD Script language. If you had Visual Basic execute this script, it would produce the same results as the model.

_____ **19.** Make and examine a JScript (for the Java language) from Gold_Model.

_____ **20.** Make and examine a Python script from Gold_Model.

Examining such scripts, knowing what they do, is a good way to begin learning how to write scripts. Of course, scripts can be, and frequently are, written directly with a text processor, rather than generated from a model.

Exercise 7-7 (Minor Project)

Making Buffers for Solving the Wildcat Boat Problem

Do your work for this exercise in

___IGIS-Arc_*YourInitialsHere*\Wildcat_Boat_Data. If you have any doubts about the correctness of the Wildcat_Boat_Data folder delete it from your folder and copy it from IGIS-Arc_AUX. You will use the data sets you make here in the next exercise.

_____ **1.** Start ArcCatalog. Using the `Multiple Ring Buffer` tool, make a `20-meter` buffer around the Streams PGDBFC. Call it Streams_mpp (for multipart polygon) and put it in the *Area_Features* PGDBFD. Make the `Field Name NoBuild`.

_____ **2.** Use the `ArcToolbox > Data Management Tools > Features > Multipart to Singlepart` tool, explode Streams_mpp into Streams_buf.

_____ **3.** Using the `Multiple Ring Buffer` tool, make a `300-meter` buffer around the Sewers PGDBFC. Call it Sewers_mmp and put it in the *Area_Features* PGDBFD. Make the `Field Name Build`.

_____ **4.** Using the `Multipart To Singlepart` tool, explode Sewers_mpp into Sewers_buf.

_____ **5.** Delete Streams_mpp and Sewers_mpp. Close ArcCatalog.

Exercise 7-8 (Project)

Finding a Site for the Wildcat Boat Facility

You will recognize this problem as the one you solved by manual means at the beginning of Chapter 1. Throughout the text, you have experimented with the data. In Exercise 7-7, you made the buffers around the streams and sewers that were required. In this exercise you finish the work of finding suitable sites. To restate the problem:

Wildcat Boat Company is planning to construct a small testing facility and office building to evaluate new designs. They've narrowed the possibilities down to a farming area near a large lake. The company now needs to select a specific site that meets the following requirements:

❑ Must reside on soils suitable for construction of buildings. (The value of the soil suitability must be 2 or 3.)

❑ The site should not have trees (to reduce costs of clearing land). A regional agricultural preservation plan prohibits conversion of farmland. The other land uses (urban, barren, and wetlands) are also out. So the land cover must be "brush land" (which has Landcover code 300).

❑ A local ordinance designed to prevent rampant development allows new construction only within 300 meters of existing sewer lines.

❑ A recent national water quality act requires that no construction occur within 20 meters of streams.

❑ To provide space for building and grounds the site must be at least 4000 square meters in size.

The data for this exercise is in

____IGIS-Arc*YourInitialsHere* /Wildcat_Boat_Data. This folder contains the personal geodatabase feature classes that describe the area from which the site must be chosen. Your assignment is to present a map showing *all the areas* that meet the requirements stated in the preceding list. Those areas should be contained in a single feature class named Final_Sites in your

___IGIS-Arc*YourInitialsHere*\Wildcat_Boat_Data folder

1. Start ArcCatalog. Show ArcToolbox. Make or confirm a folder connection with
___IGIS-Arc_*YourInitialsHere*\Wildcat_Boat_Data

2. In ___IGIS-Arc*YourInitialsHere*\Wildcat_Boat_Data\

Wildcat_Boat.mdb\Area_Features

you should find the following PGDBFCs:

      Landcover

      Sewers_buf

      Soils

      Streams_buf

Using the four preceding data sets and ArcToolbox > Analysis Tools > Overlay > Union, make a PGDBFC named Union_mpp in

___IGIS-Arc*YourInitialsHere*\Wildcat_Boat_Data\

Wildcat_Boat.mdb\Area_Features.

Union_mpp contains multipart polygons. You need to make sure each set of boundaries defines an un-subdivided area.

____ 3. Use the `Arctoolbox > Data Management Tools > Features > Multipart To Singlepart` tool to explode Union_mpp into Union_Lc_Se_So_St.

____ 4. Carefully, very carefully, examine the table of Union_Lc_Se_So_St. Ignoring the size criterion of 4000, write the relevant fields, values, and operators (like = or <> or >=) you must use to extract the sites meeting the criteria in the table that follows. (Some values have been filled in for you.)

Fields	Operators	Values	(source PGDBFC)
		300	Landcover
			Sewers_buf
			Soils
NoBuild	=	0	Streams_buf

____ 5. Write the expression that would extract the suitable polygons from Union_Lc_Se_So_St:

_____

(*After* writing your expression, check it against Figure 7-23.)

____ 6. Using `ArcToolbox > Analysis Tools > Extract > Select`, make a PGDBFC named Sites_A. Look at the attribute table of Sites_A to be sure that the values you specified for the various fields are proper. How many sites are there? _____

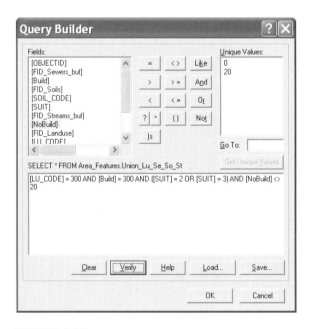

FIGURE 7-23

___ 7. Display Sites_A, checking to be sure that it makes sense. You will see some small areas that obviously will not be candidates. But in the center note the two polygons that share a common boundary. Neither by itself is large enough to meet the size criterion, but together they do. Use Identify to understand why they separated, and explain.

_____.

___ 8. You need to dissolve the boundary between those adjacent areas and make them a single polygon. When faced with a geographic problem like this, you can be pretty sure that, among the hundreds to tools supplied by ArcGIS, there will be one to solve it. Use the ArcToolbox Index to find a Dissolve tool that works on geodatabases. Locate it. What is the path?

_____

___ 9. Start the `Dissolve (management(not arc))` tool. Explore the `Help` files of the tool to understand what it does.

In this case, since all the polygons of Sites_A meet our criteria for the Wildcat Boat facility, you can just use dissolve without specifying any fields, to make Sites_B.

___ 10. Using the `Dissolve (management)` tool, make the `Input Features` Sites_A and the output Sites_B. Compare the Geography of Sites_A with that of Sites_B. Note that the lines that separated some of the polygons are gone. Since, again, the tool makes multipart polygons by default, explode Sites_B to make Sites_C. How many polygons are there in Sites_C? _____

At this point each of your polygons should have satisfied all criteria except that the area must be greater than 4000.

___ 11. Use the `Select` tool on polygons in Sites_C to make Final_Sites, such that Final_Sites contains only those polygons whose `Shape_Area` is equal to or greater than 4000. Look at the Geographics. How many polygons are left now? _____. Look at the attribute table. How many fields (columns, attributes) are left? _____. Final_Sites is the solution to the problem that you have been working on since Chapter 1!

___ 12. Launch ArcMap. Add as data Final_Sites. Use a bright green color. Also, from Line_Features add Sewers as black lines. Zoom to full extent. Use the `Measure` tool to verify that all no part of any polygon is more than 300 meters away from a sewer line.

___ 13. Add Streams as blue lines. Verify that no part of any polygon is closer than 20 meters to any stream. Zoom to the extent of the Sewers. Your map should look like Figure 7-24***.

___ 14. Compare your final map with the map you made when you worked Exercise 1-1. Give yourself a grade, based on 100, evaluating how well you did at solving this problem by hand. _____. If you didn't do very well don't be too unhappy about it. If problems like this were easy to do well by hand society wouldn't be spending billions of dollars on computer-based GIS.

___ 15. Make a layout of the results of this exercise, using your own judgment as to what to include and how to present the various layers. Print your map, preferably in color. Close ArcMap.

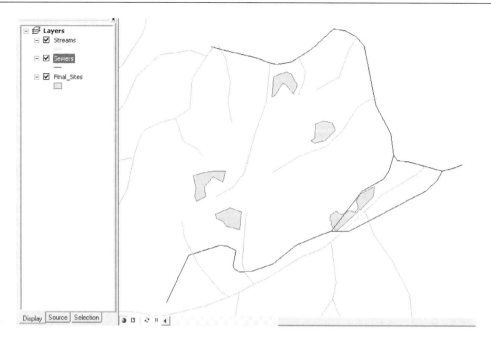

FIGURE 7-24

Exercise 7-9 (Project)

Solve a Revised Wildcat Boat Problem

As often happens when some time has gone by since a problem was posed (in this case you were busy learning GIS), the client suggests additional requirements:

1. The cost of each site should be available.

2. Unless the area is more than 7000 square meters the eccentricity ratio (the perimeter divided by the square root of the area) must be no greater than 4.6, so that the site will not be too narrow or strangely shaped.[7]

Requirement #1 should be easy to take care of. You recall that the Landcover feature class had a field called COST_HA which stood for cost per hectare.[8] So a simple computation (like dividing the area by 10,000 and multiplying by the cost per hectare) should give us the cost of each site. The difficulty, which you can verify by looking at the table of Final_Sites, is that you no longer have the COST_HA field. It went

---

[7]This ratio (perimeter to square root of area) provides a measure of how "eccentric" a plane figure is. A circle, which has the least perimeter for enclosing a given amount of area, has an eccentricity ratio of about 3.5. A square's ratio is 4.0. A rectangle that is three times as wide as it is tall has a value of approximately 4.6.

[8]Recall, a hectare (abbreviation HA) is 10,000 square meters. For example, a square 100 meters on a side is a hectare. The area is very close to 2.5 acres.)

away when you did the Dissolve. The last time you saw COST_HA was in the table for Sites_A. So you need to find a way to preserve the COST_HA field but still dissolve boundaries between smaller, adjacent polygons that meet our size criterion when they are merged together. This calls for a closer look at the Dissolve tool.

Understanding Dissolve

If you use Dissolve to make feature class Y from feature class X, and you don't specify any Dissolve Fields, then no lines that separated polygons in X will show up in Y. Further the attribute table of Y will contain only the fields OBJECTID, Shape, Shape_Length, and Shape_Area.

However, suppose you specify a Dissolve Field in the Dissolve tool window. Then a boundary line between each pair of adjacent polygons of X is erased in Y if, and only if, the two attribute values for that field are the same for each polygon. In any event, the specified field is retained in the attribute table of Y.

For example, suppose X has an attribute (field) titled Owner. Assume there are two polygons (one might be growing wheat and the other corn) and they are adjacent—sharing a common border. Assume now that the Dissolve tool, specifying a Dissolve Field of Owner, is run on X to produce Y. If the Owner attribute value for each is the same (say 'Brown') then the boundary line will not appear in Y and the attribute table will reflect a single record, with Owner 'Brown.' However, if 'Brown' owns one polygon and 'Smith' owns the other then the boundary line between them will be present in Y and there will be a record in the attribute table for each polygon.

___ 1. Start ArcCatalog. Look at the geography and the table of Union_Lc_Se_So_St. Use `Dissolve (management)` to bring up the Dissolve window. For `Input` use Union_Lc_Se_So_St. For output use Dissolve_All. In `Dissolve_Fields` leave all fields unchecked. Examine the geography of Dissolve_All. Note that all boundary lines are gone. All the attributes except the basic ones that come with any PGDBFC are gone.

___ 2. Again use `Dissolve (management)` to bring up the `Dissolve` window. For `Input` use Union_Lc_Se_So_St. For `Output` use Dissolve_None. In `Dissolve_Fields` check all fields. To be sure each polygon is separate, run the `Multipart To Singlepart` tool on Dissolve_None, making Dissolve_None_SPP. Examine the geography and table of Dissolve_None_SPP. It should be virtually identical with Union_Lc_Se_So_St, with no boundary lines gone and all the attributes the same.

___ 3. Finally, use `Dissolve (management)` to bring up the `Dissolve` window. For input use Union_Lc_Se_So_St. For output use Dissolve_Suit. In `Dissolve_ Fields` check SUIT. Run the `Multipart To Singlepart` tool on Dissolve_Suit, making Dissolve_Suit_SPP, to be sure each polygon is separate. Examine the geography of Dissolve_Suit_SPP using `Identify`. If two adjacent areas had the same suitability (say they were both 2), the boundary line between them is erased. If the suitabilities were different, the boundary line was retained. Examine the table of Dissolve_Suit_SPP. Note that the attributes are mostly gone, but the attribute field SUIT remains.

So, Dissolve does several things at once: If you check a field then that field is preserved in the attribute table on the output. If two adjacent areas have different values in that field, the boundary line between them is preserved. If two adjacent areas have the same values in that field, the boundary line between them is erased.

_____ **4.** Look at Sites_A. Use Identify on Sites_A to verify that the COST_HA is the same for the areas on both sides of the boundary lines that you dissolved earlier.

_____ **5.** Housekeeping time: Use ArcCatalog to delete Sites_B, and Sites_C. You can also rid yourself of any feature classes that end in mpp or whose names begin with Dissolve. Close ArcCatalog.

Making New Sites Including the COST_HA Field

_____ **6.** Start ArcMap. Show ArcToolbox. Bring up the Dissolve (management) tool. Pull the vertical slider bar all the way down and notice that, lurking at the bottom of the window is a check-box labeled Multipart (optional). You could conjecture that, if you uncheck that box the tool will make single-part polygons, thus saving you from the trouble of exploding the multi-part polygons. Uncheck the box! Remake Sites_C from Sites_A, using Dissolve, but this time put a check in the COST_HA box, to preserve this field in the Sites_C table. Use extraction to make Sites_D by applying the "equal to or greater than 4000" criterion to the Shape_Area of Sites_C. Look at the table of Sites_D and verify that you do indeed have the COST_HA field. Also verify that the boundaries you wanted to lose are gone. (There is the one little worry that two adjacent polygons might have different land cover values, and that by running the Dissolve tool we might have left some troublesome boundaries. Here is where visual inspection gets into the process. Examine Sites_D to assure yourself that no connected polygons exist.)

_____ **7.** Find the Options button on the attributes of Sites_D window. Add a field named SITE_COST to the Sites_D table. Its type should be Float, because its values will be calculated with an expression involving the floating-point numbers of Shape_Area. If any of the operands of an expression is floating-point, then the results must be also.

_____ **8.** Add Sites_D to the map. Open its attribute table.

_____ **9.** Right-click the SITE_COST column head. Select Calculate Values. Ignore the warning. In the Field Calculator window, part of the calculation (SITE_COST =) is already done for you. You simply need to supply the rest. Since our Shape_Area is in square meters and our cost is in hectares, calculate SITE_COST as

Shape_Area / 10000 * COST_HA

_____ **10.** Run statistics on the table. What is the cost of the average site? _____

Considering the Eccentricity Criterion

We have one more selection to perform. Recall that the eccentricity may not be more than 4.6

_____ **11.** Add another field to the table of Sites_D. Name it Eccentricity. Since the number will contain decimal places and we are not sure of its range, specify the Type as Float. OK.

_____ **12.** Right-click the Eccentricity column head; pick Calculate Values. Ignore the warning. Write the expression you need to use in the following blank: _____.
Put the expression in the Field Calculator window. Then check it against Figure 7-25. When you have it right, press OK.

The Attributes of Sites_D window should have eccentricities ranging from just over 4 to about 6 and a half.

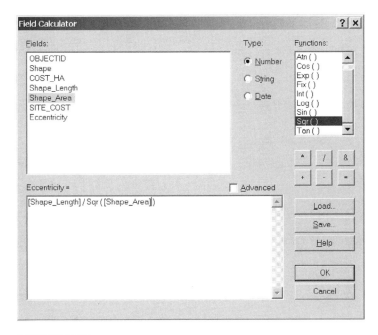

FIGURE 7-25

___ **13.** Write the SQL Expression that would let you use Sites_D to make a PGDBFC called Final_NEW_Sites, that satisfies this additional requirement. _____ _____. Wait to look at Figure 7-26 until you have written your expression. Extract the proper polygons. How many are there?

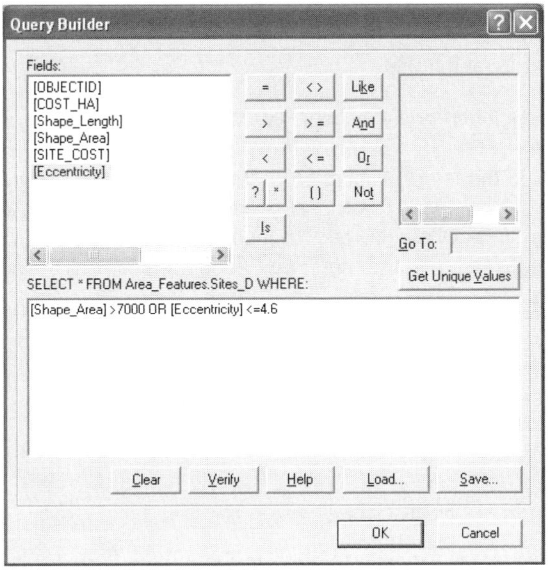

FIGURE 7-26

___ **14.** Display FINAL_NEW_SITES, showing the sites in bright yellow. Provide graphic data from the other data sources to show the relative location of the sites. Show the following:

❑ The sewers

❑ The streams

❑ The lines of the sewers buffer emphasized in red, width 3

❑ The lines of the streams buffer emphasized in black, width 2

❑ The polygons of land cover, with land cover code 300, emphasized and others in subdued colors

❑ The polygons of soils, with suitabilities 2 and 3 emphasized and the others in subdued colors

___ **15.** Zoom in on the group of sites. With the Identify tool, examine the resulting map to assure yourself that the criteria have been met. How many areas are there? _____

___ **16.** Make a layout of the map. Print the layout, using a color printer if possible.

Exercise 7-10 (Project)

Making a Model of the Wildcat Boat Solution

___ **1.** Start ArcCatalog. Make a folder named Wildcat_Boat_Model in

___IGIS-Arc_*YourInitialsHere*. Make a folder connection to Wildcat_Boat_Model.

From [___] IGIS-Arc_AUX\Wildcat_Boat_Data

copy the Wildcat_Boat.mdb into

___IGIS-Arc_*YourInitialsHere*\Wildcat_Boat_Model.

___ **2.** Make a toolbox in the folder Wildcat_Boat_Model. Highlight the newly created toolbox and choose File > New > Model. A Model window opens. Close the Model window. Expand the Toolbox icon. In the Catalog Tree, change the name Model to Wildcat_Boat_Model. Right-click the new name. In the menu that appears, click Edit. The Wildcat_Boat_Model window appears.

___ **3.** Construct a solution to the Wildcat_Boat problem, *placing the Intermediate Data and Final_Sites in*

___IGIS-Arc_*YourInitialsHere*\Wildcat_Boat_Model\

Wildcat_Boat.mdb\Area_Features.

Use the following tables for hints to building the model. Periodically run the model and check Area_Features so that you know things are operating properly.

	FROM	WITH	MAKE
__	Streams	Multiple Ring Buffer	Streams_mpp
__	Streams_mpp	Multipart to Singlepart	Streams_buf
__	Sewers	Multiple Ring Buffer	Sewers_mpp
__	Sewers_mpp	Multipart to Singlepart	Sewers_buf
__	Landcover	Union	
__	Sewers_buf		
__	Soils		
__	Streams_buf		Union_mpp
__	Union_mpp	Multipart to Singlepart	Union_Lc_Se_So_St

Run the model to this point so that the extraction will have the proper data available.

__	Union_Lc_Se_So_St	Extract > Select	Sites_A
__	Sites_A	Dissolve (COST_HA)(singlepart)	Sites_C
__	Sites_C	Extract > Select (4000)	Sites_D
__	Sites_D	Fields > Add Field[9] (Site_Cost)	Sites_D (2)
__	Sites_D (2)	Fields > Calculate Field[10]	Sites_D (3)
__	Sites_D (3)	Fields > Add Field (Eccentricity)	Sites_D (4)
__	Sites_D(4)	Fields > Calculate Field[11]	Sites_D (5)
__	Sites_D (5)	Extract > Select	Final_Sites

Exercise 7-11 (Review)

Checking, Updating, and Organizing Your Fast Facts File

The Fast Facts File that you are developing should contain references to items in the following checklist. The checklist represents the abilities to use the software you should have upon completing Chapter 7.

Important note: This checklist is on the CD-ROM that accompanies the book. It is available in Microsoft Word format. Rather than typing or writing by hand the text that follows, you can copy and paste it into your Fast Facts File from the CD-ROM file.

____ In buffers inside and outside are distinguished

____ To start ArcInfo Workstation

---

[9]Since the model runs automatically, you can't use the Add Field in the Options window of the attribute table. So look for the Add Field tool in ArcToolbox

[10]Since the model runs automatically you can't get the Calculate Field window by right clicking a column head. So look for the Calculate Field tool in ArcToolbox

[11]Unfortunately, there is no square root capability in the ArcToolbox field calculator. But you can work around that by simply squaring both sides of the equation and then asking the "less than" question. That is, is $P*P/a < 4.6*4.6$.

___ The WORKSPACE command

___ The LIST command

___ To load the Buffer Wizard

___ To bring more than one geodata set into ArcMap simultaneously

___ The buffer wizard indicates areas inside the buffer by BufferDist value of

___ When you use calculate to change cells in records and some records are selected

___ The Multiple Ring Buffer tool allows the user to

___ When one polygon data set is overlaid with another, the attribute table of the resulting data set

___ The difference between Union and Intersect is

___ When you overlay two polygon geodata sets, the number of both _____ and _____ is likely to increase in the polygon attribute table.

___ To form a new data set from attributes of an existing data set

___ SQL means

___ When a point coverage is overlaid with a polygon coverage

___ When a line coverage is overlaid with a polygon coverage

___ When a point or line coverage is overlaid the overlay coverage is always

___ Identity and Intersect are different in that

___ To join a table to a geodatabase feature class

___ To convert multipart features to single-part features

___ Building a model of a set of GIS operations has several benefits:

___ To build a model in a toolbox

___ Models can be exported to three different scripting languages:

___ Dissolve makes new feature classes by

Spatial Analysis Based on Raster Data Processing

OVERVIEW

IN WHICH you move into an entirely new realm of GIS analysis and geo-processing: solving problems with a method of data representation—called raster, grid, or cell-based—that is dramatically different from what you are used to.

A Different Storage Paradigm[2]

You have met rasters before in this book. Most images are stored in raster format. The COLE_DEM of Chapter 1 stored elevations in floating-point raster format. And MSU_SPNORTH stored landcover in integer raster format. What is new here is that you will do GIS analysis using raster representation.

To be candid, it would take another book the size of this one to cover the immense subject of raster processing. Up until a few years ago, it was almost completely divorced from vector GIS processing. ESRI has done a good job in taking down the barriers between the two. In terms of GIS data model development, raster processing (or grid processing, or cell processing, which are other terms for it) preceded vector processing. It was, at one time, about the only way to get large amounts of geographic data into the memories of the then small computers. Vector processing, with its greater capacity for precision, tended to supersede it in terms of popularity for solving GIS problems. It is now understood that raster processing has some properties that make it better for some applications than vector processing. Specifically, analysis is why GIS analysts use it now—instead of just its advantages in speed of computation, and, sometimes, conserving computer memory. Since you now have considerable background in the basics of GIS, and since this is the only chapter that is devoted to raster processing, the descriptions of the basics of raster processing will be succinct.

Raster is faster but vector is corrector.

Berry, JK, 1995[1]

[2]You actually have met rasters before in this text. As you have probably inferred by now, in GIS "paradigm" is used to describe a method of storage of geographic data. In the larger view it is a set of concepts and practices that constitutes a way of viewing reality, usually for a particular discipline.

---

[1]In GIS World, 8(6):35-35. But the matter is more complex than that as you will see. For one thing it depends on the application. For another, *everything we do in GIS, vector or raster, is an approximation.*

The major package in Arc9 that deals with rasters is called Spatial Analyst.[3] The menu of the Spatial Analyst toolbar has more than ten choices. Included are as follows:

- ❏ Distance
- ❏ Density
- ❏ Interpolate to Raster
- ❏ Surface Analysis
- ❏ Cell Statistics
- ❏ Neighborhood Statistics
- ❏ Zonal Statistics
- ❏ Reclassify
- ❏ Raster Calculator
- ❏ Convert

One of these choices is a Raster Calculator which lets you use a "Map Algebra"[4] to calculate new grids, making use of hundreds of functions, commands, and operators that are part of the ESRI GRID software. Also, if you look in ArcToolbox, you will find that the Spatial Analyst Tool Chest has 18 toolboxes:

- ❏ Conditional
- ❏ Density
- ❏ Distance
- ❏ Extraction
- ❏ Generalization
- ❏ Groundwater
- ❏ Hydrology
- ❏ Interpolation
- ❏ Local
- ❏ Map Algebra
- ❏ Math
- ❏ Multivariate
- ❏ Neighborhood
- ❏ Overlay
- ❏ Raster Creation

---

[3]The package that dealt with raster processing in ArcInfo versions before ArcGIS was called GRID. Most GRID commands have either been superseded by Spatial Analyst commands or are available through Spatial Analyst.
[4]The idea of "Map Algebra" was developed by Dana Tomlin and presented, in 1990, in a book entitled *Geographic Information Systems and Cartographic Modeling*.

□ Reclass

□ Surface

□ Zonal

Within these toolboxes are more than 150 tools. It is true that there is considerable overlap in these software tools. They represent an evolution of raster (GRID) processing techniques developed over decades. But even considering these overlaps, one cannot deny the conclusion: Spatial Analyst is *big*!

Facts About Rasters

Each raster data set must have a name, just as a coverage or shapefile or geodatabase featureclass must.

A raster database is composed of raster data sets. A raster data set may have multiple bands. If so, Spatial Analyst operates on band 1.

Rasters can be either geographic data or image data. We mostly discussed the structure of image data in Chapter 4. Here we will concentrate on geographic data.

A raster should encompass a study site. The raster representing the study area is called a Cartesian matrix, usually with rows parallel to the x-axis of a coordinate system. Columns are parallel to the y-axis.

A geographic raster is a rectangular matrix of square cells. Each cell is a member of a row and a member of a column. All the cells of a given raster are the same size. See Figure 8-1.

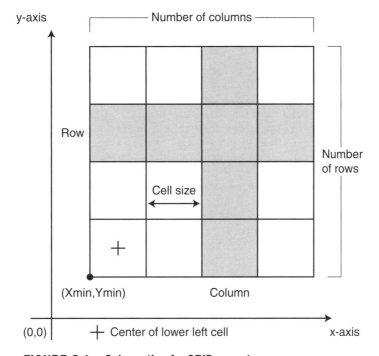

FIGURE 8-1   Schematic of a GRID or raster

Raster cells may be almost any size. Choosing the cell size that is best for both representing the features of the Earth and for analysis is a science and an art. It is not unusual to have raster composed of hundreds of millions of small cells. If you choose a cell size that is half of what you were previously considering, the new raster will have four times as many cells. This may or may not mean four times the storage requirement (depending on the compression technique used) and four times the processing time.

Coordinate Space

A rasters' coordinate system may be in real-world coordinates or "image space." For those in geographic space, cells are referenced primarily by an (x,y) location in map coordinate space, not by row and column numbers, as previous versions of grid processing were. The location referenced for a given cell is the center of that cell. For the overall raster, the geographic reference is usually the center of the lower left (southwest) cell. Cells are square in map coordinates.

Each cell represents a specific location on the surface of the Earth. While each cell of a raster is square, the surface area of the Earth it represents may not be. Such real-world areas should be almost square, however. If the northing span of a raster is too great, the representation suffers. For example, take a UTM zone whose "base" is at latitude 40 degrees and "top" is at latitude 48 degrees. Suppose the cells are 100 meters on a side. Let's say a cell at the base would represent 10,000 square meters. A cell at the top of the zone, admittedly more than 500 miles away, would represent only about 8,800 square meters!

Rasters may be described as integer or floating-point. All the cells in an integer raster contain integers or, if not, are designated NoData. All the cells in a floating-point raster contain floating-point numbers, or, if not, are designated NoData.

Rasters with Integer Cell Values

Rasters with integer cell values are used to store what is variously called discrete data or categorical data. Such data sets describe a phenomenon or object that exists at the location of that cell.

Each cell in an integer raster belongs to a *zone*. The zone number must be an integer—positive, negative, or zero. Cells in a raster that have the same value all belong to the same zone. So a zone consists of all the cells with a given value. Each cell belongs to some zone or other, even if the zone is composed of only one cell. Cells in a zone may be adjacent, but may well not be. See Figure 8-2***.

An integer raster[5] has associated with it a value attribute table (VAT). The table may have many fields but the two primary fields are Value and Count.

There is a record in the VAT for each zone. The raster shown in Figure 8-2, consists of eight zones. The zone numbers are 21, 22, 23, 24, 25, 26, 27, and 29. Obviously, sets of cells in the same zone may be connected or not.

The raster shown Figure 8-2, which we will call Basic_Raster, would have a VAT as shown in Figure 8-3. It indicates, for example, that Zone 21 consists of six cells. Figure 8-2 shows you that the cells with value 21 are not all connected.

------

[5]And some floating-point rasters have VAT's as well, as you will see.

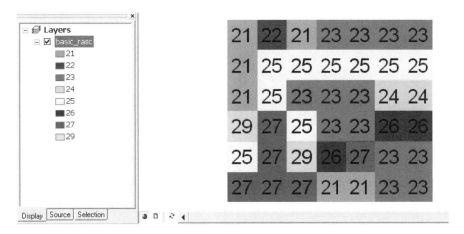

FIGURE 8-2   Basic_Raster with integer values indicating categories

Attributes of basic_rasc

ObjectID	Value	Count
0	21	6
1	22	1
2	23	13
3	24	2
4	25	9
5	26	3
6	27	6
7	29	2

Record: 14 4 [1] ▶ ▶| Show: All   Selected   Records (0 out of 8 Selected.)

FIGURE 8-3   Basic_Raster Value Attribute Table (VAT)

There is a term for sets of cells that have the same value (i.e., are members of the same zone) and that *are* connected. That term is "region."[6] Regions are subsets of zones: They are connected cells of the same value. With a command called RegionGroup, the cells in each region can be identified. Regions can be defined in two different ways. A region of Type IV I will define as one that consists of all the cells that have the same value and are connected by sharing a side; thus, each cell may have four (IV) neighbors. A region of Type VIII consists of cells that all have the same value and are connected by sharing a side *or touching at a corner*; thus, each cell may have eight (VIII) neighbors. (There are only these two types.)

Figure 8-4*** shows the regions of type IV. Note from 8-2 that Zone 23 has 13 cells. If RegionGroup is applied to Basic_Raster, and Type IV regions are asked for, the result looks like Figure 8-4 with attribute table shown in Figure 8-5. Note that the 13 cells of Zone 23 are divided into three groups; the region numbers are 4, 6, and 16. The Link number tells the value in the original raster.

---

[6]If you needed proof that raster models and vector models grew up independently, you have only to look at the definitions of region. For vector coverages, a region is defined as "a coverage feature class that can represent a single area feature as more than one polygon." Those polygons could be, and frequently are, disconnected.

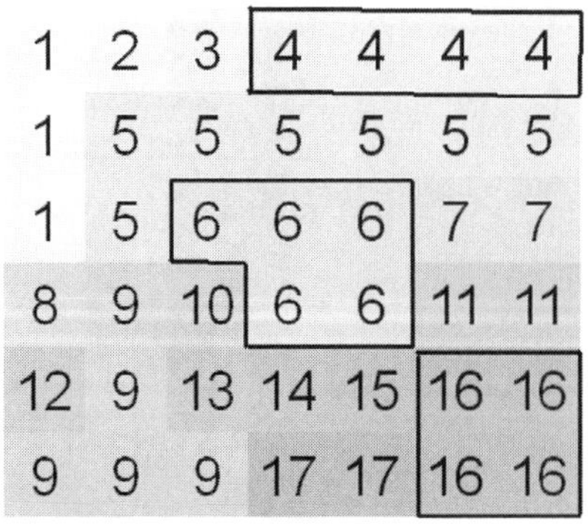

FIGURE 8-4   Regions of the
Basic_Raster (neighbors sharing sides,
not corners)

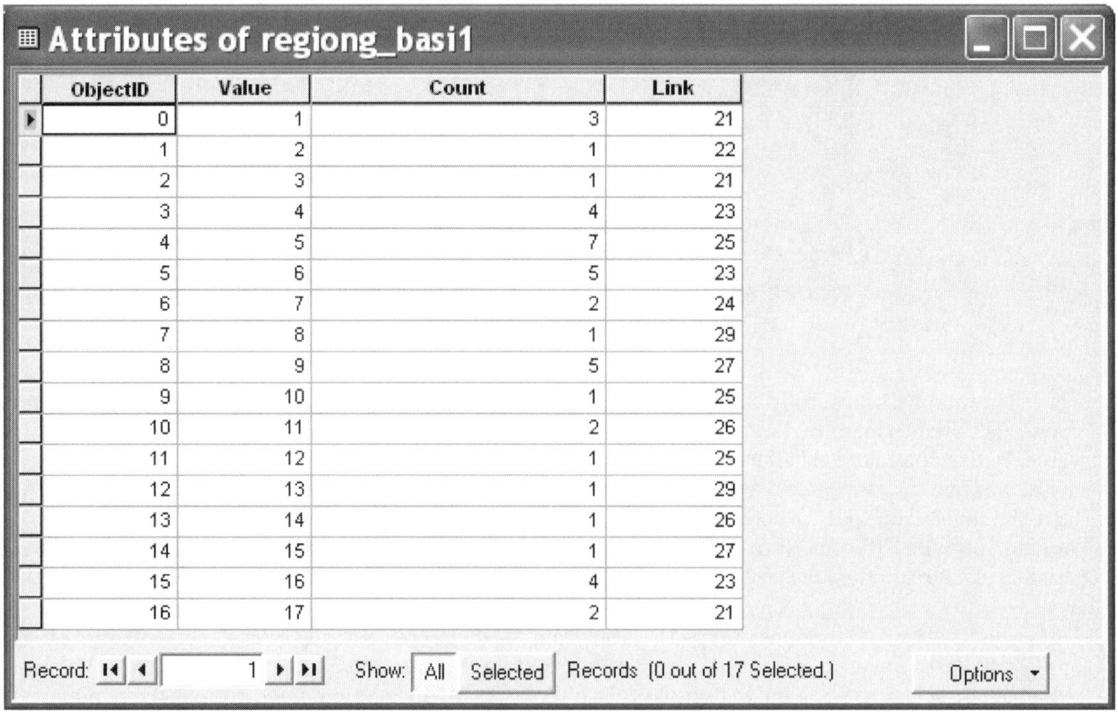

ObjectID	Value	Count	Link
0	1	3	21
1	2	1	22
2	3	1	21
3	4	4	23
4	5	7	25
5	6	5	23
6	7	2	24
7	8	1	29
8	9	5	27
9	10	1	25
10	11	2	26
11	12	1	25
12	13	1	29
13	14	1	26
14	15	1	27
15	16	4	23
16	17	2	21

Record: I◀ ◀   1 ▶ ▶I   Show: All Selected   Records (0 out of 17 Selected.)   Options ▼

FIGURE 8-5   Attribute table of the type IV regions of the sample raster

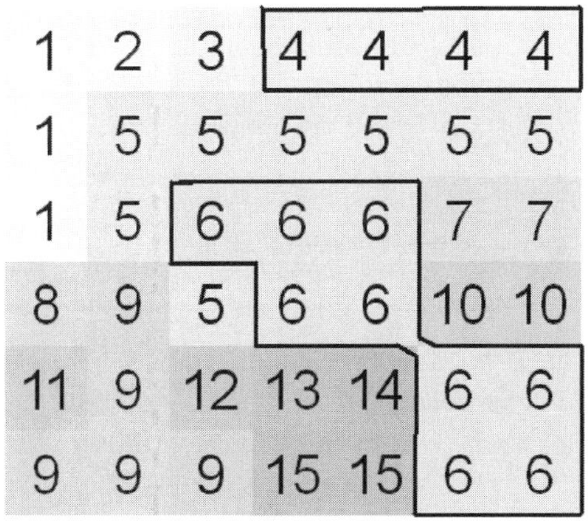

**FIGURE 8-6   Regions of the Basic_Raster
(neighbors sharing sides and corners)**

If RegionGroup is applied to Basic_Raster, and Type VIII regions are asked for, the result looks like Figure 8-6***. Note that the 13 cells of Zone 23 are divided into only two groups; the region numbers are 4 and 6.

When rasters are composed of integers, those integers generally represent categories. They are quite frequently just nominal data. It is useful if the categories are present in the table as well, so they can be used in selection. For example, suppose the Basic_Raster numbers referred to zoning. Perhaps the categories would be as follows:

21 Heavy Industrial

22 Light Industrial

23 Commercial

24 Parkland

25 Agricultural

26 Multi-Family Housing

27 Single Family Detached

29 Single Family Attached

Then a table might be created that looks like Figure 8-7.

As mentioned, cells may contain NoData. How NoData cells are handled during analysis operations depends on the particular analysis tool and/or decisions made by the user. Sometimes NoData is ignored. Sometimes NoData prevents the operation from occurring. There is not such a thing as a "NoData zone." NoData is not represented in the attribute table, but when the raster is displayed you can indicate how you want NoData symbolized.

	OID	Value	Count	ZoningDesc
▶	0	21	6	Heavy Industrial
	1	22	1	Light Industrial
	2	23	13	Commercial
	3	24	2	Parkland
	4	25	9	Agricultural
	5	26	3	Multi-Family Housing
	6	27	6	Single Family Detach
	7	29	2	Single Family Attach

Record: ◄◄ ◄ [1] ► ►I   Show: All | Selected   Records (0 out of 8 Selected.)

FIGURE 8-7   A raster attribute table enhanced with an additional column

Rasters with Floating-Point Values

Rasters with floating-point cell values are used to store what is called continuous data. Such data may also be referred to as nondiscrete data, field data, or surface data. The numbers are stored in a format that allows many significant digits and great range. A given cell may well have a different value in it than any other cell in the raster. There is no concept of zone (most of the zones would probably consist of a single cell) and, therefore, usually, no value attribute table. The values in a floating-point raster usually represent magnitude, elevation, distance, or relationships of cells to other nearby cells in that or some other raster. A major use for floating-point rasters is to represent surfaces, such as elevation of the Earth's surface, or surfaces derived from point data such as rainfall over an area.

What Is Raster Storage and Processing Good For?

To start with, rasters can be used to solve the same sorts of problems that vectors can. And as computers have become more powerful (bigger memories, higher speeds, lower costs, improved compression techniques), rasters are rivaling their vector cousins in solving problems like the Wildcat Boat facility siting problem. Raster methods may suffer, when tightly zoomed in during display, from what is inelegantly called the "jaggies" (see Figure 8-8***), where polygonal and linear boundaries are shown in stair-step fashion. This actually may not be a bad thing in most cases, since the precise lines separating polygons usually imply a level of accuracy that does not exist. In human-made phenomena (like parcel boundaries or building footprints), there are such precise lines, but they may not be represented in exactly the right place. In natural phenomena (e.g., the edge of a forest), there is usually little chance that the precise one-dimensional line accurately represents an actual boundary. Here the jagged boundary is a nice reminder that we don't know precisely where the boundary is—at least partially because there isn't a precise boundary.

A main objection to using raster methods on the kinds of problems we have used vector methods to solve is that the minimum units for line and point representation all have to be the same size. So a tree and a highway wind up being the same width. But, as has often been stated in this text, GIS only serves up an approximation of the real world, and the proof of its utility is in whether it promotes comprehension of the real world, not whether it mimics the real world.

Another problem is that one size does not fit all. You probably wouldn't want to store moose habitat at the same cell size as soil type. Adjustments can be made during analysis—generally toward using the largest cell size—but again there are compromises.

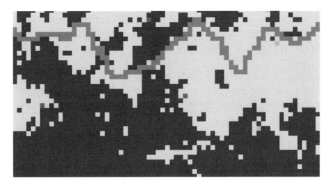

FIGURE 8-8   Raster representation creates jagged lines and polygons

Another use of rasters—an important use: A raster can pretend it is a surface. A surface is a mathematical construct such that, for every value of x and y on a horizontal Cartesian plane, there is a single z value that indicates a displacement from that plane. Any change in x and/or y, no mater how tiny, may result in a change in the z value. In pretending to be a surface, a raster cell clumps a whole bunch (an infinite number) of x-y points together and gives them a single z value. The surface is therefore lumpy. Thus, like much of GIS, it provides an approximation to the hypothetical (mathematical) world, which is endeavoring to provide an approximation of the real world.

Rasters can be effectively used in simulations. Given a raster of a forest fire covering a certain area, you could plug in other rasters that showed wind speed and direction, slope, concentrations of fuel, and so on. From this you could make predictions about where the fire would go next. The fact that each cell relates geographically to its neighbors in only one of two ways provides tremendous modeling and computer advantages.

Another simulation use is to represent, over time, the effects or material as they emanate from a source such as a pollution point. Further, floating-point rasters can be used to represent costs of moving or transporting objects across a surface.

Rasters and Features

As you work through this chapter, you will get a feel for the differences between vector and raster representation. Here I simply want to let you know that you can convert from one data model to the other—albeit with loss of information each time.

Let's consider operations in which the inputs are feature classes and the outputs are rasters: the simple conversion from feature data set to grid data set. To do these conversions, you can directly use the spatial analyst toolbar, use the raster calculator (available through the spatial analyst toolbar), or use ArcToolbox.

If you convert a point feature class to a raster, each cell that has a point within its boundary takes on the value of that point. If more than one point exists within the boundaries of a given cell, the software randomly picks one of them to represent in the cell. In terms of attributes, the VAT value is taken from the feature attribute table field that is specified by the user. If that field is an integer number it becomes the VALUE attribute. If the user-specified field is a floating-point number that number becomes VALUE in the VAT. If the user-specified field contains text then an additional field is added to the VAT, containing the field name, and the VALUE field becomes a randomly assigned integer.

If you convert a line feature class to a raster, each cell that contains any part of any line takes on the value of the line, according to the field specified by the user. The rules for what goes in the VAT are the same as for points, discussed previously. So a line in a feature class becomes a set of connected cells. Each adjacent pair is connected by either a side or a corner.

If you convert a polygon feature class to a raster, each cell that contains any part of a given polygon takes on the value of that polygon, according to the field specified by the user. The rules for what goes in the VAT are the same as for points, discussed previously. Therefore a feature-class polygon is represented by a (usually connected) set of cells.

Rasters: Input, Computation, and Output

It would be less than candid to suggest that using spatial analyst for analysis is anything but complex. A large number of ideas, approaches, techniques, and data types are in play here. Let's return to the most fundamental level of computer operation. A computer reads bits, stirs bits, and writes bits. That is, there is input, computation, and output.

For any operation in SA, the inputs can be

❏   One or more rasters

❏   One or more feature layers

❏   Parameters specified at the time of the operation

❏   Settings provided by an spatial analyst environment

For any operation in spatial analyst the outputs can be

❏   A raster

❏   A feature layer

❏   Statistics

You have a good appreciation for feature layers by this time. You are learning about rasters: what a raster is and some of its characteristics. Parameters and statistics have been part of many of the operations you have applied to vector data sets. Environment settings will be discussed in the Step-by-Step section of this chapter.

Where Raster Processing Shines: Cost Incurred Traveling over a Distance

Calculating costs (or cost-weighted distance) has applications in many fields. If you were siting a business that needed to be accessible, you could use the ArcMap Spatial Analyst Cost Distance tool to measure distance to important points in a road network. If you were doing animal migration studies, you could use it to measure migration routes. You could also use cost distance calculations to determine sites for new roads, pipelines, or transmission lines. In fact, this approach has been used in Alaska as part of a project to determine the impact of building pipelines. You will work a problem on this topic in the Step-by-Step section of this chapter.

It is important to mention another ArcMap tool designed specifically for moving across distances to points connected by linear features, such as roads or railroads. The ArcMap Network Analyst allows you to consider additional factors, including one-way streets and costs of turning at intersections. Network Analyst is very powerful, and if your problem or application is specific to street or road networks, you should become familiar with this extension, which was introduced with ArcGIS version 9.1 Chapter 9 contains an exercise using Networks Analyst.

Proximity Calculation with Rasters

GIS is meant to help humans in understanding both natural and human-generated phenomena. A fundamental concept in both arenas is the idea of closeness. Raster analysis can play a key roll. We will focus on the following:

❑ The importance of distance in geographic analysis and synthesis

❑ How to use ArcGIS Spatial Analyst to find distance from anywhere to places of interest

❑ The relationship between distance and cost

❑ How to calculate the cost effects of other factors besides distance

❑ How to find the shortest or least expensive path from anywhere to places of interest

❑ How to determine which is the closest place of interest and the place of interest that can be reached with the least cost

Human Activity, Cost, and Distance

Life forms on our planet may be divided into plants and animals. A major difference between them (although there are exceptions) is that plants do not move (much) and animals do. In fact, for humans in particular, moving themselves and their artifacts, such as clothing, weapons, and stereo equipment, is a major activity.

Frequently, a major concern about a given human activity is its cost—cost in terms of energy, time, money, fuel, suffering, or other parameters. Moving from one location to another is a frequent human activity and one whose cost we often want to minimize. We usually move because we want to be in a new location or leave an old one for some reason. (There are exceptions to this: a walk in the woods for soul healing; a jog for exercise.) The subject of this discussion is using GIS to

❑ Reduce the cost of going from one place to another

❑ Lower the cost of transporting artifacts

❑ Minimize the cost of building linear structures, such as roads, bridges, and tunnels

The costs of moving from one place to another can be complex to calculate. In cases where the distance is far, the trips are repeated frequently, or the cost is critical (as in the cost in time of getting an ambulance

459

to the scene of an accident), it may well be worth doing the computations. In this discussion, you will learn how to deal quickly and simply with the complexity of performing the "cost of moving" calculation.

What are the factors that determine the cost of moving things, including ourselves? The one that perhaps comes to mind most quickly is "distance." It costs more to go from Punxsutawney, Pennsylvania, to Redlands, California, than to go from your living room to your bedroom. Or from your home to your office. So we begin with a consideration of distance, distance of the simplest sort: straight-line distance (on a Cartesian plane) from any point to a given point.

Euclidian Distances on the Grid

Distance is measured between two points. The two points used when measuring distance on a raster (or grid) are the centers of two specific cells. That is, in the case of the raster, the measurements are made from cell center to cell center. See Figure 8-9.

If the cell size is 50, the distance shown would be 200, spanning all or parts of five cells.

For another example, if you had a grid that had cells of size 10 and 1000 columns, the longest east-west distance measurement would be 9990.

Measurements are made from cell center to cell center, regardless of intervening cells or the angle of the line connecting the two cells of interest. The distance is calculated by the law of Pythagoras, who, despite having lived 200 years before Euclid, determined that the hypotenuse of a right triangle is the square root of the sums of the squares of the other two sides.

As illustrated in Figure 8-10, the area of the square on the hypotenuse is the sum of the areas of the squares on the other two sides.

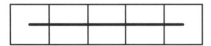

FIGURE 8-9   Distance is measured from cell center to cell center

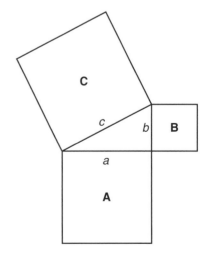

FIGURE 8-10   The Law of Pythagoras (the Pythagorean Theorem)

When you put the law of Pythagoras (also known as the Pythagorean theorem) together with René Descartes[7] great invention you can calculate the distance between two points whose coordinates are (x_1, y_1) and (x_2, y_2) as:

$$(x_1 - x_2)^2 + (y_1 - y_2)^2 = c^2$$

where c is the straight-line distance between the points. See Figure 8-11.

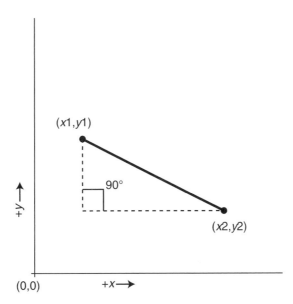

FIGURE 8-11   The straight-line distance between two points

Euclidean Distance and the Spatial Analyst

Proximity, the closeness of one point in space to another, is most easily expressed by the Euclidean straight-line distance between the two points. The term "Euclidean" comes from the geometry that was first formally developed by Euclid around 300 BCE.

ArcGIS Spatial Analyst allows you to automatically make a grid in which each cell in the grid contains the straight-line distance from itself to the closest cell of a set of source cells in another grid that represents the same geographic space. (Actually, because this distance is represented on the two-dimensional grid surface—differences in elevation being excluded—we could refer to the values in the cells as "Cartesian distance" after the mathematician Descartes,

---

[7]Descartes invented the x-y coordinate system on the 2-D plane in the 1600s.

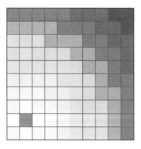

FIGURE 8-12

Figure 8-12*** shows a grid whose cells are composed of Euclidean distances from each cell to the red source cell. The lighter the color, the smaller the distance.

Proving Pythagoras Right

If you're like most people, particularly those who are spatially minded, you (vaguely) know the law of Pythagoras, but probably not why it is true. So, for those who are interested, here is a proof—one that will appeal to those interested in graphical, rather than strictly algebraic, matters. (If you just want to take my word for the validity of the law, you may skip the remainder of this discussion.)

Draw a square on a piece of paper. Draw a smaller square within the first, rotated so that its corners touch the sides of the first. Call the line segment from a corner of the big square to a point where the smaller square touches it "a." Call the segment from the corner of the big square to the other touching point "b." Label all of the line segments around the big square with "a" or "b," as appropriate. Label each of the sides of the smaller square "c." See Figure 8-13.

Now compute the areas. The area of each of the four triangles is $\frac{1}{2} ab$. So the total area of all four triangles is $2ab$. The area of the smaller square is, of course, c squared. What is the area of the larger square? The length of each side is $a + b$, so the area of the square is $(a + b)$ times $(a + b)$. When you multiply it out, that's:

$$a^2 + 2ab + b^2$$

Note now, by looking at the figure, that the area of the larger square is the same as the area of the smaller square plus the areas of the triangles, so we have

$$a^2 + 2ab + b^2 = c^2 + 2ab$$

Subtracting $2ab$ from each side of the equation gives the Law of Pythagoras:

$$a^2 + b^2 = c^2$$

Multiple Source Cells

Many distance problems involve several source cells instead of a single one. Perhaps you are flying along in your private plane, and because of an ominous new noise coming from the engine, you develop a strong interest in knowing the distance to the closest of several airports in the area. The airports might be

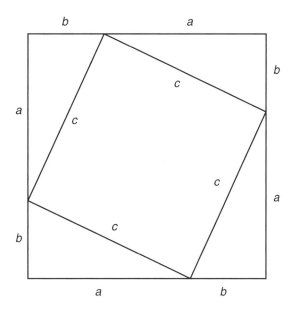

FIGURE 8-13   Re-proving the Pythagorean Theorem

recorded as source cells. If you had time to fire up your portable GIS, connected to your GPS, no matter which cell you were in, the distance to the closest airport would be shown.

Another example: Suppose you have a neighborhood of houses that need to be connected to a power line. The power line is represented as a line of source cells; houses may be built on any other cells. You want to find the shortest distances from each house to the power line so you can calculate the minimum length of wire required. Also, due to electrical requirements, there is a maximum distance the wire can be strung, so you will also need to identify the cells that are more than that distance away.

Problems of this sort may be solved with the Euclidian Distance tool, using as input a source grid with multiple source cells.

Finding the Closest of Many Source Cells

The grid of source cells that serves as input to Euclidian distance may consist of several single cells, clumps of cells (contiguous cells), a linear structure of several cells, or any combination of these. For each cell in the output grid, the straight-line distance to the closest source cell is calculated. This happens automatically. You simply apply Euclidian distance and the newly generated grid contains the closest distance.

Excluding Distances Beyond a Certain Threshold

If the problem warrants it, you may limit the search for the closest source cell so that it does not exceed a given value. To do this, you select an option that says "if you have to look further than this to get to a source cell, just forget it." Such a number may be called a limit, a cap, a cutoff value, or a threshold.

Other Factors That Influence Cost

We began this discussion on proximity saying our primary concern was the cost associated with moving (moving being defined in a very general sense) from one place to another. We also said that distance between points on the Cartesian plane was a major factor in the cost, and we have spent much of the preceding discussion discussing how to obtain that distance.

Frequently, however, distance is not the only, or even the principal, cost of moving from one place to another. An extreme example is of a person living in New York City who wants to visit a second person who lives nearby. In fact, the two people live in 33rd-floor apartments, toward the middle of parallel long-block streets, and their apartments share a common back wall. They actually live only a few feet from one another. But to travel from one apartment to the other requires two elevator rides, encounters with doormen or buzzer systems, and a not-inconsiderable walk (or cab ride). Clearly, the calculation of simple Euclidean distance is not the only tool we need to determine cost.

As a second example, suppose you want to take a hike. In between your starting and ending locations lie both swamp and dense forest. By taking a longer route, you can walk through pasture. The time and difficulty of the route you take will be a function of both the distance and the difficulty of making your way through various environments.

We can generalize these problems by creating a "cost surface." We develop this cost surface so that each grid cell contains a number that indicates the cost of going through a unit of distance in that cell. If it is three times as hard to go through the forest as it is to go through the pasture, we might assign values of 1 in the pasture cells and 3 in the forest cells. If a certain place in the swamp poses a threat of people being eaten by alligators, we might assign a very high cost (say 10,000) to traversing the distances in the swamp cells.

A third example: While off-road vehicles are becoming increasingly popular, it is still less expensive to travel by automobile from one city to another by using paved roads. The cost of traveling a straight line in this case, in terms of speed, danger, and possible bouts with the law, is prohibitive.

Consider Figure 8-14***. The time to go from A to B is greatly reduced by going around the volcanic mountain rather than over it. While the shortest path between points A and B is a straight line, it is not the fastest path, because the cost for crossing over the mountain is much greater than going around it.

In short, Euclidian distance has the limitation of assuming that the cost of crossing distances in any cell on the way to a source cell is the same as the cost of crossing distances in any other cell. You will now learn how to indicate that some cells are more expensive to cross than others. You will also learn how to find the most inexpensive route (the least-cost path) even though it may not be the shortest.

The Cost-Distance Mechanism

Spatial Analyst provides the Cost Distance tool for calculating the least cost of getting from any cell to a source cell. Two input grids are involved: the source grid and a cost surface grid, each of whose cells indicate a price of traversing each unit of distance in that cell.

The overall cost of the path from a given cell to the closest source cell is calculated on the basis of the least-cost path between the two cells. Calculating the total cost involves summing up the costs of crossing all the individual cells along the least-cost path. The cost of crossing an individual cell is found by multiplying the distance across that cell by the value found in the geographically equivalent cell in the cost-surface grid. Thus, the accumulated cost found in the output grid is the sum of the products formed by (a) the distance across each cell and (b) the value (the "price per unit distance for going through") of each cell.

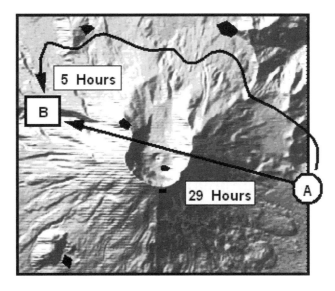

FIGURE 8-14   A straight line, while the shortest, is not always the fastest

The Cost-Distance Calculation

If you want to generate a grid similar to that made by Euclidian distance, but one that will place in each cell the least cost of traveling from that cell to the source cell, you can use the cost distance calculation. Cost distance operates on a source grid and a cost grid to produce a cost-distance grid. See Figure 8-15***.

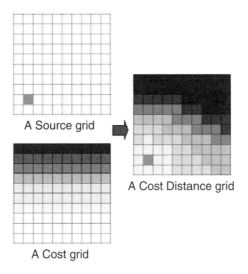

A Source grid

A Cost grid

A Cost Distance grid

FIGURE 8-15   Making a cost-distance grid

The source grid you already know about. The cost grid has darker colors for the cells that it is more expensive to traverse. The cost distance grid that is formed shows the cost to travel between a given cell and the source cell. Here the darker color indicates a greater cost in traveling between the source cell (red) and every other cell. Notice particularly the northwest and southeast corners of the resulting raster. Even though they are the same distance from the source cell, it is clear that it is more expensive to travel from the northwest than from the southeast. The reason is that southern cells are cheaper to cross than northern ones.

When you apply the Cost Distance tool, the resulting grid shows up as a map calculation.

Path Calculation in Euclidian Distance and Cost Distance

It is worth mentioning that the straight-line distance from a given cell produced by Euclidian distance and the least-cost path generated by cost distance are calculated in completely different ways. Euclidian distance calculates the direct, straight-line distance from each cell center to the closest source cell center, regardless of how this path slices through intervening cells. The cost distance path, however, must pass through cell centers, frequently generating a sequence of line segments that change direction at the center of each cell.

Therefore, even when crossing a cost surface where each cell cost is 1, so that a straight line would be the least-cost path, cost distance will usually produce a slightly longer path—because the path proceeds from cell center to adjacent cell center. This sort of "connect the dots" behavior will not result in a much longer path than the straight-line distance, but observing it illustrates the difference in the ways the two output grids are calculated. In cost distance, the path proceeds from one cell center to the center of one of the eight adjacent cells, always with the aim of minimizing the cost of getting to the source cell. In other words, except for a few special cases (e.g., due north, straight southwest), the path generated by cost distance will not be a straight line, but will zigzag so as to take in the cell centers along the way. See Figure 8-16.

The Euclidian distance direction is shown by a solid line. A possible path generated by cost distance is shown by a dotted line, under conditions of a uniform cost grid. If the cost is not uniform, then the cost distance path will be different. For example, look at Figure 8-17.

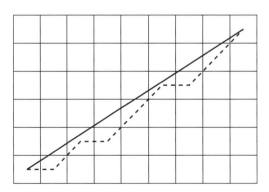

FIGURE 8-16   The cost-distance path goes from cell center to cell center

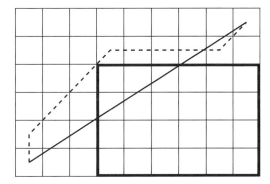

FIGURE 8-17   A cost-distance path tries to avoid areas of high cost

The EucDistance direction and path is represented by a solid line. A possible path generated by CostDistance is represented by a dotted line, under the condition that high costs exist in the cells within the outlined box.

Understanding How Total Costs Are Calculated

Calculating cost distance is somewhat involved. Take as an example a cell that costs $1.15 to cross—maybe there is a toll both there. What is the cost of traveling from the center of such a cell directly to the southern edge? See Figure 8-18. The length of this segment is 5 (half the total distance of 10 across the cell). The cost associated with the cell is 1.15 per unit of distance. The cost of this segment, then, is the product of the distance and the cost per unit distance: 5 * 1.15 = 5.75.

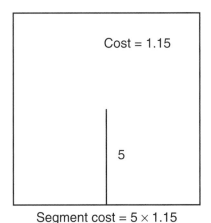

Cost = 1.15

5

Segment cost = 5 × 1.15

FIGURE 8-18   Cost of traversing a cell from its center to a side

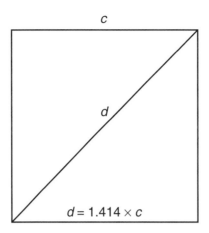

FIGURE 8-19   Cost of traversing a cell from a corner to the opposite corner

When the path goes from cell corner to corner, the distance is longer than if the path goes from the side of a cell to the opposite side. How much longer? Multiply the side-to-side distance by the square root of 2, which is approximately 1.41421356237, which we will round to 1.414. See Figure 8-19.

For example, the cost to go diagonally across a cell that had a toll of $1.15 would be

$$10 * 1.414 * 1.15 = 16.261$$

Each possible path from the source cell to each other cell is either calculated or considered in some way. That can be a lot of paths. This type of calculation may take place several times for each cell, there can be hundreds of thousands of cells, and the computer may make many attempts for each cell to find the least-cost path.

Getting More Information: Paths and Allocations

You have learned that you can calculate the shortest distance between two places using the EucDistance request and you can find the least-cost path with the CostDistance request, but there are still a couple of major questions: Which source cell is closest to (or least expensive to reach from) each given cell in the grid, and what is the direction or path from each given cell? That is, how would you identify the closest source cell and how to get there?

To answer these questions, first for straight-line distance, you can use two new parameters in the Euclidian Distance tool.

Direction and Allocation Grids for Euclidian Distance

The Euclidian Distance tool allows you to specify two new grids, which we will call a direction grid (DirGrid) and an allocation grid (AlloGrid). DirGrid's cells each contain the direction to the nearest source cell, in degrees clockwise, based on due north as 360 degrees (or 0 degrees). AlloGrid's cells contain the value of the nearest source cell. (These rasters will be created for you, with the names you specify.)

Direction and Allocation Grids for Cost Distance

The cost distance calculation is similar to the Euclidian distance calculation in that both can be used to generate direction and allocation grids. Both calculations allow caps to be placed on the maximum values in the primary grids they generate. Cost distance differs from Euclidian distance in three major ways:

❏ It uses a distance-weighting grid

❏ It differs in the way in which the direction grid points along the path

❏ It calculates the least-cost path rather than the shortest path

Because the path generated by cost distance goes from cell center to cell center, each cell in the direction grid contains information that indicates to which of the eight neighbors the path goes next. This information is coded according to Figure 8-20.

Or, in words:

1 means the next cell in the path is to the east (90 degrees).

2 means the next cell in the path is to the southeast (135 degrees).

3 means the next cell in the path is to the south (180 degrees).

4 means the next cell in the path is to the southwest (225 degrees).

5 means the next cell in the path is to the west (270 degrees).

6 means the next cell in the path is to the northwest (315 degrees).

7 means the next cell in the path is to the north (360 degrees).

8 means the next cell in the path is to the northeast (45 degrees).

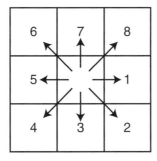

FIGURE 8-20   Direction code for the next cell in the path

A Major Application of Raster Processing: Hydrology

Because of the regular nature of the areal unit in raster processing, a number of applications, like simulation of a forest fire or an oil spill, are made much simpler. Hydrologic modeling is one such application so we look at it in some detail, both for its own sake and because it illustrates the power of spatial analyst. An important proviso: If you want to do serious hydrologic investigations, you should use a package designed for the purpose, such as those provided as separate packages by ESRI.

You may not have thought a lot about how water flows over the surface of the Earth, or how to simulate it with a computer, so let's explore the following:

❑   The fundamental elements of hydrologic modeling and analysis

❑   How to determine the direction of water flow from cells in a study area

❑   The importance of eliminating sinks from the drainage area

❑   The importance of including the entire drainage area in the study area

❑   How to determine flow accumulation in the cells of a grid

❑   How to find the distance, both upstream and downstream, from a given cell

❑   The methods used to delineate and order stream segments

❑   How to generate watersheds for streams and points

We enter into a discussion of these topics in the text that follows. Then, in the Step-by-Step section of this chapter, you simulate a hydrological system.

Basic Surface Hydrology

First a caveat: Hydrologic analysis is a complex subject. The concepts and tools presented to you here are, in themselves, not sufficient to undertake hydrologic analysis or modeling. Real-world situations frequently do not conform to the assumptions and conditions that underlie the examples presented here. However, the concepts discussed here will help you understand the basic principles of surface hydrologic analysis.

Surface hydrologic analysis (as opposed to underground hydrologic or groundwater analysis) seeks to describe the behavior of water as it moves over the surface of the Earth. Most simply, this type of analysis includes the following:

❑   Obtaining a mathematically correct representation of the surface of the area to be analyzed, considering the elevation of the surface at a given point to be the value of a grid cell at that point

❑   Determining the direction water would flow from each cell on the surface

❑   Determining to which adjacent cell water would flow when each cell is doused with a given amount of water

❑   Finding those cells that get considerable flow accumulation and delineating them as creeks, streams, and rivers, either persistently or when flooding occurs

❑ Developing a network of these creeks, streams, and rivers; determining a hierarchy of them; and classifying them as to volume, relative to their upstream tributaries

❑ Determining the areas (watersheds) that feed into given creeks, streams, and rivers and determining the outlets (pour points) of these watersheds

❑ Determining into which watershed and which water entities a given quantity of liquid (such as a polluting spill) might flow

What we will discuss here are the basic hydrologic tools available in Spatial Analyst. There are also Hydrologic Modeling tools available in other ESRI products.

How Spatial Analyst Performs Hydrologic Analysis

In Spatial Analyst, most hydrologic analysis is done by generating new grids. This operation is usually accomplished by entering formulas in the Raster Calculator or by specific tools in ArcToolbox.

Basic Surface Hydrology Concepts

The concepts are described in terms of the tools you can use to examine surface hydrology.

❑ The FlowDirection calculation determines the direction of flow from each cell of a surface grid. The grid generated by FlowDirection must be well behaved. The sort of analysis I am describing specifically excludes land areas that contain lakes or ponds. The assumption is that all the water placed on the grid will ultimately exit the grid at one or more low points on its edge.

❑ Assuming that the study area involved does not contain lakes or ponds, one of the ways the grid can be ill behaved is to contain a cell that is lower than its surrounding neighbors; such a cell is called a sink. Sinks distort the analysis; to find them, use the Sink calculation. (Editing grids with sinks is beyond the scope of this lesson.)

Another requirement of the grid is that the cells of primary interest—for example, the mouth of a river near a town that might flood—must include all the "uphill" cells. That is, all the cells that constitute the drainage basin for the cells of interest must be considered. The FlowAccumulation calculation may be configured to compute the amount of water that flows into each cell from all of its uphill cells.

❑ Stream networks are characterized by small creeks flowing into larger ones, these flowing into small streams, and so on. It is useful to speak of the "order," or relative size, of such water entities. The smallest creeks are labeled order 1. Larger entities have larger integer numbers. The StreamOrder calculation handles the process of assigning order numbers to streams. Both of the two principle methods for numbering streams (Strahler and Shreve) are available.

❑ The Mississippi River has a watershed consisting of all the land that supplies water to it. The smallest creek also has a watershed that consists of all the land that supplies water to it. The creek's watershed may be contained in the Mississippi's watershed, so the delineation of watersheds (or drainage basins, catchment areas, and contributing areas, as they are also called) is not trivial, either in concept or calculation. The WaterShed calculation assigns cells to such areas.

In addition to these calculations, an important operation that precedes surface hydrology analysis is the generation of a surface grid that gives the elevation at every cell. There are several ways to do this, as previously discussed. One way is to use the Interpolate to Raster tool in the Spatial Analyst menu.

Calculating Flow Direction

The primary data source for hydrologic operations in ArcGIS is a grid of flow direction. This grid is formed by the FlowDirection calculation based on a surface of elevation; for our discussion here, we will call the resulting grid "DirOfFlow." Each cell in the DirOfFlow grid contains an integer number; these numbers are powers of 2: 1, 2, 4, 8,16, 32, 64, and 128. (Just why these numbers were chosen, rather than 1, 2, 3, etc., has a historical and computer component, which will be discussed.) Each number indicates a direction, as shown by Figure 8-21.

The idea is, simply, that the precipitation that falls, or otherwise appears, on a given cell flows to an adjacent cell. To which of the eight adjacent cells? The one indicated by the number and the arrow in Figure 8-21, which points in the direction of the steepest descending slope.

For example, consider the simple grid shown in Figure 8-22***. The numbers in the cells indicate elevation.

The range in altitude is from 100 to 91, sloping gradually from east to west and a bit from north to south. When the Flow Direction calculation is applied to this grid, the resulting grid looks like Figure 8-23.

Note that water flows from each cell to the nearest neighbor cell so that the water flows down the steepest slope, except from the cell with lowest elevation in the southwest, where it flows off the grid.

Normally, cells along the edge of the grid are treated as any other cells in the grid, except that if none of the five adjacent edge cells have lower elevation than the edge cell under consideration, the flow will be directly off the side of the grid. But an option in the Flow Direction calculation allows water to flow directly off the edge cells of the grid. If this option is used the flow from edge cells is off the edge of the grid, regardless of the presence of adjacent lower cells. Thus, the grid in Figure 8-24 would be generated instead.

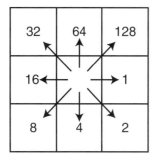

FIGURE 8-21   Direction code for the flow of water

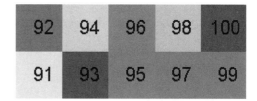

FIGURE 8-22   Elevation values in a tiny raster

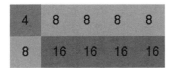

FIGURE 8-23   Flow direction following steepest slope

FIGURE 8-24   Flow allowed directly off edge cells of the raster

The Ultimate Destination of Water Is Off the Raster Area

The lowest point on the grid must be on an edge. This requirement is not as stringent as it sounds. If you think of any rectangular piece of real estate, it will have depressions in it, which will fill with water under the right circumstances. Exhibit a network of valleys that will hold linear bodies of water, at least one of which will flow off the edge of the grid or be a combination of depressions and networks of valleys. As already indicated, the ArcGIS hydrologic tools presented here do not work with lakes. They are strictly for stream networks. Lakes, which would constitute sinks, are not allowed.

It is worth remarking on the rather strange choice of numbers used to indicate flow direction. You've learned that water flows from any given cell to one of the eight adjacent cells. In the previous exercise on proximity, the directions were indicated simply by the integers 1 through 8. Why, then, are we dealing with numbers such as 32 and 64?

One conjecture is that, in the early days of hydrologic analysis, which correspond to the early days of computers, central processing unit speeds were slow and storage space in memory was at a premium. It was efficient to indicate direction with a single bit (a 1 or 0) in each position in a computer byte. Those positions correspond to columns in the base 2 number system. Those columns are designated 1, 2, 4, 8, and so on. The precedent set at that time endures in the hydrologic modeling field today.

Flow Accumulation: Drainage Delineation and Rainfall Volume

Once you have a grid that indicates flow direction, a number of other interesting and useful calculations are possible. In particular, you can determine the locations of all the linear bodies of water, and you can determine, from slope and elevation, those areas where water may accumulate during times of intense precipitation. This is accomplished with the ArcToolbox Flow Accumulation tool.

Basically, the value in each cell in the resulting grid contains the sum of the amount of water that has fallen on all the grid cells upstream from it. The intent is to simulate the flow, or potential flow, of water to form creeks, streams, and rivers. If each cell is presumed to have one unit of water (say, an inch in depth) to contribute—under a condition of "uniform rainfall," you can think of the number in a given cell as the number of cells upstream from that cell. To illustrate, examine the elevation surface in Figure 8-25. Note that the low points are in the middle of the south edge (elevation 1) and the west edge (elevation 3). All around the rest of the grid the elevations are 9 or somewhat less.

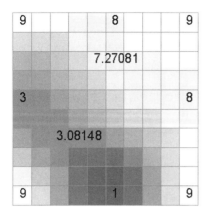

FIGURE 8-25 Elevations of a raster

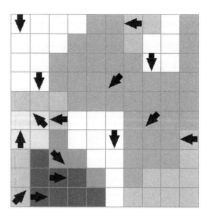

FIGURE 8-26 Some directions of flow, given the elevations of the raster of Figure 8-25

From this, you can produce a raster showing the direction of flow, using the ArcToolbox Flow Direction tool. Some arrows have been scattered on the grid to show flow direction. See Figure 8-26.

Now, applying the Flow Accumulation tool to the flow direction grid produces a grid that shows, for each cell, the water that accumulates due to adding up the accumulations from the cells "above" it. Figure 8-27 depicts some of these accumulation values.

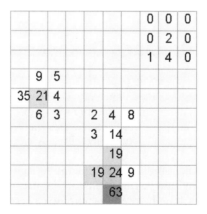

FIGURE 8-27 Accumulations of water, based on elevations and flow directions of the two previous figures

The largest accumulation is in the south, which had the lowest elevation. Another point of considerable accumulation is in the middle of the western side. If you look at the flow grid and the accumulation grid, you can get an idea of where and why the stream channels developed.

Note that some cells have the value 0, indicating that no cells are uphill of them. Note also that most of the cells accumulate very little water, but some accumulate a great deal of it—just as you might expect, since most of the land around us is uncovered by water but there are numerous creeks and streams. Finally, if you add up the values of the southern and western pour points (63 and 35), you get 98. This makes sense, since of the 100 cells total, 98 of them are above the two pour points.

The grid in the figure was developed without an input weight raster. Each cell, therefore, received the same amount of rain and was presumed to receive the same amount of water. You can change that by specifying a number for each cell in the study area. Consider a input weight raster that looks like Figure 8-28. This weight grid represents a gradation in rainfall, which was heaviest in the north.

If you consider that this was a rainfall event, and that the values in the grid cells, which vary from almost nothing to more than 1.5, constitute inches of rainfall, you can see that much more rain fell in the north than in the south across the study area. The total amount of rainfall is approximately the same as in the previous example, but there the rainfall was distributed uniformly.

Now you can apply the Flow Accumulation calculation with this weight grid. The results, shown in Figure 8-29, indicate that considerably more water volume showed up at the western pour point than before, because the rain was lighter in the south. In fact, with the weight grid, about as much water flows west as south. Recall that with no weight grid, almost twice as much flowed south as west.

This grid is the result of applying the Flow Accumulation calculation using the previous weight grid.

You can see from this example that hydrologic modeling can be a complex operation with many variables and parameters.

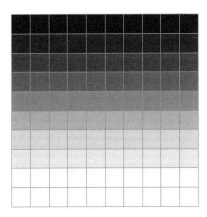

FIGURE 8-28   A weight raster: more rainfall in the northern part of the grid

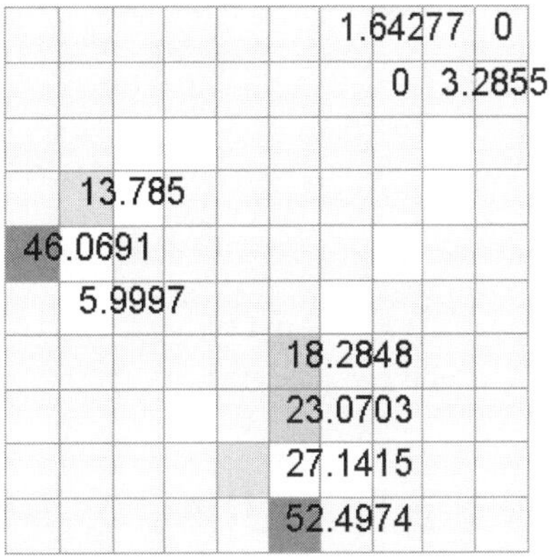

FIGURE 8-29   Flow accumulation taking into account the weight raster

FIGURE 8-30   Downstream flow length

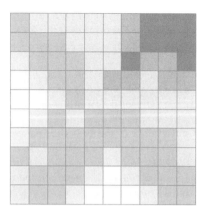

FIGURE 8-31   Upstream flow length

Calculating the Length of a Potential Linear Water Body

The length of a potential creek or stream is a useful thing to know when modeling. You can apply the Flow Length calculation to the direction of flow raster to show either the length of the flowing water from each cell upstream or downstream. Upstream flow length for a given cell is the distance, totaled from cell to cell, from the given cell to the origin of the longest path of water (the top of its basin) coming into that cell. Downstream flow length from a given cell is the distance from that cell to the pour point for the water passing through the given cell. See Figure 8-30, where darker shades indicate larger flow length.

The upstream flow length looks like Figure 8-31, where darker shades indicate longer flow lengths.

The Input Weight Raster in the Flow Length tool operates in precisely the same way as the weight grid does (impedance, cost surface) in our previous discussion of proximity. It multiplies the length through each given cell by the value in the geographically equivalent cell in the weight grid. The weight grid provides the cost or impedance for water to flow through each cell. Thus, you could simulate the fact that water flowing through forested land takes longer to cover a given distance than water flowing over rock.

You can use the output of the Flow Length tool to find the length of the longest flow path in a given basin. This is one of the values needed to calculate a more sophisticated hydrologic quantity, "time of concentration" for a basin. You can use flow length grids to create distance-area diagrams of hypothetical rainfall/runoff events, using the optional weight grid as an impediment to downslope movement.

Assigning Identities to Streams

The most basic hydrologic unit (outside of the individual cell) is the stream segment. Generally, streams segments (also called links) run between intersections in the linear network. A segment consists of all the cells between the junctions of two or more streams or between junctions and the pour points. (The cell that is the junction is considered to belong to one of the streams.) Figure 8-32 illustrates this numbering. The integer number is only a nominal, identifying value.

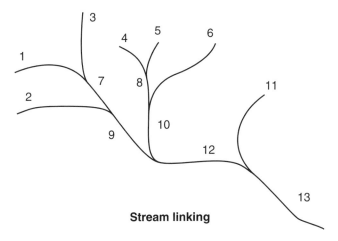

Stream linking

FIGURE 8-32   Numbered stream links

On a raster, ArcGIS places the same unique identifying number in all the cells of a given stream segment. In the discussion of the Flow Direction and Flow Accumulation calculations, every cell was considered a contributor to the creeks, streams, and rivers that developed ("Into each cell some rain must fall."). But you do not want to define all the cells in the study area as part of the water network. Instead, you can delineate specific stream channels. In other words, all of the study area contributes to the total amount of water to be dealt with, but only a small part of the study area carries most of that water. That area is known variously as the water network or the stream channels. This area is defined by including only those cells with flow accumulations greater than a chosen value; that value is called the cell threshold. Figure 8-33***, an illustration of how stream segments are numbered, shows five stream links. Each cell has a flow accumulation of greater than 7.0. These cells are considered to make up streams. Each stream segment is uniquely numbered, as shown by the color coding.

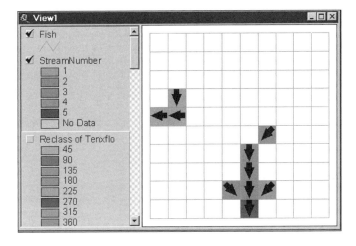

FIGURE 8-33   Stream Channels

Vector vs. Raster Representation

The two preceding figures show the difficulties involved in representing the virtually infinite, three-dimensional environment in the memory of a computer, necessarily using only the most fundamental discrete symbols: 0s and 1s. In vector mode, a stream is represented by one-dimensional arcs; the arcs have no width, only length. If quantities like flow, width, or velocity are to be included, they must be part of the attribute table.

In raster mode, a stream is represented by a sequence of adjacent cells. These cells are two-dimensional—they cover area. The area each cell covers, in basic hydrologic analysis, is the same, whether a mountain creek or a major river is being represented. Again, the geographic representation is only an approximation; even information about quantities such as width must be carried along separately.

This confluence of vector representation and raster representation in storing and displaying information about streams illustrates the challenges of using a computer to represent natural phenomena. Next, an attempt is made to represent the relative "size" of streams and stream channels.

Assigning Orders to Stream Links

You can attach an order number (integer value) to each stream segment or link. Generally, streams with lower numerical values have a size that can carry a smaller volume of water, but this is not always the case, as you will see.

Two ways of determining stream order number have been devised (one by A. N. Strahler in 1957 and the other by R. L. Shreve in 1966, as discussed below). In both methods, the smallest originating streams are numbered 1, from the source, up to the first intersection.

In the Strahler method, when (any number of) streams of the *same order* merge at a point, the downhill stream takes on an order number that is the original stream order number plus 1. For example, if a stream of order 3 merges with another stream of order 3, the resulting stream is order 4.

In any other case of stream merging, the order number of the downhill stream retains the order number of the larger uphill stream. So, if a stream of order 3 is joined by a stream of order 2, the resulting stream is still of order 3. Figure 8-34 illustrates the Strahler method.

When two streams merge according to the Shreve method, the order numbers of the uphill streams are added together to produce the order value of the downhill stream. So, when merged, two streams of order 3 produce a stream of order 6. An order 3 stream joined by an order 2 stream produces an order 5 stream. Figure 8-35 illustrates Shreve ordering.

To generate stream orders in ArcGIS, you use the Stream Order tool, which allows you to choose either the Shreve or Strahler method.

Watersheds and Pour Points

A basin is an area that drains water and other substances carried by water to a common outlet as concentrated drainage. Other common terms for a basin are watershed, catchment, and contributing area. The contributing area is normally defined as the total area contributing water flow to a given outlet, also called a pour point.

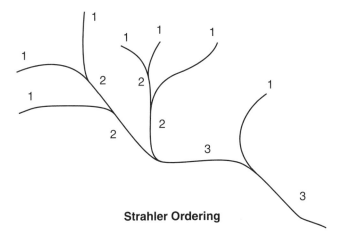

Strahler Ordering

FIGURE 8-34   The Strahler method of stream ordering

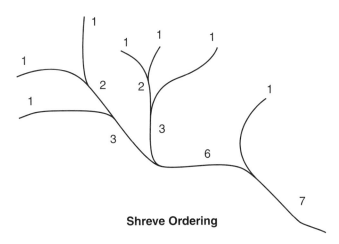

Shreve Ordering

FIGURE 8-35   The Shreve method of stream ordering

A delineation of these areas is the output of the Basin tool calculation. The geographic line between two basins is referred to as a basin boundary or drainage divide. Such a line, as you might imagine, runs along ridge tops and other lines of relatively higher elevation.

An outlet, or pour point, is the point at which water flows out of an area. It is the lowest point along the boundary of the basin. The cells in the source grid are used as pour points above which the contributing area is determined. Source cells may be features such as dams or stream gauges for which you want to determine characteristics of the contributing area.

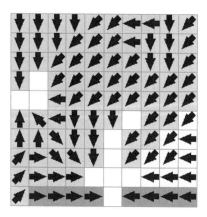

FIGURE 8-36   The five watersheds correspond to the five stream links

In Figure 8-36, the five watersheds correspond to five stream links. The arrows indicate the direction of flow from each cell; the cell shade indicates to which watershed the cell belongs.

Another ArcGIS tool that calculates areas that contribute water to a given point is the ArcToolbox Watershed tool. The distinction between them will be made clear when you work the exercises.

Spatial Analysis Based on Raster Data Processing

____ **Open up your Fast Facts text or document file.**

Basic Raster Principles and Operations

____ **1.** Start ArcCatalog. Make a folder named Raster_Experiments in your

___IGIS-Arc_*YourInitialsHere* folder. Also make sure that the Spatial Analyst extension is turned on (You will also have to turn it on in ArcMap.)

First let's create a tiny, trivial, basic raster to work with. There are several ways to create a raster. We could create a point shapefile, as we did in the project where we mapped the computer terminals, and then convert that to a raster. We will do that in a future project. Another method would be to use the ASCII to Raster tool in ArcToolbox > Conversion Tools > To Raster. This needs a text file, which you should create. (When I say "we" will use a certain tool you have, of course, noted that I make the suggestions and you do the work.)

____ **2.** Using a text or word processor, make a text file as shown in the following. Call the text file Forty_Two_Cells.txt and save it in the Raster_Experiments folder.

```
NCOLS 7
NROWS 6
XLLCENTER 4150
YLLCENTER 1150
CELLSIZE 100
NODATA_VALUE -32768
21 22 21 23 23 23 23
21 25 25 25 25 25 25
21 25 23 23 23 24 24
29 27 25 23 23 26 26
25 27 29 26 27 23 23
27 27 27 21 21 23 23
```

Minimize, but do not close, Forty_Two_Cells.txt.

We haven't defined a coordinate system for our raster, so a standard Cartesian system will be assumed. In the text file, NCOLS and NROWS are obvious. XLLCENTER is the x-coordinate of the *center* of the cell in the lower left corner. YLLCENTER similarly. CELLSIZE also is the length of a cell side. The NODATA_VALUE can be set to whatever you want. Obviously, it would be a real mistake to set it as a number that might actually be found in your data. Here it is set to the smallest number that can be represented by a 2-byte (16-bit) integer. The numbers following the first six lines are the values the cells are to take on, in "row-major order"—meaning all the values on the first row, followed by all the values on the second row, and so on. The lines in the text file do not have to correspond to the rows in the raster. Only the order is important. That is, the file

```
NCOLS 7
NROWS 6
XLLCENTER 4150
YLLCENTER 1150
CELLSIZE 100
NODATA_VALUE -32768
21 22
21 23 23 23 23 21 25 25 25 25 25 25 21 25 23
23 23 24 24
29 27 25 23 23 26 26 25 27
29 26 27 23 23 27 27 27 21 21 23 23
```

would work just as well.

____ **3.** Start ArcMap. In ArcToolbox locate the ASCII to Raster tool and start it. For the Input ASCII file, browse to Raster_Experiments and select Forty_Two_Cells.txt.

____ **4.** Click the Output Raster text box.[1] The text box probably shows the correct path, but you will want to change the name. Press End on the keyboard. Backspace up to the backslash (\) and type Basic_Raster. (If this doesn't work correctly, navigate to Raster_Experiments and type Basic_Raster for the name.)

____ **5.** The Output Data Type should be INTEGER. OK the ASCII To Raster window. Check out the dialog box in the ASCII To Raster window; close it. Basic_Raster is added to the map.

____ **6.** Use the mouse cursor to determine the extent of the raster. Round your readings off to the nearest 10.

```
Y_Maximum _____.
X Minimum _____. X Maximum _____.
Y Minimum _____.
```

____ **7.** Look at the Table of Contents. What is the smallest value? ____. The largest? _____. Use the Identify tool to check out some cell values. Restore and resize the text file, Forty_Two_Cells.txt. Verify that the values you find in the map are in the appropriate position relative to the text file that created them—for example, the two cells with 24 are in the third row, cells 6 and 7. Close Identify. Close the text file.

---

[1] If you read the right pane, you see that raster and GRID refer to the same idea.

___ 8. Bring up the Layer Properties window of Basic_Raster. Under Symbology > Show, pick Stretched. Examine the window, then click Apply and OK.

This "stretches" values over a color ramp. The default is black to white, so you get a grayscale representation of the values—low (black) to high (white).

___ 9. Open the attribute table of Basic_Raster and put it where you can see it and the map. Using the table, pick a record and highlight it. Verify that the Count field in the selected record in the table represents the number of cells highlighted. How many cells have value 23? _____. Deselect all records and close the attribute table.

It would be nice if each cell could be labeled with its value. That capability is not a part of ArcMap per se, but there is available an unsupported tool, called CellTool, that does the job. It is downloadable from an ESRI Web site.[2] It is also on the CD-ROM that accompanies this book. You can add the CellTool with the following steps.

___ 10. In ArcMap: Tools > Customize > Add from file. Navigate to

___IGIS-Arc_AUX\esriCellTool.dll.

Click it to bring it into the text box and click Open. When CellTool appears highlighted in the Customize window, click its box to turn it on and it will appear as a toolbar. Close the Customize window.

From this point on, CellTool will be available just as any other toolbar through View > Toolbars. Looking at the status bar (or ToolTips), record in the spaces provided the functions of the five icons on the Cell Tool toolbar. See Figure 8-37.

_____

_____

_____

_____

_____

FIGURE 8-37

___ 11. If you click the Value icon and then an interior cell, you will see the values of that cell and the eight around it. (If you get an error, just OK it and try again.) If you drag a box that includes a number of cells, you will see the values of all of them. Experiment. Click the Clear Graphics icon. Experiment some more. Clear Graphics.

---

[2]ESRI Technical Support says: The "Cell tool" is an ArcObjects developer sample that you can download from: http://arcgisdeveloperonline.esri.com/ArcGISDeveloper/default.asp?URL=/ArcGISDeveloper/Samples/SpatialAnalyst/CellTool/CellTool.htm. Please note that because this is a developer sample, it is provided as-is and is not supported by ESRI. The full disclaimer may be found at:
http://arcgisdeveloperonline.esri.com/ArcGISDeveloper/default.asp?URL=/ArcGISDeveloper/TOC/Samples.htm.

Important note: The Cell tool works only on the raster at the *top* of the Table of Contents. Should you fail to clear graphics, and then create a new raster, the graphics from the previous raster will remain—showing wrong results for the new raster.

The Raster Calculator—Integer Rasters

___ **12.** Using `View > Toolbars`, make the `Spatial Analyst` toolbar active. The functions available on the toolbar's menu were mentioned in the Overview of this chapter.

The most fundamental choice on the menu of this toolbar is the Raster Calculator. With few exceptions, you can use this choice to perform any operation on rasters. Many of those operations are made more simply and efficiently from other choices on the menu or from ArcToolbox. But we will start with the Raster Calculator so you know where to turn when other tools don't seem to apply. `ASCIIGRID` is the Map Algebra tool that is equivalent to ASCII_to_Raster, to make rasters out of text files.

___ **13.** Display the menu from the `Spatial Analyst` toolbar. Click `Raster Calculator`. (If Raster Calculator is grayed out turn on the Spatial Analyst Extension.) Make another raster using the same ASCII text file (Forty_Two_Cells.txt) as before, but do it with the Raster Calculator. Type the following expression into the calculate box:

```
ASCIIGRID(___IGIS-Arc_YourInitialsHere
\Raster_Experiments\Forty_Two_Cells.txt )
```

Check your work with Figure 8-38.

___ **14.** Click `Evaluate`. A layer called Calculation should appear on the map. Using the Table of Contents, the `Identify` tool, the attribute table, and/or the CellTool, verify that Calculation is the same as Basic_Raster.

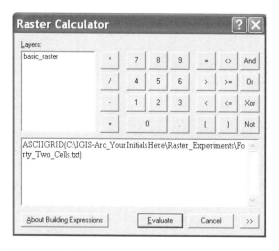

FIGURE 8-38

Arithmetic Calculation

The Raster Calculator lets us do fundamental operations on rasters. One of the simplest such operations is adding them together. When two rasters that represent the same geographic space and have the same cell sizes are added together, each cell of the resulting raster contains the sums of the cells in equivalent locations.

____ 15. Start the `Raster Calculator` again. Notice that the layers in the Table of Contents are also shown in the `Raster Calculator Layers` window. This makes it easy to do calculations with them. If you double-click a layer name, it appears in the computation pane.

____ 16. `Double-click` Basic_Raster in the Layers part of that window. It shows up in the computation pane with square brackets around it. Click the plus (+) sign in the Raster Calculator window. Now double-click Calculation. See Figure 8-39.

____ 17. Click `Evaluate`. Calculation2 appears on the map. Notice from the Table of Contents that the cell values are doubled based on the two other rasters' values. Use the `CellTool` to see the sums. Don't forget to `Clear Graphics` after you look.

____ 18. Multiply Basic_Raster by Calculation, to make Calculation3. Check a product or two out with the `CellTool`. Clear Graphics.

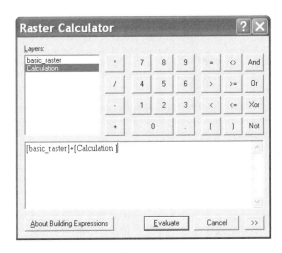

FIGURE 8-39

Boolean Operations

Rasters can be produced that indicate "truth values" or Boolean[3] values, in cells. A "1" stands for TRUE. A "0" stands for FALSE. Suppose you were interested in knowing which cells in Basic_Raster were greater than or equal to 25.

---

[3]You might wonder why they aren't called Boolian values. It's because the person who developed the concept of Boolean algebra was George Boole. Fortunately, no one wanted to take the responsibility of changing his name, just to conform to standards of English.

____ **19.** `Calculate [Basic_Raster] >= 25`. The result, Calculation4, will have a 1 (for TRUE) in those cells that were 25 or more. A zero will appear in the other cells, as you can see from the Table of Contents, the Identify tool, and the attribute table. (The `CellTool` will say ND for NoData, but that's wrong.)

The Boolean operators are as follows:

AND (shown in the expression as &),

OR (the inclusive OR, shown in the expression as |),

XOR (exclusive OR, shown in the expression as !), and

NOT (shown in the expression as ^).

The Boolean operators may be used in general expressions as well. Zero values are assumed to be FALSE. Nonzero values are considered TRUE.

____ **20.** Subtract 25 from Basic_Raster to make the Layer Calculation5. Look at the attribute table of the result. How many values of zero are there? _____. Now execute the expression `Not [Calculation5]` to make Calculation6. What values does it consist of? _____. How many values of one are there? _____. Where are the zeros, compared to Basic_Raster?

When you put a formula into the Raster Calculator, *you must put blanks around the operators, like +, ≥, and NOT*.

____ **21.** Show the preceding statement to be true by again writing an expression to add Basic_Raster and the original Calculation. Double-click Basic_Raster. Instead of clicking the + sign in the calculator window, press + on the keyboard. Then double-click `Calculation`. Press `Evaluate`. You get a cryptic, unhelpful error message: `Failed to evaluate the calculator equation`. The reason is that the + sign did not have blanks around it. When you press the + in the calculator window, a blank is automatically put on either side of the operator. Click OK to dismiss the `Error` window. Insert blanks around the + and press `Evaluate` again. Now it works.

Floating-Point Rasters

If the following math stuff leaves you scratching your head, don't worry about it. The point is that Spatial Analyst can do a lot of esoteric operations on rasters. Just go with the flow, even if the details escape you.

____ **22.** Bring up a new map without saving changes to the former one. Find and start the `Create Random Raster` tool in ArcToolbox. For the `Output` raster, browse to the Raster_Experiments folder and call the new raster Ran_Ras. Save. Leave the `Seed value` for the pseudorandom number generator blank.[4] For the output cell size, use 100. For the Output extent, use the four numbers you wrote down above for Basic_Raster. OK the window. Close the window.

---

[4]You can look on the Web for "pseudorandom number generator" to see 50,000 or so Web pages. The seed is the number that starts the process off. If you don't specify the seed, it is taken from the computer's clock, which counts the number of seconds that have elapsed since the January 1, 1970. Computers cannot actually generate true random numbers, as you would if you repeatedly threw a die. Computers are completely deterministic, so with a given seed and a given program you will always get the same sequence of random digits. Hence the word "pseudo."

_____ **23.** Ran_Ras consists of 42 cells, whose values are approximately uniformly distributed between 0 and 1. For Symbology pick a `Color Ramp` that goes from light to not too dark. Use the `CellTool` to display the values. Clear the values.

_____ **24.** Suppose you wanted to have integer numbers between 1 and 10 put in the cells, instead of 0 to 1. You could use the `INT function` on Ran_Ras (typing the INT portion):

`INT([Ran_Ras] * 10) + 1`

How many values of "1" did you get? _____. How many of "7"? _____.

A second floating-point raster demonstration: The cosine, say, x, of an angle between zero and a right angle is a number between 1 and 0. The arc cosine function finds the angle whose cosine is x. So we could use our Ran_Ras raster to make a raster of random angles, using the Acos function.

_____ **25.** In ArcToolbox locate the Acos tool. Where is it?

_____

_____ **26.** Start the tool. Make the input Ran_Ras. Name the output `Angles`. Check out the results. You may have expected to see angles between 0 and 90. But this function, which you will discover if you check the Help file, produces resulting angles expressed in radians, not degrees. There are 2*pi radians in the 360 degrees of a full circle. So one radian is about 57.296 degrees. To convert the Angles raster to degrees, use the `Raster Calculator`:

`[Angles] * 57.296`

Now check out the values. What's the smallest? _____ The largest? _____

Actually you could write a single expression entirely in the Raster Calculator that would produce the angles in degrees from a new map. You would start with the Map Algebra function RAND (instead of the ArcToolbox tool Create Random Raster), and use the Map Algebra ACOS as well. Better that you go on to Project 8-2, where you solve the Wildcat_Boat problem with rasters.

Exercise 8-2 (Project)

Solving the Original Wildcat_Boat Problem

The raster data model has some unique strengths. It is particularly suited to modeling the environment. Its advantages are that each area (cell) is homogeneous and is the same size and shape. Further, each area has four nearest neighbors and four next-nearest neighbors.

Raster data modeling trades geographic specificity for

❑ Speed

❑ Storage economy

❑ Simplicity

❑ Large group of tools

I want to introduce raster processing using the problem we solved in the last chapter regarding the location of the Wildcat_Boat facility. This particular problem is probably better solved with vector processing, but solving it using grid processing techniques is very instructive.

We'll start by setting up a folder for your work and then converting the relevant geodatabase feature classes to rasters. ESRI refers to its raster data structure as GRID, so the two words will be used interchangeably.

——— **1.** Start Catalog. Find the folder named

[___] IGIS-Arc\Spatial_Analyst_Data. Highlight it and choose `Edit` > `Copy`.

Find ___IGIS-Arc_*YourInitialsHere* and highlight it.

Choose `Edit` > `Paste`. Make a folder connection to

___IGIS-Arc_*YourInitialsHere*\Spatial_Analyst_Data\WC_Boat_SA.

Setting the General and Raster Environment[5]

The bounding rectangle of vector feature data sets pretty much takes care of itself, once you set the extents. Not so with raster data. The raster is a rectangle, and when you combine rasters, you want them to represent the same area. So we will set the output extent—in this case to the extent and projection of the vector PGDBFC Soils.

——— **2.** Start ArcMap. Dismiss ArcToolbox if it is present. Choose `Tools` > `Options` > `Geoprocessing` > `Environments`. Under General Settings, make the `Current Workspace` ___IGIS-Arc_ *YourInitialsHere*. Also make this the `Scratch Workspace`.

——— **3.** For the `Output Coordinate System`, browse to

___IGIS-Arc_*YourInitialsHere*\Spatial_Analyst_Data\

WC_Boat_SA\Wildcat_Boat_SA.mdb\Area_Features\Soils

`WGS-1984_UTM_Zone_18` should appear in the lower text box, and the upper text box should read: `As Specified Below`. Scroll down to `Output Extent` and again browse to Soils. The four bounding values should appear. Write them here (to the nearest one-tenth meter).

Top _____   Bottom _____

Left _____   Right _____

——— **4.** Scroll down further and click `Raster Analysis Settings`. Click the `Cell Size` drop-down menu and pick `As Specified Below`. Type in `10.0`. OK the `Environment Settings` window. OK the `Options` window.

---

[5]If you exit ArcMap and then start it up again, you have to set the Environment again.

____ **5.** Save the blank map, which you will call Preserve_Environments.mxd, in WC_Boat_SA. (When you quit ArcMap, the Environment settings go away. You can restore them by resuming a map containing them.)

____ **6.** Quit ArcMap. Restart ArcMap, double-clicking Preserve_Environments.mxd. Recheck the `Environment Settings`. OK the `Environment Settings` window. OK the `Options` window.

Converting Features to Rasters

The next task will be to convert the vector-based feature classes to raster-based geodata sets. You can do this (and many other Spatial Analyst operations) using the `Spatial Analyst` toolbar.

____ **7.** Select `View > Toolbars` and make sure `Spatial Analyst` has a check beside it. Move or dock the `Spatial Analyst` toolbar wherever you want it.

____ **8.** Add Soils as data from

___IGIS-Arc_*YourInitialsHere*\Spatial_Analyst_Data\WC_Boat_SA\

Wildcat_Boat_SA.mdb\Area_Features\Soils

____ **9.** On the `Spatial Analyst` toolbar, click the drop-down menu arrow. Then select `Convert > Features To Raster`. Look at the `Features To Raster` window. Note that for `Input features`, you can either select Soils from the drop-down menu or you can browse to a data set. Just use the Layer Soils indicated in the drop-down window. For `Field`, pick SUIT. If the Output cell size is not 10, make it 10. For the Output raster, browse to

___IGIS-Arc_*YourInitialsHere*\Spatial_Analyst_Data\WC_Boat_SA

and type Soils_rr10m (rr just meaning raster resolution) in the `Output raster` text box. Click Save. OK the `Feature To Raster` window.

Creating a raster in this way will put a GRID data set in the folder you specified and also add a layer to your ArcMap Table of Contents. Note that the GRID data set does not go into the personal geodatabase.

____ **10.** Arrange it so that the feature class soils is at the top of the table of contents and raster layer Soils_rs10m is just below it. Make the Soils layer `Hollow` or `No Color`, so you can see the raster, and also the polygon boundaries on top of the squares. Zoom up on the northernmost point of land in the northwest quadrant of the map. See Figure 8-40***.

From this you can get a pretty good idea of the rule for assigning a value to a cell: The cell gets the value of the polygon that exists at the center of the cell.

____ **11.** Zoom up some more and measure the length of a side of a cell. _____.

____ **12.** Right-click the Soils_rr10m name in the Table of Contents and choose `Properties > General`. Under Name, change the name from Soils_rr10m - soils_rr10m to just Soils_rr10m. Click Apply. Click Source. How many columns and rows are there? _____. _____. What is the

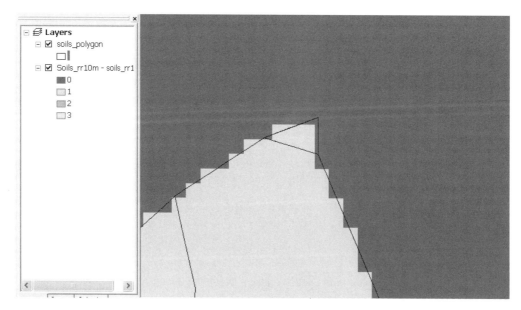

FIGURE 8-40

Cellsize (X,Y)? _____. _____. What is the Uncompressed Size of the raster? _____KB. What is the Format? _____. Scroll down through the list to get an idea of the information you can get about a raster data set. Click OK.

___ **13.** Repeat the creation of a Soils GRID data set, but this time use a cell size of 5 meters. Call the data set Soils_rr5m.

___ **14.** Change the name from Soils_rr5m - soils_rr5m to just Soils_rr5m.[6] Click Apply. Click Source. How many columns and rows are there? _____. _____. What is the Cellsize (X,Y)? _____. _____. What is the Uncompressed Size of the raster? _____KB. Notice that this is about four times as large. Twice as many rows and twice as many columns makes four times as many cells. Click OK.

___ **15.** Compare Soils_rs5m with the polygon feature class by flipping Soils_rr5m off and on. Note how this change in resolution results in an improvement in precision. You can see that the edges of the squares of the raster more nearly conform to the polygon boundaries. So the area covered by the raster is more nearly congruent to the polygons than at 10-meter resolution. To see this clearly, flip the 5-meter layer on and off at different levels of zoom.

___ **16.** Minimize ArcMap and ArcCatalog, and, using Windows Explorer or just My Computer, navigate to

___IGIS-Arc_*YourInitialsHere*\Spatial_Analyst_Data\WC_Boat_SA\Soils_rs10m.

---

[6]Change the name of raster data sets whenever you get tired of seeing the name repeated.

Right-click the folder Soils_rs10m and, with Properties > General, determine the Size of the contents of the *folder*. It is _____ kilobytes (KB). Now determine the Size of Soils_rr5m. _____KB. Close the window.

What is interesting about the comparison of the sizes of these two rasters is that, while the 5-meter resolution raster contains four times as many cells as the 10-meter resolution one, the data set is not four times as large (which would be 300 percent larger). In fact, it is only about 56 percent larger. So it is not the case that doubling the resolution quadruples the data set size—far from it! This happy fact is due to the data compression techniques discussed in Chapter 4.

___ 17. Bring back ArcMap. Since neither of these data sets is very big and didn't take very long to create, let's increase the resolution—say, to 2 meters. Make Soils_rr2m.

___ 18. Add the PGDBFC LANDUSE[7]

(from ___IGIS-Arc_*YourInitialsHere*\Spatial_Analyst_Data\WC_Boat_SA\Wildcat_Boat_SA.mdb\ Area_Features)

to the map. Using 2-meter resolution, make a raster (GRID) data set of Landuse, called Landuse_rr2m. Use LU-CODE as the field to make the "value variable" of the raster. Turn off all layers except the Landuse pair. Make the Landuse Symbol No Color or Hollow. Look at how closely the cells approximate the Landuse boundaries.

___ 19. Open the attribute table of Landuse_rr2m. How many cells of value 100 are there? _____. How many square meters does each cell represent? _____. So, how much area is occupied by the land use with LU_CODE equal to 100? _____.

___ 20. Just as a check, open the attribute table of the PGDBFC Landuse. Click Options. Using Select By Attributes, select those records with LU_CODE = 100. Display only selected records. Run statistics on Shape_Area. What is the sum? _____. Compare this number with the area you calculated in the previous step. The result should be confirming. Clear the selected features and records.

___ 21. Save the map as WC_Boat_Area_Map.mxd in WC_Boat_SA. Dismiss ArcMap.

Next we will convert (create rasters for) the PGDBFCs Roads, Streams, and Sewers. To do this, we should verify the Environment variables, so that we can be sure extents of all the data sets will be the same.

Creating Rasters for Sewers, Streams, and Roads

We could create rasters for the line feature classes—Sewers, Streams, and Roads—in the same way as above, but it is also possible to do it with ArcToolbox. There are two advantages. First, we don't have to load the PGDBFC into ArcMap, and second, we can easily set or check environment variables, important for determining the extent of a GRID.

The extent of a vector data set is pretty much determined by the data. But GRID data sets are rectangles, and if you convert a vector data set to raster, it is important to set the extents to the same as other data

---

[7]We called this dataset Landcover when we worked with these datasets before. Land use and land cover are technically not the same, but are often used interchangeably.

sets of the study area. Since the soils PGDBFC covers the entire study area (and the sewers data set, for example, does not), we earlier set the extent of rasters that we are about to create to that of the soils data set. Here we check just to make sure.

____ **22.** Start ArcMap with WC_Boat_Area_Map.mxd. Open ArcToolbox. Press the Index tab and type in Feature to Raster. Feature to Raster (conversion) is highlighted. Press Locate. Where is the tool in the ArcToolbox tree?

_____ > _____ > _____. Double click the tool.

In the Feature To Raster window that opens you find a text box for Input Features.

There are two ways to load that text box. The first is to use a layer in ArcMap. The second is the data set on which the layer is based. Click the drop-down menu arrow. This shows the current ArcMap layers that one might convert. You could load Sewers into ArcMap and it would appear in the drop-down menu box. Instead, browse to

___IGIS-Arc_*YourInitialsHere*\Spatial_Analyst_Data\ WC_Boat_SA\ \Wildcat_Boat_SA.mdb\Line_Features\Sewers

and click Add. For Field, choose DIAMETER. Browse to make the Output raster Sewers_rr2m in ___IGIS-Arc_*YourInitialsHere*\Spatial_Analyst_Data\WC_Boat_SA.

If the Output Cell Size does not say 2, type in 2.0. Do not OK the window yet.

____ **23.** Press the Environments button. Check that the General Settings are as you specified earlier. If not, reset them, as in Steps 2 through 4 above, **except under Raster Analysis settings set the cell size to 2**. OK the Environment Settings window. OK the Feature To Raster window. The raster Sewer_rr2m will show up in ArcMap.

____ **24.** Add the Sewers PGDBFC and make it a bright line of width 2. Make sure every other layer is turned off except the two that relate to sewers. Zoom to the extent of the feature layer Sewers. Look at it on top of Sewers_rr2m. Zoom way up on a place where the sewer lines intersect. Note how a raster data set represents lines. Use its attribute table to determine how many cells it took to represent the 45-inch pipe. _____. The 60-inch pipe. _____.

____ **25.** Using the same approach as above, make a raster of Streams, called Streams_rr2m. For Field (which will become Value in the attribute table), pick STRM_CODE and use 2.0 again for Output Cell Size. Change the name to simply streams_rr2m. From the attribute table of streams_rr2m, what are the two values that were put in for STRM_CODE? _____, _____.

____ **26.** Use the same approach a third time to make Roads_rr2m, using RD_CODE.

Look at the attribute tables of each of the three raster data sets. What are the numbers you find for Value and Count?

	Values	*Counts*
Roads_rr2m	_____	_____
Sewers_rr2m	_____	_____
Streams_rr2m	_____	_____

Buffering with Spatial Analyst (Maybe)

Recall that the site for the Wildcat_Boat facility has to be within 300 meters of a sewer line. And the site must be at least 20 meters away from any stream. Both the sewer lines and the streams are represented by square cells that are 2 meters on a side, connected either at a side or a corner. To make a buffer a raster representation of these linear entries, we need to fatten up the zones. There is a Spatial Analyst command, EXPAND, which is reputed to be the raster equivalent of the vector procedure BUFFER. We get to EXPAND through the Raster Calculator, which gives us access to all Spatial Analyst commands and functions, but requires that we type a command using proper syntax, compared with pointing and clicking, which we normally use.

___ **27.** From the Spatial Analyst drop-down menu, pick Raster Calculator. The version of the EXPAND command we want to use has the following syntax:

- ❏ EXPAND (name of the raster, then the number of cells to expand by
- ❏ The word LIST
- ❏ A comma-separated-value list of the VALUES of the zones we want expanded)

Let's use EXPAND first to make the Streams raster fatter by ten cells (at 2 meters per cell), so the command will look exactly like:

```
EXPAND ([Streams_rr2m],10,LIST,1,2 )
```

By typing and double-clicking, put this into the Raster Calculator window. See Figure 8-41. Press Evaluate and wait. The status bar will indicate progress toward making the new raster. When it is finished, a layer named Calculation will be added to the map. Turn off all layers except Streams_rr2m and Calculation. Pull Streams_rr2m above Calculation. Zoom up some to verify that a buffer has been built around the streams of code 1 and code 2.

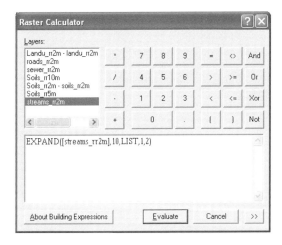

FIGURE 8-41

_____ **28.** In the same way, use

```
EXPAND ([Sewers_rr2m],150,LIST,45,60)
```

to generate an expansion of 150 two-meter cells around the 45- and 60-inch-diameter sewer pipes.

_____ **29.** The resultant raster is called Calculation2. Drag the Sewers PGDBFC above it in the Table of Contents. Examine them together. To say things don't look right is to understate the case. Take some measurements. You will notice that in some places the edge of the "buffer" is much more than 300 meters from the sewer line.

The problem with EXPAND is that it adds on the number of cells regardless of whether they touch side to side, or corner to corner. If they touch side to side, the increment is 2 meters. But if they touch corner to corner, the increment is about 2.828 meters (the diagonal of the square). So the moral of this story is that EXPAND is not a good equivalent of BUFFER. You can check the width of Streams_rr2m to verify this fact. You'll find that when a line runs horizontally or vertically, the distance to the edge of the raster zone is about 20 meters. When the line is diagonal, the "buffer" is more like 30 meters.

Buffering—Plan B

One nice thing about ArcGIS is the ability to slide easily from one data model to another. Let's take advantage of the fact that you created vector buffers of both sewers and streams when you solved the Wildcat Boat problem in the last chapter.

_____ **30.** Using Spatial Analyst toolbar > Convert > Features To Raster, start to make a new raster from

___IGIS-Arc*YourInitialsHere*\Wildcat_Boat_Data\
Wildcat_Boat.mdb\Area_Features\Streams_buf.

The location where you want to put the raster buffer is

___IGIS-Arc_*YourInitialsHere*\Spatial_Analyst_Data\WC_Boat_SA

Make the cell size 2.0 and the Field value NoBuild.

Try the name Streams_buf_rr2m. Your result should look like Figure 8-42.

_____ **31.** Try again. The name "Strm_buf_rr2m" meets the <= 13-character requirement. For the Field value, use NoBuild.

FIGURE 8-42

_____ **32.** The GRID will be added to the map. The Legend may be strange. If so, fix it by going into the Properties of Strm_buf_rr2m > Symbology > Classes > 1. Click Apply, then on OK. Check out its attribute table. What is value? _____, What is count? _____.

_____ **33.** Make Sewr_buf_rr2m from Sewers_buf. For the Field value, use Build.

_____ **34.** The GRID will be added to the map. The Legend may be strange. If so, fix it by going into the Properties of Sewr_buf_rr2m > Symbology > Classes > 1. Click Apply, then on OK. Check out its attribute table. What is the value and count? _____, _____.

Reclassifying the Data

When we solved the Wildcat_Boat problem with vector-based GIS, we combined landuse, soils, the sewer buffer, and the stream buffer. We will now do the same with the rasters. We have

```
Landuse_rr2m
Soils_rr2m
Sewr_buf_rr2m
Strm_buf_rr2m
```

The strategy will be to reclassify the values of each of these rasters. We'll make the value "one" if the zone meets the requirement, and zero or NODATA otherwise.

For Landuse_rr2m this means changing all the values to "0," except "300," which we change to "1."

For Soils_rr2m we will change values of "0" and "1" (which both indicate unsuitability) to "0." We will change the values "2" and "3" to "1."

For Sewr_buf_rr2m we will replace the "300" with "1," leaving NODATA alone.

For Strm_buf_rr2m, the matter is a bit different. The values of 20 we want to replace with NODATA, while we give the current NODATA values a "1."

Once we have done this, we merely add the four rasters together, so that the sum at each location is a 0, 1, 2, 3, or 4. From the way we constructed the reclassification, a cell must have a 4 in it to be eligible to be a part of the areas that would allow the Wildcat_Boat facility.

_____ **35.** On the Spatial Analyst toolbar, pick Reclassify. For the Input raster, browse to Landuse_rr2m in WC_Boat_SA. Press the Unique button so all the values of landuse are shown. The Reclassify window shows columns titled Old Values (100, 200, and so on) and suggested New Values (1, 2, and so on). Click the New Value 1 and type 0. Do the same for 2, 4, 5, and 6. For the value 3, put in a 1. For the name of the Output raster, browse to

___IGIS-Arc_*YourInitialsHere*\Spatial_Analyst_Data\WC_Boat_SA

and put in the name rcl_landu. See Figure 8-43. Click OK.

_____ **36.** Reclassify Soils_rr2m, calling the resulting raster rcl_soils. Again, you may note that some of the categories of soil suitabilities are shown as ranges—such as 2-3. You want each category

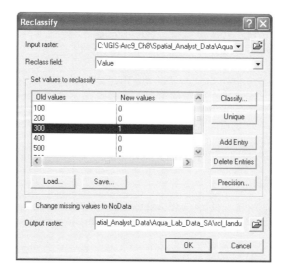

FIGURE 8-43

to have a unique value, so punch that button. Change the New values of 1 and 2 (which indicate unsuitability) to 0, and the values 3 and 4 to 1. Click OK.

___ **37.** Start to reclassify Sewr_buf_rr2m, but notice that the new value is just what you want (1 for 300), so just fix up the name, rcl_sewr, and click OK.

___ **38.** Finally, reclassify Strm_buf_rr2m, so that 20 becomes 0, and NoData becomes 1. Call the result rcl_strm. Click OK.

___ **39.** Use the Identify tool to look at each of the four reclassified data sets: rcl_soils, rcl_landu, rcl_sewr, and rcl_strm. Make sure that the Values are 0 or 1, and that they indicate the correct places.

Adding the Rasters with the Raster Calculator

Now you may perform addition on the four rasters whose names begin with rcl. You will use the Raster Calculator.

___ **40.** From the drop-down menu on the Spatial Analyst toolbar, click Raster Calculator. By double-clicking the raster names and single-clicking the plus (+) operator, create the expression shown in Figure 8-44. Click Evaluate. A layer named Calculation will appear in the Table of Contents. Its values will be 1, 2, 3, and 4. Of course, 4 indicates those areas suitable in all respects. Check it out with the Identify tool.

___ **41.** Once you are convinced that the Layer Calculation is correct, it is a good idea to make it a permanent data set. Otherwise, it goes away should you close ArcMap.

___ **42.** Right-click Calculation and select Make Permanent. The Make Calculation Permanent window appears. The Output raster name in the window is not exactly the same as the name in the

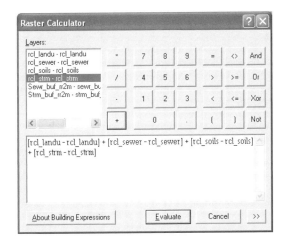

FIGURE 8-44

Table of Contents, but it is synonymous with it.[8] Change the name to SUM_LuSoStSw, and make sure it is going into the WC_Boat_SA folder with all the other rasters. Click Save.

___ **43.** Remove the temporary calculation from the Table of Contents. Add as data SUM_LuSoStSw.

Converting Zones to Regions

You know that a *zone* consists of cells that all have the same value. So the areas that are candidates for the Wildcat_Boat facility are all connected in a way—part of a single zone. You may recall that a raster *region* consists of disconnected sets of cells that are in the same zone. So it is the regions of zone 4 that we want as distinct entities. We can form these using the Region Group tool in ArcToolbox or the REGIONGROUP function typed into the Raster Calculator.

___ **44.** Locate the Region Group tool in ArcToolbox. Where is it?

_____ > _____ > _____

___ **45.** While you are in the toolbox, count the number of sub-toolboxes in the Spatial Analyst toolbox. _____. That is to say, this is a *big* chunk of software, all by itself.

___ **46.** Start the Region Group tool. Browse to SUM_LuSoStSw for the Input raster. Make the Output raster name in the window RG_WC_Boat, in WC_Boat_SA.

Recall that the term "connected cells" can have two meanings. Cells can be connected by their sides, in which case each cell has up to four neighbors. Or cells can be considered connected if either their sides or corners touch, in which case each cell has up to eight neighbors.

The tool lets you put in a LINK value in the new raster, which is the old zone value. You will need that, since the candidate areas are all of zone 4.

---

[8]Students from Missouri can verify the true name of the raster by using the Source tab in the Table of Contents, and then following the tree (it will go something like Documents and Settings\<username>\LocalSettings\Temp).

___ **47.** For Number Of Neighbors To Use, select EIGHT. Leave Add Link Field To Output checked. Click OK.

___ **48.** RG_WC_BOAT will be added to the map. It consists of over 50 distinct zones, formed by all the regions of all the zones in SUM_LuSoStSw. Observe the map. Open the attribute table of RG_WC_BOAT. Notice, by looking at Count, that some of these zones are quite small; remember that each cell contains only 4 square meters. Notice that the Link value in each record is 1, 2, 3, or 4.

As a last step, to isolate the sites, you will use the ArcToolbox command Extract by Attributes. You will want the Link value to be 4. You will also want the areas to be at least 2000 square meters. How many cells would that be? _____

___ **49.** Locate the ArcToolbox command Extract by Attributes. Where is it?

_____>_____>_____

___ **50.** Start the tool. The Input raster should be RG_WC_BOAT. Browse to make the Output raster FinalSites_SA. Press the SQL (Structured Query Language) button. Construct the following expression[9] (noting the blanks around the operators):

Link = 4 AND Count >= 500

and click OK in Query Builder window.

___ **51.** Verify that the Extract By Attributes window looks like Figure 8-45 and click OK.

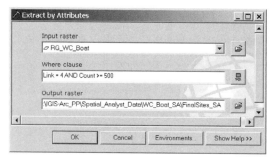

FIGURE 8-45

Turn off all layers except FinalSites_SA. Examine the sites and compare them with those you found using the vector approach in Chapter 7. How many sites are there? _____ Then add as data the Roads, Streams, and Sewers data sets from

___IGIS-Arc_*YourInitialsHere*\Spatial_Analyst_Data\WC_Boat_SA\ Wildcat_Boat_SA.mdb. Save the maps as Final_Sites.mxd.

___ **52.** Show each linear feature with significantly different symbology. Switch to Layout View and add other map components such as title, legend, and so on, as you wish. Print the layout.

_____

[9]The Verify button may show that there is an error in the expression. There isn't. Just OK the window.

Solving a Wildcat_Boat Problem with Different Requirements

Raster processing, by its nature, allows a lot of flexibility. Suppose the Wildcat_Boat facility requirements could be varied somewhat. Perhaps the 300-meter rule might be waived if the builders are willing to put in a septic tank. This is not as desirable a solution, but it is a possibility. Suppose also that agricultural land is deemed as okay, and forested land is no longer absolutely forbidden, but not encouraged. As for soils, you would like to favor the very suitable soils (type 3) over the moderately suitable (type 2) soils. The streams requirement is still firm, however.

When you solved the WildCat Boat problem with vector data, the study area was partitioned into two categories: where building could take place and where it couldn't. What you could do easily now, with raster processing, is to display areas of greater or lesser degrees of suitability. You do this by going back to reclassification. Keep in mind that you are going to combine the values of the zones. In the last exercise you simply added the four variables together (each variable being either 0 or 1) and took only those zones with value 4 as the sum as acceptable. But consider the flexibility you now have. You could reclassify the sewers—making the value 10 for areas within the 300 meters and 5 outside that distance. You could make brushland and agricultural land have value 10 and forested land have value 7.

Regarding Soils: You could use 6 for moderately suitable and 10 for very suitable. What about the prohibition regarding building on unsuitable soil? You could reclass those areas with a small negative number (say, −100), so the zone would show a negative value after addition takes place. What about the streams requirement? You could reclass the NOBUILD area as −100 also, while the permissible areas could have value 0.

If you added these four up, you could have zones with value 30 for the most suitable, and smaller values for less suitable areas. For example, a zone depicting an area outside the 300-meter sewer zone, on forested land that was only moderately suitable for building would be 18. (You would know that any area that came out negative was prohibited.)

You could also decide that the land use issue was twice as important as the soils or sewers issue, so you could weight that variable. Your function might look like this:

```
rcl_soils + (2 * rcl_landu) + rcl_sewr + rcl_strm
```

You would, of course, have to rethink the maximum and other values to form conclusions as to what was ultimately desirable and what wasn't.

—— **1.** In ArcMap, open Final_Sites.mxd. Make a new map by reclassifying the variables.

—— **2.** Print the new map, using text to indicate the assumptions made.

Making Surfaces with IDW, Spline, Trend, Nearest Neighbor, and Kriging

I will now introduce a very big subject with a very small exercise. The subject is basically this: Given a finite set of points with known values, say, "z" values (e.g., altitudes, pressures, pollution levels), what values could reasonably be assigned to the remaining (infinite number) of points that are in the same area? That is, how would one interpolate between the known points?

This is an important area, because all we can really measure with a high degree of precision are conditions at a point. Think about altitude, temperature . . . We can be very sure what the situation is at a given point (in space and time). We have to infer what conditions exist in the vicinity of that point. So what follows is a quick and very cursory look at the tools in ArcGIS that let you do that. An entire course could be built around the techniques these tools represent and the statistics involved. The Help files are some help, but to make real use of these tools, you need to do a lot of reading, or engage a statistician.

(You have, of course, seen one approach to developing a continuous surface from a set of points: TIN. What is different about these surfaces you are about to explore is that they have no sharp line breaks—that is, they are differentiable, in the calculus use of the word—and they are represented by mathematical functions.)

---- **1.** In ___IGIS-Arc_*YourInitialsHere* make a folder named Experiment_with_Interpolation

---- **2.** From ___IGIS-Arc_*YourInitialsHere*\Trivial_GIS_Datasets, copy over three shapefiles:

Known_Altitudes.shp
Known_Populations.shp
Only_points.shp

to this new folder.

---- **3.** Start ArcMap and add as data Known_Altitudes.shp.

---- **4.** *Consider the shapefile Known_Altitudes.shp:* Make the map units and the display units Meters. Look at the attribute table. Each number is the altitude in feet of the particular point. What is the range of altitudes: Lowest? _____. Highest? _____. Label the map using Altitude as the label field. Very roughly, on average, how far from each point is its nearest neighbor? _____ meters. How would you characterize the lay of the land if the numbers represented hundreds of feet of altitude. High areas, low areas?

Close the attribute table.

The question now is, interpolating from these known altitudes, at known positions, what are the altitudes at other nearby points we might randomly select. There are a number of tools, that use different statistical techniques, that can make very good, educated guesses at the answers to this question.

___ 5. Display ArcToolbox. Choose Spatial Analyst Tools > Interpolation. Disregarding the Topo to Raster tools (read about them if you want), list the remaining tools:

_____, _____, _____

_____, _____

___ 6. We'll start by using the Spline tool. Pounce on Spline. Read the Help panel. In the Help panel, click the Help question mark. Click Learn More About How Spline Works. Read the first three paragraphs. (You are welcome to read more, but, without statistical and calculus knowledge you are likely to just go: "huh?") Close the Help window. Click the down arrow on Input Point Features and choose Known_Altitudes. For the Z value field, use Altitude. Put the Output raster in

___IGIS-Arc_*YourInitialsHere*/Experiment_with_Interpolation

and call it simply Spline. Make the output cell size 5.0. Leave the rest of the text boxes as they are—after reflecting that REGULARIZED, Weight, and Number Of Points are options that would affect the surface you are about to create. (Again, all the tools, with their parameters, are intended to make guesses at the values between the points. Those guesses are affected by the options chosen by the user.) Click OK. Close the Spine window when the Close button appears.

___ 7. The raster Spline has now been added to the map. If necessary, move Known_Altitudes to the top of the Table of Contents so you can see the point locations and the associated elevations. From the Table of Contents, what are the values related to the range of altitudes that the generated surface exhibits?: Lowest? _____. Highest? _____. Range? _____.

___ 8. Examine the Properties of the raster. Turn on Map Tips for the raster (under the Display tab).

___ 9. Start the Identify tool (specifying Spline) and check some (Pixel) values. Check close to points and between points. Recall from the Help file that the generated surface is supposed to pass through all the points. Also notice that you get a "Class Value." This number comes from the Table of Contents. Turn off Map Tips.

___ 10. *For all the steps that follow, use a cell size of 5.0.* Always move the Known_Values shapefile to the top of the Table of Contents, above all the rasters.[10]

___ 11. Create a surface, called IDW, with the IDW (Inverse Distance Weighted) tool, after reading the first three paragraphs of the help file. By flipping IDW off and on, you can compare it with the results of Spline. You see a fair amount of difference. Eyeball a location between two of the known points and get its altitude with Spline. _____. Then pick approximately the same point with the IDW surface. _____. You see why I use words like "guess" and "reasonable estimate" to describe the outputs of these tools.

___ 12. Create a surface, called Ntrl_Ngbrs, with the Natural Neighbors tool. Read the Usage Tips. Explore. What is one obvious difference between Ntrl_Ngbrs and the other two presentations? _____

---

[10]Actually, these rasters don't look like rasters initially (no jags) because ArcMap initially displays them smoothed out, but they are rasters, as you can see from the Properties. Also, if you select Classified or Stretched under Show in Symbology, you can restore the raster-looking appearance.

_____ **13.** Create a surface with the Trend tool. Call it Trend1. Your immediate impression may be that something went wrong, because you just get stripes! What has happened here is that the process fitted a plane, a flat surface, through the space occupied by the set of points. The angle of the plane was set so that the sum of the squared distances between the plane and the points was minimized. If you identify cells at either end of a given stripe, you'll see what I mean. You might get a better picture if you change the Symbology to Stretched.

_____ **14.** **_Create another surface with the Trend tool, calling it Trend2:_** The difference? You'll put in 2 for the Polynomial Order instead of 1. This allows for a surface whose defining equation can contain squared terms, rather than just linear ones. Again go to Stretched Symbology to view it. Still, the surface makes no pretense of going through the points.

_____ **15.** Make Trend3, with the value 3 for the polynomial order, and explore.

_____ **16.** Make a surface with the Kriging tool, after reading the first paragraph of the Help that shows up when you click Learn More About How Kriging Works. You will also notice that you have lots of options in the Kriging tool window—Ordinary vs. Universal, Spherical vs. Circular vs. Exponential vs. Gaussian vs. Linear—and so on. The lesson here is that, as you start your career in GIS (after all, you are almost done with this book) when you need to create a surface you should seek help from people or documents that will let you know the most appropriate tool to use and how to use it.

Points and Density

While we are on the subject of making surfaces and areas out of points, let's consider the Density tool. This procedure takes the values that are associated with points and, with a raster, spreads them out over the landscape around the points. For example, suppose you have data that places all the population of each county in a state at the points designated as courthouses. Obviously, everyone in a county doesn't live at the courthouse.

_____ **17.** Start a new map. Add as data Known_Populations.shp.[11] Label each point with the Population value. Make the map units and the display units Kilometers. What we will do is spread the population at each courthouse over an area of several hundred square kilometers, just to illustrate how the Density tool works.

_____ **18.** Find Point Density (sa) in ArcToolbox. Start the Point Density tool. The Input Point Features would be Known_Populations and the Population field Population. Make the Output raster Density_test, in

___IGIS-Arc_*YourInitialsHere*/Experiment_with_Interpolation

using an output cell size of 5.0.

For the neighborhood, pick Rectangle. For the Neighborhood Settings, make a square from Map Units (not Cell Units), 40 kilometers on a side. How many square kilometers will there be in the neighborhood? _____. OK.

---

[11]This is basically the same file as Known_Altitudes.shp, but we are not planning on making a surface, although one could consider a surface of population—with the z being, say, the number of people per acre.

_____ **19.** Move Known_Populations to the top of the Table of Contents. Examine the Density_test raster with the Identify tool. Click in the square which has a population of 3200. How many people are there per square kilometer? _____ If you didn't get 1600 for the number of square kilometers in each neighborhood, go back and rethink the previous step. Click some other neighborhoods, including the ones with populations of 1600 (which should have one person per square kilometer) and 1000.

Thiessen, Voronoi, Dirichlet (and, of course, Decartes)

While we are on the subjects of points and their surrounding areas, let's look at the creation of a set of polygons, each created from a point, that have the following property: Every interior point of each polygon is closer to its generating point than to any other point. This idea has "been invented" by at least three different people (and was used informally by Decartes), giving rise to Thiessen polygons, Dirichlet domains, and Voronoi cells. The polygons tessellate a portion of the Cartesian plane. Unlike the procedures used above, no value at the point is considered. Only the position of the point is important. In Arc-Toolbox you find the tool among the Coverage tools:[12]

ArcToolbox > Coverage Tools > Analysis > Proximity > Thiessen

_____ **20.** With ArcToolbox > Conversion Tools > To Coverage > Feature Class To Coverage, convert Only_Points.shp to a coverage of the same name.

_____ **21.** Using the Thiessen tool with the coverage ONLY_POINTS, make an output coverage named T_D_V.

_____ **22.** Start a new empty map in ArcMap. Add as data the point component of ONLY_POINTS and the polygon component of T_D_V. Put the points at the top of the Table of Contents. Label with the ONLY_POINTS-ID field. Use the Measure tool to convince yourself that the arcs lie halfway between pairs of adjacent points, by measuring from the first point to the second point, clicking, and then measuring back to the arc. The segment distance should be one-third of the total distance.

_____ **23.** With the drop-down menu of the Spatial Analyst toolbar, convert the coverage T_D_V to the raster ThieDiriVoro. Use T_D_V-ID for the Field and 5.0 for the cell size. Put the point coverage at the top of the Table of Contents. Turn off the T_D_V coverage, leaving the ThieDiriVoro raster on. Use the Identify tool on the raster to verify that the raster zone is the same as the point value. Purists may use identify to assure themselves that the number of cells times 25 is about equal to the area reported by the coverage.

_____ **24.** Close ArcMap.

---

[12]For anyone who tells you that geodatabases cover everything in GIS and that coverages are passé, the quick answer is: Thiessen polygons?

GRIDS: Distance and Proximity

Making a Grid Showing Straight-Line Distances to a Single Place

In the following steps, you will see how ArcMap Spatial Analyst makes a grid of Euclidean (straight-line) distances from each cell to a single source cell.

_____ **1.** If ArcMap is already running, start up a new map. Otherwise, start ArcMap. Choose Extensions from the Tools menu, and make sure the Spatial Analyst box is checked.

_____ **2.** Open (rather than Add Data) a map file named Proximity_startup.mxd, which you will find in

_____IGIS-Arc_*YourInitialsHere*\

Spatial_Analyst_Data\Proximity_Data_SA.

A map with a line coverage called Fishnet opens. Click the Display tab. Note that "Fishnet arc" appears as a 10 × 10 matrix of transparent squares. Each square is 10 units (say the units are kilometers) on a side, so the square represents a space 100 kilometers on a side. Fishnet will serve as a backdrop for the next several raster data sets.

_____ **3.** Click the Add Data button. In the Add Data dialog box, navigate to the

_____IGIS-Arc_*YourInitialsHere*\Spatial_Analyst_Data\Proximity_Data_SA directory, and click the grid named Onecell. Click Add. Onecell is a raster consisting entirely of No Data values, except for a single cell with a value of 999. Use Identify to verify this. Open, examine, and close the attribute table for Onecell. Change the color of OneCell to solid black. See Figure 8-46.

_____ **4.** Bring up the Properties of Onecell. Under Source, notice that the cell size is 10 and there are 10 rows and 10 columns in the grid. Scroll down to look at the rest of the information, though it's pretty uninteresting, given that the entire raster consists of one cell that isn't NoData. Under Symbology, leave Unique Values highlighted. Cancel the Layer Properties window.

By using Spatial Analyst, you will be able to make a raster whose cells are the distances from the center of each cell in the grid to the center of the source cell (Onecell).

_____ **5.** *Prepare to calculate the Euclidean distance:* From the Spatial Analyst Toolbar drop-down menu, choose Distance > Straight Line. A Straight Line window opens. For Distance to browse to

_____IGIS-Arc_*YourInitialsHere*\

Spatial_Analyst_Data\Proximity_Data_SA\Onecell.

(From now on in this exercise you will be working in the Proximity_Data_SA folder, so I will only tell you the name of the data set.)

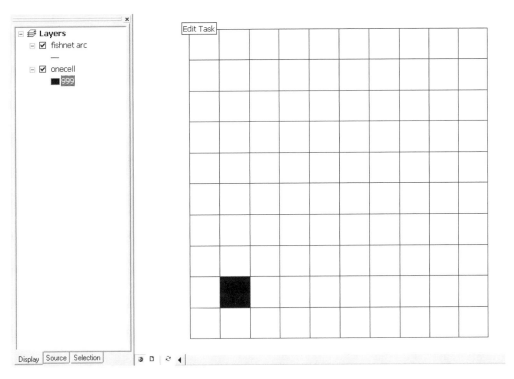

FIGURE 8-46

The Output Cell Size should say 10. For the Output raster, browse to the Proximity_Data_SA folder and type the name `MapDist1`. The type is ESRI GRID. Press Save in the Save As window. OK the Straight Line window.

_____ **6.** Spatial Analyst will calculate a grid with the distances and add that data set to the map. Change the order of entries, so that from top to bottom they are ordered: Fishnet, Onecell, MapDist1. See Figure 8-47***.

_____ **7.** *Explore the grid with the Identify tool:* Click the Identify tool and make MapDist1 the displayed data set. Run the mouse over the grid and note that the cursor position (both X and Y between zero and 100) is given by the numbers on the status bar of the ArcMap window. Now click one of the cells. The cell's value, which is the distance from it to the source cell, is displayed. Set Map Tips on (Properties > Display > Show Map Tips) to look at the values of MapDist1. Using the Identify tool or Map Tips, query several more cells.

_____ **8.** *Understand the grid values by checking some other cells:* Each cell represents the straight-line distance from the center of that grid cell to the center of the source cell. Verify that the cell on the new grid that corresponds geographically to the source cell on the Onecell grid has a value of zero.

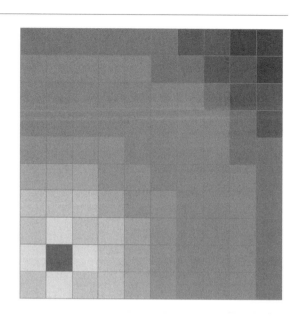

FIGURE 8-47

Note that if you go three cells to the right of Onecell, the distance is 30. If you go from that cell up four cells (that is, 40 units), the (diagonal) distance to the source is 50. That is reassuring, because Pythagoras tells us that the sum of the squares of the two sides of a right triangle equals the square of the hypotenuse (900 plus 1600 equals 2500, whose square root is 50).

___ **9.** *Verify distances using the Measure tool:* First, click a cell and use the Identify tool to obtain its value. Now click the Measure tool and verify that the distance from the center of your chosen cell to Onecell is approximately the same as the value displayed in the Identify Results window.

___ **10.** *Determine the distance from a source cell to a diagonal neighbor:* Use the Identify tool to check the distance from the cell in the lower left-hand corner of the raster to the source cell. What is it? _____. Use the measure tool to determine the distance from the center of the lower left cell to the center of OneCell. _____.

The distance should be 14.142136, which is approximately 10 times the square root of 2. ($10^2 + 10^2 = 200$; the square root of 200 is approximately 14.142136.)

___ **11.** In the Spatial Analyst toolbar, there is a text box named Layer. Using the drop-down menu, put MapDist1 in that text box. Then click the Histogram button at the right of the toolbar. By looking at the Table of Contents and using the Identify tool on the Histogram, how many cells are about 101 to 113 units away from Onecell? _____. Dismiss the Histogram and the Identify Results window.

From this basic understanding of Euclidian distance we can add some capabilities. The new features that you will work with are: (a) more than one source cell in the raster and (b) a cap, or cutoff value, so that no value greater than the cap will be recorded in any cell.

Examining Many Source Cells and the Capping Distance

____ **12.** Turn off Onecell and MapDist1. Add the grid data set Manycells to the map. Notice the presence of several cells containing data, while the rest of the grid shows No Data. There is still a single source cell in the southwestern portion of the grid, but there is also a line of source cells in the northeast and a clump (cluster) of source cells in the northwest. What are the values of the three zones? _____, _____, _____. Change the colors of all the source cells to black.

You will generate a grid containing distances to the nearest source cells, provided that the distance is no further than 30 units. You can limit the search for the closest source cell so that it does not exceed a given value, thereby excluding cells in the output grid that are more than a certain distance away from any source cell. You are telling the software "if you have to look further than this to get to a source cell, just forget it." Such a number may be called a limit, a cap, a cutoff value, or a threshold.

____ **13.** Use the same Spatial Analyst tool as you did in the previous steps to create the new distance raster. Browse to make its name DistManycells. In the text box for Maximum distance, type 30. The Output cell size should be 10. The type is ESRI GRID.

The grid of source cells that serves as input to this tool may consist of several single cells, clumps of cells (contiguous cells), a linear structure of several cells, or any combination of these. For each cell in the output grid, the straight-line distance to the closest source cell is calculated. This happens automatically. You simply apply the straight-line distance tool, as you did in the previous steps, and the newly generated grid contains the closest distances.

____ **14.** Arrange the entries in the Table of Contents so that from top to bottom they are ordered Fishnet, Manycells, then DistManycells. See Figure 8-48***.

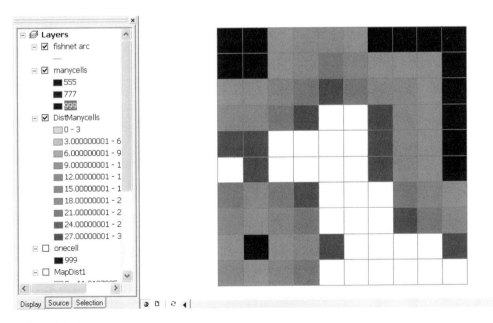

FIGURE 8-48

___ **15.** Examine the distance grid, making sure that DistManycells is the active data set in the Identify tool. Verify that the distances shown by cells in the new data set are indeed associated with the closest source cells in Manycells. The cutoff value of 30 on the distance resulted in some No Data cells in the new data set. Verify by using the Measure tool that the No Data cells are more than 30 units away from any source cells. Note that 30 is the largest value in the legend of DistManycells.

Developing a Grid with Cost Distance

The following steps will illustrate the principle of calculating cost that is based on both distance and a cost surface. First, a quick review question:

___ **16.** What would be the cost of the path going horizontally through a cell (cell size 10, cell cost 0.85)? Answer: 8.5. How about the cost of moving through that same cell, going diagonally? _____.

___ **17.** Turn off all visible entries except Fishnet. In the steps that follow, keep Fishnet at the top of the Table of Contents to delineate the cells of the displayed grids.

The Tolls data set is an artificial cost surface grid data set. The surface is divided into rows of cells; all the cells in a given row have the same value.

___ **18.** Add the cost surface raster, Tolls, from

___IGIS-Arc_*YourInitialsHere*\Spatial_Analyst_Data\Proximity_Data_SA

Imagine that the center of each cell contains a toll booth. Depending on which row the cell is in, the toll booth charge varies from 25 cents to $1.15 to pass through *one unit of distance* of the cell. In the Tolls grid, all the cells in the bottom row have values of 0.25; the top row cell values are 1.15. So, to cross a single cell in the top row in an east-west direction would cost $11.50 (1.15 dollars times 10 distance units). (Recall that it costs more to cross the cell in a diagonal direction because the distance is longer.) Use the Identify tool to examine the values of cells in the rows. See Figure 8-49***.

___ **19.** *Examine the source grid:* Move Onecell to the top of the Table of Contents and turn it on. Recall that Onecell was the source grid you originally worked with that contained a single source cell.

___ **20.** *Apply the cost weighted distance tool using the Spatial Analyst drop-down menu:* Choose Distance > Cost Weighted. Distance To should be Onecell. The Cost raster should be Tolls. Call the Output raster TCostOnecell (the total cost of the path from the center of each cell to the center of the source cell in the data set Onecell). Click OK.

___ **21.** Arrange the entries (Fishnet, Onecell, TCostOneCell, Tolls) and look at the new legend. Each cell contains the cost of moving across the cost surface grid from that cell to the source cell. You can get a general idea of the relative costs by looking at the colors of the grid.

___ **22.** Using the Identify tool, examine the total cost of traversing the cells due north of the source cell (pick the cell on the top row directly above the source cell). Now look at the total cost of

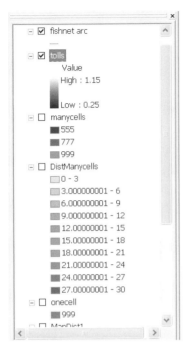
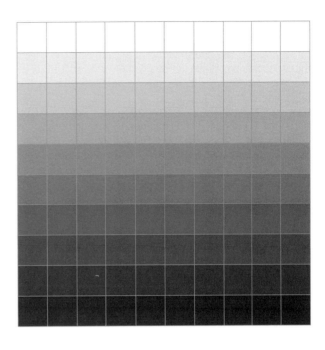

FIGURE 8-49

traversing the cells due east of the source cell (pick the cell on the right column directly to the right of the source cell). It is obviously cheaper to come from the east (cost is about $23.50) than from the north ($60). The overall range of costs is from 0 (from the source cell to the source cell) to approximately $82. Again use the Identify tool to display the values of the Toll cells that are traversed. To come from the east, one need only traverse cells with costs of 0.35 per kilometer or less. To come from the north, the cell costs are 1.15, 1.05, 0.95, and so on. The most expensive trip is from the northeasternmost cell—it is furthest away and in an area where the cost per cell is highest.

_____ **23.** ***Check out the contouring capability of the Spatial Analyst.*** In the Spatial Analyst Toolbar there is a text box named Layer. Enter TcostOneCell in that text box. Click the Contours button. Then click various places on the raster. The lines that appear are estimates of constant cost. That is, from any point on a given contour line, it should cost about as much to go to Onecell as from any other point on that line. Of course, this won't be exact, since the cost jumps incrementally each time you cross a cell boundary, but you see the idea. Use Identify to check out one or two lines.

_____ **24.** Erase the contour lines by pressing the Select Elements button, dragging a box around the contour lines, and pressing the Delete key or Delete button.

Creating Direction and Allocation Grids

You learned that you can calculate the shortest distance between two places using the Euclidian Distance request and you can find the least-cost path with the Cost Distance request, but there are still a couple of major questions: (1) Which source cell is closest to (or least expensive to reach from) each given cell in the grid, and (2) what is the direction or path from each given cell? To answer these questions, first for straight-line, do the following.

Going back to the Euclidian (straight line) calculation, you will be able to make not only a grid showing the distance to a source cell, but you will make two additional grids. One of these will show, in each cell, roughly the direction one must travel to get to a source cell. The second will show, in each cell, the number of the closest source cell. You will use the two parameters in the euclidian distance request that were left unchecked in previous examples.

_____ **25.** Click the New Map File icon. Open Proximity_startup.mxd. Add the layer Twocells to the map.

_____ **26.** On the Spatial Analyst drop-down menu, click Distance, then Straight Line. The Distance To raster should be Twocells. The cell size should be 10. Max Distance should be 50. Check the boxes of Create Direction and Create Allocation. For the Direction grid, browse to Proximity_Data_SA folder and enter the name Dir. For the Allocation grid, do the same, but enter Allo for the name. Name the Output raster EucTwocells. Click OK.

_____ **27.** Arrange the layers in the Table of Contents this way:

    Fishnet
    Twocells
    Allo
    Dir
    EucTwocells

    Turn on Fishnet, Twocells, and Allo; turn all other layers off.

_____ **28.** Using the colors of Allo and the legends of Allo and Twocells, notice that like cells are allocated to (clustered around) the nearest source cell, and (using Identify) that the allocated cells take on the value of that source cell. Cells greater than 50 distance units away from a source are set to NoData. Note that there are several of these.

_____ **29.** Turn off Allo and turn on Dir. Using the Symbology Properties, set the Layer Properties Show box to Stretched. Examine Dir with the Identify tool. Verify (by looking at Pixel value) that the direction in each cell appears to be the compass direction you would travel from that cell to reach the source cell. The direction 360 is used to indicate north; the value 0 is placed in those cells that correspond to the source cells. Figure 8-50*** has arrows drawn to show the value of several cells.

    Figure 8-51*** shows the values in the cells.

(You have used the Cell tool that made these arrows and values, available from

http://arcgisdeveloperonline.esri.com/ArcGISDeveloper/default.asp?URL=/ArcGISDeveloper/Samples/SpatialAnalyst/CellTool/CellTool.htm.

This is an unsupported ESRI product. The CellTool is also on the CD-ROM that accompanies this book.)

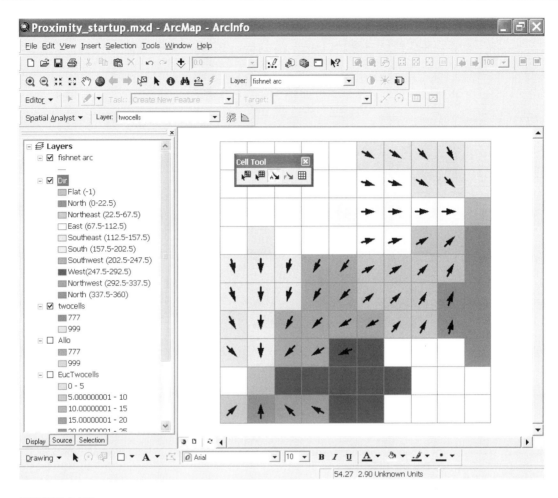

FIGURE 8-50

Using Cost Distance to Make Direction and Allocation Grids

Solve the following problem in a new map. Basically, you are repeating the previous exercise, but using cost distance instead of Euclidian distance.

_____ **30.** Start a new map with Proximity_startup.mxd. Add TryThis as a source grid and Tolls as a cost grid.

_____ **31.** _Using_ `Spatial Analyst toolbar > Distance > Cost Weighted`, generate the Accumulated Cost raster, Direction raster, and Allocation raster. Make Cc, which is the output, the cost of going from any cell to a source cell in TryThis. Make Dd the direction grid—sometimes called a backlink grid, because it can be used to find the least-cost path from any cell _back_ to the source cell. Make Aa the allocation grid. Use Tolls for the Cost raster. Cap the cost at a maximum of 30. Click OK.

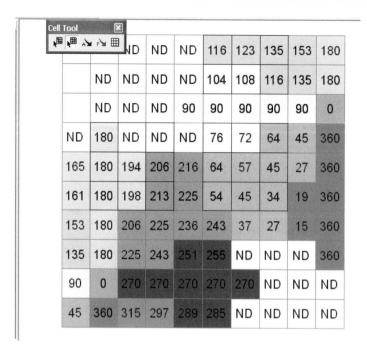

FIGURE 8-51

_____ **32.** Collapse the legends. Place the entries in the Table of Contents in the following order: Fishnet, TryThis, Aa, Dd, Cc, Tolls. Turn off all except Fishnet and TryThis. Examine TryThis with Identify, and the Table of Contents (which you will have to re-expand), so you know what is where.

_____ **33.** Start the Identify tool and set it so that it shows the values of Cc, even though that layer is turned off. Position the cursor in the cell whose center is at location approximately (85, 15). The cost (Pixel value) from that cell should be about 21. Examine other cells in the accumulated cost grid. What is the cost from (5,55)? _____

_____ **34.** Turn on and expand the allocation grid Aa. Examine Aa with Identify. Verify that the cell at (85, 15) is allocated to (generates the least-cost path to) a source cell whose value is 50. Determine where the cells of value 50 are. To what cell value is the cell at (5,55) allocated? _____

_____ **35.** _**Compare the allocation grid (Aa) with the source grid (TryThis):**_ Display TryThis with a grayscale ramp. Choose Properties > Symbology > Classified. Set the number of classes to 8. Repeat the process for grid Aa. Turn both entries on and everything else (except Fishnet) off. Now, by toggling Aa off and on, you can see the cells in Aa that are clustered around the source cells in TryThis. Since the two entries have the same grayscale scheme, Aa seems to "grow out of" TryThis. However, because of the cost surface Tolls, there are more cells clustered toward the southwest. This is the essence of the weighted allocation operation. Turn off Aa.

____ 36. Turn Dd on and expand its legend. Explore the values of some of its cells with Identify. The values tell toward which of the eight nearest neighbors the path to the desired cell runs. The code is as follows:

1	East	90
2	Southeast	135
3	South	180
4	Southwest	225
5	West	270
6	Northwest	315
7	North	360
8	Northeast	45

The code is not particularly intuitive.

____ 37. ***Reclassify the direction code to degrees:*** From the Spatial Analyst menu choose Reclassify. For the Input raster, use Dd. The Reclass field should be Value. Click the Unique button to show all values. Call the new raster Dd_Deg. Type in the new values from the list that follows. The new values are in degrees, based on the compass direction north as 360. Not all the old values listed will be represented in the Reclassify window (there is no 2); just type values for those that are.

Old Value	*New Value*
0	NoData
1	90
2	135
3	180
4	225
5	270
6	315
7	360
8	45
No Data	NoData

Click OK to generate the new data set called Dd_Deg. Move it to the top of the Table of Contents under Fishnet and turn it on. Turn all other layers off. Verify that to head toward the source cell from cell (85,15), you would first go to cell (75,5). That is, cell (85,15) will have the value 225, indicating a direction of southwest. Trace the rest of the path from cell (85,15). Notice how it takes a southern route despite the fact that the shortest distance (the straight-line distance) is due west. Why is this?

_____.

Calculating a Least-Cost Path

In the following steps, you will find the least-cost path between a starting cell and a destination cell. You may have noticed that you could figure out the least-cost path by investigating the direction grid in the last step. If so, you will have certainly noticed that doing so was a total nuisance. Using the Shortest Path tool will generate a line shapefile that will show you the shortest path (or least-cost path) from a specified cell to a target cell.

Shortest Path can handle multiple cells in both the starting grid and the destination grid, but this exercise will demonstrate the request using only one cell in each. If it has not become clear by this point, I should state the obvious: The cost (and/or the path) of going from A to B, as calculated by the grid manipulations you have been directing, is the same as the cost (and/or the path) of going from B to A. There are certainly instances in which this assumption is not warranted. One example would be if A is at the bottom of a mountain and B is at the top. Another example is driving in a city that has one-way streets. But the tool can be useful anyway.

The goal of this exercise is to use Shortest Path to create a *shapefile* that shows the least-cost path between the single cell in a grid called Single_cell and the cell in the grid Onecell. You could, of course, determine this path by exploring the destination grid, but Shortest Path generates a Shapefile with the path delineated.

_____ **38.** Start with the New Map File Proximity_startup.mxd. Add the grids Onecell, Tolls, and Single_cell. Arrange them in the following order: Fishnet, Onecell, Single_cell, Tolls. Make Onecell bright red. Make Single_cell bright green. Make Tolls stretched.

Shortest Path requires several inputs. The tool itself operates on the grid containing the starting point, but Shortest Path uses the rasters that you generate ahead of time with the Cost Weighted tool. These rasters are the cost grid and the direction grid. (You won't need an allocation grid.) So, you need to first use the Cost Weighted tool.

_____ **39.** *Make a least-cost grid and direction grid for Onecell:* In the Cost Weighted window, the Distance To should be Onecell. The Cost raster is Tolls. Call the direction grid Onecell_dir. Name the Output raster Onecell_cost. Click OK. Collapse the legends.

_____ **40.** Using the drop-down menu on the Spatial Analyst toolbar, pick Distance, then Shortest Path.[13] In the window, For Path To, browse to Single_cell. Browse to Onecell_Cost for the Cost Distance raster. Browse to Onecell_Dir for the Cost Direction raster. For the Path Type, use Best Single. The output is a *shapefile*; name it Least_Cost_Path.shp and plan to put it in Proximity_Data_SA. Click OK.

_____ **41.** Fishnet and Least_Cost_Path should be at the top of the Table of Contents and turned on. Single_cell, Tolls, and Onecell should also be on. Turn all other entries off.

_____ **42.** Examine the new layer Least_Cost_Path. Note the somewhat circuitous route that was calculated as the least-cost path runs from cell center to cell center. Is the path surprising in light of the Tolls cost surface grid?

_____ **43.** Exit ArcMap.

---

[13]In this case Shortest Path means Least-Cost Path.

Putting the Tools Together: Site a Regional Park

The power of GIS is most evident when you use a number of tools in concert. This is also the stage where the creativity comes in—and sometimes the frustration as well, because frequently, even if data sets are available, they are not in the form you can most easily use.

This exercise uses the Shortest Path tool, which calculates and tries to minimize costs according to the "price you pay" in traveling between a source and a destination. This price has many names, among them: cost surface, friction, resistance to flow, resistance to travel, and impedance.

Suppose that the county of San Bernardino, California, would like to build a regional park somewhere near the city of Redlands. One of the criteria for the park location is that it must be accessible to those who might drive some distance using freeways or interstate highways.

For this exercise, accessibility is defined very specifically: The less time it takes to drive from any freeway offramp, the more accessible the park is deemed to be. The source grid, therefore, will consist of cells containing Off_ramps.

Your approach will be to create a cost surface, expressed in minutes required to travel. You will derive travel cost weights from speed limits on the non-freeway roads in the area. The cost surface, or weighting grid, will be the inverse of the speed (divide the speed into 1.0, so you get, say, hours per mile, rather than miles per hour). The grid will be based on the time it will take a car to traverse each given cell in the study area.

Here are the steps:

1. Set up the Environment.

2. In ArcMap, add the shapefiles STUDY_AREA and ROADS.

3. Join a table that augments the ROADS attribute table with road types and speed limits.

4. Select the roads records with road type Offramp.

5. Make a grid called Offramps of these selected roads.

6. Select all the other road type records.

7. Make a grid called Roads_raster of the selected records.

8. Reclass Roads_raster with appropriate speed costs to make a raster called Speed.

9. Calculate a Cost raster, Mn_per_Ft, as an inverse function of Speed.

10. Use Offramps and Mn_per_Ft to calculate Drive_Time, the driving minutes to each cell.

Setting Things Up

___ **1.** Start ArcMap. Add the data files ROADS.shp and STUDY_AREA.shp from

___IGIS-Arc_*YourInitialsHere*\Spatial_Analyst_Data\
Proximity_Data_SA\Park_Data.

Make sure that ROADS is at the top of the Table of Contents. Save the resulting map as Site_Park.mxd in the Park_Data folder.

___ **2.** Examine the Site_Park entries and set some options: The study area is a simple, single polygon representing the Redlands, California, area. ROADS is a line file. Each single feature in ROADS is an amalgamation of all the roads of a given type in the study area.

___ **3.** Change the mapping units and the display units of the data frame to Feet.

___ **4.** Set Environment Variables: `Tools > Options > Geoprocessing > Environments > General Settings`. Leave the coordinate system undefined. Make the `Output Extent Same as Dataset` "STUDY_AREA.shp." Under `Raster Analysis Settings`, set the cell size to 150 (under As Specified Below). Click OK, and Ok again.

Now that setup is complete and you begin thinking about the problem, a couple of issues might come to mind.

(a) How are you going to identify the Off_ramps?

(b) How are you going to exclude cells that have no roads, and cannot be traversed? Recall that your previous experience has been with grids in which any cell could be traversed.

As to (a), you will be able to select the ramps based on values in the Roads table, which indicate the road types. For (b), once you have identified cells with road segments, you can set their impedance based on speed limits, which will also be available in the Roads table. For the cells without roads, you can set a high impedance, or cost, so that it would be prohibitive to drive through very many of them.

___ **5.** Open the Attributes of Roads data set table. Some of the data massaging has already been done for you. The roads have been clipped to the study area, and the number of roads has been reduced to 12 sizable networks (from 29,326 road segments in the valley). The 12 road networks were derived by merging similar road types based on their Census Feature Classification Codes (CFCCs). Roads is a feature-based layer, not a grid data set.

The assumption has been made that all roads of the same CFCC will have the same speed limit. (The speed limit data is in a table that you will later join to the Attributes of Roads table).

___ **6.** ***Add the cfcc.dbf table to the map:*** Open it. Notice the different road classifications and the speed limits associated with each.

___ **7.** ***Join the two tables:*** Right-click ROADS in the Table of Contents. Choose `Joins and Relates`. Click Join. In the Join Data window, you want to use the `Join Attributes From A Table selection`. The field that the join will be based on is CFCC in both layers (that is, window items 1 and 3). The table to join (item 2) is CFCC.dbf. Click OK. The data contained in CFCC.dbf is added to the Attributes of Roads table. Close CFCC.dbf and resize the now-enhanced Attributes of Roads table so you can see all its fields.

The Roads table identifies the Off_ramps with a code (A63). You want to make a grid containing source cells that represent road segments with the A63 designation; you want to assign No Data to all other cells. You will select the Off_ramps and then make a raster data set of the selected features

___ **8.** ***Select the Off_ramps:*** With the Attributes of Roads table active, select the offramp record. Minimize the table so you can see the Off_ramps selected in the map, highlighted in cyan.

_____ 9. ***Prepare to create the source grid.*** From the Spatial Analyst drop-down menu, choose Convert > Features To Raster. The Input Features should be ROADS. Make the Field CFCC.CFCC. The field chosen here is only informational; the integer value of the one record in the grid will be 1. If necessary, set the Cell Size by typing 150 (recall that the units are feet). To create the name, browse to

___IGIS-Arc_*YourInitialsHere*\Spatial_Analyst_Data\
Proximity_Data_Spatial Analyst\Park_Data

and type as the name Interstate_off_ramps. You will be reminded that GRID names cannot exceed 13 characters. So use Off_ramps instead. Click Save, then OK.

_____ 10. Notice that Off_ramps shows up in the Table of Contents. Click the Display tab in the Table of Contents. Make the Table of Contents order, from the top, ROADS, Off_ramps, and Study_Area. Have Off_ramps on, ROADS and STUDY_AREA off. Zoom up on a region of colored cells, so you can see the individual pixels. Measure the cell size. _____ Feet.

_____ 11. Turn on the ROADS Layer. By panning and zooming, you should assure yourself that the selected Roads (Off_ramps) did, in fact, create cells in the Off_ramps grid.

_____ 12. Open the Off_ramps attribute table. How many cells are there representing the Off_ramps? _____. All the other cells are set to No Data. What is the CFCC code? _____.

Preparing to Create a Cost Surface

You have just made the source grid. The next step is to create a cost surface grid that represents "resistance to travel" by assigning travel costs to every cell in the study area. In those cells where there are roads, you will base the travel cost on speed limits. In areas where there are no roads, you will simply assign a high cost based on a very low speed limit (5 miles per hour), primarily to keep cars on the roads. It would be reasonable to ask, "Why not just assign a super-large impedance to off-road cells?" You won't do this, because you want to allow every cell in the study area the possibility of access. If you assign an impossibly high impedance value to cells that don't have roads in them, you exclude those cells from consideration for the park because they could not be accessed unless a road ran through them. Presumably, if a group of cells is to become a park, a low-speed access road would be built. You do, however, want to completely exclude freeways to force travel on local roads and to prevent a path from crossing the freeways. In the following steps, you will select highways, county roads, and local roads to make a grid based on their speed limit values. You will then reclassify the non-road cells to give them a speed limit of 5 mph.

_____ 13. ***Select the proper roads and create the Speed raster:*** Make the Attributes of Roads table visible. Clear any selections. Now select the Interstate, US Highway, County Road, Local Road, and Not A Public Road records. Do not include the Offramp records. Prepare to convert these features to a raster. In the Features To Raster window, do not browse for Input features. Instead, use the drop-down menu. Pick ROADS. For the Field, pick CFCC.SPEEDLIMIT and type 150 as the cell size. Browse to Park_Data and name the Output raster SpeedLimit. Click Save. Click OK. Once the SpeedLimit GRID is built and added to the map, pan and zoom to check it out with the Identify tool. Click the Off_ramps layer off and on to assure yourself that they are not included in SpeedLimit. In the Attributes of Roads table, clear the selections.

The SpeedLimit data set is a raster representation of the selected roads; the values in the cells are the speed limits. Notice from the map that the cells between the roads were assigned the No Data value. Let's now make a raster in which the speed in each cell is the same as the speed limit in that cell, except for Interstate cells (which we will make low because we don't want our path to the park to try to cross an interstate). The No Data cells we will set to 5 mph. To do all this, we will use the Reclassify tool.

____ **14.** ***Make a GRID with speeds in the cells:*** Go: Spatial Analyst > Reclassify. The Input raster should be SpeedLimit. Press the Unique button to be sure that you get all the different values of the cells. The Old Values are 1 (for private roads), 30, 45, 50, 65, and No Data. The suggested new values are 1, 2, 3, 4, 5, and No Data. You do not want these suggested values. Mostly you want the new values to be the same as the old values, except for the No Data value, which you want to be 5, and the Interstate value (65), which we'll make 2. Type in the new values: 1, 30, 45, 50, 2, and 5. Browse and call the new Output raster Speeds (in Park_Data, of course). Click Save, then OK.

____ **15.** ***Examine the cells containing the speed by panning and zooming, using Identify:*** Use the Identify tool (operating on the Speeds GRID) to verify that the data cells that were formerly NoData cells have a value of 5 and that other cells have appropriate values. Turn off the Speed_Limit, Off_Ramps, and the Study_Area entries, if they are on. Move the Roads data set to the top of the Table of Contents. Turn on the Speeds data set. Next, zoom in on an area that includes part of the interstate where the ramps are. Notice how the speed limit cells follow along with the local roads. Notice also that all the cells between the roads now have the 5 mph value. Zoom in some more to see the relationship between the feature data set Roads and the GRID data set Speeds. When you are finished looking, choose Full Extent from the map menu.

____ **16.** ***Examine the Properties of the Speeds GRID:*** Right-click the Speed layer name (called Speeds–Speeds, for some reason) and click Properties > Source. How many rows are there? _____. Columns? _____. What is the uncompressed size of the raster? _____. What is the format? _____. Dismiss the Layer Properties window.

Building a Cost Surface

Now you have the raw material (those speeds at which those going to the park may drive) for building a cost surface. What numbers do you put in the cost surface cells? You cannot put miles per hour (mph), because then higher speeds would imply greater resistance to flow—the opposite of what you want. You could use the reciprocal, hours per mile. To get this reciprocal, you simply divide "1" by miles per hour, which gives you hours per mile. For example, 50 mph is 0.02 hours per mile; 25 mph is 0.04 hours per mile. Since 0.04 is a larger amount than 0.02, and that larger amount represents a slower speed, 0.04 is what you want to see in your cost grid.

But there is another consideration: It is important to express the cell costs in terms of the units of the grid and the problem statement. You are interested in drive times in minutes. And the grid is measured in feet. The numbers you have been given to work with are miles and hours, but what you want, strange as it may seem, is minutes per foot! How do you convert hours per mile (hpm) to minutes per foot? Multiply hpm by a conversion factor: 0.0113636, which is the number of minutes it takes to travel a foot at the rate of one hour per mile.[14]

---

[14]One hour per mile is 60 minutes per 5280 feet. 60/5280 equals 0.0113636—your conversion factor. So, for example, 50 miles per hour is 1/50 hours per mile. 1/50 * 0.0113636 = 0.000227272 minutes per foot. If a cell is 150 feet across, say, from east to west, it therefore takes 150 * 0.00022727 or 0.034 minutes to cross it at 50 mph. (Or you can just take my word for it.)

___ **17.** ***Create the travel cost surface:*** From the Spatial Analysis toolbar, open the Raster Calculator. Enter the expression that will convert mph into minutes per foot for every cell in the grid:

```
1.0 / [Speeds - Speeds] * 0.0113636
```

being sure, in constructing this expression, to *leave blank spaces* on either sides of the operators * and /.

This may seem like a lot of trouble just to divide the speeds into 1.0 and multiply by a constant, but the arithmetic operations take place over the entire study area, composed of more than 200,000 cells (522 * 387). You get a lot of computation for your expression. Click Evaluate.

___ **18.** ***Make the cost layer permanent and examine it:*** The calculation that appears in the map as a result of the raster calculation is temporary and will go away unless you make it permanent. Right-click the name and choose Make Permanent from the drop-down menu. Make the new name Min_per_Ft and put it in the Park_Data folder. Add Min_per_Ft to the map and remove the temporary calculation from the Table of Contents. Since the numbers are small enough to require exponential notation (for example, E-04), the software assumed you wanted to see them as a "continuous raster." But we know that there are only a small number of unique values. So change the symbology: Right-click the Min_per_Ft layer name, choose Properties > Symbology > Unique Values. Click OK.

You now have the ingredients necessary to run the Cost Distance request. The source grid is called Off_ramps; the cost surface is Min_per_Ft.

___ **19.** Choose Spatial Analyst toolbar > Distance > Cost Weighted. For Distance To, use the little yellow file folder (not the dropdown menu) to browse to Off_ramps in the folder Park_Data. The Cost raster is, of course, Min_per_Ft, browsed to in the same way. The next field says Maximum Distance, but that is misnamed in this case, since it is minutes we are producing. Put in 60 for the maximum number of minutes. For the Output raster browse to Park_Data and then put in the name Drive_Time. Click Save, then OK.

Depending on the speed of your computer, this may take a while to calculate. Why does it take so long? Realize that almost one-fifth of a million cells must have values calculated for them. You will know that Cost Distance is finished when the new data set is added to the map.

___ **20.** Turn all entries off except Drive_Time and Roads. Keep the Roads data set above Drive_Time in the Table of Contents. Examine the Drive_Time layer. Look carefully at the Drive Time data set values. While many areas are within a 10-minute drive of an offramp, others are an hour away.

Of course, drive time is only one factor to consider in siting the park. Probably the areas closest to the Off_ramps are the most developed and therefore least suitable for a park.

Improving the Understandability of the Map

You can get a better idea of travel times by redoing the legend of Drive Time. The current legend simply breaks the drive times up into equal intervals, resulting in such unwieldy ranges as 9.065 minutes to 18.131 minutes. A better set of classes can be defined, and a better legend can be created.

_____ **21.** Through Drive_Time Properties, bring up Symbology. In the Layer Properties window, select Show Classified. Specify 10 classes, then press Classify. In the Classification window, make the break values 3, 5, 10, 15, 20, 25, 30, 40, 50, and 60. Click OK. Back in Layer Properties pick a color ramp you like. Under Label, type in the top value of each range: 3, 5, and so on. Click Apply, then OK.

The new legend uses more appropriate divisions of driving time.

_____ **22.** Close all legends except Off_ramps and Drive_Time. Make the symbol for Off_ramps solid black. From the top to the Table of Contents, make the layers Roads, Off_ramps, and Drive_Time. Turn them on and everything else off. Explore the resulting map.

It should be apparent that the mathematical massaging you did earlier (calculating minutes per foot) to make the ultimate results show up as minutes was worthwhile.

_____ **23.** Save the project with a new name: Park_Sites_*YourInitialsHere*.mxd. Close the file. Exit ArcMap.

Exercise 8-7 (Project)

Watershed Analysis

To recap from the Overview and prepare you for the next exercise, let's recap the various steps in determining a model of surface water flow.

It starts with elevation information, since water (unless extremely provoked) runs downhill. In this exercise the layer is called Elevpts. From Elevpts we will create ElevSurface. From ElevSurface we will create FlowDir, which will define, for each cell in the raster, the adjacent cell into which the water will flow. We can also create a raster that shows the maximum downhill slope for each cell, called ElevDrop.

From FlowDir we can determine three other rasters:

❑ UniqueBasins—A basin is an area where the elevations and slopes are such that the ultimate drainage all winds up in the same place

❑ FlowAccu—A measure of how much water arrives in each cell, given that every cell initially gets one unit of water

❑ AnySinks—Showing any areas inside the study area where water could not drain to an adjacent cell

From FlowAcc we can determine StrmChannels by saying that if a cell has more than a prescribed amount of water in it, it is part of a stream. StrmChannels lets us produce two new rasters:

❑ StrmOrder_StM—In which each stream is assigned an integer value that is an indication of its size, in terms of the volume of water in it. The Strahler Method (hence the StM in the name) is used.

❑ StrmID—In which each stream is assigned an integer value that is unique within the study area.

Finally, to show the utility of this information, we will consider that stream pollution has been discovered, indicated by a raster called POLLUTION_PTS, from which we will determine SuspectAreas.

In this exercise, you will create an elevation surface from a large number of points, generate a flow direction grid, and then check it for sinks. You will also create a drainage network, assign stream order to linear water bodies, and identify watershed basins.

___ 1. Start ArcMap. From the

___IGIS-Arc_*YourInitialsHere*\Spatial_Analyst_Data\Hydrology_Data_SA

folder add Elevpts.shp. Examine its attribute table. SPOT is the elevation. What is the lowest? _____. Highest? _____. How many points are there? _____

___ 2. *Use the elevation point data set to make a terrain surface grid:* From the Spatial Analyst toolbar menu, choose Interpolate To Raster. Pick Spline. The Input points should be Elevpts. The Z value field should be SPOT. Accept the default cell size. In the Output raster specification, browse to the Hydrology_Data_SA folder, which you will use for all your work here, and type ElevSurface for the name. Accept the other defaults. Click OK. ElevSurface should be added to the map.

Examining the Surface with Various Spatial Analyst Tools

___ 3. From the Spatial Analyst drop-down menu, pick Surface Analysis > Contour. From the drop-down menu of Input Surface, select ElevSurface. Under Contour Definition, you will be shown the altitude minimum and maximum. Accept the contour interval of 100, but use a base contour of some even hundred number under the altitude minimum, say, 1500. For output features browse to

___IGIS-Arc_*YourInitialsHere*\Spatial_Analyst_Data\Hydrology_Data_SA

and name the feature ElevContours.shp. Click OK. Examine the contour lines added to the map and the conformance with the surface ElevSurface.

___ 4. From the Spatial Analyst drop-down menu, pick Surface Analysis > Slope. Make the input surface ElevSurface. Make the output measurement percent slope. Note that the output will be a raster—a temporary one that will go away when you leave ArcMap. (If you want to keep the raster, you can give it a name in any location you browse to.) Click OK. The Slope grid will cover up the ElevSurface grid. Notice that the slope is greatest (like over 60 percent—100 percent is 45 degrees) where the contour lines are closest together.

___ 5. From the Spatial Analyst drop-down menu, pick Surface Analysis > Aspect. Make the Input surface ElevSurface. Again, a raster is placed on the map that covers up the previous ones. As you know, aspect is the direction the slope faces. What you see is something of a hodgepodge—a riot of colors. Examine it further in the next step.

___ 6. In the Layer text box on the Spatial Analyst toolbar, pick Aspect of Spline of ElevSurface. Click the Histogram button. Comparing the colors of the histogram with those of the Table of Contents for Aspect of ElevSurface, in what direction would you say most of the slopes faced? _____

___ 7. From the Spatial Analyst drop-down menu, pick Surface Analysis > Hillshade. Make the input surface ElevSurface. Click OK. Turn off all layers except ElevSurface and Hilshade of ElevSurface.

By flipping the Hillshade layer off and on, you can get an idea of what the area would look like with sun in the northwest, 45 degrees above the horizon, and, from ElevSurface, why it would look that way.

____ **8. *Generate a flow direction grid:*** Navigate to ArcToolbox > Spatial Analyst Tools > Hydrology and start the Flow Direction tool. Do not browse for the Input surface raster! Rather, click the drop-down menu arrow at the end of the text box. It should say ElevSurface. Make the name of the resulting data set just FlowDir, by fixing up the last characters in the text box. Set the option not to force water off the edge of the grid. (Click the `Output drop raster` text box and read about it. Create such a raster, called ElevDrop.) Click OK.

____ **9.** FlowDir and ElevDrop will both be added to the map. From the Table of Contents you can determine the maximum and minimum slopes based on ElevDrop. What are they? _____, _____. Remove ElevDrop.

____ **10.** By looking at the Table of Contents for FlowDir, and by clicking a few cells with the Identify tool, verify that the cells contain the codes for each direction: east (1), southeast (2), south (4), southwest (8), and so on. Recall Figure 8-21. You may be able to note visually that much of the water in the study area tends to drain toward the west (16), with other directions being toward the southwest, and northwest (values of 8 and 32, respectively). You can also note that, in small areas, water flows in lots of different directions.

____ **11.** Open the attribute table of FlowDir. Run Statistics on the Count column. How many cells are there total? _____. Sort the Count column into descending order. Select the first three records, which indicate flows to the west (16), southwest (8), and northwest (32). Look at the map. Run Statistics. How many cells fall into these three categories? _____. Clear all selections.

As mentioned previously, water needs to ultimately flow to the edge of the grid for this model to work properly. If it flows to internal cells from which it cannot exit because the surrounding cells are all of greater elevation, the model gives wrong answers.[15] You can check for this situation by applying the Sink tool to the Flow Direction data set, to make the raster AnySinks.

____ **12. *Identify sinks:*** Turn off all layers. Run the Sink tool on FlowDir. Call the Output raster AnySinks. Once it is added to the map, you will be able to see many spots where sinks have been identified. Open its attribute table. How many sinks have been identified? _____

Only if the data set consisted entirely of NoData values would you be assured that there were no sinks in the study area. But, unfortunately, this is not the case. Maybe there are real sinks here. Or perhaps the surface generated by the Interpolate to Raster has sinks in it where none exist in reality. Or maybe the elevation data was wrong. In any event, you need a sinkfree surface to proceed.

To fill sinks, you would normally use more extensive Spatial Analyst or Hydrologic Modeling tools. This is a time-consuming process and requires knowledge you may not have at this point. Filling sinks is an iterative process, meaning that filling one set of sinks may generate others. That is, there are ways of filling sinks effectively, but they are beyond the scope of this text. So you will use a "repaired" layer, called ElevSurface2, which has all the sinks filled for you.

---

[15]One other problem in modeling the environment rears its head. In the real world there are such things as sink holes, which connect the surface water with the under surface water (ground water). Here we choose to ignore this complication.

_____ **13.** Start a new map. From the

_____IGIS-Arc_*YourInitialsHere*\Spatial_Analyst_Data\Hydrology_Data_SA

folder add the raster ElevSurface2. Re-create the flow direction data set with ElevSurface2, making the output FlowDir2. Run the Sink tool on FlowDir2, naming the resulting data set SinksNow. The raster should consist entirely of NoData. The attribute table for SinksNow should be empty. Remove the SinksNow data set.

Recall that the Flow Accumulation tool computes the amount of water that flows into each cell from *all* the upstream cells. In the absence of a weight grid, the request assumes that one unit of water will come from each cell and will flow into one adjacent cell. The one unit of water for a given cell does not show up in the accumulation of water for that cell. Therefore, some cells that have no water draining into them will have an accumulation value of zero. As you saw earlier, some cells at the bottom of the drainage can have very large values. The output grid of accumulation can be used as a measure of runoff for hypothetical rainfall events. The optional weight grid could be a surface of rainfall values interpolated from weather station measurements and modified with a model to adjust for loss from evapotranspiration and soil absorption. With a weight grid, the Flow Accumulation tool would return an estimate of actual runoff.

_____ **14.** *Calculate flow accumulation:* Use the Flow Accumulation tool on the FlowDir2 layer. Do not use an Input weight raster. Name the new data set FlowAccu. From the Table of Contents, what is the highest value of flow accumulation? _____ Compare that with the number of cells in the study area, recalling that each cell contributes one unit of water. Notice how the map delineates the accumulations of water, and therefore the streams, and perhaps rivers.

Determining the Stream Channels

You can use the Flow Accumulation raster to identify a drainage network. We will identify those cells that have high accumulated flow values. The cells that contain the most water are the stream channels. We can create a drainage network of any detail by choosing those cells that have more than a certain mini-mum threshold value. In this case, say that a stream exists at a given cell if the flow accumulation in that cell is equal to or greater than 140—the equivalent of 1 unit of rain over at least 140 other cells that drain to that cell. Let's say that a cell either is or is not in the drainage network, so we will make a raster consisting of binary values: 0 or 1. We can do this with the Raster Calculator.

_____ **15.** In the Raster Calculator, build and evaluate this expression:

```
[FlowAccu] >= 140
```

Make the calculation permanent: Right-click the name, pick Make Permanent, and call the re-sult StrmChannels. Add StrmChannels to the map; remove the calculation. Turn off all layers except StrmChannels. The cells that constitute the stream channels have values of 1; the rest of the cells have values of 0.

Calculating Stream Order

The volume of water flowing through a stream is a function of many things, including the stream's width (number of cells) and its depth. For that and other reasons, the number of "width cells" that depict a stream is not a good indicator of its size or volume. As previously discussed, one way to get an idea of stream size is to assign a stream order number, which indicates the relative volume of water in a stream segment. In this step, you will use the Strahler ordering.

_____ **16.** Use the StreamOrder tool on the StrmChannels layer and the FlowDir2 layer: Make this data set active with the name to StrmOrder_StM.

The Strahler method is more conservative than the Shreve method—that is, the numbers tend to be smaller. In Shreve, every time one stream joins another, the order number goes up. Not so with Strahler. Only when two streams of the same magnitude join does the order number increase, and then only to identify the downstream by a number greater by 1 than those above. Notice that there are only four categories of the Strahler layer, even though you had a large number of stream segments. With Shreve, the largest order is almost 40, as you can prove to yourself if you choose to by making StrmOrder_ShM.

Numbering Each Stream Individually

_____ **17.** To individually designate each stream segment, use the StreamLink tool. This tool operates on the StrmChannels data set, and it uses FlowDir2. Call the new data set StrmIDs, leave it on, and turn all other entries off. An amazing number of individual stream segments have been produced—more than 200. Zoom in on a portion of StreamIDs and use the Identify tool to examine a few assigned numbers—found in the (Pixel) Value field. Click the Zoom To Full Extent button to see the complete map. Open the attribute table of StrmIDs. Select the one with Value 189. Check it out on the map. Clear selections. Close the table.

Identifying Basins

Basins can be defined simply by the flow directions of water that falls on the study area. The flow directions are themselves defined by the elevations and slopes.

_____ **18.** Start the Basin tool. Note that the only input is FlowDir2. Call the output UniqueBasins. How many are there? _____

_____ **19.** With the Spatial Analyst toolbar drop-down menu, convert the raster data set StrmChannels to the feature class shapefile Strms.shp. Make sure the Output geometry type is _Polyline_. Browse so you can put it in the Hydrology_Data_SA folder. Display Strms.shp with bright red. Turn all layers off except Strms.shp and Unique_Basins. It is pretty clear how the basins are delineated.

_____ **20.** Set the Unique Basins layer at 85 percent transparency. (Wait! Wait! Don't tell me![16]) In the Table of Contents put Strms.shp at the top, UniqueBasins next, and ElevSurface2 third. Turn those three on and every other layer off so you can get a picture of why the basins are the way they are and why the streams form as they do.

Finding Pollution Culprits

Each of the streams has associated with it a watershed. The WaterShed request finds all the upstream cells that flow down to a given point—perhaps a point where pollution is discovered.

Specifically, the WaterShed request finds the up-gradient cells of a specific set of cells; those cells may or may not be cells in stream segments. It is possible to find the watersheds of points where pollution has been found by monitoring devices. The POLLUTION_PTS data set (points at which pollution has been found) indicates pollution in streams at two points.

---

[16]That's under Display in Layer Properties.

____ **21.** Add as data the raster POLLUTION_PTS. Turn off all other layers. In the POLLUTION_PTS attribute table, select both records so you have a better chance of spotting the points on the map. If that doesn't work, turn the symbols of the two cells of the POLLUTION_PTS raster (550 and 675) bright red.

____ **22.** Use the Watershed tool to determine those areas that drain to the points. The Input flow direction raster is FlowDir2. The pour point data comes from POLLUTION_PTS. Make the Pour Point field Value. Call the result SuspectAreas. Move POLLUTION_PTS above SuspectAreas in the Table of Contents. Open the attribute table of SuspectAreas. How many cells might be involved in the search for the location that is polluting 550? _____. 675? _____.

____ **23.** Make a polygon shapefile of UniqueBasins. Call it Basins_Outline. Make its color Hollow and the width of its outline 3 picas, in black. Turn off all layers except Basins_Outline and FlowDir2. With Basins_Outline at the top of the Table of Contents, note that all flows are away from the boundaries of the basins, except for a few cases when the flow is parallel to the boundary.

The distinction between basins and watersheds is a subtle one. Also, they are not completely distinct. In terms of ArcMap, basins are determined only by flow direction, which is determined by the elevation surface, That is, topography is the primary ingredient in basin determination. The water shed calculation is based on the aggregation of cells that feed a particular point, usually on a stream.

____ **24.** Save the project. Make a layout and print a map that consists of:

Basin_Boundaries (with dark outlines of hollow polygon basins)

StrmOrderStM (with a symbology that goes from light blue to dark blue)

____ **25.** Close the project and exit ArcMap.

Exercise 8-8 (Review)

Checking, Updating, and Organizing Your Fast Facts File

The Fast Facts File that you are developing should contain references to items in the following checklist. The checklist represents the abilities to use the software you should have upon completing Chapter 8.

Important note: This checklist is on the CD-ROM that accompanies the book. It is available in Microsoft Word format. Rather than typing or writing by hand the text that follows, you can copy and paste it into your Fast Facts File from the CD-ROM file.

____ To create a raster or GRID from a text file

____ Row-major order means

____ The Symbology "stretch" option

____ The two fields in a VAT for a bare bones raster are

____ A zone is

___ A region is

___ The Cell tool

___ Rasters are of two types:

___ To do arithmetic computation with rasters use the

___ The Boolean operators are

___ When using the Raster Calculator, you must put blanks around

___ To keep environment settings

___ To convert features to rasters

___ When the cell size is halved, the storage requirement does not necessarily go up by a factor of four because

___ EXPAND does not work well as a buffer because

___ Reclassification

___ To convert zones to regions

___ SQL is

___ Extract by Attributes allows the user

___ The ways of creating a surface from a set of points are

___ The Density tool

___ Thiessen polygons

___ To create rasters for elevation, slope, aspect, and hillshade

___ Straight-line distances on a grid are calculated

___ Cost of moving across a grid is calculated

___ Distances on a raster or GRID are measured from cell center to

___ The threshold distance is

___ A Direction GRID is

___ An Allocation GRID is

___ A Cost Surface GRID is

___ CFCC means

___ To join two tables

___ Flow direction is the basis for much of hydrologic investigations; it means

___ Two ways of delineating stream order are

CHAPTER 9

Other Dimensions, Other Tools, Other Solutions

OVERVIEW

IN WHICH we examine the third spatial dimension in GIS, time and GIS, address geocoding, network analysis, and linear referencing.

Ignored Third Dimensions: The Temporal and the Vertical Spatial

So far in our GIS work, though it pains me to say it, we have had all of the disadvantages and none of the advantages of the fact that the world (and everything in it) resides in four dimensions. The disadvantages have come about partly because (1) the Earth is approximately spherical—requiring all that projection complication in moving from three spatial dimensions to zero, one, or two dimensions—(2) three-dimensional stuff is just harder to deal with, so we have contented ourselves with "flatland" in which nothing is quite right, and (3) it's hard enough just to get a data set right at a given moment or period in time—never mind historical or anticipatory data sets.

But now we take on both the third spatial dimension and the time dimension, at least briefly. After moving up a dimension (first 2-D to include the 3-D vertical and, second, 2-D to include the time dimension), we'll move down a dimension (2-D to 1-D), where we will consider networks—roads, pipes, wires, rivers, and such—and how locations along them are recorded. While these networks exist in 4-D (what doesn't?), we'll look specifically at data structures that are made of linear (1-D) elements.

I present the Overview and the Step-by-Step sections together for each topic. When we say 3-D GIS, we usually mean three spatial dimensions. The title of the next section is meant to convey that idea. But in the section after that, which is "3-D: 2-D (Spatial) Plus 1-D (Temporal)," we will take up looking at what happens to two-dimensional data over time.

3-D: 2-D (Spatial) Plus 1-D (Spatial)

This Overview will be short—because

❑ ArcGIS 3D Analyst Extension is so rich in capabilities that covering it in detail in an introductory course is out of the question.

❑ Many of the capabilities of ArcScene and ArcGlobe, the constituents of 3D Analyst, fall into the "a picture is worth a thousand words" category of explanation.

The size of 3D Analyst Extension is testified to by the fact that ESRI has a 375-page manual[1] on the subject, and OnWord Press has a 200-page text on 3-D modeling using the extension.[2]

Throughout the text I have hinted at the possibilities of three-dimensional GIS. Before you start into the exercises, let's look at a list of things you can do with this 3-D visualization and analysis extension. One caveat: There is a lot of overlap between the capabilities of Spatial Analyst and 3D Analyst. For example, both deal with surfaces. Spatial Analyst is more concerned with analysis and 3D Analyst emphasizes display. But these two extensions share some tools and work as a team.

ArcScene

With ArcScene you can[3]

❑ Create surfaces with a number of tools and techniques.

❑ Analyze surfaces in a variety of ways.

❑ Drape raster images over surfaces.

❑ Drape vector features over surfaces.

❑ View three-dimensional surfaces in perspective from multiple observer points.

---

[1]*Using ArcGIS 3D Analyst*, from ESRI Press. Also available in a folder named ESRI_Library\ArcGIS_Extensions in the documentation that came with the system.
[2]*Data in Three Dimensions: A Guide to ArcGIS 3D Analyst* by Heather Kennedy. Thomson Learning/Delmar Learning/OnWord Press, 2004. www.delmarlearning.com. Full disclosure: Heather Kennedy is the author's daughter.
[3]Some of these capabilities require ArcMap.

❑ View scenes differently by changing the shading, transparency, and illumination properties of 3-D layers.

❑ Change the "Z" visual component of 3-D scenes (exaggerate the vertical).

❑ Create surface models from rasters and TINs.

❑ Make queries on raster values.

❑ Get instantaneous information about the elevation, slope, and aspect of TINs.

❑ Create contours of surfaces.

❑ Find the steepest path on a surface.

❑ See a surface under different levels of illumination.

❑ Calculate the volume between a horizontal plane and three-dimensional surface specified by a TIN.

❑ Determine what parts of a landscape can be seen from various vantage points.

❑ Draw lines of site across landscapes.

❑ Create a 2-D cross-section profile graph of a 3-D surface, given a line across the surface.

❑ Create 3-D features from 2-D data.

❑ Digitize 3-D features.

❑ Extrude 3-D features from 2-D features.

❑ Simulate "flying through" a landscape, using animation capabilities

ArcGlobe

The other software package in 3D Analyst is ArcGlobe—a startlingly effective piece of software that ought to be used in every elementary school in the nation. ArcGlobe is to a manual globe what ArcMap is to a paper map. It is a software lever that gives you an intellectual advantage in looking at the world.

With ArcGlobe you can

❑ Get marvelous views of the Earth, rotating and panning its surface

❑ Add large data sets which are made part of the Earth's surface, correctly placed regardless of their coordinate systems.

❑ Automatically spin the globe around its axis, looking from any vantage point.

❑ Measure distances along great circle routes.

❑ Rotate an image either around the center of the Earth or around the center point of the image—doing global or local navigation.

❑ Make animations of the images you generate.

Again, the capabilities of 3D Analyst are best experience rather than described. You do that next.

The Third Spatial Dimension

STEP-BY-STEP

A New Software Package: ArcScene

ArcScene is the package you saw briefly in Chapter 2. We will use it to look at multiple 3-D data sets simultaneously. If we want to look at an individual data set, we can use ArcCatalog, using 3D View Preview. ArcScene operates a lot like ArcMap.

_____ Open up your Fast Facts text or document file.

_____ Before starting the exercise that follows, let's get into a three-dimensional, "large view of the world" frame of mind. Start Menu > Programs > ArcGIS > ArcGlobe. In Globe Layers, turn off Countries. Under the View menu, click Toolbars and place a check mark beside Spin to bring up the Spin toolbar. In the Speed text box, type 2.0. Press the Enter key. Click the Spin Counter Clockwise button (that's counterclockwise when looking down on the North Pole—and this is the way Earth spins). Sit back and be impressed for at least three revolutions. Press the Navigate (see Navigates the scene—on the Status bar) icon. Turn on the Countries layer. While the rotation continues, use the left and right mouse buttons to position the globe so that you are looking at the middle of your county from about a thousand kilometers away.[4] Let another revolution go by. Minimize the ArcGlobe window and begin the following exercise with Step 1, knowing that, as you do, the Earth will continue to turn.

Exercise 9-1 (Project)

Experimenting with 3-D

_____ **1. Preliminaries:** Start ArcCatalog. Under Tools > Options > General, make sure Hide File Extensions is unchecked. Click OK. Also be sure that ToolTips is turned on (Tools > Customize > Options). Click Close. Under Tools > Extensions, make sure a check is in the box labeled 3D Analyst. Click Close.

_____ **2. More Preliminaries:** Under View > Toolbars, make sure 3D View Tools is turned on. Locate the toolbar and drag it into the display window (right pane). Using ToolTips, list the names of the buttons:

_____

[4]See the Distance value in the lower left corner of the window.

———————————, ———————————, ———————————

———————————, ———————————, ———————————

———————————, ———————————, ———————————

———————————, ———————————, ———————————

———————————, ———————————,

Double-click the title bar of 3D View Tools to put it back among the other toolbars.

___ **3.** *Final Preliminaries:* In ___IGIS-Arc_*YourInitialsHere* make a folder named 3-D_Spatial. Make a folder connection to 3-D_Spatial. From [___] IGIS-Arc\River, use ArcCatalog to copy the following data sets into the folder 3-D_Spatial.

- ❏ Boat_SP83.shp

- ❏ cole_dem

- ❏ cole_tin

- ❏ cole_DRG.TIF

- ❏ COLE_DOQ64.JPG

ArcScene

___ **4.** Start ArcScene from ArcCatalog using one of the buttons you identified previously. Locate the ArcScene Tools toolbar and move it onto the display area. List the buttons on the Tools toolbar.

———————————, ———————————, ———————————

———————————, ———————————, ———————————

———————————, ———————————, ———————————

———————————, ———————————, ———————————

———————————, ———————————, ———————————

———————————

Notice that while some of the buttons are the same as with ArcCatalog, some are different. Draw circles around the ones that are the same. Put the toolbar back with the others by double-clicking its header.

Chapter 9

What's 3-D and What's Not

The data set you have worked with that truly includes the vertical dimension *as spatial data* is the TIN. All other feature classes, even if they have the possibility of the third dimension, are in flatland.

_____ **5.** From ___IGIS-Arc_*YourInitialsHere*\3-D_Spatial

add as data, all at once by Ctrl-click, to ArcScene:

- ❏ Boat_SP83.shp (symbolize it with a red dot, size 4)
- ❏ cole_dem
- ❏ cole_DRG.TIF
- ❏ COLE_DOQ64.JPG
- ❏ cole_tin

A lot of visual information will appear, although it is mostly covered up by the green TIN. You are viewing it from a point southwest of the image and well above it. This is what is considered full extent in ArcScene. We will explore these data sets using the Navigate tool in ArcScene. The Navigate tool is a bit like a Swiss Army Knife—lots of different tools in one artifact. As such, it is both useful and dangerous.

_____ **6.** Click the Navigation tool on the Tools toolbar. Move its cursor into the display pane and play with it. If you drag left and right with the *left* mouse button, you can rotate the image around its center. If you drag up and down with the *left* mouse button, you can tilt the scene. If you drag (up and down) with the *right* mouse button, you can zoom the image. If you drag with the center mouse button (or the left and right mouse buttons, held down together), you can pan the image. If the center mouse button is a wheel, you can use that to zoom the image. Experiment. Throughout all this your good friend is the Full Extent button. Use it liberally.

_____ **7.** Using the Zoom In/Out tool on the Tools toolbar, zoom in on the GPS track and where it cuts through the landscape made by the Kentucky River. View the scene from the southeast. See Figure 9-1***

_____ **8.** ***Use the rotating and tilting capabilities of the Navigate Tool:*** If you rotate the scene so that you are looking at it from the southeast (along the GPS track) and from elevation zero, you will see two sorts of images. On the zero elevation level most of the data sets have merged into pretty much a single line. "Floating" above this line is the TIN, with its true 3-D representation.

We are not stuck with 2-D representation of these data sets. First we will give the DEM three-dimensional representation.

_____ **9.** Turn off all the images except the Digital Elevation Model (DEM). Open the attribute table of the DEM. Note the Value field. Close the attribute table.

The DEM, while it contains elevation information as an attribute (look at the VALUE field), was represented as flatly as everything else (except the TIN). We can alter that. First let's display it with more color and more differentiation between the heights.

_____ **10.** Bring up the Properties window of cole_dem. Choose Symbology. Set the window to Show: Classified. Make 25 classes. Right-click the color bar of the Color Ramp. Uncheck Graphic

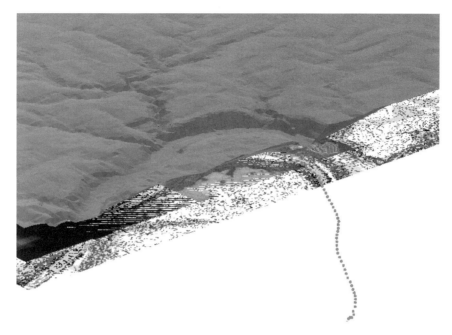

FIGURE 9-1

View. From the drop-down menu, choose Elevation #2. Click Apply, then OK. Take a quick look at the Value symbols in the Table of Contents, then collapse them so they don't clutter up the T/C. Move cole_dem around with the navigation tool to look at it.

Things are still flat—but more colorful and obvious. Now we add the vertical dimension. We actually have two sources of elevation information: the TIN, of course, and the elevation attribute (VALUE) contained in the DEM's table.

_____ **11.** Bring up the Properties window of cole_dem again. Note that there is a Base Heights tab. Select it. One of the possibilities under Height is to Obtain Heights For Layer From Surface. Pick that and choose the DEM itself as the surface to use. Slide the Layer Properties window away from the display so you can see the effect when you press Apply. Concentrate on the river bend area, using pan, zoom and navigation. One image you can see is Figure 9-2***.

_____ **12.** Just to reassure yourself that you have only changed the way cole_dem is displayed, and not the original data, add cole_dem again as data, and examine the situation with the Navigation Tool. What is going on here is that every pixel of the original cole_dem in the display has been given a height in accordance with the value associated with that pixel. The original data set remains the same. (There is a way to permanently give a data set height values; you'll learn about this later.)

_____ **13.** Right-click this last cole_dem data set and remove it. Since you have gone to some trouble to make the first cole_dem—changing symbology and base height—make it a layer file (right-click

FIGURE 9-2

on the name, then select Save As Layer File) named cole_dem_layer.lyr in the 3-D_Spatial folder you have been working with. Add cole_dem_layer.lyr to the map.[5]

Now that you know how to set base heights, you can do the same for the digital raster graphics image of the Coletown topographic map: cole_drg.

_____ **14.** Turn off cole_dem. Turn on cole_drg. Zoom to full extent. Then zoom to the river bend. Set things up so you are looking at the bend in the river from the southeast, somewhat above. Bring up cole_drg properties and select the Base Heights tab. Press the radio button in front of Obtain Heights From Surface and, for the surface, pick the TIN this time. Click Apply, then OK. You should see the DRG "draped" over the elevations of the area. See Figure 9-3***.

_____ **15.** Turn the Coletown DEM layer file back on. Zoom up on the southeast quadrant. Notice that you have a mess on your hands. In some places you see the DEM. In other places you see the DRG. Flip each on and off to see what is happening.

With 2-D images you will recall that the drawing order was controlled by the Table of Contents. With 3-D images this really doesn't work. But you can control the order in another way. Suppose that if both the DEM and the DRG are on, you want the DEM to prevail visually.

_____ **16.** Display the properties of the DRG. Select the Rendering tab. Under Effects, read the statement that suggests you have at least some control over the drawing order. Change the priority to seven, which lowers the priority, since "1" is the highest.

---

[5]In version 9.1, the layer file is not identified as such, either by its extension or by the new name you gave it. To be sure you have the layer file, you can remove both instances of cole_dem and then add back the layer file.

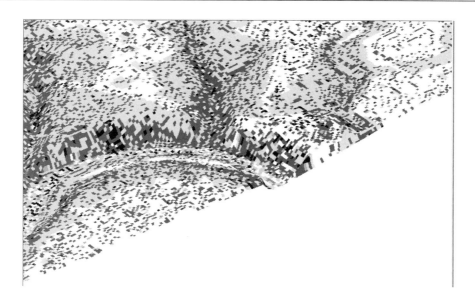

FIGURE 9-3

——— **17.** Make the display priority for the DEM two. Now as you navigate the scene you still will see places where the DRG will poke up through the DEM, but the DEM certainly predominates. Set the drawing priority of the TIN to 4 and turn it on.

——— **18.** Turn on the GPS track. Navigate to a horizontal view of the river bend from the east. Note how the GPS track is decidedly underneath the other drawn layers. Set the base height of the GPS track to the same as the DEM.

——— **19.** Finally, set the base heights of the DOQ to that of the DEM. Turn off all layers except the GPS track and the DOQ. Notice how the DOQ is draped over the landscape. You can see the boat's path through the valley.

Viewing 3-D Data with Animation

——— **20.** Call for a new ArcScene file. Add as data COLE_DOQ64.JPG. Then add COLE_TIN. Turn off COLE_TIN. Zoom to Layer of COLE_DOQ64.

We will be performing several complex operations on the data. Navigating and other operations have to be applied to every feature and every pixel, which takes CPU horsepower and time. In general, you want only the data sets in the ArcScene table of contents that you need.

——— **21.** Set the base heights of the DOQ to those of the TIN. Remove the TIN. Add as data the GPS track Boat_SP83. Set its base height at 640 (feet), which puts it 100 feet above the river (which you may or may not remember has a pool elevation of 540 feet). Make the symbol for the elevated GPS track a red dot, size 4. To set the background color, right-click Scene Layers at the top of the table of contents. Click Scene Properties. Under the General tab, choose Apatite Blue as the background color. Click Apply, then OK.

_____ **22.** Go to Full Extent. Pan, navigate, then Zoom to COLE_DOQ64, so that the scene looks like Figure 9-4***. Pretend that this is the beginning point of a helicopter ride that will take you through the gorge.

With ArcScene you can record a sequence of observer-target positions, called keyframes. You usually do this within an ArcScene session with a particular set of entries in the table of contents. ArcScene then lets you use these observer-target positions to view data sets (usually the same ones you used to make the sequence), giving the appearance of a number of consecutive snapshots of scenes. Further you can set things up so that you can "morph" from each to the next, giving the appearance of motion.

_____ **23.** Select View > Toolbars > Animation and turn it on. On the Animation toolbar drop-down menu, click Create Keyframe. In the Create Animation Keyframe window that results, click New. An Animation Track window appears. Accept the default name Camera Track 1. Click OK. Now accept the keyframe name Camera Keyframe 1. Tuck the window away as much as you can—like in the Table of Contents. See Figure 9-5.

When you press the Create button, in the next step, the software will record the observer position and the target position that has been set up.

_____ **24.** Click Create. This makes Camera Keyframe 1 and sets things up so Camera Keyframe 2 will be created next. Click Center On Target on the Tools toolbar and click the red dot that is fourth outside the DOQ. Click Zoom To Target, using that same dot. Click Create again.

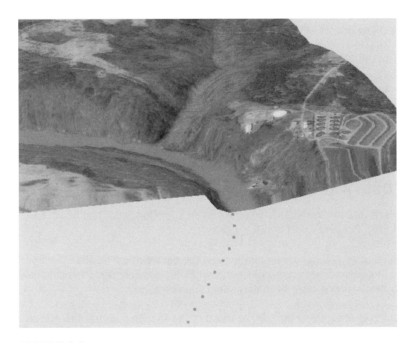

FIGURE 9-4

FIGURE 9-5

_____ **25.** Click Center On Target on the Tools toolbar and make the target the first red dot that is in the DOQ. If you then use the Navigation tool and the Zoom tool, you can get an image that looks like Figure 9-6.

_____ **26.** Click Create. This makes Camera Keyframe 3.

_____ **27.** Create a third target about four or five dots along, and position the scene so that you are flying down river. Click Create.

_____ **28.** Continue creating keyframes about every four dots, using `Center On Target` and Zoom To Target, flying down river, zooming in as close as you want to (or dare to).[6] Make the final dot one that is off the edge of the DOQ. Close the Create Animation Keyframe window.

Once you have made the keyframes, you can view the animation.

_____ **29.** On the Animation toolbar, click Open Animation Controls. On Animation Controls, click Options and set the duration to 20 seconds. Press Play (the triangle).

---

[6]Professional pilot on planned course. Do not attempt this on your own.

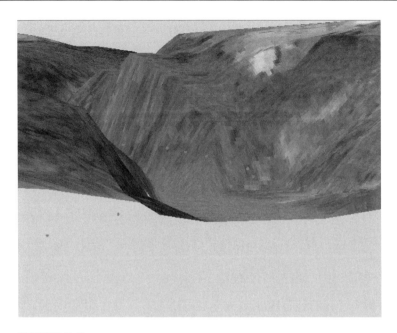

FIGURE 9-6

___ **30.** A rather neat effect occurs. However, you will probably not be happy with your flight path (or impressed with your flying ability). On the Animation toolbar, click Clear Animation and re-make the sequence until you are satisfied with it.

___ **31.** Now that you have a sequence you are happy with, save the Animation File (via a menu item on the Animation toolbar drop-down menu) as Helicopter_Ride in ___IGIS-Arc_*YourInitialsHere* 3-D_Spatial

___ **32.** Also take a look at the Animation Manager, which gives you specifics about the position of the observer and the target. Examine. Close.

Actually, what you have done is to record a sequence of observer-target position that can be used with any ArcScene data set that shows the vicinity. First let's turn off the GPS dots.

___ **33.** Turn off Scene layer Boat_SP83. Rerun the animation, noting that the guidance dots are gone.

___ **34.** Add as data COLE_DRG.TIF and COLE_TIN. Set the DRG to have the base height of the TIN. Remove the TIN. Turn off the DOQ. Rerun the simulation. Exit ArcScene.

Two-and-a-Half-D: Calculating Volumes

There is sometimes a tension, if not confusion, within a given description regarding which dimension is being addressed by a term in geometry. Is a square the four lines that meet at vertices of 90 degrees, or is a square the surface formed by those four lines? Does a cube consist of 12 lines in 3-D space, or the surfaces that those lines imply, or, thirdly, the solid defined by those planes? That is, is the cube the wire frame of the lines, the surfaces of the planes, or the volume contained within?

When we say that a road network is represented by one-dimensional elements, we have to note that the network wanders all over the two-dimensional plane. When we talk about ArcGIS doing 3-D, we are usually talking about surfaces that are 2-D at any point, but overall occupy 3-D space. What we haven't addressed are volumes in three dimensions—that is, true three-dimensional objects. For example, a coal seam is a true three-dimensional object. You can think of it as a volume that is defined by a multitude of planes, in the same way you think of a polygon as an area defined by a multitude of line segments. Frankly, ArcGIS doesn't do much with defining volumes, but it does have some capabilities.[7] The volume under (or over) a TIN can be calculated relative to some horizontal base plane. This takes place in the 3-D world, but the result might be described as 2.5-D, since one plane is the trivial, horizontal one. What follows is an example of calculation in two and a half dimensions.

Calculating a Volume with ArcGIS

Between two and three miles west of the river bend that you have looked at so often is a little hill peak that we would like to build a house on. We have to do some grading, because of the rugged terrain. We have surveyed the area and determined that if we create a flat surface at the 1020 contour elevation, we will have enough space to build and have a garden, a lawn, and a tennis court. How much dirt would we have to move to flatten it out? We will use as data cole_TIN and our surveyor's contour lines to determine the answer.

_____ **35.** Start ArcCatalog. From ___ IGIS-Arc_YourInitialsHere\Trivial_GIS_Datasets copy Four_contour_ lines.shp to ___IGIS-Arc_*YourInitialsHere*\3-D_Spatial. After looking at the geography of the lines, use Identify to determine the elevation, given by the field name CONTOUR, of the longest, and, in elevation, lowest line. _____. What is the elevation of the highest line? _____. Measuring up from the 1020 foot elevation to the top, what difference in altitude are we dealing with? _____ feet.

_____ **36.** Start ArcMap. Make sure the 3D Analyst Extension is turned on. Make sure the 3D Analyst toolbar is turned on. From ___IGIS-Arc_*YourInitialsHere*\3-D_Spatial, add as data cole_TIN and Four_contour_lines.shp. Make the symbol for the contour lines bright red, size 2. The location of interest will show up as a red dot. Find it and zoom up on the area of interest. See Figure 9-7***.

_____ **37.** Explore Four_contour_lines.shp. What is the maximum distance across, indicated by the outermost, and lowest, line? _____.

Using the 3D Analyst toolbar, we can create a TIN whose extent is roughly that of the outermost contour line.

_____ **38.** From the drop-down menu of the 3D Analyst toolbar, select Create/Modify TIN > Create TIN From Features. Put a check mark by Four_contour_lines. The height source should be CONTOUR. Under Triangulate, accept hard line. Call the new TIN Mtn_Retreat and make sure it is going into the

___IGIS-Arc_*YourInitialsHere*\3-D_Spatial

folder. See Figure 9-8. Click OK.

The new TIN should automatically appear in the ArcMap Table of Contents and in the map area. Zoom up on it, after turning everything else off. See Figure 9-9***.

---

[7]There are computer programs that operate on three-dimensional objects. Mechanical engineering, and other fields, make use of these "solids modeling" programs.

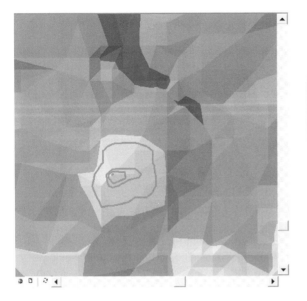

FIGURE 9-7

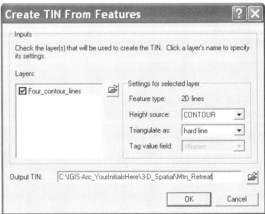

FIGURE 9-8

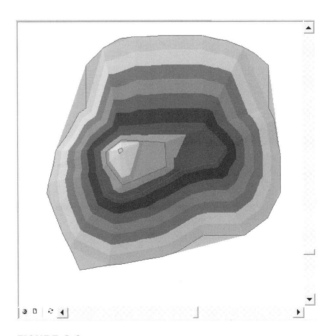

FIGURE 9-9

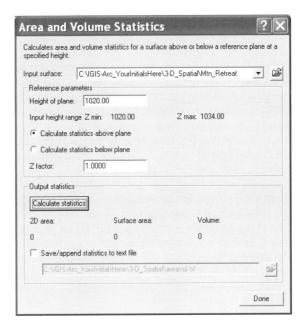

FIGURE 9-10

_____ **39.** Explore Mtn_Retreat TIN with the Identify tool, noting the elevation, slope, and aspect of various positions on the mountain.

Now you are ready to determine the volume and surface area of the volume that is under the TIN but above the 1020-foot plane.

_____ **40.** On the 3D Analyst drop-down menu, choose Surface Analysis, then Area and Volume. The input should be Mtn_Retreat. The window should look like Figure 9-10.

Note that the Height of plane is 1020 feet. The height range is from 1020 to 1034—the height of the top contour line. Leave the Z factor as 1.

Since the horizontal measurements of the data set is set to feet and the vertical measurements are also in feet, there is no need to convert. If you had horizontal units in, say, meters, as with UTM, and you wanted the vertical units in feet, you could use the Z Factor text box to compensate.

_____ **41.** Under Output Statistics, press Calculate Statistics. How many cubic feet of material reside under the TIN? _____. How many square feet of surface area does the TIN represent? _____. This is a number that represents the actual surface area, found by summing the areas of the triangles of the TIN, rather than the plane area you are used to seeing when ArcGIS reports area to you. Press Done.

Other Neat Stuff You Can Do with 3D Analyst: Viewshed and Hillshade

Suppose you are interested in knowing what part of the Coletown landscape you can see from the river. ArcMap can calculate a "viewshed." Basically, you can select a set of points and the software will show you all the surface area that can be seen from one or more of them. In our case we will use the points from the GPS track.

_____ **42.** Add Boat_SP83 from ___IGIS-Arc\River. From the 3D Analyst drop-down menu, pick Surface Analysis, then Viewshed. Make the Input surface cole_tin. The observer points should be Boat_SP83. The result will be a temporary raster, which will go away when you leave ArcMap. Make the output cell size 100 and click OK. Wait. The percentage of the calculation that is complete shows up at the bottom of the screen as a blue bar. Realize that ArcMap is calculating a ray from each point in the GPS track to each 30 x 30 square in the entire Coletown TIN. You would probably have time to hand-calculate how many such pixels there are in the 6.7-mile-by-8.6-mile TIN. Or you could just go get some coffee.

_____ **43.** The Viewshed of Boat_SP83 will appear in the Table of Contents. Collapse all Table of Contents entries and move the viewshed to the top. Turn on only the viewshed and cole_tin. Zoom to the Visible area. Look at the properties of the viewshed. Under Display you will see that it is 20 percent transparent. Choose 40 percent so you can better see the TIN underneath. Change the Visible symbol to a light yellow. Change the Not Visible symbol to black. The yellow pixels are those that could be seen from some point on the GPS track. The black ones are hidden. Turn on the GPS track.

_____ **44.** Turn off the viewshed. Zoom to the GPS track. Pan the map to the east so that it resembles Figure 9-11***. From the 3D Analyst drop-down menu, pick Surface Analysis > Hillshade. For the input surface choose cole_tin. Note that the default has the light source coming from 315 degrees (northwest) and at an angle above the horizon of 45 degrees. Leave everything the same except output cell size, which you should set at 30. Click OK.

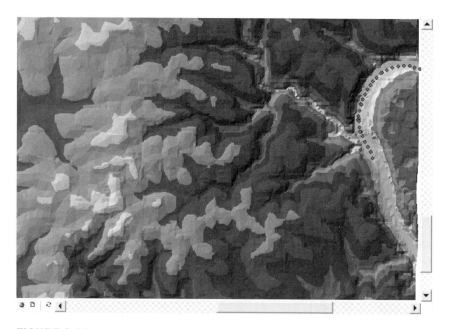

FIGURE 9-11

Once this process completes, a Hillshade of cole_tin should appear at the top of the Table of Contents and a grayscale raster will appear. Note that everything is pretty bright except the southwest-facing slopes of the river gorge.

___ **45.** Rerun Hillshade (being careful to specify cole_tin), lowering the sun altitude to 30 degrees. Note the reduced amount of light everywhere and the deeper shadows.

___ **46.** Rerun Hillshade, putting the sun overhead at 90 degrees. There are no shadows, of course, but the image reflects different amounts of illumination on steep slopes. Dismiss ArcMap.

Adding Data to ArcGlobe

___ **47.** While you have done the preceding steps, ArcGlobe has been running in the background, spinning the Earth. Restore the ArcGlobe window. Stop the spin. Zoom to Full Extent. Move the globe around so you can see the United States. Zoom in on it.

___ **48.** Using the same procedure as with ArcMap, add the following data sets, one at a time, to ArcGlobe.[8] After each one, use Zoom to Layer. Plan to wait for the computations to take place.

[___] IGIS-Arc\Kentucky_wide_data\KY_streams_spf.shp[9]

Change the width of the streams to 0.3. Make the symbol color blue.

[___] IGIS-Arc\River\Lexington.mdb\Roads

Change the symbol color to black.

[___] IGIS-Arc\River\ cole_DEM

[___] IGIS-Arc\River\ cole_drg.tif

[___] IGIS-Arc\River\ COLE_DOQ64.JPG

[___] IGIS-Arc\River\Boat_SP83.shp

Use a red, square symbol, size 20.

___ **49.** Experiment with the Zoom control, using the icon on the Tools toolbar. Notice that the image remains centered, allowing you to keep an area of interest in the middle of the map, even though you can't see it. Look at the image at *about* 10, 20, 50, 100, 500, 1500, and 3000 kilometers away.

---

[8]This is one place where you don't have to worry much about the warning regarding different coordinate systems, for two reasons: (1) one of ArcGlobe's features is that in converts data on the fly to the globes coordinate system and (2) with the large environment we are dealing with here, small problems with accuracy won't usually be relevant.
[9]This is a big data set. The whirling globe on the status bar tells you that the data set is still loading. If the whirling globe stops and you still don't see the data set, click on the image itself. This is a good idea generally if nothing seems to be happening. ArcGlobe sometimes needs a little encouragement.

___ **50.** Save the 3000 kilometers-away map as

___IGIS-Arc_YourInitalsHere\3-D_Spatial\Globe_Map.3dd

___ **51.** Using the operating system, determine the size of Globe_Map.3dd. _____. This small size should tell you that ArcGlobe functions like ArcMap: The map file only serves to point to the data sets and how they are drawn. If you send someone a 3DD file without sending the underlying data, you might as well not send anything.

What's interesting here, other than the startling graphics and the ability to zoom from a space ship view or moon view of Earth, is that you can add your own data to the globe—something not yet possible with other Earth view programs that you find on the Internet.

___ **52.** Start a new ArcGlobe file. Turn off the Countries layer. Add as data

[___] IGIS-Arc\River\Lexington.mdb\Roads (use red lines, width 3)

[___] IGIS-Arc\Other_Data\Crystal_Lake_GPS.shp (use a red square, size 18)

[___] IGIS-Arc\Image_Data\Pilgrim.tif

___ **53.** Order the Layers so that you see the GPS track on top of the orthophotoquad.[10] Zoom to the Pilgrim Layer. The GPS track outlines Crystal Lake, Michigan, and goes through the town of Frankfort, on the western coast of Lake Michigan. Zoom out so you can determine the general location.

___ **54.** Zoom out so you are looking from about 500 kilometers away. Position the globe so you can see both the GPS track in the north and Lexington Roads in the south.

___ **55.** Use the Measure tool to determ=ine the approximate distance from Lexington, KY to Frankfort, MI. _____. (Don't forget the measurement units.)

___ **56.** Click on the Navigate icon. Find the Navigation Mode button (between Pan and Full Extent). Read the explanation on the status bar.

Until you change the sense of this button, the Navigate tool operates in "globe navigation mode," with which you are familiar. When you move the cursor, the globe revolves around the center of the Earth. When you go into "surface navigation mode," the center of the map becomes the center of rotation.

___ **57.** Click the Navigation Mode button so that it no longer appears pressed in. Experiment with the Navigate cursor. You can actually look at the Earth from the inside. With Zoom, move out from Earth some 5000 kilometers. Use Navigate again and note the effect: The center of rotation stays locked at the point on the surface where it was before.

___ **58.** Zoom to Lexington roads. Experiment again with the Navigate cursor. Change the mode back to globe navigation. Again move the cursor around.

---

[10]Sometimes you can see the GPS track only at some levels of zoom and if you flip the TIF layer off and on again. Refreshing each layer (right-click on the Layer name, then click Refresh) sometimes works also.

_____ **59.** Add another GPS track as data to the map:

[___] IGIS-Arc\Other_Data\Mystery_Location _1.shp (use a red square, size 18). Where is it? _____. What is it?[11] _____.

_____ **60.** Turn on the Countries layer. Add another GPS track as data to the map:

[___] IGIS-Arc\Other_Data\Mystery_Location _2.shp (use a red square, size 18). Zoom out. Use the Identify tool on the Countries layer. Which ones are represented? _____. Describe the trip. _____. Open the attribute table. In what direction was the car traveling? _____

_____ **61.** Dismiss ArcGlobe.

---

[11]*Hints:* Hot. Dangerous.

The Time Dimension | OVERVIEW

3-D: 2-D (Spatial) Plus 1-D (Temporal)

In this section, we look again at three-dimensional GIS. But the third dimension is time rather than a spatial dimension.

Time is usually considered an enemy of GIS. Time, among other things, ensures that a GIS is always out-of-date in some respect or other. The world around us changes constantly—due to both natural effects and human action. The bits in a computer's memory persist in their states unless changed. Time marches on, gradually making our data obsolete or forcing us to update continually. Such updates are time-consuming and expensive. Imagine, for example, that a major feature of your database consisted of orthophotos, obtained from aircraft. How often do you re-fly the area so the photos reflect new development or changes to the environment? And at what cost?

Another issue with time and GIS is that very little attempt is usually made to preserve states of the database with an eye to analyzing changes later. If a piece of property is subdivided, or a road built, the affected databases may well be updated, but the time at which the update happens is usually lost, as far as any easy access by those who might want to compare "what is" with "what was."

Again we face the issue of continuous versus discrete. The movement of time is a continuous phenomenon. Incremental changes happen within the unfolding of time. Are the changes large enough to warrant modification of the database. Consider a house in an historic district. The paint on the house deteriorates slowly. When is the record of the appearance of the house changed from excellent to good to fair to poor? Question 1 is how and how often do you update your database. Question 2 is what does it take to trigger a change. Suppose the house is repainted a different color. Suppose an extension is added. Suppose the house is demolished.

These are merely the complications time poses for our two-dimensional GIS. Looking to the future, can GIS be harnessed to allow analysis of changes to the environment over time? Certainly, old maps can be scanned in and interpreted. Such data can be compared with more recently installed GIS data. Those interested in using GIS for history should read Anne Kelly Knowles' book *Past Time, Past Place* (ESRI Press, 2002).

But consider: Old maps have dates on them. Suppose you are interested in making comparisons between conditions now and those ten years ago in a venue that has been using, and continually updating, a GIS database. How would you know what changed when?

Since the beginning of this book you have grappled, at least theoretically, with representing the continuous world in the discrete memory of a computer. Now another continuum, time, is tossed in. Your reaction, and, for the most part, the reaction of the "GIS industry," has been to ignore the topic. After all, it's hard enough to get things right in two dimensions. But this issue, along with the problem of information quality control in general, is one that will have to be faced. For a theoretical treatment of the temporal issues related to GIS read Gail Langran's book *Time in Geographic Information Systems* (Taylor & Francis, 1993).

The Time Dimension

STEP-BY-STEP

Let's run through a quick example of how files of the same variety of data, taken at different times, might be compared. Just as an example of one way of doing the comparison, we'll make use of the topology tool available to us in geodatabase feature data sets. We'll use the buffer tool as well.

Suppose we are interested in knowing what new streets were built in a city in the United States over approximately a decade. We have available a feature data set derived from the 1994 TIGER/Line files from the 1990 census. We also have more current files: roads from the 2000 census. Suppose we place both files in a personal geodatabase's feature data set. Then we could make a display showing both files. There would, of course, be a great amount of overlap. We can identify that overlap using topology. If we could then delete the streets that are (almost) coincident, we would be able to see the new streets that were added. You do that in the following project.

Exercise 9-2 (Project)

Looking at Infrastructure Changes Occurring over Time

——— **1.** Start ArcCatalog. Copy the folder [___] IGIS-Arc\Historical_Data to your ___IGIS-Arc_*YourInitialsHere* folder.

——— **2.** Create a folder connection to this new Historical_Data folder.

——— **3.** Inside the Historical_Data folder you will find the Geodatabase Roads.mdb. It contains the feature data set Lexington_SP_North_83_Feet. Within that are the feature classes Lex_Roads_1994 and Lex_Roads_2002.

——— **4.** Highlight the feature data set Lexington_SP_North_83_Feet. Using Ctrl-C and Ctrl-V, make a copy of it in Roads.mdb. Make a second copy, also in Roads.mdb. Rename the copies Lex_Backup_A and Lex_Backup_B. Anytime you use topology operations, you need backups of your original data, because those operations can move objects. Two backups may be overdoing it, but it is unusual to have too many backups. Often one does not have enough.

——— **5.** Preview the Geography of each feature class: Lex_Roads_1994 and Lex_Roads_2002. You can note, as you flip back and forth, that a fair number of roads were added in the eight years from 1994 to 2002.

_____ **6.** Look at the table of each feature class. How many road segments are represented in the 1994 roads file?[12] _____. In the 2002 file? _____.

_____ **7.** Look at the `Metadata` of one of the feature classes. What is the Geographic Coordinate system name? _____. Under Details, what are the Units? _____.

We are going to attempt to identify those road segments that are congruent, so we can ignore them and be left with only those segments that are not part of both data sets. You will recall that it is inadvisable to ask if two numbers that are computed by a machine are equal. Applying the same principle, we will ask if two segments are almost congruent—that is, their vertices are almost identical. We will do this by using topology. As you will recall, a topology may exist on the feature classes that are within a feature data set.

_____ **8.** Highlight the feature data set Lexington_SP_North_83_Feet. Choose File > New > Topology > (read the window) > Next. For the name, accept the default. But do not accept the default for the cluster tolerance. Read the description of the cluster tolerance. Let's assume that two road segments are coincident if all their vertices fall within 200 survey feet of each other. Click Next.

_____ **9.** In the New Topology window, select both feature classes. Click Next.

It would probably be a safe bet that the 2002 TIGER files are more accurate than the 1994 files. So we should set things up so it will be the 1994 lines that will move in case they are not coincident with the 2002 lines.

_____ **10.** Set the rank for Lex_Roads_1994 to 5. Click Next. Add the rule `Must Not Overlap` **With**. In the `Add Rule` window use `Lex_Roads_1994` for one feature class and `Lex_Roads_2002` for the other. Click OK, then Next. Check the `Summary` and click Finish.

_____ **11.** After the new topology has been created, select Yes to validate it. Be prepared to wait for a bit. Think of all the road segments the computer has to look at, and maybe reposition. It takes time.

_____ **12.** Lexington_SP_North_83_Feet_Topology should appear in the Catalog Tree, along with the Roads feature classes. If you click on it with the Preview tab pressed, you will see a lot of pink, representing the road segments that were coincident. Those that weren't coincident are not shown. Check the drop-down menu in the Preview text box at the bottom of the display. You will notice that there is no table associated with the topology.

_____ **13.** Launch ArcMap. Add as data Lex_Roads_1994 and Lex_Roads_2002 from

___IGIS-Arc_*YourInitialsHere*\Historical_Data\
Roads.mdb\ Lexington_SP_North_83_Feet.

_____ **14.** Make the 1994 roads black, width 1. Make the 2002 roads bright green, width 1. Arrange it so the 1994 roads are drawn last—that is, place its entry at the top of the Table of Contents. You can see that most development occurred along the southern and eastern edges of the city. Running "topology" on the data sets has snapped the features of the data sets together, where

---

[12]These files were modified somewhat: records in which the FEATURE NAMES were blank were deleted.

appropriate. Now individual road segments can be identified, so the new ones can be picked out. Not so good is this: If you zoom up on some areas, you will notice that roads that are probably meant to be the same show up in both data sets.

Frankly, older TIGER/Line data left a lot to be desired, so there are segments such that at least one vertex in a feature was more than 200 feet away from vertices in the same feature in the other data set.

_____ **15.** Let's further improve the presentation. It would be nice if we could use the topology data set to erase the 2002 roads that are coincident with the 1994 roads. Bring the `Topology` toolbar into view. Add the

Lexington_SP_North_83_Feet_Topology

to the map. (Don't add the associated feature classes—they are already there.) In the Table of Contents, make sure Lex_Roads_2002 is on, as well as the topology. Turn off Lex_Roads_1994.

_____ **16.** Choose `Selection > Select By Attributes`. Select those roads in Lex_Roads_2002 with FENAME equal to 'Strawberry', putting `Strawberry` in single quotes. Zoom to the selection and then clear selected features. Label the features in Lex_Roads_2002 with FENAME. Note that portions of Strawberry in Lex_Roads_2002 are flagged by the pink of the topology as overlapping Lex_Roads_1994.

_____ **17.** ***Erase parts of Strawberry:*** Start the Editor and then make the `Fix Topology Error` tool active on the `Topology` toolbar. Click on a portion of Strawberry that is overlapped by the pink topology line. It will turn black. Right-click, then click Subtract. In the Subtract window, pick Lex_Roads_2002. See Figure 9-12***. Click OK. Both the feature and the topology disappear. Erase a few more road segments in the area, by "subtracting" the segments that are overlaid with the "topology error" lines. This would be a long process for the thousands of segments in the city. How could we do it differently?

_____ **18.** One possibility for erasing all the Lex_Roads_2002 identified by topology would seem to be to select a large number of them at once. Drag a large box that encompasses several of the features identified by the topology. They will turn black. Now right-click. Unfortunately, the Subtract option doesn't exist when we right click. If you check the Help files, you will discover that multiple subtracts are not implemented. Turn off the labels of Lex_Roads_2002.

We certainly don't want to go through Lex_Roads_2002 subtracting segment by segment. Let's try another approach: buffering the Lex_Roads_1994, which have been snapped to the Lex_Roads_2002 where both exist, and then erasing, with an analysis tool, those Lex_Roads_2002 that lie within the buffer.

_____ **19.** With the Multiple Ring Buffer tool in ArcToolbox, start to buffer Lex_Roads_1994 by 200. Call the output Buffer_1994, and make sure it goes into

___IGIS-Arc_*YourInitialsHere*\Historical_Data\
Roads.mdb\ Lexington_SP_North_83_Feet.

The rest of the Multiple Ring Buffer window is OK as is. Run the tool. Explore the results.

_____ **20.** Use ArcToolbox > Analysis Tools > Overlay > Erase (analysis) to erase those features in Lex_Roads_2002 that lie within Buffer_1994, calling the result No_Coincident_Roads.

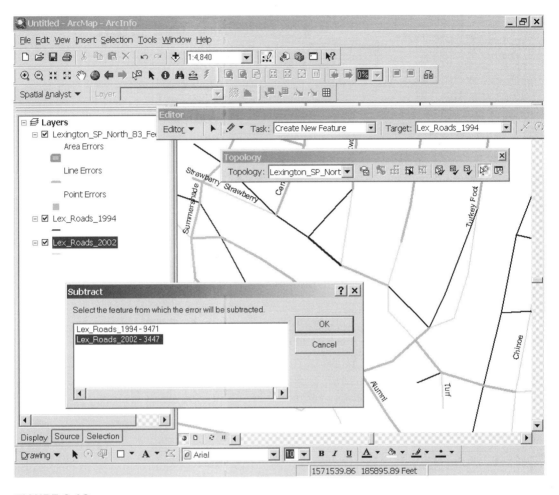

FIGURE 9-12

_____ **21.** No_Coincident_Roads appears in the Table of Contents. Turn off all other layers. What appears should be the remaining roads. Zoom up on various areas to inspect. Flip Lex_Roads_2002 on and off to see the context.

_____ **22.** While our results are not perfect, you can see that we have managed to use two historical data sets to provide new information. How many segments are represented in No_Coincident_Roads? _____. What is the approximate total length of road segments (in feet, use Shape_Length) that were added? _____. How many miles is that? _____.

Address Geocoding OVERVIEW

A Second Fundamental Way of Defining Location

As discussed in detail earlier, a location on the surface of the Earth may be referenced in a spherical coordinates system by a ray, emanating from the center of the Earth, defined by an angle of latitude measured from the equator and an angle of longitude, measured from the prime meridian. This method is "perfect" in that no error is introduced by the system itself; our accuracy is limited only by our ability to measure. Latitude and longitude descriptions may be applied anywhere on Earth. (I also discussed earlier how these two angles may be converted into myriad other pairs of coordinates through processes of projection.)

A large part of the Earth that is most interesting to us consists of human-made infrastructure. Almost always such infrastructure includes streets and buildings. Further, the most usual way to describe a location in such areas is with a street address. Such an address is a text string: a sequence of letters, numbers, and other characters.

Outside of all addresses being text strings, addresses appear in a wide variety of formats. Some examples:[13]

> 26376 Alpine Lane, Twin Peaks, CA 92391, USA
>
> 76-20 34th Avenue, Jackson Heights, Queens, NY 11372, USA
>
> 305 W 100 S, Salt Lake City, UT 84119, USA
>
> N84W 16301 W Donald Ave, St Charles, IL 60175, USA
>
> Rua Aurora 735 01209001 Sao Paula, Brazil
>
> Wendenstrasse 403, 20537 Hamburg, Germany

Such addresses are usually quite precise—locating a structure to within several meters if you have a good map of the area. But, by themselves, they tell us nothing about the location with respect to most other locations, except perhaps for other structures located on the same street. For purposes of mapping and analysis, we often want to turn addresses such as these into latitude-longitude coordinates, or other coordinate pairs.

Making use of addresses requires either personal or institutional knowledge of the location, or requires a map. To use a map, one must first find the right map—not necessarily an easy task—and not one guaranteed of success if the address is in a sparsely populated area.

---

[13]Courtesy of *Geocoding in ArcGIS*, published by ESRI.

Suppose we have this problem: A municipality wants to determine the spatial distribution of burglaries. It has records of the addresses of the crimes. One way to do this would be to put a map on a wall and mark each instance, based on the address. Person A might look at a police report and call out the street address. Person B would look at an alphabetical list of streets and announce the row-column zone that the street was in (e.g., C-6). Person C could then locate the street on the map and make a guess at the position of the structure along the street by knowing the house number. In other words, this is not a particularly simple operation. What would be required to perform the equivalent process with a GIS?

First we would need some sort of base map showing streets that could be referenced by spatial coordinates. Next we would need an extensive database that related locations of structures to the coordinate system of the base map. Finally, we need some automated way to parse[14] addresses, such as those exemplified in the preceding list, so that they may be matched against the database. Such a parser must be sophisticated enough to derive the "true" address from any of a large number of text strings that might represent an address in varying forms (with variation in blanks, commas, hyphens, letter case, and so on).

TIGER/Line Files

Determining geographic coordinates from addresses is a tall order. Let's begin by restricting ourselves to the United States, and the forms of address used there. You have already met TIGER/Line files, developed by the Bureau of the Census. If the terms behind the acronym[15] TIGER leave you confused, consider the predecessor description of such files: DIME, standing for Dual Independent Map Encoding. The idea behind "dual" is this: Each record in the file contains two geographic references. One is a range of address; the other is latitude-longitude coordinates. So this sort of file ties together (a) locations specified by text strings with (b) locations that are specified by geographic coordinates.

Each record in the TIGER file specifies a geographic line (called a chain). For streets in urban and suburban areas, the line usually represents a single block. At the ends of chains are intersections. See Figure 9-13***.

Each record also contains the street name (Ninth, Jenkins) and street type (Lane, Drive, etc.) together with any prefixes and suffixes (NW, E, and so on). The referenced line has a beginning point and an ending point. Each such point is defined by a latitude and longitude pair, in the NAD 1983 coordinate system. Further, at the beginning of each line, two structure numbers (street address numbers) are specified—one for the structure on the left and one for the structure on the right. The same is true for the ending point. Finally, if the line curves, its path is prescribed by vertices—familiar to you from your previous work with linear representations in shapefiles, coverages, and geodatabase feature classes.

So you can see that the TIGER database has all the ingredients that are logically necessary to convert an address into coordinates that will be "spatially close" to the actual coordinates of the structure at the address. A record in TIGER also contains lots of other information, useful in taking the census and making use of census results. It is worth mentioning that the TIGER file is a "flat" file. This means that there is no hierarchy—such as putting the records of a given state together and then requiring the user to know which state file a record came from. Each record contains all the information necessary to locate in geographic space the chain to which it refers.

---

[14]Parse: To break down a sequence of letters or numbers into meaningful parts based on a set of rules.
[15]Topologically Integrated Geographic Encoding and Referencing.

FIGURE 9-13

Following is what the primary record in the TIGER line files looks like.

The primary TIGER record (called a Type 1 record) is 228 characters long. Broken up into logical pieces, here is a typical record. Each piece is explained in the key that follows the record.

```
10021 31922071 AN Martin Luther King     Blvd  A41
301    313    300    342
01004050840508        2121067067
9122491224     46027460270003  0003   310 307
-84490912+38047734
-84489656+38048778
```

Record type: 1

File version: 0021

Record ID: 31922071

Single side: Blank, meaning data exists on both sides of the chain

Source code: A (see www.census.gov/geo/www/tiger for explanation)

Feature direction prefix: N, meaning north

Feature name: Martin Luther King

Feature type: Blvd (Boulevard)

Feature direction suffix: none

Census Feature Classification Code (CFCC): A41, meaning local street

Address number at FROM node, LEFT side: 301

Address number at TO node, LEFT side: 313

Address number at FROM node, RIGHT side: 300

Address number at TO node, RIGHT side: 342

Impute flags (four digits): 0100 (see www.census.gov/geo/www/tiger for explanation)

Zip code on LEFT: 40508

Zip code on RIGHT: 40508

State code on LEFT: 21 (Kentucky)

State code on RIGHT: 21 (ditto)

County code on LEFT: Fayette

County code on RIGHT: (ditto)

County subdivision on LEFT: 91224

County subdivision on LEFT: 91224

Codes such as census tract and block numbers that provide keys to census data: 46027460270003 0003 310 307 (see www.census.gov to use this information)

Longitude of the FROM node: -84490912 (read as 84.090912 West)

Latitude of the FROM node: +38047734 (read as 38.047734 North)

Longitude of the TO node: -84489656 (read as 84.489656 West)

Latitude of the TO node: +38048778 (read as 38.048778 North)

Precision of the Geographic Coordinates in TIGER Files

Consider that the 90-degree angle from the equator to the North Pole (latitude) is about 10,000 kilometers. So:

A degree of latitude is about 111 kilometers.

A tenth of a degree of latitude is 11.1 kilometers.

A hundredth is 1.11 kilometers or 1110 meters.

A thousandth is 111 meters.

A ten-thousandth is 11.1 meters.

A hundred-thousandth is 1.11 meters.

A millionth is 0.111 meters.

The precision of the recorded number (e.g., 38.047734), where the last digit is in the "millionths" position, might suggest that the position is known to about the nearest tenth of a meter (i.e., 0.111 meters). Here is where your knowledge of the difference between accuracy and precision comes in. TIGER data is rarely good to 10 centimeters accuracy. Ten meters might be more like it. Sometimes the last two digits are zeros in the files, which implies an accuracy of the nearest ten-thousandth of a degree, or about 10 meters—roughly 33 feet.

Address Locators

The second piece required for converting a set of addresses to a set of latitude-longitude coordinates is a mechanism for parsing the addresses so that queries to a TIGER-style database[16] will be recognized by the database. With ArcGIS this takes the form of creating an "address locator." You kick the process off in ArcCatalog by finding Address Locators in the Catalog Tree, expanding it, and double-clicking on Create New Address Locator. This opens a window that allows you to pick an address locator style. The idea here is to provide information about the general format of addresses in the region of interest. You have more than three dozen different styles to pick from. Your choice depends on the structure of your reference data and what the form of address in the region of interest looks like.

Once the choice of address locator is made, you will encounter a dialog box that will let you specify a number of parameters that will allow the form of your input addresses to be matched with the form of the locations in the reference database. If you are only typing one or a few addresses into the address locator, you can monitor whether or not you were successful. But if you have a large file of addresses to display, the problem gets a little stickier. Some addresses may pass muster and be represented on the map; others may be rejected. Still others may find close, but not exact, matches. These latter ones are assigned a score as to the closeness of the match, based on a number of factors. Successfully converting a large file of addresses requires both some tweaking on your part and something coming close to artificial intelligence on the part of the software. In the Step-by-Step section, you will go through the process of converting addresses.

---

[16]Many agencies and companies have taken the basic TIGER files and enhanced them.

Address Geocoding | STEP-BY-STEP

Exercise 9-3 (Project)

Experimenting with Addresses and Coordinates

_____ **1.** Using the operating system of your computer, open the text file

[___] IGIS-Arc\Address_Geocoding\
Clinton_County_OH_Type1_1994_TIGER_Records.txt

Clinton_County_OH_Type1_1994_TIGER_Records.txt is a raw TIGER file. Notice it is composed of only letters and digits—228 of them in each record. I show you this so you can understand that the basic TIGER files are simply character files. They contain geographic information, but to plot this information, you must convert the files into GIS files (e.g., shapefiles, geodatabase feature classes).

_____ **2.** Locate the sixth record from the top of the file. Verify that it describes the 4000 block of state highway 729. Verify that the longitude of the FROM node of that block is 83.613958 WEST. What is the latitude of the TO node of that block? _____.

_____ **3.** Looking at that same record you can discover that the zip code on the left of the highway is the same as the zip code on the right. In what zip code does this block of highway lie? _____.

_____ **4.** Start ArcCatalog. Copy the folder [___] IGIS-Arc\Address_Geocoding into ___IGIS-Arc_*YourInitialsHere*. Make a folder connection to the new folder.

_____ **5.** Start ArcMap. Add as data

___IGIS-Arc_*YourInitialsHere*\Address_Geocoding\Lexington_Roads.mdb\Lex_Roads_DD83\
Lex_Roads_2002.

You will see the road network of Lexington, Kentucky. This data set was derived from the totally character-based TIGER files of the 2000 census. The geographic component, present in the TIGER files as only character-based information giving longitude and latitude, has been converted to (geo)graphic coordinates so it can be plotted. Some of the other character-based information remains.[17]

---

[17]If you are interested in geographically augmented street data for your city, it is available free from the ESRI Web site. It comes in zipped-up shapefile format, so getting to it takes a few steps, but keep in mind the word "free." By the way, you can also unzip files for free with Windows XP.

—— **6.** Make the `display units` Decimal Degrees. In round numbers, what is the longitude of Lexington? _____. The latitude? _____. Now make the `display units` Feet.

—— **7.** Open the attribute table. How many records are there? _____. (Hint: click on the icon that gives you the last record.)

—— **8.** Look at the field names. Here you see the usual sorts of items: Feature Identifier (FID), the FROM and TO nodes, and the length (meaningless, because it is in decimal degree representation, which has no consistent scale). You also see the TIGER/Line ID (TLID), which is a unique identifier for the road segment—unique nationwide, actually.

Recall that we want to use this data set to find a mappable geographical point by specifying the address of the building at that point, For this we need the street name (here in the field FENAME); its type, such as avenue, road, and so on (here in the field FETYPE); direction prefixes and suffixes (FEDIRP and FEDIRS); and the numerical address ranges for each segment (i.e., block), which we will examine in detail shortly.

—— **9.** How many unique feature types (e.g., Rd, Ave, Blvd) are there in this feature data set? _____. Try to identify what each of the abbreviations means. (*Hint:* Use `Selection > Select By Attributes`. Construct the beginnings of an expression

`[FETYPE] =`

and then ask for Unique Values.)

Finding the Geographic Position of an Address "Manually"

To further understand address ranges, let's look at a particular block.

—— **10.** Using `Select By Attributes` (from Selection on the Main menu), locate all street segments whose FENAME is `Chinoe` (shin-o-ee). Zoom to selected features. Now, using `Add To Current Selection`, find those streets with FENAME of `Cooper` OR FENAME of `Cochran`. Again zoom to selected features. The map should look like Figure 9-14***. In the attribute table, show only selected records. Inspect. Dismiss the table.

The block of Chinoe we are interested in is the one between Cochran and Cooper.

—— **11.** Use the Identify tool to examine the block of interest. What is the From Address on the right-hand side? _____. The From Address on the left-hand side? _____. The To Address on the right-hand side? _____. The To Address on the left-hand side. _____. In which direction of Chinoe are the house numbers increasing? ___ North or ___ South. The designation of the left or right side of a road is determined by moving forward along that road in the direction of increasing structure numbers. So, on which side of the road would you expect to find the house with number 525? _____. About what percent of the way along the block, going south from the intersection of Chinoe and Cochran, would you expect to find this house? _____.

—— **12.** Make sure that the Drawing toolbar is visible. At the right-hand end there is a drop-down menu that allows you to change the marker color. Change the color to red. Near the left end is a drop-down menu that lets you select the sort of graphics to be drawn. Pick `New Marker`. Place the cursor crosshairs over the spot on Chinoe where you think 525 is, and click. Press Esc to unselect the marker.

FIGURE 9-14

_____ **13.** Save the map file in ___IGIS-Arc_*YourInitialsHere*\Address_Geocoding with the name 525_Chinoe_Road. Dismiss ArcMap

Making an Address Locator

_____ **14.** Back in ArcCatalog find Address Locators > Create New Address Locator and double-click. In the window you will find all sorts of styles of address locators. What we want in this instance is the US Streets (GDB).[18] Double-click. In the sizable window that comes up, put in the name Lexington_Streets. For Reference Data, use, of course,

___IGIS-Arc_*YourInitialsHere*\Address_Geocoding\Lexington_Roads.mdb\ Lex_Roads_DD83\Lex_Roads_2002.

Look at the Fields area. You will see that all the defaults are appropriate. Under Output Options, make the side offset 30 Feet. In the Output Fields area, put a check by the four items. Click OK. Now the item Address Locators in the Catalog Tree will contain Lexington_Streets, qualified (using a prefix) by your username. (Address locators are private. If someone else logs onto this computer with a different username, he or she will not have access to your address locators.)

---

[18]GDB—Abbreviation for geodatabase.

Chapter 9

Finding the Geographic Position of an Address "Automatically"

_____ **15.** Restart ArcMap. Open the map 525_Chinoe_Road. Change the color of future markers you might draw to black. Choose Edit > Find (or click the binoculars on the Tools toolbar). Select the Addresses tab. To choose an address locator, browse to the one you just made in the Catalog Tree under Address Locators. For `Street Or Intersection`, type

```
525 Chinoe Road
```

and press Find. The Find window should show that such an address was found by giving details in the lower part of the window. Note that the structure is on the "R" side and that the Pct_along is 24.5. Look at how these results compare to your answers you wrote earlier when you found the position manually.

_____ **16.** Move the Find window out of the way so you can see Chinoe Road. Right-click the text that contains the found information. From the drop-down menu, pick `Add As Graphic(s) To Map`.

_____ **17.** A new marker symbol (black) should appear, near the one you placed earlier. Dismiss the Find window. Zoom way up so that the block of interest fills the screen. In the Data Frame Properties window, make sure the display units are in Feet. Measure the distance from the street to the black marker. It should be about 30 feet. Measure the distance from the red marker symbol to the black one. _____. If it is more than a couple of hundred feet, try to determine why.

TIGER Files and ZIP Codes

_____ **18.** Zoom to full extent. Using Select By Attributes, find those records in Lex_Roads_2002 in which the ZIPL is not equal to the ZIPR. Dismiss the selection window and minimize the attribute table so you can see the map, with the selected segments.

What I had planned on showing you by this step was a clever way to delineate zip code areas. After all, when a street segment has one zip code on one side and another on the other, the street is obviously a dividing line between the two areas. While you can sort of get an idea of the zip code areas, really what I have shown you instead is how wretched the data set is. Clearly, the zip code data in this set of TIGER records leaves a lot to be desired. If you want, you may peruse the fields to see what sorts of problems give rise to the unfortunate map, but the main lesson is that there are data sets out there, from reputable sources like the U.S. Bureau of the Census, that are shot through with errors. User beware!

To see the actual zip codes of Lexington, add the shapefile

___IGIS-Arc_*YourInitialsHere*\Address_Geocoding\ZIP_Codes_in_405.shp

to the map.[19] Compare the zip code boundaries with the selected elements.

More to Know—More Information Available

There is much, much more to know about address geocoding, but you have the basic idea. Addresses in text form can be used to create geographic coordinate points. The next step, which we won't undertake,

---

[19]Zip codes data sets for the entire United States are available in the data sets that came with your ArcGIS suite. Look for a folder named ESRIDATA\census.

would be to convert files of addresses. To refer back to the example I used in the Overview, suppose you have a file of thousands of addresses on crime reports. You might like to see where, on a map, these events occurred. You can submit this file to the software and get a map of the events.

Problems occur when an address yields up no geographic point. Or when an address presents you with two or more possible points. The address geocoding software is sophisticated. For example, when there are multiple possibilities, the software provides you with a "score" that tells you how good a match the input address is to what's in the database.

To learn more, read *Geocoding in ArcGIS*, which exists as a paper manual available from ESRI or as a PDF (Portable Document Format) file.[20] The filename is Geocoding_in_ArcGIS. You may find it in a folder named ESRI_Library\ArcGIS_Desktop.

---

[20]Software for displaying such files is the Acrobat Reader, from Adobe Systems Incorporated.

Analysis of Networks | **OVERVIEW**

In ArcGIS Version 9.1, ESRI introduced the Network Analyst in ArcGIS Desktop. The capability to do network analysis has been present in ArcInfo for years, but it had to be accessed through typed commands, rather than point-and-click activities.

The sorts of networks that Network Analyst deals with are primarily composed of roads. You might think of network analysis as sort of a simulation of a vehicle, confined to streets, whose driver wants to get from A to B by the shortest route or in the least time—staying on the roads, of course. Or network analysis can develop service areas, considering the network, such as assigning voters to the nearest polling place.

To analyze a network, one needs a network to analyze—and this is no mean feat. First you have to have all the sort of data that is, very roughly, supplied by TIGER-like files. Then the actual lengths, in linear units—not degrees of latitude and longitude—of segments must be developed. Connectivity of segments must be ensured. Since time is usually as important as distance, the times to traverse each segment needs to be produced. While traffic signals and turns from one segment to the next aren't reflected in distance calculations, they very much affect time. So tables, for each intersection, that describe averages of how long it takes to turn right, turn left, or go straight are needed. And the 99 percent of you who have driven a car know that how long it takes to drive a particular block can depend on what time of day it is, whether there is construction ahead, and whether there is an accident. In other words, a truly effective network needs to be sensitive to the time of day, and, better yet, sensitive to traffic conditions. The need for "dynasizsm" of this sort is well beyond our scope here. But in the basics of network analysis are contained in the next section.

Experimenting with Routes and Allocations

____ **1.** Start ArcCatalog. Copy the folder Network_Analyst_Data from

[____] IGIS-Arc to your

____IGIS-Arc_*YourInitialsHere* folder

____ **2.** Start ArcMap. Choose Tools > Extensions and make sure that there is a check mark by Network Analyst. (If you are using a version of ArcGIS earlier than 9.1, you will not find the Network Analyst Extension.) Under View > Toolbars also make sure the Network Analyst toolbar will be shown.

____ **3.** On the Network Analyst toolbar, find the Network Analyst window button. Click to display the pane.

What you will see in the following steps is made possible by data and a network data set that were constructed with considerable effort. We will not look at the construction effort nor the building of the network database, called streets_nd.ND. For further information, take the Network Analyst tutorial that comes from ESRI. The data for this exercise was derived from that tutorial data.

____ **4.** Add the data set streets_nd.ND from

____IGIS-Arc_*YourInitialsHere*\Network_Analyst_Data\Route_&_Allocate\Network.

You will be asked if you want to also add all participating feature classes. You do.

____ **5.** Turn off all layers except Streets. You will see the streets of the San Francisco area. On the Network Analyst toolbar, note that the network data set is streets_nd.

____ **6.** Under Selection > Set Selectable Layers, place a check by Streets only.

____ **7.** Open the Streets attribute table. Make it occupy the full screen. How many street segments are there? _____. Notice that the fields suggest that this data set was derived from TIGER/Line files. Notice also that several fields have been added. You will see Full_Name, street length in meters, FT_Minutes (some sort of average time for traversing the street, moving from the from node to the to node), and TF_Minutes (same sort of thing, but traversing the opposite direction).[21] Notice also the field Oneway, which contains either FT (meaning traffic may proceed from the FNode to the TNode) or TF. Minimize the attribute table.

_____ **8.** Choose Selection > Select By Attributes. Make the layer Streets. In order to find the intersection of Pine and Fillmore Streets, create *exactly* the following query, being aware that every character is case-sensitive:

"Full_Name" = 'Fillmore St' OR "Full_Name" = 'Pine St'

and press Verify. If it works, press Apply. Now choose Zoom To Selected Features.

_____ **9.** Obviously, from all the selections, there are Pine and Fillmore streets scattered all over the Bay Area. You are interested in the two that intersect, so zoom up on those. See Figure 9-15***. Then zoom up tightly on the intersection itself.

_____ **10.** Use the Identify tool with the Streets layer to find the block of Pine Street immediately west of the intersection.

_____ **11.** On the Network Analyst toolbar drop-down menu, click New Route. The Table of Contents fills up with lots of features associated with the about-to-be-created route. The Network Analyst window will show that, so far, there are no stops, routes, or barriers.

_____ **12.** On the Network Analyst toolbar, find the Create Network Location button and click. Using the mouse cursor, which now contains a flag, place the crosshairs midway down that block of Pine Street, and click. This will put Stop 1 in position. Choose Selection > Clear Selected Features.

To make Stop 2, you will use the Find capability of the Address Locator name SanFranStreets.

_____ **13.** Choose Edit > Find. In the Find window, click Addresses. For the Address Locator navigate to

___IGIS-Arc_*YourInitialsHere*\Network_Analyst_Data\
Route_&_Allocate\SanFranStreets

and Add. For Street Or Intersection type *exactly*

Cabrillo St & 5th Ave

where the & specifies an intersection. Press Find.

_____ **14.** Two possibilities appear: the one we want (score 100) and one on 45th Avenue—obviously not of interest. Right-click on the first candidate and pick Zoom To Candidates(s) And Flash. You probably didn't see the flash because the Find window was in the way. Move it and click Flash Candidate Locations. Use Identify to assure yourself that this is the right intersection.

_____ **15.** Right-click the candidate again and pick Add As Network Location.

_____ **16.** Right-click on Stops (2) and select Zoom To Layer so you can see the map that contains both stops.

---

[21]These two numbers might be used to indicate one-way streets (by making the "wrong way" a very large number), but the one-way issue is handled in a different way: by the field Oneway.

FIGURE 9-15

_____ **17.** Using the same procedure as above, place Stops 3, 4, 5 and 6 at

❑ Teddy Ave and Delta St

❑ Clement St and 18th Ave

❑ 245 Brazil Ave

❑ Fillmore St and Pine St (near the starting point)

_____ **18.** Dismiss the Find window. Zoom to the map containing the stops.

Now that the stops are established you can have Network Analyst create a route.

_____ **19.** Click the Solve (Run the current analysis) button. Right-click on Routes (1) and Zoom To Layer. The route will be shown. Click the button to give you the Directions window. How long is the trip? _____ How many minutes will it take? _____. Peruse the Directions window. Click on the word Map next to a direction to see a detailed map of the described turn. Dismiss the Directions window.

_____ **20.** Expand Stops in the Network Analyst window. Click each graphic pick to select it and note its position on the map. Notice that you are doing a lot of extra driving to visit these stops. See Figure 9-16***. You could make the trip more efficient if you switched some of the stops.

_____ **21.** You can reorder the stops (confine yourself to numbers 2, 3, 4, and 5 only) by dragging a graphic pick to a different position in the list. Try changing the order in this way and re-solving. What is the most efficient trip you can manage? _____Miles. _____Minutes.

Not only does Network Analyst provide a very efficient algorithm for finding the shortest (or fastest) path that visits a number of stops in which the order is predefined. It also at least attempts to solve the famous

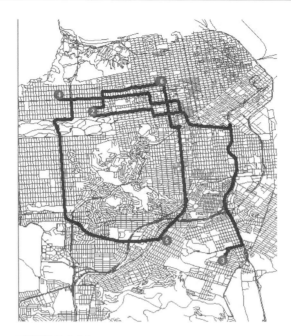

FIGURE 9-16

"Traveling Salesman" problem, which minimizes trip length or time.[22] You can have Network Analyst decide on the order of the stops.

____ **22.** In the Network Analyst window, at the top right, is a button that will give you layer properties. Under Analysis Settings, check the box Reorder Stops To Find Optimal Route, while preserving the first and last stop. Also you will notice that you can choose to have the impedance (the factor that Network Analyst tries to minimize in selecting a route) be either minutes or meters. You can also notice that you can restrict the route not to go the wrong way on one-way streets—which seems like a good idea, so check that box. Click Apply, then OK.

____ **23.** Solve the network again. With the Directions window, determine if the new network analysis improved on your last effort for an efficient trip. (If not, blame the one-way streets.)

Zoom up on one of your stops. With the Select/Move Network Locations tool, drag the stop a few blocks away. Resolve the network. Notice how the route is changed to go through the moved point.

There is one other button on the Network Analyst toolbar that deserves your attention. Find and press the Build Entire Network Dataset button. Answer No when asked if you want to build the network data set. What you should realize from this is that a lot of effort goes into building the database that allows a user to do all the nifty operations that you just performed.

---

[22]There is no known solution for ordering a set of points so that the path that encounters each one of them once (and then returns to the starting point) is the shortest possible, except for examining every possible path, which is computationally out of the question for a large number of points. But one can get very close to an optimal path by a variety of techniques, which is probably what this software does.

Finding the Shortest Route to a Facility

Suppose there is an automobile accident and routes from the closest hospitals for ambulance service are needed. Network Analyst can provide this information almost instantly.

____ **24.** Add as data to the map the data set Hospital.shp, found in

___IGIS-Arc_*YourInitialsHere* \Network_Analyst_Data\Route_&_Allocate\Layers

Zoom to that layer.

____ **25.** In the Network Analyst toolbar drop-down menu, click New Closest Facility. Note that headings for Facilities, Incidents, Closest Facility Path, and Barriers appears in the Network Analyst window.

____ **26.** Right-click on Facilities (0). Choose Load Locations. In the window that appears, choose Hospital from the Load From drop-down menu. Click OK. Notice that several hospitals appear on the map with the symbol indicated in the Table of Contents under Closest Facility > Facilities > Located. These are, of course, in the same locations as the Hospital layer you added earlier.

____ **27.** Click the Layer Properties button (on the Network Analyst Window) and select the Analysis Settings tab. Under `Facilities To Find`, change the number to 3. Travel From should be set at Facility To Incident. One-way streets should be respected. Click Apply, then OK.

____ **28.** Click Incidents (0). Click on the Create Network Location button on the Network Analyst toolbar. Move the cursor onto the map.

Pretend that this cursor, rather than being pushed around by your mouse, is directed by an automatic signal coming from a vehicle whose airbag has just deployed. The car knows where it is because it has an onboard GPS receiver. That location is identified with a marker. Figure 9-17*** shows the location with a dark solid dot.

____ **29.** Click on the map at approximately the location shown by the dot. Click Routes (0). Click Solve. The path from the three closest hospitals to the accident should appear. Also, the names of those hospitals will show up if you expand Routes (3).

____ **30.** Click on one of the hospital names under Routes (3). One path will be selected. Press the Directions button. How long will it take an ambulance to reach the scene of the accident? _____. How about for the other two hospitals? _____. _____.

Allocating Territories to Facilities

Assume you want to do an analysis of how well fire stations can respond. Perhaps a new standard is that every point in the downtown area of the city should be reachable within 1.5 minutes from some fire station.

____ **31.** Under Layers remove all layers except Streets, Streets_ND_Junctions, and Streets_nd. Make Streets a black line of width 1. Add as data Fire_Station.shp. Choose Zoom To Layer.

____ **32.** On the Network Analyst toolbar menu, choose New Service Area. Right-click on Facilities in the Network Analyst window and load the Fire_Station locations. How many are there? _____.

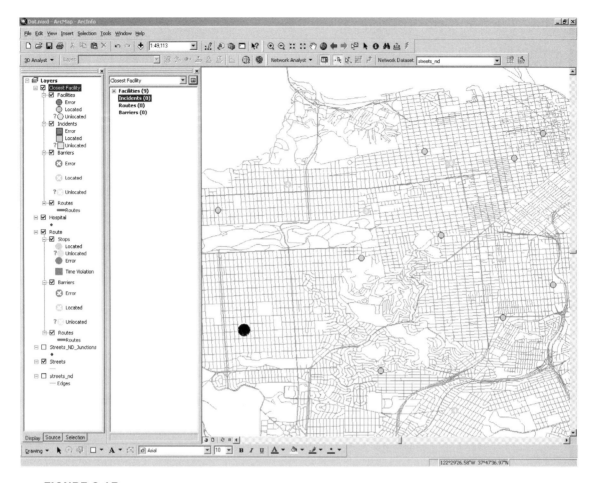

FIGURE 9-17

___ **33.** Click Service Area Polygons (0). Bring up Layer Properties > Analysis Settings. The Default Breaks field specifies the extent of the service area by calculating how far, along each possible route from the fire station, a vehicle could reach in the time allotted. Set it at 1.5. Respect one-way streets. Click Apply. Under the Polygon Generation tab, click on Generalized. (If solving the problem with generalized polygons doesn't take too long, you can come back and ask for the detailed polygons.) Click Apply, then OK.

___ **34.** Press the Solve button and be prepared to wait. The software is computing, for each fire station, the 1.5-minute range along each possible path. That's a lot of computing.

___ **35.** Expand Service Area Polygons. Click on Station 22 to highlight the name and the polygon. Select Zoom To Selected Features. Notice the area that is covered by the station in the 1.5-minute time allotment, as well as the area left uncovered. See Figure 9-18***.

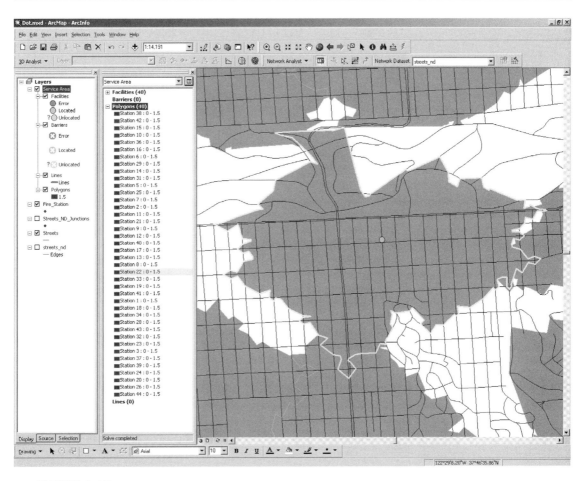

FIGURE 9-18

Linear Referencing | OVERVIEW

Managers of artifacts with long linear structures (e.g., highways, pipelines, railroads) are generally unconcerned with the precise two-dimensional geographic coordinates of the entity they deal with. A locomotive is pretty well locked onto the railroad's linear structure—unlike, say, an airplane, which can operate in three dimensions. As a result, many of the coordinate systems that have grown up around linear structures are different from those we have been talking about up to now. Terms like road miles, river miles, and rail miles are in common use in those industries. The average citizen who uses the Interstate Highway System will probably be familiar with the tiny green signs with numbers on them that indicate the number of miles from some origin—frequently a state border.

We are in the habit of representing linear structures with lines drawn between junction points (e.g., nodes). (Of course, these structures are portrayed on a two-dimensional field, but the smallest part of each such line is one-dimensional [a vector], so we call them one-dimensional, or linear.) The problem with simply representing such a structure—let's take a highway, for example—from intersection to intersection is related to attributes. The requirement, so far in your studies, is that for any GIS feature (whether point, line, or polygon), it must be homogeneous in all its attributes. For example, all of a given cadastral polygon is owned by one entity, the taxes are paid on all of it or none of it, and so on. If different attributes apply to two different parts of the feature, then you need two features (e.g., if different people own different parts of a piece of land, then you need more polygons).

Imagine that you have a stretch of highway that is 2 miles long between intersections A and B with other roadways. Among the attributes you want to store for this length of road are speed limit, pavement type, pavement quality, political jurisdiction, and number of lanes. The difficulty is that the speed limit changes four times and the road goes from four lanes down to two and then back to four. It crosses a county boundary. Part of it is blacktop and part concrete. And repairs have taken place on different segments of the road at different times. To represent this "traditionally" in ArcGIS, we would have to have a plethora of features. Every time an attribute changed (e.g., the speed limit changed from 55 to 45), a new feature would have to be declared. So the single-line feature or arc that represented the 2 miles between intersections might have to turn into dozens of short features. The complications created then—e.g., how would you find the distance from A to B?—are considerable.

The invention that turned out to make attribute representation tractable for such linear features is called "linear referencing." The fundamental idea is that you can have several sets of attribution information that accompanies a single linear feature. What makes this possible is the concept described earlier in this section: The thing that fixes the point at which a change takes place is a number—a distance—that is related to the origin of the feature, in the style of, say, road miles.

To recap, linear referencing lets you store geographic information without explicit x-y coordinates. Instead, a measure (distance) along a linear feature is provided (which is itself, of course, defined by x-y

570

coordinates). Linear referencing is a mechanism that allows you to associate multiple sets of attributes to portions of linear features.

One way to view linear referencing is "features within features." As such, some new terminology and tools are necessary to understand and operate linear referencing. We address those in the Step-by-Step section that follows.

Linear | **STEP-BY-STEP**
Referencing |

The first concept to grasp is that of a "route." A route is a linear feature, probably made up of several or many other linear features, that has a unique identifier and has a measurement system stored with its geometry.

Exercise 9-5 (Project)

Experimenting within Linear Features

---

1. Start ArcCatalog. In the Catalog Tree find the folder

[____] IGIS-Arc\Linear_Referencing.

Copy it into ____ IGIS-Arc_*YourInitialsHere*.

---

2. In ____IGIS-Arc_*YourInitialsHere*\Linear_Referencing\Pittsburgh.mdb\PITT_Roads_Routes

preview the Geography of Just_Roads. Using the Identify tool, click on a few features. You will notice that you that you get some standard information (e.g., feature name) and also at least the possibility of a route identifier (called ROUTE1), a beginning mile point, and an ending mile point. Look at the attribute table of Just_Roads. How many road segments are there? _____.

---

3. In the Catalog Tree, click All_Routes. Switch back to the Geography display, which thins out a lot compared to Just_Roads. These are road features that have been combined and designated as "routes"—all under the name of ROUTE1. Each route is composed of sets of features from Just_Roads.

---

4. Again use Identify. Notice that you get completely different results from previously. The shape is still polyline. But usually the Shape_Length is way longer. When you click on a feature, the line that flashed is usually lengthy—composed of many of the features you saw before. Look at the table. How many routes are there? _____.

---

5. Lastly, display Some_Routes from the Catalog Tree. Some_Routes is a subset of All_Routes that we will use for demonstration purposes. How many of Some_Routes are there? _____.

From the preceding you can see that routes are features. But they are features with remarkable, complex, and useful characteristics, as you will see from the steps that follow.

6. Start ArcMap. Add two data sets from

___IGIS-Arc\Linear_Referencing: (1) County and (2) Cities, making sure that Cities is at the top of the Table of Contents. From the pgdbfd PITT_Roads_Routes, add Some_Routes. Make the line feature bright green, with a width of 2.

7. Open the Some_Routes attribute table. Find the route 30000030. Select it in the table and note its location in the northeast quadrant of the map. See Figure 9-19***.

You have identified Some_Routes as a feature. Next you will Identify it as a route, but you have to add a button to a toolbar to do it.

8. Choose Tools > Customize > Commands. Find Linear Referencing in the Categories list. Drag Identify Route Locations to the Tools toolbar, where it becomes a button. Close the Customize window.

9. Make the Identify Route Locations tool active and click on the selected route. What are the minimum and maximum values of the Measure of the route? _____. _____. Of how many parts does the route consist? _____. Click on different points of the route so you can determine, based on the Measure value, in which direction (northwest or southeast) the Measure increases? _____. Dismiss the Identify Route Location Results window.

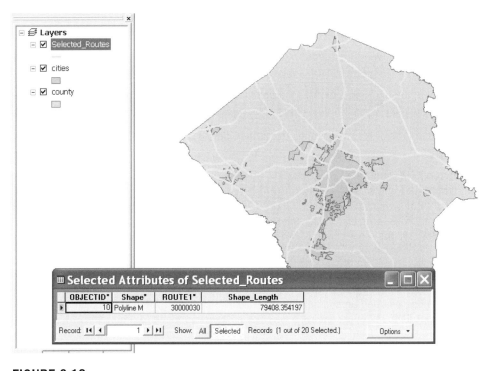

FIGURE 9-19

Suppose you know that a call box has malfunctioned at measure 3.35. You want to find that point on the map.

_____ **10.** Click Find (the binoculars) > Route Locations. For Route Reference, pick Some_Routes. The Route Identifier is, as always, ROUTE1. Load Routes. Pick the route that you have been working with: 30000030. The `Type` should be `Point`. Put in the Location 3.35 and press Find. Information about the route should appear in the bottom panel of the window.

_____ **11.** Right-click the highlighted text selection. Flash the route. Flash the route location that was found. Draw the route location. Label the route location. Click the Select Elements pointer and place it over the route label; click once and pause. Drag the label around, noting that it is a callout box. Press Delete to remove the box. Select the drawn location by dragging a box around it; delete the drawn location. Close the Find window. Minimize the table.

_____ **12.** Add the table accident.dbf as data from the Linear_Referencing folder. Open the table. How many accidents are recorded? _____. With Options > Select By Attributes, select the records with ROUTE1 = 30000030.[23] Show the selected records. How many are there? _____. Right-click the gray box to the left of a record and click the Identify box that comes up. Note that the record contains a MEASURE field that indicates the position along the route where the accident occurred. Lots of other data about the accident is also recorded. Close the Identify Results window.

_____ **13.** Move the table around so you can see the map, and particularly Route 30000030. Notice that no points along the route where these accidents occurred are shown. That's because the table is just that: a table. To see the locations of the accidents, you can use the table to make a layer. (This layer will exist only in memory and will go away once you close ArcMap, unless you save the map, in which it is saved [only with the map]. Of course, you may specifically save it as a layer file by right clicking and choosing save as LayerFile.)

_____ **14.** Show All Records in the table. Under `Options,` choose `Clear Selection`. Close the table. Choose `Tools > Add Route Events`. Set up the window so it looks like Figure 9-20

_____ **15.** Examine the map. Now you see the accidents depicted in the table—put there by a layer called `Accident Events`. Click the Source tab.

For reasons whose logic or history escapes the author, points or segments within a route are termed "events."

_____ **16.** Open the `Accident Events` attribute table. Again use Select By Attributes to select Route 30000030. Show only selected records. Resize and rearrange the table so you can see the map. Notice that the accidents on our favorite route are highlighted. Sort the MEASURE attribute values, smallest to largest. Click the gray box to the left of the record with MEASURE value 7.23. The record should turn yellow. The point on the map should turn yellow. Display All Records. Choose `Selection > Clear Selected Features`. Dismiss the table.

_____ **17.** Add and open the pavement.dbf table. How many different pavement events are there? _____. Select those in Route 30000030. How many? _____. Notice that there is a beginning mile point and an ending mile point field. Sort BEGIN_MP is ascending order. That should also put END_MP in ascending order. The segments connect but do not overlap. Display All Records. Clear all selections.

---

[23]You may have noticed that sometimes a field name in a query is enclosed in square brackets and sometimes in double quotes. It depends on what database management system is in use for the layer or table.

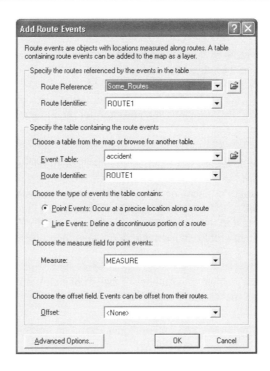

FIGURE 9-20

_____ **18.** Make a layer from the Pavement.dbf table. This time, of course, the event table should be "pavement" and it is a line event. The From-Measure should be BEGIN_MP and the To-Measure END_MP. Always the route identifier is ROUTE1. Click OK. Click the Source tab.

_____ **19.** Turn off Some_Routes. Make the Pavement Events layer bright red. Open the attribute table. Using Options, select those pavement events on Route 30000030 and show selected records. Yellow-fy the record of the segment that runs from mile point 4 to mile point 7. Zoom in and look at it on the map. Examine the RATING attribute, which is a value from 0 to 100, indicating the quality of the roadway. What is the range of RATING values for all selected records? High _____. Low _____.

_____ **20.** Show all records. Minimize the `Attributes Of Pavement Events` window. Close the Some_Routes table. Save the map as Pittsburgh_Routes in ___IGIS-Arc_*YourInitialsHere*\Linear_Referencing.

Intersecting Route Events

Just as you could combine a set of polygons with another set of polygons to create a third set of polygons with appropriate (geo)graphic and attribute information, you can also combine the graphic and linearly referenced attributes of routes.

Suppose someone has suggested that there is a correlation between auto accidents and road conditions on Route 30000030. You want to combine the accident data with road condition data.

_____ **21.** If necessary, start ArcMap with Pittsburgh_Routes, as saved in the previous step.

_____ **22.** Open the attribute table of Some_Routes. What is the Shape_Length of 30000030? _____. Close the table.

The length of a route, considered as a feature, is, of course, in the units of the coordinate system. But, as you noted previously, the event measures are in miles. Just for the sake of confidence, let's compare one against the other.

_____ **23.** Restore the attribute table of Pavement Events. Select 30000030. Display Selected Records.

Nowhere in the table is the length of each segment, but we can fix that, since we have the beginning and ending mile point number. With this few records (six), you could easily verify that there are no overlaps (a requirement for valid segmented data) and determine the total length in miles covered by the segments. But let's add a field to the table whose value is the segment length.

_____ **24.** Under Options in the table, bring up the Add Field window. Call the new field Seg_Len. Make its type Float. Set the precision (the number of digits possible) at 6. Set the scale (the number of digits to the right of the decimal) at 2. Click OK.

_____ **25.** Right-click on the field name Seg_Len and click Calculate Values. Ignore the warning.

The segment length is calculated at the beginning mile point minus the ending mile point. Because this number might be negative, we will take its absolute value.

_____ **26.** Calculate Seg_Len as Abs([BEGIN_MP] - [END_MP]). The calculation will take place only for selected records. Click OK. By looking at the table, verify that Seg_Len does indeed contain the positive difference between the beginning and ending points.

_____ **27.** Run Statistics on Seg_Len. What is the Sum? _____. Divide the length of the route in feet that you found previously by the number of feet in a mile (5280). The result should be reassuring. _____.

_____ **28.** Using Options > Select By Attributes (Method: Select From The Current Selection), obtain those segments of 30000030 that have RATING >75. How many miles are in these segments? _____.

_____ **29.** Select segments of 30000030 and RATING <=75. How many miles? _____.

The intersection of point events and line events has a lot in common with its polygonal counterpart. The graphic result will consist of those points and segments that occupy common space. The attribute table will reflect both event tables. We will do the intersection for all route segments and all accident event points. (There are certainly more efficient ways to do this, as far as number of calculations is concerned, but this is the most straightforward.)

The process will be to intersect the accident and pavement event layers to produce a new event table, consisting of records of pavement event segments where the segments contain accidents. Then this table will be made into a (temporary) event layer and displayed.

_____ **30.** Display All Records of Pavement Events. Clear Selected Records. Close that table. Show Arc-Toolbox. In Linear Referencing Tools, right-click on Overlay Route Events and select Open.

_____ **31.** In the Overlay Route Events window, click the down arrow at the end of the text box. Select Pavement Events. The next four text boxes should be filled in for you. For the Overlay Event table, pick accident Events in the same way.

_____ **32.** Keep the path the same but change the name of the Output Event Table to Accidents_and_ Pavement.dbf. Accept the rest of the table. Click OK. You may have to wait a bit, since thousands of point events are being matched up with hundred of segments. Close the window when the deed is done.

What you have now is an attribute table named Accidents_and_Pavement.dbf. To see the graphics, you must make a layer. You did this before with the Tools menu. Since you have ArcToolbox open, You may use the tool there.

_____ **33.** In Linear Referencing Tools, right-click on Make Route Events Layer and select Open. In the Make Input Route Features text box, take the only choice offered: Some_Routes. The Route Identifier field is, of course, ROUTE1. The Input Event Table, from the drop-down menu, should be Accidents_and_Pavement.dbf. Again, the Route Identifier field is ROUTE1. The rest of the window should be filled in for you. Click OK.

_____ **34.** Hide ArcToolbox. Turn off Accident Events and Pavement Events. Open the Accidents_and_ Pavement.dbf. Events _layer_ attribute table. Select the records in route 30000030. How many are there? _____. Show only those records.

_____ **35.** Right-click on the gray box to the left of some record. Notice the large number of fields and values in the Identify Results box. Dismiss the Identify Results window. Under Options, click Select By Attributes. In the window, under Method, pick Select From The Current Selection. For the SQL expression, use RATING <= 75.

_____ **36.** How many accidents occurred on the stretch of road where the rating is less than or equal to 75? _____. From your previous calculation, how many miles were involved? _____.

_____ **37.** By using subtraction, determine of the number of accidents that occurred on segments with a RATING > 75. _____. From a previous step, copy the length of road involved: _____. Based on accidents per mile, what is your conclusion as to whether the condition of the road was related to the number of accidents? _____.

What's Not Covered Here

As with all the sections in this chapter, we have barely scratched the surface. There is the issue of calibrating routes. And, as with everything we do, editing is a major issue. To make use of linear referencing, you should refer to ESRI's publication _Linear Referencing in ArcGIS_, which exists as a paper manual available from ESRI or as a PDF file from Adobe Systems Incorporated. The filename is Linear Referencing in ArcGIS. You may find it in a folder named ESRI_Library\ArcGIS_Desktop. In addition to a Quick Start Tutorial (from which the data for this section of this book are adapted), the topics covered there are Creating Route Data, Displaying and Querying Routes and Events, Editing Routes, and Creating and Editing Event Data.

Checking, Updating, and Organizing Your Fast Facts File

The Fast Facts File that you are developing should contain references to items in the following checklist. The checklist represents the abilities to use the software you should have gained upon completing Chapter 9.

Important note: This checklist is on the CD-ROM that accompanies the book. It is available in Microsoft Word format. Rather than typing or writing by hand the text that follows, you can copy and paste it into your Fast Facts File from the CD-ROM file.

___ The two ESRI software packages that deal with 3-D spatial data are

___ To automatically spin the globe

___ To turn on 3D Analyst

___ The data set type that is naturally, truly three dimensional is

___ Base Heights refers to

___ In 3D Analyst, drawing order is controlled by

___ Keyframes are

___ One way to produce animation is

___ To get to the animation controls

___ To calculate volumes

___ To make a TIN from line features

___ A viewshed is a scene that shows

___ To add your own data to ArcGlobe

___ To change the center of rotation in ArcGlobe from the center of the Earth to another target

___ When dealing with two data sets and topology, the one with the greater number indicating the rank is likely to move _____.

___ To erase features flagged as topology errors

___ An address locator is

___ Distances obtained directly from TIGER files are not useful because the coordinate system

___ TIGER files define location in two different ways, which are

___ To make an address locator

___ The reason for an offset in an address locator is

___ Network Analyst has capabilities that allow the user to determine

___ To use Network Analyst, one needs ArcMap version

___ Best route is defined as

___ To create a network location (called a stop)

___ You can manually reorder the stops by

___ The traveling salesman problem

___ To move a stop

___ Finding closest facilities is done by

___ Allocating geographic areas to facilities is done by

___ Routes are features that are composed of

___ To find a particular location on a particular route

___ A route event is

___ Just as polygons can contain lines, linear route events can contain

___ Intersecting route events means

___ Intersecting route events produces a table, which, in order to have its geographics displayed, must be converted to

Afterword: From Systems to Science

Michael F. Goodchild[1]

In the Preface, Michael Kennedy listed three objectives for this book, the last of which was to provide a "basis for the reader to go on to the advanced study of GIS or to the study of the newly emerging field of GIScience, which might be described as the scientific examination of the technology of GIS and the fundamental questions raised by GIS." This idea of a "science behind the systems" goes back to the early 1990s. The GIS software industry was flourishing, courses were being instituted in universities, and GIS was being adopted as an essential tool in a wide range of applications. Nevertheless several prominent academic geographers began asking awkward questions. If GIS was simply a tool, then why was so much attention being devoted to it? After all, universities didn't feel the need to teach courses in word processing or spreadsheets. GIS was clearly useful to powerful interests, but what was it doing for the poor and the marginalized? Wasn't GIS simply "nonintellectual expertise," a bag of tricks?

By the middle of the decade, a consensus began to emerge that there was indeed "more to it," that the use of GIS raised some very interesting and fundamental questions, and that there were important research issues to be resolved if the design of GIS was to be improved in the next generation. In addition to training in the latest version of GIS software, a course in GIS ought to be an education in some very basic principles that would still be true ten or twenty years into the future. The term "geographic information science" was widely adopted in the titles of books and degree programs and the names of journals and conferences, though other terms such as "geomatics" were also used, particularly outside the United States.

So what exactly are these questions, issues, and principles? Many of them have been raised or at least hinted at in the pages of this book, but here are some of the more important and challenging:

❏ The problem of uncertainty. Every GIS database attempts to represent some selected aspect of reality in the geographic world, but none succeeds exactly because reality is almost infinitely complex, but the storage capacity of a digital computer is always limited. Choices must always be made about how much detail to include, how much to generalize or approximate in the interests of creating a database at reasonable cost, and what to leave out. In short, a GIS database always leaves its users with some degree of uncertainty about the real world. How should that uncertainty be described and measured, and what impact does it have on the results of GIS analysis? Over the past two decades, a very valuable body of knowledge has accumulated on the answers to these questions.

[1]National Center for Geographic Information and Analysis, and Department of Geography, University of California, Santa Barbara, CA 93106-4060, USA. Phone +1 805 893 8049, FAX +1 805 893 3146, Email good@geog.ucsb.edu.

❑ Representation choices. The reader has encountered several different approaches to the representation of geographic phenomena in GIS—shapefiles, coverages, geodatabases, TINs, rasters, vectors—and there are many others. Over 300 formats are recognized by software packages that specialize in conversions. Yet many types of phenomena are still difficult to capture in GIS. There are still only very rudimentary techniques for representing three-dimensional data, or time-dependent data on dynamic phenomena, or data on flows and interactions. Many GIScientists specialize in this area, developing new data models and formats that will likely form the foundation of future generations of GIS software.

❑ Public participation. GIS is complex technology, and it can take several courses at university level to master it. But if it is to be truly useful in everyday life it needs to be accessible to people who cannot afford to take the time to read books, sit through lectures, and work through complex lab exercises. A branch of GIScience called *public-participation GIS* (PPGIS) examines its use in public decision-making and in community planning. Instead of a single database, PPGIS allows each stakeholder to develop a database that expresses his or her own views. This is particularly important in decisions that involve indigenous cultures who may have entirely different ways of thinking about their local environments.

❑ Spatial data infrastructures. Geographic data is essential in many aspects of human activity, from day-to-day way-finding to disaster management and from logistics to forestry. Many of the feature classes that the reader will have encountered in this book are useful for many different purposes, and it makes sense if they can be effectively shared. The term spatial data infrastructure describes all of the arrangements society makes to produce and share geographic data. In the United States, the National Spatial Data Infrastructure was initiated under President Clinton in 1994 and has grown into a complex of committees, format standards, data warehouses, and training programs that together seek to improve access to geographic data. Similar strategies have been adopted in other countries, and research continues into better approaches to achieve these objectives and into the impediments that stand in the way.

These four are only a few of the topics that make up the research agenda of GIScience. An excellent source of more information is the Web site of the University Consortium for Geographic Information Science (www.ucgis.org), which has published perhaps the best-documented research agenda and also organizes conferences and workshops and offers scholarship programs for students. Another good source is the book *Geographic Information Systems and Science* by Longley, Goodchild, Maguire, and Rhind, also published by Wiley.

This book has covered a large amount of material, and its exercises will have brought the reader to an advanced state of expertise in today's technology. But what are the wider implications of today's technology, and where is it headed? How accurate are the results it produces, and what broader impacts will it have on society and the decisions it makes? These are all important questions, and they are the stuff of GIScience. Like all good books, this one will have succeeded if it leaves the reader with as many questions as it answers.

Index

Index

Index

Index